PLASMA PHYSICS

PLASMA PHYSICS

Proceedings of the 1997 Latin American Workshop
(VII LAWPP 1997),
held in Caracas, Venezuela, January 20–31, 1997

Edited by

PABLO MARTÍN
Universidad Simón Bolívar
Caracas, Venezuela

and

JULIO PUERTA
Universidad Simón Bolívar
Caracas, Venezuela

Reprinted from *Astrophysics and Space Science*
Volume 256, Nos. 1–2, 1997

KLUWER ACADEMIC PUBLISHERS
DORDRECHT / BOSTON / LONDON

A C.I.P. Catalogue record for this book is available from the Library of Congress.

ISBN 0-7923-5527-X

Published by Kluwer Academic Publishers,
P.O. Box 17, 3300 AA Dordrecht, The Netherlands.

Kluwer Academic Publishers incorporates
the publishing programmes of
D. Reidel, Martinus Nijhoff, Dr W. Junk and MTP Press.

Sold and distributed in the U.S.A. and Canada
by Kluwer Academic Publishers,
101 Philip Drive, Norwell, MA 02061, U.S.A.

In all other countries, sold and distributed
by Kluwer Academic Publishers,
P.O. Box 322, 3300 AH Dordrecht, The Netherlands.

Printed on acid-free paper

Printed in the Netherlands

TABLE OF CONTENTS

Instabilities, Shock Waves and Plasma Spectroscopy

Astrophysical and Space Plasmas

Magnetic Confinement

Pulse Power Experiments in Plasmas

PREFACE

In 1982, the First Latin American Workshop on Plasma Physics was held in Cambuquira (Brazil). It has been a long way since this first meeting up to the actual one, the VII LAWPP held in Caracas, Venezuela. However, the structure of this first Meeting has remained unchanged. The first week of the workshop is always devoted ot short courses in the different fields of Plasma Physics, and the presentation of contributed and review papers is carried on during the second week.

In relation to the preceding meetings, the emphasis on Plasma Processing was increased and a short course in dusty plasma was introduced too. This area of plasma has become more important every year, even though the research in this field is at incipient stage in Latin-America. However we feel that its importance will increase in future years and we should keep step.

A large scope of fields in Plasma Physics has been covered in this meeting, however, several areas were left aside, because the field of Plasma Physics has become so broad that it cannot be covered in such a short period of two weeks. However, we have kept some equilibrium between Basic Plasma Physics and Applications. On the other hand, some of the most important plasma-related problems were presented. The actual situation on Plasma Fusion through Tokomak confinement has been analyzed, other alternatives of magnetic confinement devices (mirror machine, stellarators, etc.) were left aside. Prof. Malcolm Haines presented two excellent minicourses and review talks in z-pinch and inertial confinement. His contributions during the meeting and also in several stages of the preparation of the workshop are greatly appreciated. The excellent lectures given by most of the Invited Speakers and detailed here are also very esteemed.

The contributed papers presented here cover an ample spectrum of the plasma physics areas, including some papers in plasma processing and chemical plasmas, which were absent in previous LAWPP meetings.

Participants from twenty different countries were in this meeting, and nearly half of them were from beyond the Latin-American continent, showing that the LAWPP meetings are truly international.

All the sessions of the workshop were hel at the campus of the University Simón Bolívar, located in a suburban area of Caracas called Sartenejas Valley. The site is surrounded by several green areas and not far away from Avila Mountains, which can be seen from the nearby hills.

Astrophysics and Space Science **256:** ix–xi, 1998.

Prof. Freddy Malpica Pérez, Rector of the University Simón Bolívar, was the main speaker during the Inaugural Session at the beginning of the second week of the Workshop, devoted to the presentation of the Contributed and Review Papers. The vice-rector, Prof. Osmar Issa Issa, addressed the Opening Session at the beginning of the minicourses.

We also thank the financial support from other authorities of the University, as Prof. Carlos Pérez García (Dean of Postgraduate Studies) and Prof. Gianfranco Passariello (Dean of Research and Development). Architect Alejandro Chataing Roncajolo has also been very important as General Coordinator of all the activities of the VII LAWPP, as well as Mr Mauro Varela.

We also thank the several international and government agencies, that have contributed with funds, or other means to the development of this Workshop. Particularly we appreciate the support of CLAF (Centro Latinoamericano de Física), ICTP (International Centre of Theoretical Physics), The British Council and the sponsorship of the IAEA (International Atomic Energy Agency). As a whole, the main funds contributions came from the University Simón Bolívar, but the financial support of CONICIT (Concejo Nacional de Investigaciones Científicas y Tecnológicas) was also very important and appreciated.

We would also like to thank the Universidad Central de Venezuela, the Universidad del Zulia and the Universidad de Los Andes, for its encouragement and support in several ways to the development of the meeting, and to their students for their attendance during the Workshop. Finally thanks to the Ministry of the Secretary of the Presidency of the Venezuelan Republic for its contribution during the publication process.

We feel that the meeting has been very positive in the establishment of strong relations and collaborations among the physics researchers of Latin-America.

Finally the contributions of all the invited speakers are very much appreciated. Their lectures presented here give an excellent overview of the state-of-the-art in several areas of Plasma Physics. As such, it will be very useful for post-graduate students and researchers in this area of Physics.

PABLO MARTÍN
CHAIRMAN VII LAWPP
Depta. Física,
Universidad Simon Bolivar,
Apdo. 89 000,
Caracas 1080 A,
Venezuela

SPONSORS

- Universidad Simón Bolívar, USB
- Consejo Nacional de Investigaciones Científicas y Tecnológicas, CONICIT
- Centro Latino Americano de Física, CLAF
- International Center for Theoretical Physics, ICTP
- The British Council
- Ministry of the Secretary of the Presidency of the Republic of Venezuela
- Universidad Central de Venezuela, UCV
- La Universidad del Zulia, LUZ

INTRODUCTION

Old Greeks used to think that the matter of the universe was composed by four essential elements: earth, water, air and fire. Clearly the first three elements correspond to the three main states of the matter: solid, liquid and gas. However the fourth state (we are not considering here more sophisticated states as superconductivity, superfluity, etc.) the fire was forgotten until this century, when some physicist as K. Birkeland and I. Langmuir, etc, studied this important state that was denoted as plasma in 1923. One of the principal characteristics of this state is that the main constituent particles of the state were charged particles instead of neutral ones.

The discovery of this fourth state in Latin-America came much later in the second part of this century. We could say that the beginning of the study of the Plasma Physics in Latin-America was in 1949 during the government of Juan Domingo Perón in Argentina, when experiments with electric discharges in great scale were performed by Ronald Richer (Austrian physicist), looking for controlled thermonuclear fusion. In Brazil the Plasma Physics activities began in the 1950's at the University of Sao Paulo and it is now the most active country in Latin American. In about the same period work in the area of gaseous discharges started at the National Institute of Nuclear Research (ININ) in Mexico. In Chile these activities began in the 1960's (about 1966) with the installation of a shock tube at the University of Chile (FCFM). About ten years later in 1974 the first plasma physics group was created at the Universidad Simón Bolívar, Caracas, Venezuela and a DP machine was build there later. A plasma chemistry group began also activities at the Universidad Central, in Caracas in the 1970's. Colombian plasma physics began activities during the 1980's in the Universidad de Antioquia (Medellín), Universidad del Valle (Cali) and Universidad Nacional Sede Manizales (Manizales).

The development of Plasma Physics in each Latin-American country happened very independent. There was very little contact among the plasma physics groups in each country. However it is interesting to point out the strong collaboration of the Argentinean plasma physics researchers (Naum Friedenreich and Enrique Di Stefano) in the creation of the Chilean plasma physics group, as a result of the flight away of several researchers from the Juan Carlos Onganía dictatorship. The main relations of the Latin-American plasma physics groups were with the U.S.A. universities and colleges as well as with some European countries such as the United Kingdom (mainly Imperial College in London), Germany and France.

Astrophysics and Space Science **256:** xiii–xv, 1998.

In the 1980-decade (1982) is when the first Latin American meeting started, in Cambuquira, Brazil. The other Latin American Workshops were successively: 1987, Medellín (Colombia); 1988, Santiago, (Chile); 1990, Buenos Aires, (Argentina); 1992, Mexico City, (Mexico); 1994, Fox de Iguazu, (Brazil) and 1997, Caracas, (Venezuela).

The number of non-Latinamerican participants have increased with every new meeting and, in the VII LAWPP, there were participants from twenty different countries, with twelve countries from outside Latin-America. The meeting at Fox de Iguazu was held in coincidence with the 1994 International Conference in Plasma Physics, thus the international collaboration there, was very strong.

The scope of Plasma Physics is very broad. Thus electron plasma in a solid can be as low temperature as a few Kelvin degrees, and in the opposite sides plasma in fusion devices can reach temperatures as high as hundred of millions degrees. In the length scales, distances of the order of nanometers are significant in the etching integrated circuits and pellet manufacture for inertial fusion with laser in holhraums. In the opposite sides distances of the order of kiloparsecs are important for galactic plasmas. Thus, the selection of the invited speakers for the VII LAWPP was not easy. A criterion could be to select the people according to the field of interest of the Latin American plasma physics groups, but this idea is only attractive in appearance due to the following reasons. First, the creation of each Latin-American plasma physics groups is not always motivated by the interest of the Latin-American economy. Most of the time the field of research was derived for that of the field of the Ph.D. thesis of a given researcher, done in an USA or European laboratory. On the other hand, there are some fields where there is not any development in Latin-America, which seem very important, such as dusty plasmas, etc. Furthermore other areas of Plasma, as Plasma Focus, have experimented a strong development, somehow unusual, compared to their interests within the international community. There is another point to be considered and it is that these meetings represent a good opportunity to look the main stream of problems of the international plasma physics community. We feel that the selection of invited speakers should be a compromise among all the aspects of the problems here displayed.

Looking in detail tokamaks are actually in operation in Brazil, University of Sao Paulo and in Mexico, Comisión de Energía Nuclear (tokamak Novillo). Thus several invited lectures on this theme were presented including a general scope on confinement and instabilities. The problems of internal barriers, plasma rotation, design of magnetic divertors and edge problems were also considered. Field reversed configurations were also analyzed.

Inertial confinement fusion has been also treated widely including non-linear growth of Raleigh-Taylor instabilities, recent holhraums compression results, the proposed National Ignition Facility, old and new ideas on ICF, etc.

Excellent lectures on dense z-pinches have been also presented including numerical simulations and recent experimental results in MAGPIE and IMP fiber z-pinches. Basic problems of Plasma Kinetic Theory were also analyzed. The fast ignitor concept was also presented. Very clear and complete lectures on plasma Astrophysics and Cosmology have been presented including reviewed ideas on the expansion rate of the early universe, JETS, filaments etc. Numerical simulation in Astrophysics, ICF and plasma physics in general, including application, are described in a very clear way. An extensive presentation and a general review of dusty plasmas have been also carried out.

We are sorry that some of the lectures presented in the VII LAWPP do not appear finally in this book, because the papers were not sent or arrived late to be included.

Finally a very important part of the book contains the contributed papers to the VII LAWPP. The classification of the papers has been done according to the themes treated. The extension of each paper was also limited to about four pages. It seems to me that this book will be beneficial not only for the Latin-American plasma physics groups, but also for all the international plasma physics community.

PABLO MARTÍN

DENSE Z-PINCHES

M.G. HAINES
Blackett Laboratory, Imperial College, London SW7 2BZ, U.K.

Abstract. The basic underlying physics of the Z-pinch is reviewed; the Bennett relation, the snow-plough model, particle orbits, heat losses to electrodes, equilibrium profiles, radiative collapse, and finally stability in many regimes. Recent experimental results from Imperial College using the MAG-PIE and IMP generators on fibre Z-pinches are presented, together with X-ray results from wire arrays from SATURN and PBFA-Z at Sandia National Laboratory.

1. Introduction

Historically the Z-pinch was one of the earliest plasma devices, and was studied with a view to attaining thermonuclear fusion. There was early disappointment because of MHD instabilities that developed. Together with the perception of end-losses to the electrodes it led to the development of toroidal pinches with the addition of a toroidal magnetic field for stabilising. As is well known, the subject then branched into the reversed field pinch where the toroidal and poloidal field components are comparable and into the tokamak where the toroidal field is dominant but the plasma β is lower.

However, several factors have led to a revival, even a renaissance, in Z-pinches. First was the development of pulsed power (particularly at AWE Aldermaston) enabling a current of $>$ 1 MA at a voltage $>$ 1 MV for a pulse time $\sim$ 150 ns. The terawatt power input rendered heat losses to electrodes acceptable or even negligible. The high energy density needed to satisfy the Lawson conditions for 100 ns or to create an intense X-ray source in a high Z material could be achieved with this pulsed power. Secondly, it was discovered that the ideal MHD theory for stability was not applicable to a fusion Z-pinch, but other regimes, in particular the Vlasov large ion Larmor radius regime, were relevant. Lastly, it was realised that the conventional approaches to fusion, e.g. the tokamak, lead inevitably to a low power density, and hence to an expensive fusion reactor, because of the limitation of the magnitude of the magnetic field produced by a coil to 5–10 tesla. A new study of the dense Z-pinch as a compact, high power density source could therefore be justified. At Imperial College another objective of trying to attain radiative collapse to very high density at currents above the Pease–Braginskii value is also being addressed.

Astrophysics and Space Science **256:** 1–12, 1998.

2. Basic Physics

2.1. THE BENNETT RELATION

Bennett (1934) derived an expression for pressure balance in a Z-pinch which turns out to be valid for any current or pressure profile. From

$$\frac{\partial p}{\partial r} = -J_z B_\theta = -\mu_0 \frac{J_z}{r} \int_0^r J_z r dr$$

it can be shown that

$$Nk\,(T_i + ZT_e) = \int_0^a 2\pi p r dr = \mu_0 \pi \int_0^a \left[r J_z \int_0^r J_z r dr \right] dr = \frac{\mu_0}{8\pi} I^2,$$

where N is the ion line density and I the total current. Thus, in an experiment if N and I are known, a so-called Bennett temperature can be calculated.

2.2. THE SNOWPLOUGH MODEL

Early Z-pinch experiments were almost exclusively compression pinches in which gas in a cylindrical vessel with electrodes at each end was energised by an applied voltage. This led to a skin current which formed a magnetic piston. A simple snow-plough model in which the magnetic piston sweeps up also the gas in its path leads to a pinch time proportional to $\rho^{1/4}$ where ρ is the initial gas density, and a final particle energy proportional to the ion mass, the rate of rise of current $\dot{I}$ and $\rho^{-1/2}$. Other simplified models such a shock model or slug model gave similar scalings, but now more physics can be included in numerical simulations.

2.3. PARTICLE ORBITS

In the same way as Galeev and Sagdeev's (1968) consideration of trapped and passing particles led to a more complete understanding of tokamak transport properties, so particle orbits in a Z-pinch can lead to a better understanding of the physics. Figure 1 illustrates the two basic types of particle orbits: the off-axis (conventional) guiding centre drift orbit in which the E_r/B_θ, ∇B and curvature drifts are in the $\pm Z$ directions, and the singular orbits within one gyro radius of the axis. Haines (1978) showed that the singular flows of electrons and ions each can constitute half the total current, but diamagnetism redistributes the current density over the whole pinch cross-section.

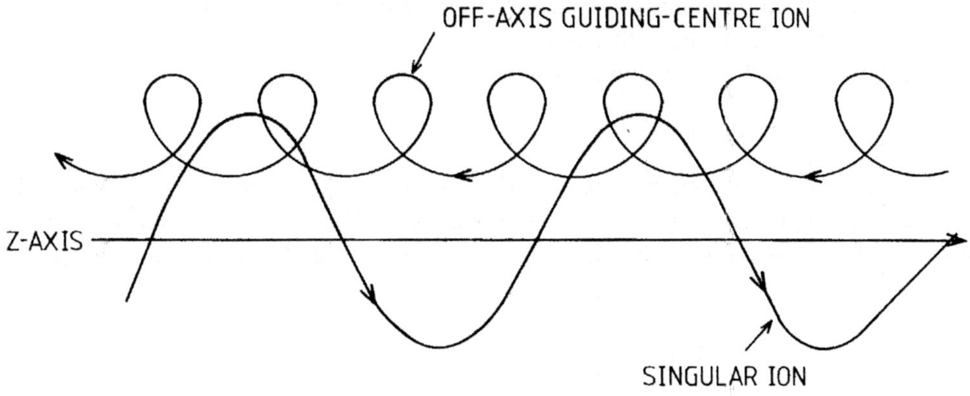

Figure 1. Two basic types of charged particle orbits in a Z-pinch: the off-axis guiding centre motion, and the singular snake-like motion within one Larmor radius of the axis.

2.4. Joule Heating under Pressure Balance

By considering the rate of heating by Joule heating together with bremsstrahlung losses, Haines (1960) showed how the temperature T and current I varied in time under pressure balance at constant radius. In the limit where bremsstrahlung is negligible it is found that the current must vary as $t^{1/3}$ where t is the time, and a self-similar current profile having the property of a weak skin effect where the surface current density is 1.66 times the axial value occurs. The final steady state when the Joule heating balances bremsstrahlung occurs at a unique current, the Pease–Braginskii current $I_{\rm PB} = 0.43(\ln\lambda)^{1/2}$ MA found independently in 1957 (Pease, 1957;, Braginskii, 1958).

2.5. Heat Losses to Electrodes

A model of a steady state Z-pinch under pressure balance with Joule heating balanced by heat flux to the electrodes and bremsstrahlung has been solved analytically (Haines, 1961) in two extreme cases, $\omega_e\tau_{ei} = 0$ and ∞, where $\omega_e\tau_{ei}$, the product of the electron cyclotron frequency and electron-ion collision time, is called the Hall parameter. Surprisingly the same scaling laws are found, the numerical coefficients differing by only 30%. Thus we can have confidence in the results being robust.

A simple estimate for $\omega_e\tau_{ei} = \infty$ is based on the heat flux being essentially the enthalpy flow by the current density. This gives for power balance

$$IV = \frac{5}{2}\frac{kT}{e} I,$$

where V is the voltage. The resistance of the pinch column is

$$\frac{V}{I} = \frac{z_0}{\pi a^2 \alpha T^{3/2}},$$

where the electrical conductivity σ is $\alpha T^{3/2}$, α being approximately constant, z_0 is the length of the pinch and a the radius.

For the $\omega_e \tau_e = 0$ case we write instead

$$IV = \pi a^2 \lambda T^{5/2} T / z_0,$$

where the thermal conductivity κ is $\lambda T^{5/2}$, λ being approximately constant and ∇T has been represented as T/z_0. Both models give $V \propto T$ and $I \propto \pi a^2 T^{5/2}/z$ in absence of bremsstrahlung, while a full calculation yields for T_e (°k) and $\omega_e \tau_{ei} = \infty$

$$T_e = 4.64 \times 10^3 V \left(1 - \frac{I^2}{I_{\rm PB}^2}\right),$$

$$T_e^{5/2} = 2.04 \times 10^7 \frac{z_0}{\pi a^2} I \left(1 - \frac{I^2}{I_{\rm PB}^2}\right).$$

2.6. Fusion Conditions

The above relations can be used to show for $I/I_{\rm PB} \ll 1$ the Lawson product

$$n\tau_E = 1.71 \times 10^{-11} N T^{3/2},$$

where the energy confinement time τ_E is defined by

$$\tau_E = \frac{\text{energy content}}{\text{power input}} = \frac{\frac{3}{2} Nk\,(T_e + T_i)}{IV} z_0.$$

Thus, for a given $n\tau_E$ and $T (= T_e = T_i)$ the parameters N, V, I (from the Bennett relation) and $z_0/\pi a^2$ are determined. This only leaves τ_E, n, z_0 and a to be found; the above relations mean that only one of these is free to be chosen, the other three being determined (Haines, 1978). Pulsed power technology dictates that τ_E should be about 100 ns. This then leads to a current of 1 MA at a steady voltage of $\sim$ 100 kV (over 1 MV is needed to arrive at this point) and a pinch length of about 5 cm, radius of 20 μm and density 5×10^{21} cm^{-3}. Of relevance to stability, the low line density means that the ion Larmor radius is 0.1 to 0.3 times the pinch radius. Thus directly using the Z-pinch for fusion leads to a very compact magnetically confined system.

2.7. Radial Profiles

We have already discussed a self-similar weakly skin current profile which arises in a Joule-heated pinch with uniform temperature. Coppins et al. (1988) found another self-similar profile which arises when the radial thermal conduction is negligible and leads to a centrally peaked current and temperature. This will be relevant to experiments with the gas-embedded pinch.

2.8. Lower Hybrid Turbulence

Near the surface of the Z-pinch the electron density is low and ends to zero, yet the classical current density would be finite. This would lead to a high drift velocity which would trigger a velocity-space instability, especially the lower hybrid instability (see Krall and Liewer, 1971) since the current density is orthogonal to the magnetic field. This micro-turbulence will lead to a reduction in $\underline{J}$ and hence in the $\underline{J} \times \underline{B}$ force, so causing a radial expansion or ablation of the low density surface plasma. Radial streak camera images of optical emission show indeed a rapid expansion from a fibre pinch (Chittenden et al., 1994). Chittenden (1995) has simulated this using Drake et al.'s (1984) formula for the effective anomalous collision frequency.

3. Radiative Collapse

We have discussed how with energy and pressure balance the pinch current is the Pease–Braginskii value. A simplified model shows that because σ is $\alpha T^{3/2}$ the Joule heating J^2/σ balancing bremsstrahlung loss per unit volume $\beta_b n^2 T^{1/2}$ where β_b is a constant leads to $J \propto nT \propto p$. Pressure balance means $JB \sim p/a$ and hence $Ba \propto I$ is a constant.

A zero-dimensional model of a pinch with variable radius $a(t)$ and current $I(t)$ satisfying the Bennett relation shows how a linearly rising current will lead to a collapse to zero radius at $I = 3I_{\mathrm{PB}}$. But this requires an infinite voltage. Coupling $I(t)$ to a realistic circuit leads to a finite but spectacular collapse of a Z-pinch to 10^4 times solid density followed by an expansion to a Pease–Braginskii equilibrium (Haines, 1989). A time-dependent 1-D simulation by Chittenden and Haines (1990) which includes electron degeneracy effects on the electron pressure and on the conductivity and which also includes opacity effects and cut-off when the radiation angular frequency is below the very high value of ω_p the plasma frequency leads to a greater compression. This is illustrated in Figure 2. However, all of this supposes stability which may not be true. This we consider next.

4. Stability

4.1. The Universal Diagram

It is a classical example of ideal MHD stability theory that the Z-pinch is unstable to the $m = 0$ and $m = 1$ instabilities, commonly called the sausage and kink modes respectively. It is possible to stabilise the $m = 0$ in a fixed boundary model with an axially peaked current and pressure profile that satisfies the Kadomtsev criterion (Kadomtsev, 1966) for interchange modes. A conducting wall can stabilise long wavelength modes but not short wavelength modes.

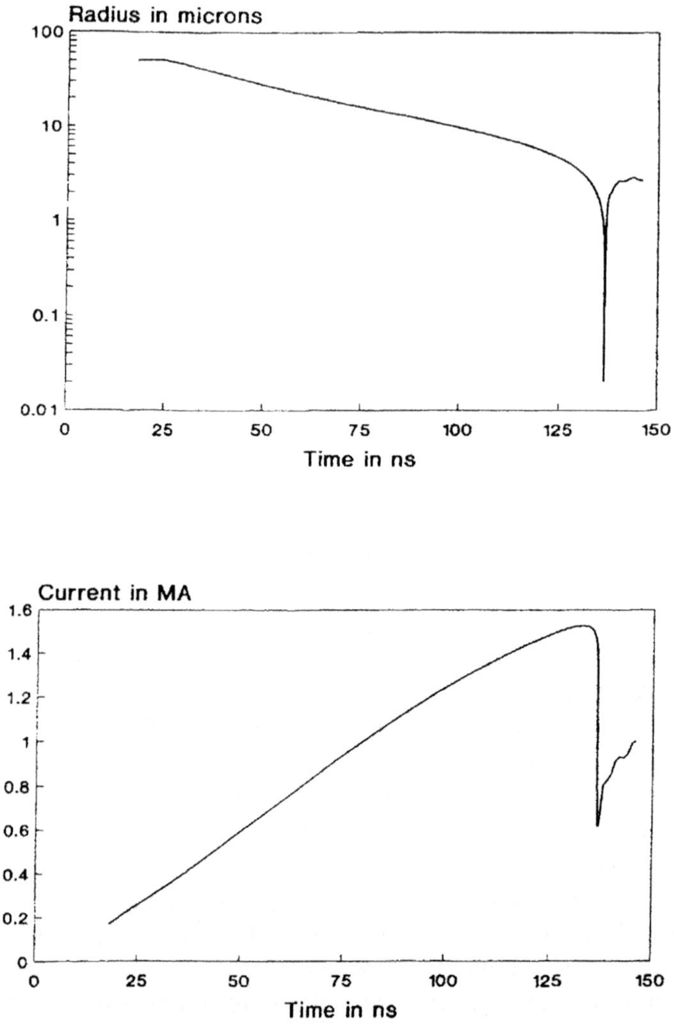

Figure 2. Radiative collapse of a Z-pinch is illustrated by the evolution of the pinch radius and current with time, the collapse occurring at about 141 ns.

However, ideal MHD theory is not valid for many Z-pinches of interest because it neglects resistivity, viscosity, finite ion Larmor radius effects, pressure anisotropy, and the Hall and electron pressure effects in Ohm's law. The Lundqvist number, $S \approx \mu_0 \sigma v_A a$, is the relevant dimensionless parameter for resistive effects on MHD modes. Here v_A is the Alfvén speed. When the Bennett relation is satisfied S becomes $3.9 \times 10^{23} I^4 a/N^2(A^4 m^3)$. Theory indicates that for $S < 10^2$, resistive effects dominate and there is no exponential growth of MHD modes (Culverwell and Coppins, 1990). The condition for isotropic viscous damping and for anisotropic ion pressure effects to be important is $v_A \tau_i / a = 2.1 \times 10^{39} I^4 a/N^3$ to

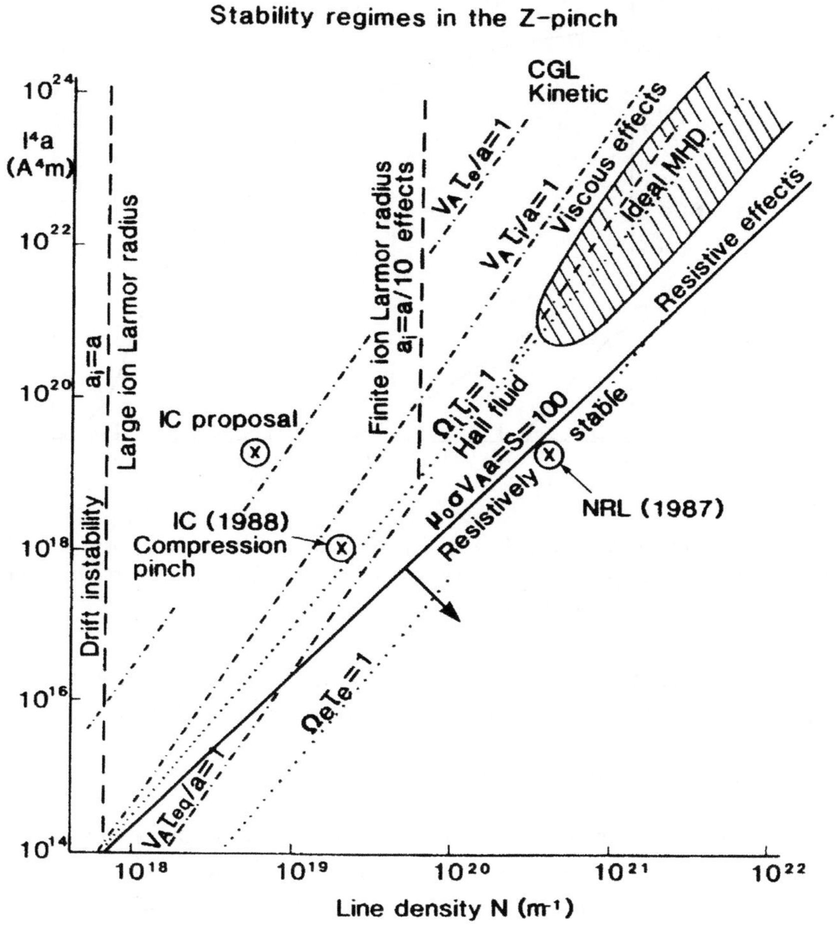

Figure 3. Regimes of validity of various stability models are shown in a plot of I^4a *versus* N for a Z-pinch under pressure balance.

be larger than unity. Finite ion Larmor radius effects are important when $a_i/a = 8.1 \times 10^8 N^{-1/2} > 0.05$ when $\omega_i \tau_i = 3.6 \times 10^{30} I^4 a/N^{5/2} > 1$. All of these conditions can be encapsulated in Figure 3 which is a plot of I^4a *versus* N. Only in the wedge shaped hatched region is MHD stability valid (Haines and Coppins, 1991).

4.2. Gas-Embedded Z-Pinch Stability

In nearly all Z-pinches the dominant mode that is found is the $m = 0$. The gas embedded Z-pinch, initiated by laser or e-beam ionisation on axis is an exception. Results by Choi et al. (1988) show that only an $m = 1$ kink develops, and that

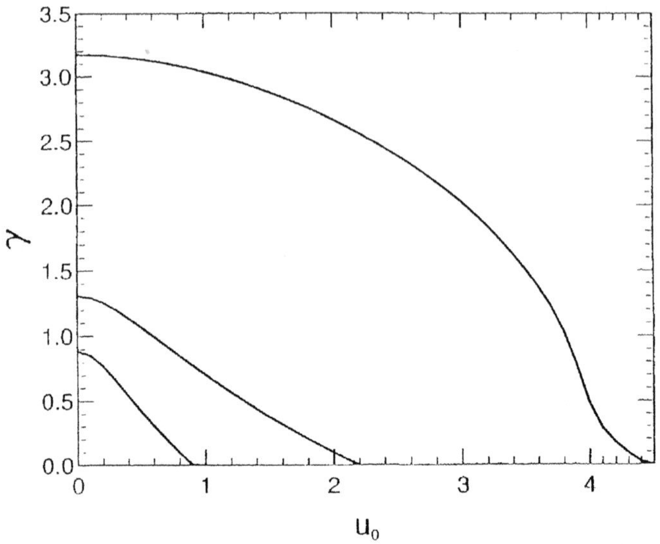

Figure 4. The effect of sheared axial flow on Z-pinch stability is shown in a plot of normalised growth rates of the three fastest growing modes *versus* u_0, where $u_z(r) = u_0(1 - r^2/a^2)$. This is for free boundary modes for $m = 0$ and $k_a = 10$.

the early displacement (eigenmode) is a maximum on the axis, in agreement with theory. This is a good example of experiment supporting theory.

4.3. Large Larmor Radius Effects

Two numerical models, an initial value code and a variational code, have been used by Arber et al. (1994, 1995) to solve the linearised Vlasov equation for ions in a fluid electron background. Maximum reduction in growth rate for a pinch with a parabolic pressure (density) profile is found at $a_i/a \sim 0.1$; for $m = 0$ a 70% reduction, for $m = 1$, 80%. The next question is whether under this condition there is non-linear saturation of the modes. Arber (1996) has shown that this does not occur, but with his model unrealistically large values of the electron drift velocity occur at the plasma-vacuum interface. The model is very sensitive to this, and further work including the onset of lower hybrid edge turbulence is required.

4.4. Sheared Axial Flow

Arber and Howell (1996) have developed a linear stability theory with an equilibrium axial flow velocity $u(r)$ added to ideal MHD. Figure 4 shows the reduction in growth rate with amplitude of u. However, high k modes are always unstable, and too great a value of u (e.g. a Mach number > 4) leads to the triggering of Kelvin–Helmholtz instabilities. Both 2-D and 3-D non-linear simulations show, however, that the modes saturate at an amplitude of about 20% with a shear Mach number 1.25 to 2 (Chittenden and Lucek, 1997).

4.5. Additional Axial Magnetic Field

An additional axial magnetic field can stabilise the pinch provided there is sufficient shear for Suydam's criterion to be satisfied. In a dense Z-pinch realistic values of magnetic field $\sim 10^4$ tesla can however only be provided by employing a compressing liner. In this case (e.g. Felber, 1982) it is possible to conceive an isolated fibre on axis developing into a stable θ-pinch.

5. Experimental Z-Pinch Configurations

Figure 5 illustrates nine methods of producing Z-pinches: the original compression Z-pinch for which the snowplough model was developed, the capillary discharge now used for X-ray laser studies, the gas embedded pinch, again with additional micro-discharge, a wire-array pinch as used for intense transient X-ray sources, a gas-puff pinch, a fibre pinch, multiple shell pinch and a plasma on wire pinch. In addition these can be modified by the addition of sheared axial flow, rotation, or additional axial magnetic field.

6. Imperial College Experiments with the MAGPIE Generator

The MAGPIE generator can supply up to 2 MA at a maximum voltage of 2.4 MV with a pulse length of 200 ns to a fibre pinch. The main purpose of these experiments is to study radiative collapse and fusion conditions in hydrogen and deuterium fibres. Preliminary experiments have been carried out on mainly 33 μm diameter carbon or CD_2 fibres. The main features of the results have been (i) initial surface ablation and coronal MHD instabilities resulting in transient ($\sim$ 1 ns) intense X-ray bright spots from $m = 0$ necks near the surface of the ionising fibre. The spots bifurcate and propagate axially with a high velocity. A 2-D MHD code with an ionisation package has successfully modelled this, showing that the effect is an ionising MHD axially propagating front, triggered by the $m = 0$ mode. (ii) There follows a quiescent period during which there is an evolution of a slowly growing $m = 0$ mode. (iii) This is followed by the growth of distinct density islands along the axis and the onset of a disruption at about 160 ns characterised by the production of an intense hard X-ray burst and, in the case of CD_2 fibres, a neutron burst with a yield of $\sim 10^9$. The neutrons are not isotropic in origin but have axially peaked distributions in both directions.

Recently, simultaneous interferometry, schlieren shadowgrams and Faraday rotation measurements have been made that demonstrate the detailed structure of the coronal MHD modes and expansion. The magnetic field measurements show that over 90% of the current is carried within 1 mm of the axis.

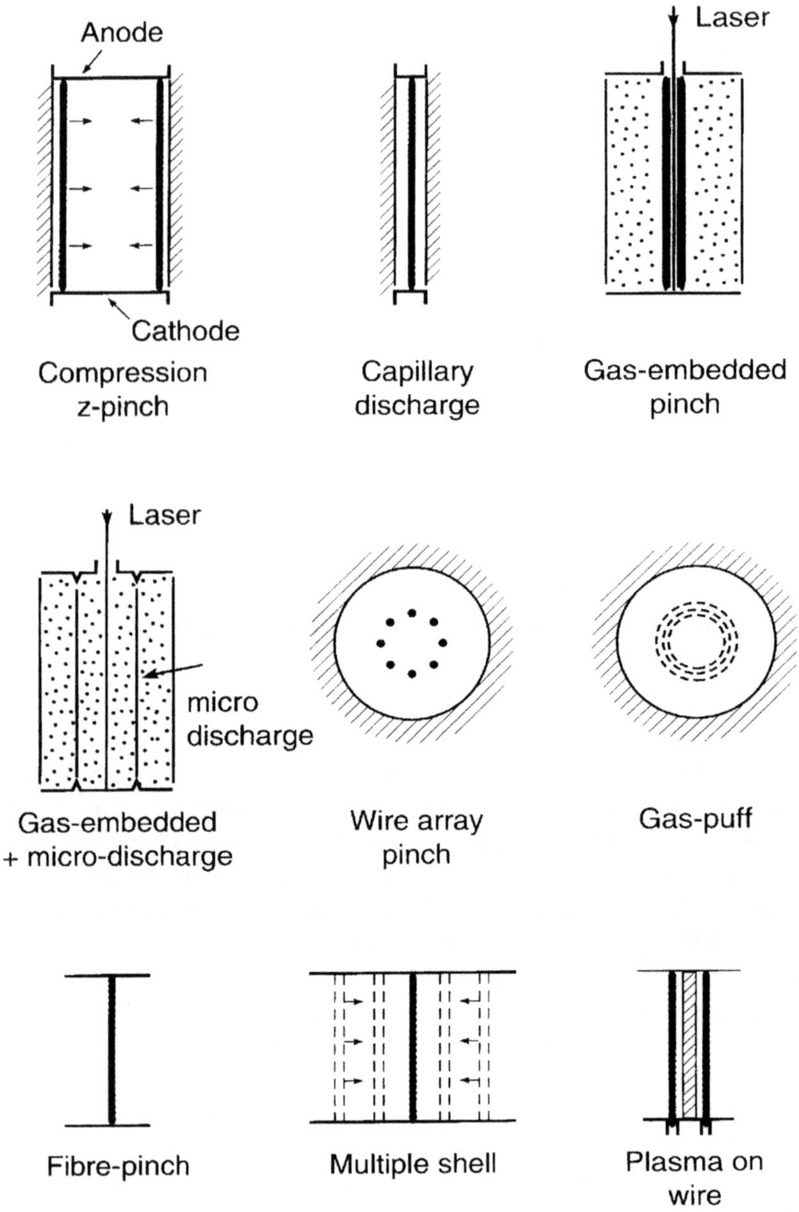

Figure 5. Nine methods of producing Z-pinches.

7. Experiments at Sandia National Laboratory

The Sandia Z-pinch programme is mainly concerned with the production of intense transient soft X-rays which can be used for example to create a hohlraum for radiation-drive experiments including spherical capsules for inertial confinement. Gas puffs, e.g. Kr or wire arrays are employed. Recent experiments with up to 136 aluminium wires (Sanford et al., 1996) and up to 120 tungsten wires (Deeney, 1997) have shown a greatly enhanced and sharper in time soft X-ray emission. With tungsten up to 85 TW of soft X-rays in 4 ns (i.e., $\sim$ 500 kJ) were generated on the Saturn generator leading to a hohlraum temperature approaching 100 eV. More recently using the larger 11.4 MJ PBFA-Z generator 1.8 MJ of soft X-rays have been produced with a wall-plug efficiency of 15% at 19.5 MA.

8. Conclusion

The dense Z-pinch is yielding exciting results as a powerful transient X-ray source; theory and simulations are leading to a better understanding of its properties, and there are several proposed schemes for improving its overall stability as a compact plasma of high energy density for fusion applications.

References

Arber, T.D.: 1996, *Phys. Rev. Lett.* **77**, 1766.
Arber, T.D. and Howell, D.F.: 1996, *Phys. Plasmas* **3**, 554.
Arber, T.D., Coppins, M. and Scheffel, J.: 1994, *Phys. Rev. Lett.* **72**, 2399.
Arber, T.D., Russell, P.G.F., Coppins, M. and Scheffel, J.: 1995, *Phys. Rev. Lett.* **74**, 2698.
Bennett, W.H.: 1934, *Phys. Rev.* **45**, 890.
Braginskii, S.I.: 1958, *Sov. Phys. JETP* **6**, 494.
Chittenden, J.P.: 1995, *Phys. Plasmas* **2**, 1242.
Chittenden, J.P. and Haines, M.G.: 1990, *Phys. Fluids* **B2**, 1889.
Chittenden, J.P. and Lucek, S.: 1997, Private communication.
Chittenden, J.P., Bayley, J.M., Mitchell, I.H., Wilson, K., Worley, J.F., Choi, P., Dangor, A.E. and Haines, M.G.: 1994, in: *Proc. 10th Int. Conf. on High Power Particle Beams* (San Diego), NTIS No. PB95-144317, Vol. 2, p. 756.
Choi, P., Coppins, M., Dangor, A.E. and Favre, M.B.: 1988, *Nucl. Fusion* **28**, 1771.
Coppins, M., Culverwell, I.D. and Haines, M.G.: 1988, *Phys. Fluids* **31**, 2688.
Culverwell, I.D. and Coppins, M.: 1990, *Phys. Fluids* **B2**, 129.
Deeney, C.: 1997, Private communication.
Drake, J.F., Guzdar, P.N., Hassan, A.B. and Huba, J.D.: 1984, *Phys. Fluids* **27**, 1148.
Felber, F.S.: 1982, *Phys. Fluids* **25**, 643.
Galeev, A.A. and Sagdeev, R.Z.: 1968, *Sov. Phys. JETP* **26**, 233.
Haines, M.G.: 1960, *Proc. Phys. Soc.* **76**, 249.
Haines, M.G.: 1961, *Proc. Phys. Soc.* **77**, 643.
Haines, M.G.: 1978, *J. Phys. D; Appl. Phys.* **11**, 1709.
Haines, M.G.: 1989, *Plasma Phys. Controlled Fusion* **31**, 759.

Haines, M.G. and Coppins, M.: 1991, *Phys. Rev. Lett.* **66**, 1462.
Kadomtsev, B.B.: 1966, in: M.A. Leontovich (ed.), *Reviews of Plasma Physics*, Vol. 2, Consultants Bureau, New York, p. 153.
Krall, N.A. and Liewer, P.C.: 1971, *Phys. Rev. A* **4**, 2094.
Pease, R.S.: 1957, *Proc. Phys. Soc.* **70**, 11.
Sanford, T.W.L. et al.: 1996, *Phys. Rev. Lett.* **77**, 5063.

COMPUTATIONAL SIMULATION OF PLASMAS

A.R. BELL
Imperial College, London SW7 2AZ, U.K.

Abstract. This paper reviews the use of computational simulation in plasma physics. It describes a range of numerical models varying from particle models of low density collisionless plasmas to fluid models of high density collision-dominated plasmas. Some applications of these models, particularly to laser-produced and astrophysical plasmas, are described.

1. Introduction

The greater part of plasma physics can be described by the four Maxwell equations and the Lorentz force equation of motion for charged particles. These equations which appear so simple are in actuality very difficult to solve for realistic problems. Plasmas are usually far from thermal equilibrium, rapidly evolving, highly non-linear and often unstable. Under these circumstances, the plasma equations can often only be solved numerically. Numerical simulation is important for both theoreticians and experimentalists. In fact, numerical modelling is essential for many experiments for which simulation forms an integral part of the difficult process of experimental diagnosis. Simulation codes are also important in designing future experiments. For example, the design of the US National Ignition Facility (NIF) has been specified with the help of large inertial confinement simulation codes, and the same applies to all large experimental facilities. The situation in astrophysics and space physics is similar since observations are always limited. Simulations codes can be used to explore the gaps in our knowledge and test theories by investigating whether they can reproduce the observations.

Plasma simulations can be divided into two main categories: kinetic and fluid. Kinetic simulations follow the plasma evolution in phase space. This is essential for many basic plasma phenomena such as wave-particle interactions. Fluid simulations, in contrast, integrate over velocity space to maintain only very basic information about distribution functions such as the mean particle velocity (fluid velocity) and temperature (mean square velocity spread). A fluid model reduces the information content of the simulation with the pay-off that much larger systems can be modelled for much longer times. Most simulations of complete experiments, such as tokamak and laser-produced plasmas, use a fluid model, and the same applies to astrophysical objects such as supernovae. Kinetic simulations usually model only one element of a complete system such as a collisionless shock in the solar wind or the absorption of laser irradiation in a laser-plasma experiment. In

Astrophysics and Space Science **256:** 13–35, 1998.

this review, I firstly discuss kinetic simulation before moving on to fluid models. Such a wide range of numerical models is used in plasma physics that it would be impossible to give a full account of them all, but my aim here is to concentrate on the core codes used to describe plasmas. Examples of applications will be taken primarily from laser-produced and astrophysical plasmas.

2. Phase-Space Models

The particle-in-cell (PIC) method is nearly the most basic form of plasma simulation. It is sufficiently close to a real plasma that a PIC calculation can be called a numerical experiment. If all the particles in a plasma can be treated classically (neglecting quantum effects) and interact through the Lorentz force then their motions are described by the equations

$$\frac{d\mathbf{v}_i}{dt} = \frac{q_i}{m_i}(\mathbf{E} + \mathbf{v}_i \wedge \mathbf{B}), \qquad \frac{d\mathbf{r}_i}{dt} = \mathbf{v}_i,$$

$$\nabla.\mathbf{E} = \rho/\epsilon_0, \qquad \nabla.\mathbf{B} = 0,$$

$$\nabla \wedge \mathbf{E} = -\partial\mathbf{B}/\partial t, \qquad \nabla \wedge \mathbf{B} = \mu_0\mathbf{j} + \frac{1}{c^2}\partial\mathbf{E}/\partial t,$$

$$\rho = \sum_{i=1}^{N} \delta(\mathbf{r} - \mathbf{r}_i)q_i, \qquad \mathbf{j} = \sum_{i=1}^{N} \delta(\mathbf{r} - \mathbf{r}_i)\mathbf{v}_i q_i,$$

where $\mathbf{r}_i$, $\mathbf{v}_i$, m_i and q_i are the position, velocity, mass and charge of the ith ($i = 1, N$) particle. These equations can in principle be solved as they stand. The problem is that the total number of electrons and ions can be excessively large. N is of the order of 10^{21} for a tokamak, 10^{20} for an ICF target, and 10^{57} for a supernova remnant. The maximum practical limit for modern computers is of the order of 10^7 particles. The way around this limitation is to calculate the motions of representative particles, or alternatively to think of the computational particles as macroparticles which are each the combination of many plasma particles. Macroparticles behave in the same was as real particles if their charge to mass ratio is the same as for the real particles. The other problem with the particle model is that each of N particles interacts with $N - 1$ other particles giving $N(N - 1) \approx N^2$ interactions in total. Computers cannot handle this large number of interactions if N is of the order of 10^7. The solution to this problem is to model only collective interactions instead of each individual binary interaction. Instead of calculating the fields on each particle directly, the fields are calculated on a spatial grid. If the computational space is 2-dimensional (2D) in x and y, it is divided into $N_x \times N_y$ computational cells of dimension $\Delta x \times \Delta y$. The number of operations in the field calculation is then proportional to either N or $N_x \times N_y$, which are both much smaller than N^2, thus making the calculation feasible for a large number of particles.

2.1. THE PIC METHOD

A PIC code (Birdsall and Langdon, 1991; Hockney and Eastwood, 1988) can be separated into two parts: the particle-mover and the field-solver. The particle-mover takes the known fields, accelerates the particles in those fields, and then moves the particles according to their new velocity. The difference form of the equations for the acceleration of each ith ($i = 1, N$) particle is

$$\mathbf{v}'_i = \mathbf{v}_i^{m-1/2} + \frac{q_i}{m_i}\mathbf{E}_i^m \frac{\Delta t}{2}, \qquad \mathbf{v}''_i = \mathbf{v}'_i + \mathbf{v}'_i \wedge \mathbf{t},$$

$$\mathbf{v}'''_i = \mathbf{v}'_i + \mathbf{v}''_i \wedge \mathbf{s}, \qquad \mathbf{v}_i^{m+1/2} = \mathbf{v}'''_i + \frac{q_i}{m_i}\mathbf{E}_i^m \frac{\Delta t}{2},$$

where

$$\mathbf{t} = \frac{q_i}{m_i}\mathbf{B}_i^m \frac{\Delta t}{2}, \qquad \mathbf{s} = \frac{2\mathbf{t}}{1 + t^2}$$

and Δt is the timestep. The superscripts m and $m \pm 1/2$ refer to the time level. For example, any quantity superscripted m is evaluated at time $m\Delta t$. The equation for $\mathbf{v}'$ steps forward half a timestep, $\Delta t/2$, the acceleration due to the electric field. The equations for $\mathbf{v}''$ and $\mathbf{v}'''$ apply the magnetic field for a full timestep Δt by rotating the direction of the velocity without changing its magnitude. The equation for $\mathbf{v}_i^{m+1/2}$ applies the electric field for a further half timestep. The electric and magnetic fields $\mathbf{E}_i$ and $\mathbf{B}_i$ are evaluated at the position $\mathbf{r}_i^m$ of the ith particle at time $m\Delta t$. Note that the equations are symmetric in time about the time $m\Delta t$. Such equations are 'centre differenced' and are more accurate than uncentred equations. $\mathbf{v}_i^{m+1/2} = \mathbf{v}_i^{m-1/2} + (q_i/m_i)\mathbf{E}_i^{m-1/2}\Delta t$ would be an example of an uncentred equation because the electric field is evaluated at time $(m - 1/2)\Delta t$ instead of $m\Delta t$. Each particle position is updated according to the equation

$$\mathbf{r}_i^{m+1} = \mathbf{r}_i^m + \mathbf{v}_i^{m+1/2}\Delta t.$$

This equation is similarly centred in time so as to give an overall high accuracy. The overall scheme is known as a leapfrog scheme. The position equation is centred at time $(m + 1/2)\Delta t$ by leapfrogging from time $m\Delta t$ to $(m + 1)\Delta t$ over a velocity evaluated at time $(m + 1/2)\Delta t$. Similarly the acceleration equations leapfrog the velocity from time $(m - 1/2)\Delta t$ to $(m + 1/2)\Delta t$ over fields $\mathbf{E}$ and $\mathbf{B}$ evaluated at time $m\Delta t$.

In a PIC code the electric and magnetic fields are calculated on a spatial grid as mentioned above. This is most easily demonstrated when there is only one spatial dimension x. Each of the particles carries a charge and a current. If the ith particle is positioned in the nth spatial cell, meaning that its x coordinate lies in the range $n\Delta x < x_i < (n+1)\Delta x$, then its charge q_i is split between the two nearest grid-points to increase the charge at each point n by an amount σ_n. A common way of splitting the charge is to use a linear weighting:

$$\sigma_n = q_i \left(\frac{(n+1)\Delta x - x_i}{\Delta x} \right), \qquad \sigma_{n+1} = q_i \left(\frac{x_i - n\Delta x}{\Delta x} \right).$$

The charges of each particle are accumulated in this way to calculate the total charge σ_n associated with each grid-point. The currents of each particle are similarly accumulated onto the grid-points to give the total currents $\mathbf{j}_n$. The 1D Maxwell equations can then be solved to update the fields. The electric potentials ϕ_n and the electrostatic field in the x direction are calculated from the difference equations

$$\frac{\phi_{n-1} - 2\phi_n + \phi_{n+1}}{\Delta x^2} = -\frac{\sigma_n}{\epsilon_0 \Delta x}, \qquad E_x = -\frac{\phi_{n+1} - \phi_{n-1}}{2\Delta x}.$$

The magnetic field and the transverse electric field (in the y and z directions) can be calculated in a variety of ways, one of which is to difference the Maxwell equations straightforwardly as a leap-frog scheme with **B** and the transverse components of **E** offset by $\Delta x/2$ in space and $\Delta t/2$ in time. The coupled equations for E_z and B_y take the form

$$\frac{E^{m+1/2}_{z\ n} - E^{m-1/2}_{z\ n}}{\Delta t} = c^2 \frac{B^m_{y\ n+1/2} - B^m_{y\ n-1/2}}{\Delta x} - \frac{j^m_{z\ n}}{\epsilon_0 \Delta x},$$

$$\frac{B^{m+1}_{y\ n+1/2} - B^m_{y\ n+1/2}}{\Delta t} = \frac{E^{m+1/2}_{z\ n+1} - E^{m+1/2}_{z\ n-1}}{\Delta x}.$$

Differencing of this kind is used in more than one spatial dimension, but in 1D a simpler form can be implemented in terms of quantities P and Q which correspond to waves propagating in the $+x$ and $-x$ directions respectively. In this method, **E** and **B** do not have to be staggered in time or space.

$$P^m_n = (E^m_{z\ n} + cB^m_{y\ n}), \qquad Q^m_n = (E^m_{z\ n} - cB^m_{y\ n}).$$

The timestep is often chosen such that $\Delta t = \Delta x/c$ in which case the electromagnetic wave moves a distance Δx in one timestep Δt, and the difference equations take a simple form:

$$P^{m+1}_n = P^m_{n+1} - \frac{\Delta t}{\epsilon_0 \Delta x} j^{m+1/2}_{z\ n+1/2}, \qquad Q^{m+1}_{n+1} = Q^m_n - \frac{\Delta t}{\epsilon_0 \Delta x} j^{m+1/2}_{z\ n+1/2}.$$

The electric and magnetic fields can easily be calculated from P and Q. As an alternative method, it is also practical to solve the transverse field equation in terms of the vector potential **A**.

When the electric and magnetic fields have been updated, the calculation cycles back to the particle-mover and each particle is accelerated in the new fields. The values of the field at the position of each particle are calculated by interpolating from the spatial grid using the same weighting process used to weight the particle charges and currents onto the grid. It is essential that the weighting procedure is the same for fields, charges and currents. Otherwise the particles experience a spurious 'self-force' which attracts them to or repels them from the nearest grid-point.

The consequence of this grid method is that the fields cannot be resolved on distance scales less than Δx. Strong binary collisions arising from the close approach of two particles are ignored. In effect, the grid method smooths out the fields over distances of Δx. Fortunately, this is unimportant for many plasma problems since the evolution of a plasma is often dominated by collective oscillations rather than binary collisions. The PIC method solves the Vlasov rather than the Boltzmann equation.

The book by Birdsall and Langdon (1991) gives an excellent introduction to the methods of solving both the particle and the field equations.

2.2. Applications of PIC Codes

The use of PIC codes has been reviewed by Dawson (1995). They have been widely used in the simulation of solar system plasmas (e.g. Ashour-Abdalla, 1985). Space plasmas have a low density and high temperature such that the collisional mean free paths are comparable with interplanetary distances and hence collisions can be neglected. Satellite measurements have shown that space plasmas are not Maxwellian, so they can only be modelled with a method which retains detailed velocity space information. A fluid model cannot model many important phenomena in space, and a PIC code is needed. One of the most successful uses of PIC codes in space plasmas is the simulation of shock fronts since the physics of collisionless shocks is dominated by non-linear waves, turbulence and wave-particle interactions. Shock fronts have a spatial structure which is relatively well defined and have a relatively limited spatial extent as required for PIC modelling for which the number of particles is limited by the speed and memory of computers. PIC codes have been used to develop very detailed models of laminar shocks which can adequately be modelled in one-dimension. However, a 1D model is less successful for shocks which propagate nearly parallel to the magnetic field and for highly supersonic shocks. These shocks are highly turbulent and unsteady, requiring simulation in two or three dimensions.

PIC codes have also been used extensively for laser-produced plasmas (e.g. Wilks et al., 1992; Gibbon and Bell, 1992; Guerin et al., 1995; Pukhov and Meyer-ter-Vehn, 1996). These plasmas have a much higher density. In some respects they are very collisional so PIC simulation appears inappropriate at first sight. However, the laser is unable to penetrate above the critical density of 10^{21}–10^{22} electrons cm^{-3}, so there is an important low density high temperature corona in which collisions are less important. Moreover the electromagnetic and electron plasma waves have periods which are very short, around 10^{-15} second, which is much shorter than the collision time. These waves are kinetic phenomena which need to be modelled in velocity space. Hence the Vlasov equation gives the appropriate description and PIC codes are widely used.

PIC codes are used to simulate the absorption of laser light (Figure 1). Collisionless absorption is strongest at oblique incidence. If the laser deviates from normal incidence by an angle θ the electromagnetic wave penetrates to a density of $n_{\text{reflect}} = n_c \cos^2 \theta$ where n_c is the critical density. An evanescent wave tunnels from the reflection density n_{reflect} to the critical density n_c where it resonantly ($\omega_p = \omega_{\text{laser}}$) excites electron plasma waves. The plasma waves undergo wave-breaking and give their energy to high energy 'suprathermal' electrons. Thus, the laser energy is absorbed into fast electrons which then travel freely into the solid target. This has been the subject of many PIC simulations. It used to be thought that collisionless absorption had to be modelled in two spatial dimensions because of the requirement of oblique laser incidence. 2D PIC calculations are very demanding computationally. However, it was shown by Gibbon and Bell (1992) that many aspects of absorption could be modelled computationally in 1D. Oblique laser incidence at an angle θ onto a stationary target is equivalent (by Lorentz transformation) to normal incidence on a target which is moving laterally at a speed $c \cos \theta$. 1D simulation is computationally much less expensive, and extensive modelling can be performed on relatively small computers.

In the context of laser-produced plasmas, PIC codes are also used to model parametric instabilities (Kruer, 1988). This is particularly important in both directly and indirectly (x-ray) driven ICF targets. These experiments last a relatively long time (nanoseconds) and the ablating plasma evolves into a large plasma corona. The electromagnetic wave propagating a large distance through the plasma is parametrically unstable to the growth of Langmuir waves (Raman instability) and ion acoustic waves (Brillouin instability). The Raman instability is dangerous because it generates very energetic electrons which preheat the target. The Brillouin instability is dangerous because it can reflect the laser light before it can be absorbed. Because the linear growth rates of these instabilities are large, their non-linear development and interaction needs to be understood. Only a PIC code is capable of treating this properly. A laser beam passing through a low density plasma is also unstable to the filamentation and self-focussing instabilities, whereby the beam tends to filament into many beams or self-focus into one narrower beam. This can be modelled to some extent with a fluid model, but it has proved to be another powerful application of PIC codes (Pukhov and Meyer-ter-Vehn, 1996).

There is great interest at present in laser propagation problems in the context of the 'Fast Ignitor' approach to Fusion in which a pellet is first compressed to high density and relatively low temperature, and then heated to thermonuclear temperatures and ignited by propagating a very intense short laser pulse into the core of the dense plasma (Tabak et al., 1994). Propagation by ponderomotive and relativistic

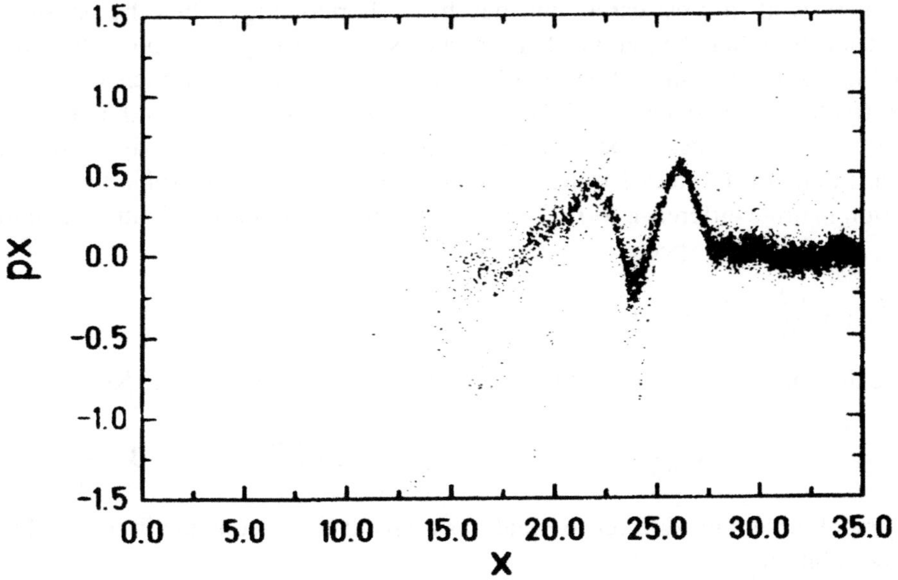

Figure 1. An example of the phase space output from a PIC code simulating the generation of plasma waves on the surface of a high density target (right) caused by laser irradiation (input at the left boundary). The waves are damped by the production of energetic electrons. Figure taken from Gibbon and Bell (1992).

channelling through the overdense, i.e. greater than critical density, plasma (Adam et al., 1995, Guerin et al., 1995) is one of the crucial elements of the scheme.

PIC codes are also important in the development of plasma-based particle accelerators such as the beat-wave accelerator (BWA) and the laser wake-field accelerator (LWFA). In both these schemes, laser pulses excite a large amplitude Langmuir wave in a low density plasma. In the BWA, two laser beams are superimposed with frequencies which differ by the electron plasma frequency and the Langmuir wave is excited by the beat between the two. In the LWFA the Langmuir wave is excited by a single very short laser pulse with a pulselength approximately equal to half the period π/ω_p of the Langmuir wave. In both cases the wave is driven by the ponderomotive force and the Langmuir wave has a phase velocity equal to the group velocity of the laser pulse. Relativistic electrons moving in phase with the wave experience a constant electric field and 'surf-ride' the wave to high energy. Because the problem is collisionless and kinetic, a PIC code is ideal for modelling the BWA and the LWFA. In fact, the original paper by Tajima and Dawson (1979) proposing these acceleration schemes was based on PIC calculations.

2.3. Vlasov Codes

PIC codes are a very robust and efficient way of modelling collisionless plasmas in phase space. They come close to modelling a plasma from first principles. They work very well in most cases, but they have shortcomings when the physics is dominated by a small fraction of the particles. For example, suppose that wave damping results in a small fraction of the electrons gaining a high energy. If this small fraction of particles is to be modelled with adequate statistics then the total number of particles has to be very large and the code executes slowly. If this statistical limitation is excessive then it becomes more efficient to solve the Vlasov equation in finite difference form. The Vlasov equation for the distribution function $f(\mathbf{r}, \mathbf{v}, t)$ of a single species of particles,

$$\frac{\partial f}{\partial t} + \mathbf{v}.\frac{\partial f}{\partial \mathbf{r}} + \frac{q}{m}(\mathbf{E} + \mathbf{v} \wedge \mathbf{B}).\frac{\partial f}{\partial \mathbf{v}} = 0,$$

can be time integrated by finite differencing it in two parts (Cheng and Knorr, 1976)

$$f' = f^m - \Delta t \left(\mathbf{v}.\frac{\partial f}{\partial \mathbf{r}} \right)^{m+1/2}, \qquad f^{m+1} = f' - \Delta t \left(\frac{q}{m}(\mathbf{E} + \mathbf{v} \wedge \mathbf{B}).\frac{\partial f}{\partial \mathbf{v}} \right)^{m+1/2}.$$

The first of these finite difference equations is an advection equation in space which has the solution

$$f'(\mathbf{r}, \mathbf{v}) = f^m(\mathbf{r}', \mathbf{v}), \qquad \mathbf{r}' = \mathbf{r} - \mathbf{v}\Delta t$$

which consists merely of shifting the distribution at each velocity $\mathbf{v}$ a distance $\mathbf{v}\Delta t$. Similarly, the second difference equation is an advection equation in velocity with the solution

$$f^{m+1}(\mathbf{r}, \mathbf{v}') = f'(\mathbf{r}, \mathbf{v})$$

which represents shifting the particles from the old velocity $\mathbf{v}$ to the new velocity $\mathbf{v}'$. If the magnetic field is zero, this constitutes a simple translation in velocity given by

$$\mathbf{v} = \mathbf{v}' - \frac{q}{m}\mathbf{E}\Delta t.$$

If the magnetic field is non-zero, then the velocity shift involves a rotation as well as a straight line advection in velocity space. This is more complicated to implement, but takes a form similar to that of the particle acceleration equation for PIC codes since the relationship between $\mathbf{v}'$ and $\mathbf{v}$ is the same as that of a particle which has a velocity $\mathbf{v}$ at time $m\Delta t$ and is accelerated by the fields to a velocity $\mathbf{v}'$ at time $(m+1)\Delta t$.

The two difference equations have an analytic solution which involves reading off firstly the value of f^m at point $\mathbf{r}'$ and calling that f', and then secondly reading off the value of f' at velocity $\mathbf{v}'$ and calling that the value of f^{m+1} at the velocity $\mathbf{v}$. This reduces to a process of interpolation, usually using cubic splines, on a grid in configuration and velocity space.

The advantage of this finite difference solution of the Vlasov equation is that computational effort is spread uniformly over the phase space grid. This contrasts with PIC codes in which the computational effort is uniformly spread over the particles instead of the grid. In cases where a small fraction of the particles is particularly important, a Vlasov code can be more efficient than a PIC code because the Vlasov code applies equal computational effort to the part of phase space occupied by those particles and to the part of phase space occupied by the majority of particles. If necessary a non-uniform grid can be used to concentrate the computational effort on any required part of phase space.

A disadvantage of Vlasov codes relative to PIC codes is that they run more slowly because the numerical equations are more complicated than the relatively simple PIC equations. Also, the advantage of Vlasov codes is also their disadvantage. They have to spend time modelling a vacuum or an unoccupied high velocity part of phase space where there are no particles and which would be ignored by PIC codes. A further disadvantage is that Vlasov codes are numerically diffusive and dispersive. A sharp feature can spread diffusively, and high k spectral components can advect at the wrong speed. This distortion is caused by 'truncation' errors introduced by the difference equations not being exactly the same as the differential equations they approximate. These truncation errors are particularly problematic where (i) there are sharp density gradients such as at the plasma/vacuum interface in laser-plasma experiments or (ii) when the motion in phase spaces tracks diagonally across the rectangular grid. Truncation errors can be overcome by the use of a very fine grid, i.e. making Δx and Δv small, but this increases the computational expense. Vlasov codes are particular slow to run in two or three dimensions, but in 1D they have advantages over PIC codes for particular problems. With the increasing power of computers, Vlasov codes are increasingly feasible since their lower speed of execution is less restricting. Recently, Vlasov codes have been used very effectively for the modelling of laser absorption on solid targets (Ruhl and Mulser, 1995).

2.4. Collisions in PIC Codes

PIC and Vlasov codes cannot model physics occurring on distance scales smaller than the spatial cell-size Δx, and therefore the cell-size has to be smaller than a Debye length λ_D since Debye shielding is an essential feature of any plasma. If $\Delta x > \lambda_D$ the plasma artificially heats up, violating energy conservation, increasing λ_D until $\lambda_D > \Delta x$. However, even if $\Delta x < \lambda_D$, the physics of interparticle collisions is unresolved. The classical distance of minimum approach b_0 at which an electron is deflected through 90° is smaller than the Debye length by a number of the order of N_D^{-1}, where N_D is the number of electrons in a Debye sphere. For a tokamak plasma $N_D \sim 10^8$ and for an interstellar plasma $N_D \sim 10^{12}$. Thus, it is impossible for PIC codes to model collisions properly for these low density plasmas. Even laser-plasmas (taking $T = 1$ keV and $n_e = 10^{22}$ cm^{-3}) have $N_D \sim 500$ making direct modelling of collisions infeasible in PIC codes.

Most PIC codes are collisionless, but collisions have been included in some on a phenomenological basis. If the plasma has a relatively high Z, the dominant electron collision process is angular scattering by ions. This is relatively easy to include in a PIC code at a basic level. The scattering rate for each electron is calculated from the electron velocity and the ion density. Each electron velocity is then deflected randomly to give the required angular scattering rate. Electron-electron scattering is more difficult to implement because of the importance of energy exchange. The electron-electron collision rates for angular scattering, velocity diffusion and frictional drag must be calculated by integrating over the local electron distribution, and then each electron's velocity randomly changed accordingly. This is a complicated procedure which is often inaccurate because of inadequate particle statistics, and energy is not easily conserved.

Although implementation is difficult, collisions have been successfully included in some PIC codes. In some special cases, even inelastic collisions and/or ionisation have been added successfully (Wallace et al., 1991, Janulewicz et al., 1996).

2.5. N-Body Codes

PIC codes have limited applicability to collisional plasmas. Phenomenological collisional models have their uses, but a particle code can only model the detailed results of collisions if binary particle-particle interactions are included. It was pointed out in Section 2 that N particles have approximately N^2 binary interactions, thus limiting N to impracticably small values. However, the N^2 interactions includes binary interactions between individual particles which may be a large distance apart. Binary interactions are only important between nearby particles. Interactions between distant particles are collective, meaning that the fields due to distant particles can be smoothed out as in PIC codes. The detailed fields of only nearby particles need calculating, and this principle is used in Tree Codes (Barnes and Hutt, 1986).

Tree codes provide a means of identifying particles which are close neighbours. At the heart of the code is the tree which is a data structure. The volume is divided into a series of boxes as shown in Figure 2. The first step in creating the tree in two spatial dimensions is to divide the overall computational box into four equally sized boxes (eight boxes in three dimensions). We can call the overall box the 'level 0' box, and its four sub-boxes 'level 1' boxes. The level 1 boxes are examined to see how many particles they hold. If, as expected, a level 1 box contains more than one particle it is further sub-divided into four level 2 boxes. The level 2 boxes are each sub-divided if they contain more than one particle. This process continues until each box contains at most one particle as shown in Figure 2. The end-product is a list of boxes and the particles inside them. The total charge in each of the boxes can be calculated, where $q_n^{(p)}$ would be the charge in the nth box at level p.

Suppose that the problem is purely electrostatic and we want to calculate the electric field experienced by one (ith) particle. The Tree can be used to identify the

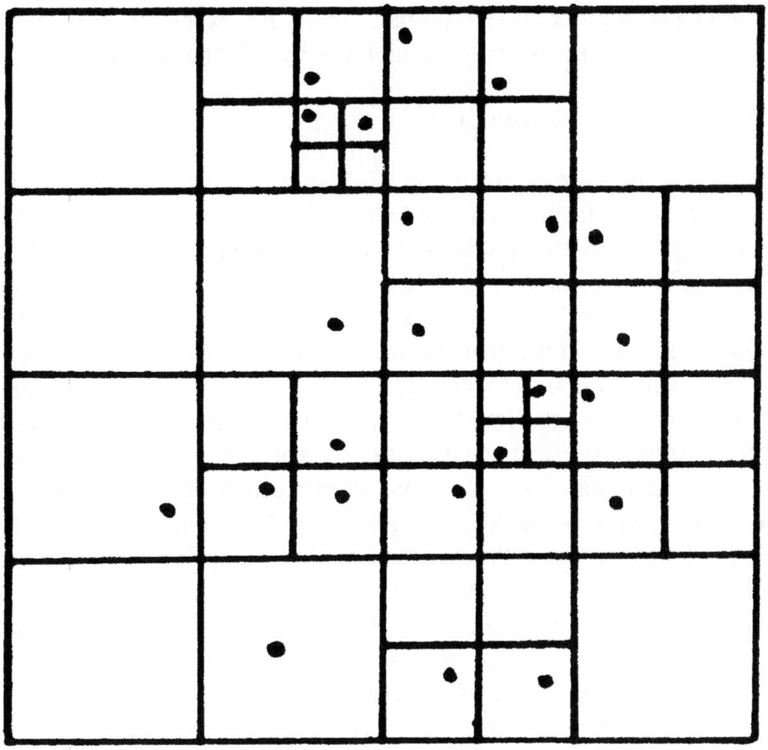

Figure 2. Tree structure which places each particle in a separate box with minimum subdivision of the space.

close neighbours of the ith particle and their binary interactions taken account of. Interaction with particles distant from the ith particle do not need to be considered individually, and it might be sufficient to consider the distant particles as lumped together into boxes at level 2. If the nth box at level 2 has a total charge $q_n^{(2)}$, then it would be necessary only to calculate the electric field at the ith particle exerted by a charge $q_n^{(2)}$ at the centre of the nth box at level 2. This is much faster computationally than calculating the electric field due to all the individual particles in that box. Particles at intermediate distances could be lumped together in boxes at level 3, and still nearer particles in boxes at level 4, and so on.

Using tree data structure, the electric field at each ith particle interacts with N_{clump} clumps of particles in variously sized boxes. N_{clump} is proportional to log N which is much smaller than the total number $(N - 1)$ of individual particles that interact with the ith particle. The total number of operations needed to accelerate the particles each timestep is then $\propto N \log N$ instead of N^2. This tree method permits N-body calculations with much increased numbers of particles. This method has been widely in astrophysics for the related problem of clusters of stars interacting with each other gravitationally (Hernquist, 1987). An excellent description of the method and its applications is given in the book by Pfalzner and Gibbon (1996).

2.6. Fokker–Planck Modelling

Few laboratory plasmas are collisionless. Collisions can be neglected when studying certain aspects of plasmas, but for other aspects they must be included. For example, although mean free paths of particles in tokamaks are of the order of a kilometre, collisions cannot be neglected for many tokamak phenomena. Most other laboratory plasmas have a higher density and a lower temperature, making collisions even more important, even to the extent that in high density laser-produced plasmas the collision mean free path can be very much less than one micron. Collisions determine the freedom of particles to move relative to each other, thus controlling transport effects such as the electrical and thermal conductivities.

If a plasma is dominated by collisions then thermal equilibrium prevails and all particles have a Maxwellian distribution at the local temperature. In this case the plasma can be treated as a fluid and a kinetic description is unnecessary. The slowest collision process is energy equilibration between electrons and ions, so electrons and ions can have different temperatures, but this can be allowed for by treating electrons and ions as separate fluids with different temperatures.

The fluid model is appropriate for a collision-dominated plasma, and the PIC or Vlasov model is appropriate for a collisionless plasma. In between these two extremes lies a regime in which collisions cannot be neglected but they are not sufficient to ensure Maxwellian distributions. Velocity space information must be retained and collisions included. The addition of collisions to a PIC code has the disadvantages described in Section 2.4, and Tree codes are relatively slow because the collisions are modelled in detail. The preferred option is to add a Fokker–Planck collision term to a Vlasov code. These 'Fokker–Planck' codes solve in finite difference form the equation:

$$\frac{\partial f}{\partial t} + \mathbf{v}.\frac{\partial f}{\partial \mathbf{r}} + \frac{q}{m}(\mathbf{E} + \mathbf{v} \times \mathbf{B}).\frac{\partial f}{\partial \mathbf{v}} = \left(\frac{\partial f}{\partial t}\right)_{\text{collisions}}.$$

The collision term on the right-hand side represents the cumulative effect of a large number of small angle collisions:

$$\left(\frac{\partial f}{\partial t}\right)_{\text{collisions}} = -\frac{\partial}{\partial \mathbf{v}}.(f\langle\Delta\mathbf{v}\rangle) + \frac{1}{2}\frac{\partial}{\partial \mathbf{v}}\frac{\partial}{\partial \mathbf{v}} : (f\langle\Delta\mathbf{v}\Delta\mathbf{v}\rangle),$$

where $\langle\Delta\mathbf{v}\rangle = \int \psi(\mathbf{v}, \Delta\mathbf{v})\Delta\mathbf{v}d\Delta\mathbf{v}$ and $\langle\Delta\mathbf{v}\Delta\mathbf{v}\rangle = \int \psi(\mathbf{v}, \Delta\mathbf{v})\Delta\mathbf{v}\Delta\mathbf{v}d\Delta\mathbf{v}$ (e.g. Boyd and Sanderson, 1969), and $\psi(\mathbf{v}, \Delta\mathbf{v})$ is the probability that a particle with velocity $\mathbf{v}$ is deflected by an amount $\Delta\mathbf{v}$ per unit time. $\langle\Delta\mathbf{v}\rangle$ is the vector rate of advection in velocity space due to collisions, and is known as the coefficient of dynamical friction. $\langle\Delta\mathbf{v}\Delta\mathbf{v}\rangle$ is the tensor rate of diffusion in velocity space due to collisions, and is known as the coefficient of diffusion. The probability ψ can be calculated by summing over all possible binary collisions using the Rutherford cross-section.

The Fokker–Planck collision term can be written in a variety of ways. Probably the most convenient uses the Rosenbluth potentials (Rosenbluth et al., 1957) G and H which are called potentials because they satisfy potential equations,

$$\nabla_v^4 G_j = -8\pi f_j \quad \text{and} \quad \nabla_v^2 H_j = -4\pi \left(1 + \frac{m_i}{m_j}\right) f_j,$$

where the differentials ∇_v are gradients in velocity space. The subscripts i and j refer to different species (electrons and ions). This is necessary because the species interact with each other. The collision term for the evolution of the ith species due to collisions with all other species then takes the form

$$\left(\frac{\partial f_i}{\partial t}\right)_{\text{collisions}} = \sum_j \Gamma_{ij} \left[\frac{\partial}{\partial \mathbf{v}} \cdot \left(f_i \frac{\partial H_j}{\partial \mathbf{v}}\right) + \frac{1}{2} \frac{\partial}{\partial \mathbf{v}} \frac{\partial}{\partial \mathbf{v}} : \left(f_i \frac{\partial}{\partial \mathbf{v}} \frac{\partial G_j}{\partial \mathbf{v}}\right)\right],$$

where Γ_{ij} is a collection of constants. The potentials are usually evaluated using the integral form of the potential equations:

$$G_j = \int |\mathbf{v} - \mathbf{u}| f_j(\mathbf{u}) d\mathbf{u} \quad \text{and} \quad H_j = \left(1 + \frac{m_i}{m_j}\right) \int \frac{f_j(\mathbf{u}) d\mathbf{u}}{|\mathbf{v} - \mathbf{u}|}.$$

The Fokker–Planck (FP) collision term consists of two terms which are a first and a second differential respectively in velocity space. The second differential represents diffusion in velocity space, and the first differential represents advection or drift in velocity space. Advection in velocity space dominates for a low intensity beam of heavy ions propagating through a low Z plasma since the ions steadily lose energy without undergoing much angular scattering. The steady loss of energy corresponds to a drift towards zero in velocity space and is represented by the first differential. In contrast, a beam of electrons in a highly ionised high Z plasma principally undergoes angular scattering with a much slower loss or gain of energy. Angular scattering corresponds to diffusion in the angle of propagation which is represented by the second differential. For most particles in most plasmas both the first and second differentials are important. A Maxwellian distribution results from the competition of the two terms since the advection tends to draw particles towards zero velocity by energy loss to other particles, while diffusion tends to spread the particles out in velocity space.

Velocity diffusion has the computationally desirable effect of smoothing out the distribution in velocity space and reducing the amount of information needed to describe it adequately. If the system is 1D in space (in the x direction) the velocity distribution is a function only of $|\mathbf{v}|$ and θ where $\cos\theta$ is the angle between the electron velocity and the x direction, $v_x = v\cos\theta$. In this case the velocity distribution can be expanded as a sum of Legendre polynomials (Bell et al., 1981):

$$f = \sum_{n=0}^{\infty} f^{(n)}(x, v) P_n(\cos\theta).$$

Isotropisation by angular scattering damps each $f^{(n)}$ at a rate proportional to $n(n+1)$, so the large n terms in the series are close to zero and the series can be truncated at a small value of n. In many cases it is sufficient to keep only the first two terms in the series, in which case the distribution can easily be generalised into two or three dimensions.

$$f(\mathbf{r}, \mathbf{v}) = f^{(0)}(\mathbf{r}, v) + \mathbf{f}^{(1)}(\mathbf{r}, v).\frac{\mathbf{v}}{v},$$

where $\mathbf{f}^{(1)}$ is a vector with components $(f_x^{(1)}, f_y^{(1)}, f_z^{(1)})$ representing particle fluxes in the x, y and z directions. At any spatial point, the electron density in the whole of the velocity space is represented by the four functions $f^{(0)}$, $f_x^{(1)}$, $f_y^{(1)}$ and $f_z^{(1)}$ which are functions of the one-dimension magnitude v of the velocity instead of the 3D velocity vector $\mathbf{v}$ of which f is a function, thus reducing the number of velocity dimensions from three to one. This represents a large reduction in information content and a large saving in computing effort. For this reason, a finite difference Fokker–Planck is much faster than a PIC + collisions code for most problems. The PIC + collisions approach is useful when the collisions are very weak and the velocity distribution strongly anisotropic requiring the inclusion of many terms in the Legendre polynomial expansion for f.

2.7. Applications of Fokker–Planck Modelling

Fokker–Planck (FP) codes have been widely used in laser-produced plasmas. The laser energy is absorbed at or below the critical density and is transported to higher density by thermal conduction. In ICF targets the laser pulse is relatively long and the ablating plasma approximately reaches a steady state in which the energy flux by thermal conduction $Q = -\kappa \nabla T$ is balanced by the enthalpy flux $J_E = 5Pu/2 + \rho u^3/2$ where u is the plasma flow velocity, P is the pressure and ρ the mass density. Ablation is transonic at the critical surface. Equating Q and J_E, and setting $u = \sqrt{P/\rho}$ gives $Q \approx 0.1 Q_{\text{free}}$ where Q_{free} is the so-called 'free-streaming' limit, $Q_{\text{free}} = n_e m_e v_t^3$, $v_t = \sqrt{kT/m_e}$ is the electron thermal velocity, and n_e is the electron density. It is called the free-streaming limit because it is notionally the energy flux carried by n_e electrons flowing freely down the temperature gradient at their thermal velocity v_t. Hence the electron heat flow is very large, entailing drift velocities which are not much smaller than the electron thermal velocity. In order to carry this heat flow, the temperature scalelength $L = T/\nabla T$ has to be in the range 10-100 mean free paths. In these extreme conditions, the electron distribution is highly non-isotropic and non-Maxwellian, and is most appropriately modelled with an FP code. Electron heat transport in laser-produced plasmas was only understood after it was modelled with an FP code (Bell et al., 1981). For a review, see Bell (1995).

Fokker–Planck modelling of laser-produced plasmas has been extended to model the way in which non-uniform laser energy deposition is smoothed by thermal conduction, the effect of thermal conduction on thermal filamentation

(Epperlein, 1990), and the generation and convection of magnetic field (Kho and Haines, 1985). In recent years, the development of high intensity very short (psec) laser pulses has led laser-produced plasmas into even more extreme regimes in which the electron distributions are even more non-Maxwellian and anisotropic (Town et al., 1995). This pushes FP codes to the limit and may require a move to PIC + collision models. FP codes have also been used to model the transport of energetic particles in tokamaks (e.g. O'Brien et al., 1996), the transport of energetic particles in the sun and the solar wind (e.g. Fletcher and Brown, 1995; Canullo et al., 1996) and cosmic ray acceleration (e.g. Bell, 1987).

3. Hydrodynamic Simulation

The methods discussed so far in this review retain detailed information of the particle distribution in velocity space. This is essential for many plasma phenomena especially non-linear waves and wave-particle interactions. However, the consequent heavy demand on computational resources severely limits the size of the problem that can be modelled. It is impossible to model a complete tokamak or a complete ICF implosion with a PIC or Vlasov code. In these, and many other cases, approximation is essential if the problem is to be reduced to a scale which can be modelled on present-day computers. Fluid codes reduce the scale of the problem by dispensing with detailed velocity space information. This reduces the six dimensions of phase space to the three dimensions of configuration space. Only the crudest velocity space information is kept such as the root mean square particle velocity (the temperature), the mean particle mass drift velocity (the fluid velocity), and the electron drift velocity relative to the ions (the electrical current). Other quantities such as the heat flow and the stress tensor can be included to add further information about the velocity distribution. Each of these quantities are essentially velocity moments of the distribution, i.e. average quantities derived by integrating over the velocity distribution. The fluid equations which connect these quantities are derived by integrating in velocity over the Vlasov equation. In some plasmas it is possible to neglect the magnetic forces but generally they must be included. The simplest magnetohydrodynamic (MHD) equations are those for an ideal MHD perfect gas in which transport effects such as thermal conduction, viscosity and electrical resistivity are ignored:

$$\frac{\partial \rho}{\partial t} = -\nabla.(\rho \mathbf{u}), \quad \frac{\partial (\rho \mathbf{u})}{\partial t} = -\nabla.(\rho \mathbf{u}\mathbf{u}) - \nabla P + \frac{(\nabla \wedge \mathbf{B}) \wedge \mathbf{B}}{\mu_0}, \quad \frac{d}{dt}\left(\frac{P}{\rho^\gamma}\right) = 0,$$

$$\frac{\partial \mathbf{B}}{\partial t} = \nabla \wedge (\mathbf{u} \wedge \mathbf{B}),$$

where ρ is the mass density, $\mathbf{u}$ is the fluid velocity, P is the pressure, and γ is the ratio of the specific heats. The first three equations describe mass conservation, momentum conservation, and adiabaticity respectively. The fourth equation expresses magnetic flux-freezing. These equations are sufficient to model the most basic aspects of magnetically confined plasmas. ICF modelling requires the inclusion of heat flow and the adiabaticity equation is replaced by a more complicated energy conservation equation:

$$\frac{\partial U}{\partial t} = -\nabla.(U\mathbf{u}) - P\nabla.\mathbf{u} - \nabla.\mathbf{Q}, \quad \mathbf{Q} = -\kappa\nabla T,$$

$$P = P(\rho, T), \quad U = U(\rho, T),$$

where U is the energy density, $\mathbf{Q}$ is the heat flow, κ is the conductivity, and T is the temperature. P and U are not necessarily simple functions of ρ and T since the plasma in general will not be an ideal gas. Collisional magnetised plasmas such as Z-pinches and laser-produced plasmas require the inclusion of diffusion of magnetic fields due to electrical resistivity so that the magnetic field equation becomes

$$\frac{\partial \mathbf{B}}{\partial t} = \nabla \wedge (\mathbf{u} \wedge \mathbf{B}) - \nabla \wedge \left(\frac{\eta}{\mu_0}\nabla \wedge \mathbf{B}\right).$$

Numerical solutions of these equations fall into two broad categories: Eulerian and Lagrangian. In the Eulerian method a computational grid is fixed in space and the fluid moves across it. In the Lagrangian method the grid is fixed into the fluid and moves with the fluid in space.

3.1. Lagrangian Method

The Lagrangian method is usually preferable when the problem is 1D. For example, the basic aspects of an ICF implosion can be modelled in 1D since every quantity is a function of the radial distance r from the centre of the implosion. The target can be considered as a series of concentric shells of plasma and the Lagrangian model tracks the position of these shells. Each nth shell has a radial position r_n and a radial velocity u_n which obey the equations

$$\frac{dr_n}{dt} = u_n, \qquad \frac{\Delta m_n}{\Delta V_n}\frac{du_n}{dt} = P_{n-1/2} - P_{n+1/2},$$

where Δm_n is the mass of the nth shell, $P_{n-1/2}$ and $P_{n+1/2}$ are the pressures each side of the nth shell, and $\Delta V_n = 4\pi(r_{n+1/2}^3 - r_{n-1/2}^3)/3$ (where $r_{n+1/2} = (r_{n+1} - r_n)/2$) is the volume of the nth shell. In the simplest adiabatic Lagrangian code, $P_{n+1/2}$ can be thought of as the pressure between the nth and $(n+1)$th shell which can be calculated via the mass density and the condition that P/ρ^γ remains constant for each element of the plasma:

$$P_{n+1/2}(t) = P_{n+1/2}(0)\left(\frac{\rho_{n+1/2}(t)}{\rho_{n+1/2}(0)}\right)^\gamma, \qquad \rho_{n+1/2} = \frac{\Delta m_{n+1} + \Delta m_n}{2\Delta V_{n+1/2}},$$

where $V_{n+1/2} = 4\pi(r_{n+1}^3 - r_n^3)/3$. These Lagrangian equations fit well into a leap-frog scheme in which the shell positions and shell velocities leap-frog over each other in time as in the PIC equations.

A Lagrangian scheme was used by the MEDUSA ICF simulation code written in the early 1970s (Christiansen et al., 1974). The code included sufficient physics to model an ICF implosion, and extra physics and diagnostics have been added over the years (e.g. Djaoui, 1995). The robust basic algorithm has served as a platform for the development of a sophisticated simulation code. Similar Lagrangian codes have been used for 1D astrophysical problems such as supernova explosions and the evolution of a supernova remnant as it expands into the interstellar medium.

Lagrangian codes are well suited to 1D problems but less so to 2D and 3D problems. Suppose an initially square spatial grid is used for a 2D problem. As the fluid moves, the grid distorts. Severe difficulties arise when one grid-point crosses the boundary between two other grid-points. The grid becomes tangled and the calculation halts. Grid-tangling can be overcome by stopping the calculation and establishing a new grid but the process of re-zoning introduces numerical errors and may not extend the calculation much further in time before the grid tangles again, particularly if vorticity is generated. For this reason, purely Lagrangian methods are not usually used in 2D and 3D, although a notable exception to this is LASNEX (Zimmerman and Kruer, 1975), the Lagrangian ICF simulation used in the U.S.A.

3.2. Eulerian Codes

In 2D and 3D the Eulerian method is usually employed for fluid simulation. In this method the computational grid remains fixed in space and the fluid is moved (advected) between cells. In Lagrangian codes, the advection equation for mass conservation

$$\frac{\partial \rho}{\partial t} = -\nabla.(\rho \mathbf{u})$$

is automatically obeyed because the mass moves with the grid. In an Eulerian code, this equation has to be solved in finite difference form. Similarly, advective terms are added to the momentum and energy equations. Advective terms appear deceptively simple, since they are only first order differentials, but they are very difficult to solve accurately without numerical instability. The simplest stable algorithm uses the donor-cell or upwind method (e.g. Gentry et al., 1966). For the 1D (in x) mass equation, the donor-cell difference equation takes the form

$$\rho_n^{m+1} = \rho_n^m + \frac{\Delta t}{\Delta x}(F_{n-1/2} - F_{n+1/2})$$

where

$$F_{n+1/2} = u_{n+1/2}\rho_n^m, \quad \text{if} \quad u_{n+1/2} > 0,$$

and

$$F_{n+1/2} = u_{n+1/2}\rho^m_{n+1}, \quad \text{if} \quad u_{n+1/2} < 0.$$

$F_{n+1/2}$ is the mass flux across the boundary between the nth and $(n+1)$th cells. At first sight it would be better to represent the fluxes by the difference equation

$$F_{n+1/2} = u_{n+1/2}\frac{\rho^m_{n+1} + \rho^m_n}{2},$$

where the density at the $(n + 1/2)$th boundary is represented by the average of the densities in the nth and $(n + 1)$th boundaries, but this differencing is unstable and the algorithm produces nonsense. The donor-cell method assumes that the density at the boundary is the density in the upwind, or donating, cell from which the fluid is flowing across the cell boundary. This method is inherently stable provided the timestep is not too large. The donating cell is either the nth or $(n + 1)$th cell depending upon the sign of $u_{n+1/2}$.

The donor-cell method is stable but it lacks accuracy. In the simplest case of constant and uniform positive velocity u, its solution is the equivalent of the solution of the differential equation

$$\frac{\partial\rho}{\partial t} = -u\frac{\partial\rho}{\partial x} + \frac{u\Delta x}{2}\frac{\partial^2\rho}{\partial x^2} + \text{terms of order } \Delta x^2, \Delta x^3 \ldots$$

The extra terms introduce errors. The dominant error term adds diffusion which causes any feature to smooth out as it advects. An example of this is shown in Figure 3 which shows the spreading of a top-hat function as it advects at constant velocity. It rapidly loses its shape. This diffusion causes the donor-cell method to be stable because it smooths out and inhibits the growth of instabilities. However, diffusion is a serious disadvantage since it means that donor-cell cannot model the discontinuities which naturally arise in fluids as a result of shocks and boundaries between different materials.

Numerical diffusion makes the donor-cell method unacceptable for most applications. The challenge is to produce an advection algorithm which can maintain discontinuities without being unstable. There is no simple solution to this problem. A number of different algorithms of varying complexity have found favour. One of the most popular is 'Flux-Corrected Transport' (Boris and Book, 1973). The philosophy of this method is that a strongly diffusive method is used initially at each timestep to ensure stability and then the diffusion is subtracted by 'correcting' the calculation. If the correction removed all the diffusion, then the total effect would be to introduce instability. This is avoided by limiting the correction in such a way as to avoid producing any new spatial maxima or minima. This works very well for shocks because it produces a steep shock front without any spatial oscillations on either side. It works less well for gentle density gradients which it can in some circumstances steepen into a series of discrete jumps which look like a 'staircase' (Woodward and Colella, 1984).

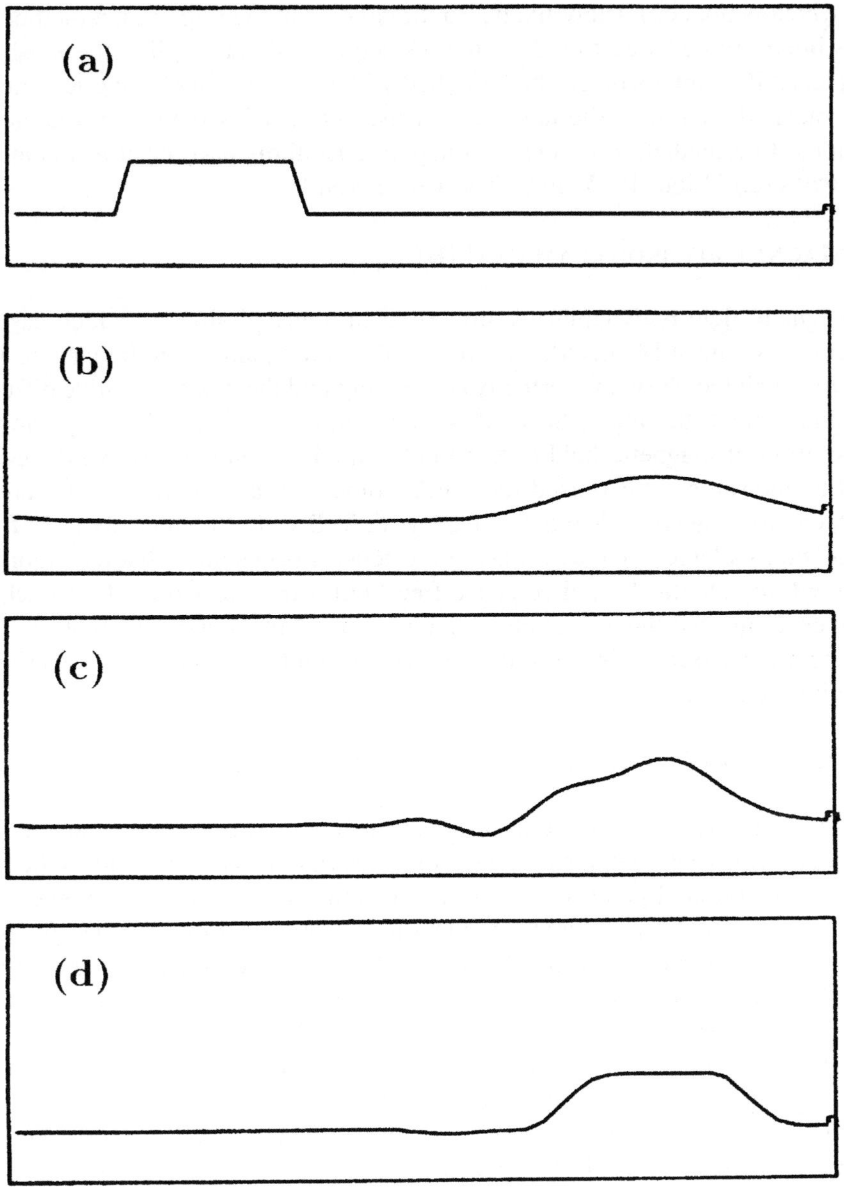

Figure 3. The difficulties of solving the advection equation: (a) a square pulse at time $t = 0$; (b) after the initially square pulse has been propagated across the grid using a donor-cell method, showing the effect of numerical diffusion; (c) the same as (b) but using the Lax–Wendroff method, showing the effect of numerical dispersion; (d) the same as (b) and (c) but with a numerical scheme using flux-corrected transport (FCT). The boundary conditions are periodic.

Lagrangian and Eulerian codes each have advantages, and it is possible to design algorithms which try to combine the best of each. Some codes use a sliding mesh which is not completely fixed into the fluid as in the Lagrangian method, but moves in an ordered way to reduce the advection without tangling the mesh. For example, an ICF implosion can be modelled with a code in which the grid contracts in the radial direction as the target contracts, but remains regular in the angular directions. Intermediate Eulerian/Lagrangian algorithms have been formulated in many ways (e.g. Liou, 1995, and references therein).

3.3. Magnetohydrodynamics (MHD)

The magnetic field is dynamically important in many plasmas in which case the magnetic force must be included in the momentum equation. In the case of ideal MHD in which the plasma is perfectly conducting and there is no resistive diffusion of magnetic field, the magnetic field is frozen into the plasma. The equation for the evolution of magnetic field is essentially an advection equation which can be solved by the same method used for mass, momentum and energy. Some care has to be taken to preserve a divergence-free field, $\nabla.\mathbf{B} = 0$. Small errors in $\nabla.\mathbf{B}$ can accumulate to a large value over many timesteps. It is desirable that the algorithm should enforce identically a divergence-free field. Lucek and Bell (1996) achieve this by ensuring that the advection of magnetic field from one computational cell to the next is performed by re-routing magnetic field lines instead of creating or destroying field lines.

3.4. Transport and Other Additional Effects

Plasmas can rarely be described as an ideal magnetohydrodynamic adiabatic perfect gas of electrons and ions. Magnetohydrodynamics is rarely ideal because resistivity is non-negligible and field-line reconnection cannot be ignored. Adiabaticity is violated by thermal conduction, radiative losses and energy input. Behaviour as a perfect gas can be violated by ionisation and by interatomic forces. Suprathermal populations of particles transport energy and make a single temperature description inappropriate. If these non-hydrodynamic effects are too strong, then a fluid description must be abandoned, but often they can be included phenomenologically at an adequate level of approximation.

Diffusive effects such as thermal conduction and resistive diffusion of magnetic field can be included through the diffusion equations for energy or magnetic field

$$\frac{\partial(CT)}{\partial t} = \nabla.(\kappa\nabla T), \quad \frac{\partial \mathbf{B}}{\partial t} = -\nabla \wedge \left(\frac{\eta}{\mu_0}\nabla \wedge \mathbf{B}\right),$$

where C is the specific heat, κ the thermal conductivity, and η the resistivity. These diffusion equations must be solved implicitly, meaning that if the time difference on the left-hand side of the equation advances from time $m\Delta t$ to $(m+1)\Delta t$, then the quantities (T or $\mathbf{B}$) on the right-hand side of the equations must be at time $(m+1)\Delta t$. Because there are unknown quantities on both sides of the equations, the equations become a set of coupled equations in $N_{\rm grid}$ unknowns where $N_{\rm grid}$ is the number of spatial grid-points. The problem reduces to the solution of a large sparse matrix equation which is best solved iteratively using such techniques as the Incomplete Cholesky Conjugate Gradient (ICCG) method (Kershaw, 1978). The diffusion equation cannot efficiently be solved explicitly (meaning that the quantities on the right-hand side are at time $m\Delta t$ instead of $(m+1)\Delta t$) because the timestep has to be very short if numerical instability is to be avoided.

Energy transport by suprathermal particles and by radiation can sometimes be treated diffusively, but often the mean free paths are long and a Monte Carlo treatment is needed in which the paths of sample particles and photons are tracked through the plasma. Radiation transport can be a particular problem because of its sensitivity to detailed opacities. Opacities, and similarly the equation of state and other material properties, are often accessed from large tables of data such as those held in the Sesame datafiles at the Los Alamos National Laboratory.

These extra items of physics often swamp the basic fluid code in terms of lines of code and execution time. The fluid code acts as a computational platform on which large physics packages can be supported.

3.5. Applications of Fluid Codes

Time-dependent fluid codes have many applications but their special strengths are seen in processes which are far from equilibrium and velocities are comparable with or greater than the characteristic sound speed. In astrophysics, energetic events such as supernova explosions (e.g. Jun and Norman, 1996), the behaviour of active galactic nuclei (e.g. Bell and Lucek, 1995), and the propagation of jets (e.g. Falle, 1994) lend themselves to hydrodynamic simulation. In the solar system, the supersonic solar wind and its interaction with earth can be modelled with MHD (e.g. Ogino et al., 1994). Energetic terrestrial plasmas such as z-pinch evolution, ICF implosions (Craxton et al., 1986), and nuclear detonations require hydrodynamic modelling.

Simulation codes are an essential part of experiments. For example, the behaviour of an ICF implosion can only be interpreted with the aid of a fluid code. A typical ICF experiment lasts a few nanoseconds, the characteristic scalelength is 10s or 100s of microns, and the target is opaque to visible light. The experimental diagnostics used in ICF are impressive and highly developed, but naturally limited. The experimental procedure consists of making whatever measurements are possible and appropriate and then matching these to the output from a simulation code. The code fills in the gaps which cannot be directly accessed by experimental diagnostics. Good agreement between the code output and the diagnostic output where the two overlap indicates that the code is reproducing the behaviour of the target and the results of the code may be believed, with an appropriate level of scepticism, even concerning phenomena which are not directly measurable. In astrophysics, the observational information can be even more limited. The central engine of an AGN at a distance of 10^{26} m has a characteristic size of around 10^{12} m, which is too small to be observationally resolvable, yet the output from the central engine can stretch 10^{22}m in the form of the observed radio jets. Numerical simulation can be used to test the validity of theoretical models. If simulation and observation coincide then maybe the simulation code approximately models the actual physical processes that are active within the AGN.

In complicated events such as an ICF implosion, the simulation code is a major repository of knowledge concerning ICF. A few decades ago, our knowledge of physical systems was contained within mathematical equations. Today it is often contained within the simulation codes.

4. The Future

Numerical simulation has progressed enormously in the last 30 years, partly due to the development of new algorithms, but mainly because of the development of computer technology. The supercomputer of yesteryear is insignificant compared with the desk-top computer of today. Where 1D simulation used to be the limit, full 3D simulation is possible now. Computational power will continue to increase, especially as massively parallel computers become easier to use. Physics is no longer divided between theory and experiment. The new topic of computational physics sits between and overlaps both theory and experiment.

Acknowledgements

This work is supported by the SILASI European Network ERB 4061 PL 95-0765 and by grants from the UK Research Councils (PPARC and EPSRC). The author's participation in the Latin-American Workshop in Plasma Physics was partly funded by the British Council.

References

Adam, J.C., Guerin, S., Heron, A., Laval, G. and Mora, P.: 1995, *J. de Physique IV* **5**(C6), 1.

Ashour-Abdalla, M. (ed.): 1985, *Space Science Reviews* **42**, 1.

Barnes, J. and Hutt, P.: 1986, *Nature* **324**, 446.

Bell, A.R.: 1987, *Mon. Not. R. Astr. Soc.* **225**, 615.

Bell, A.R.: 1995, 'Transport in laser-produced plasmas', in: M. Hooper (ed.), *Laser-Plasma Interactions – 5*, Scottish University Summer Schools Press and Institute of Physics Publishing, p. 139.

Bell, A.R. and Lucek, S.G.: 1995, *Mon. Not. R. Astr. Soc.* **277**, 1327.

Bell, A.R., Evans, R.G. and Nicholas, D.J.: 1981, *Phys. Rev. Lett.* **46**, 243.
Birdsall, C.K. and Langdon, A.B.: 1985, *Plasma Physics via Computer Simulation*, Adam-Hilger.
Boris, J.P. and Book, D.L.: 1973, *J. Comp. Phys.* **11**, 38.
Boyd, T.J.M. and Sanderson, J.J.: 1969, *Plasma Dynamics*, Thomas Nelson and Sons.
Canullo, M., Costa, A. and Ferro-Fontan, C.: 1996, *Astrophys. J.* **462**, 1005.
Cheng, C.Z. and Knorr, G.: 1976, *J. Comp. Phys.* **22**, 330.
Christiansen, J.P., Ashby, D.E.T.F. and Roberts, K.V.: 1974, *Comp. Phys. Comm.* **7**, 271.
Craxton, R.S., McCrory, R.L. and Soures, J.M.: 1986, *Sci. Amer.* **255**, 60.
Dawson, J.M.: 1995, *Phys. Plasmas* **2**, 2189.
Djaoui, A.: 1995, *J. Quant. Spect. and Rad. Transfer* **54**, 143.
Epperlein, E.M.: 1990, *Phys. Rev. Lett.* **65**, 2154.
Falle, S.A.E.G.: 1994, *Mon. Not. R. Astr. Soc.* **269**, 607.
Fletcher, L. and Brown, J.C.: 1995, *Astronomy and Astrophys.* **294**, 260.
Gentry, R.A., Martin, R.E. and Daly, B.J.: 1966, *J. Comp. Phys.* **1**, 87.
Gibbon, P. and Bell, A.R.: 1992, *Phys. Rev. Lett.* **68**, 1535.
Guerin, S., Laval, G., Mora, P., Adam, J.C. and Heron, A.: 1995, *Phys. Plasmas* **2**, 2807.
Hernquist, L.: 1987, *Comp. Phys. Comm.* **48**, 107.
Hockney, R.W. and Eastwood, J.W.: 1988, *Computer Simulation Using Particles*, Adam-Hilger.
Janulewicz, K.A., Grout, M.J. and Pert, G.J.: 1996, *J. Phys. B* **29**, 901.
Jun, B.I. and Norman, M.L.: 1996, *Astrophys. J.* **465**, 800.
Kershaw, D.S.: 1978, *J. Comp. Phys.* **26**, 43.
Kho, T.H. and Haines, M.G.: 1985, *Phys. Rev. Lett.* **55**, 825.
Kruer, W.L.: 1988, *Physics of Laser-Plasma Interactions*, Addison-Wesley.
Liou, M.S.: 1995, *J. Comp. Phys.* **118**, 294.
Lucek, S.G. and Bell, A.R.: 1996, *Mon. Not. R. Astr. Soc.* **281**, 245.
O'Brien, M.R., Cox, M., Gardiner, C.A. and Zajtsev, F.S.: 1996, *Phys. Plasmas* **3**, 2827.
Ogino, T., Walker, R.J. and Ashour Abdalla, M.: 1994, *J. Geophys. Res. Space Phys.* **99**, A6, 11027.
Pfalzner, S. and Gibbon, P.: 1996, *Many-Body Tree Methods in Physics*, Cambridge.
Pukhov, A. and Meyer-ter-Vehn, J.: 1996, *Phys. Rev. Lett.* **76**, 3975.
Rosenbluth, M.N., MacDonald, W.M. and Judd, D.L.: 1957, *Phys. Rev.* **107**, 1.
Ruhl, H. and Mulser, P.: 1995, *Phys. Lett. A* **205**, 388.
Tabak, M., Hammer, J., Glinsky, M.E., Kruer, W.L., Wilks, S.C., Woddworth, J., Campbell, E.M., Perry, M.D. and Mason, R.J.: 1994, *Phys. Plasmas* **1**, 1626.
Tajima, T. and Dawson, J.M.: 1979, *Phys. Rev. Lett.* **43**, 267.
Town, R.P.J., Rose, S.J. and Bell, A.R.: 1995, *Phys. Rev. Lett.* **74**, 924.
Wallace, J.M., Forslund, D.W., Kindel, J.M., Olson, G.L. and Comly, J.C.: 1991, *Phys. Fluids B* **3**, 2337.
Wilks, S.C., Kruer, W.L., Tabak, M. and Langdon, A.B.: 1992, *Phys. Rev. Lett.* **69**, 1383.
Woodward, P. and Colella, P.: 1984, *J. Comp. Phys.* **54**, 115.
Zimmerman, G.B. and Kruer, W.L.: 1975, *Comments on Plasma Phys. and Controlled Fusion* **2**, 85.

PLASMA PHYSICS IN ASTRONOMY AND COSMOLOGY

REUVEN OPHER
Instituto Astronômico e Geofísico, Universidade de São Paulo, Av. Miguel Stéfano, 4200 CEP 04301-904, São Paulo, SP, Brazil

Abstract. We discuss some of the major areas in astronomy and cosmology where plasma physics is important: (1) origin of stars; (2) distortions of the microwave background radiation; (3) expansion rate of the Early Universe; (4) the magnetic fields and relativistic electrons in jets; (5) the collimation of jets; (6) the origin of stellar winds; (7) the origin of filaments and clouds not gravitationally bound; and (8) the origin of cosmic rays.

1. The Origin of Stars

One major problem that exists in the formation of stars is the requirement that protostellar clouds must lose orders of magnitude of angular momentum. Stars are formed from interstellar matter which participate in the rotation of the galaxy. A typical density of the interstellar matter is ~ 1 cm^{-3}. A stellar mass of ~ 1 solar mass ($M_\odot$) then has an any angular momentum of $\simeq 10^{55}$ g cm^{-2} s^{-1}. A normal $\sim 1\ M_\odot$ star has, however, only an angular momentum of $\sim 10^{49}$ g cm^2 s^{-1}.

The prime candidate for the remover of the excess angular momentum is torsional Alfvén waves. If the radius of the cloud is r_C and the Alfvén velocity in the cloud is V_{AC}, then the characteristic time for the cloud to lose its angular momentum is $\tau = r_C/V_{AC}$. If the magnetic field is not distorted by the cloud, and the ratio of the densities inside and outside the cloud is $\rho_{CL}/\rho_{\rm EXT}$, we can write τ as $\tau_U = (r_C/V_{A\rm EXT})(\rho_C/\rho_{\rm EXT})^{1/2}$, where $V_{A\rm EXT}$ is the Alfvén velocity in the external medium.

If the cloud distorts the magnetic field out to a radius R_o with a magnetic field distribution $B \propto (R_o/r_C)^2$, the characteristic time for the loss of the angular momentum is $\tau_D \sim (r_C/V_{A\rm EXT})(\rho_C/\rho_{\rm EXT})^{1/2}(r_C/R_o)^2$, or $(\tau_D/\tau_U) \sim (r_C/R_o)^2$.

Even after the cloud loses its angular momentum, it still does not collapse. The free-fall time of a cloud is $t_{ff} = (4\pi G\rho)^{-1/2} \simeq 2 \times 10^7 n^{1/2}$ years, where n is the particle density (cm^{-3}). A cloud which has a Jeans dimension, λ_J, is a cloud in which the internal thermal pressure just balances the gravitational attraction. A cloud which has a dimension $\lambda \sim 2\lambda_J$ then should collapse in approximately a free-fall time. We have $\lambda \sim 2\lambda_J$ for $\lambda \sim 50$ pc, $T_C \sim 6000$ K, $n \sim 1$ and $t_{ff} \sim 2 \times 10^7$ years, which is *not* observed.

The velocity near the edge of the cloud should be $V_{ff} \sim 100$ km/s, which is *not* observed. A central concentration in the cloud is also not observed. These nor-

Astrophysics and Space Science **256:** 37–50, 1998.

mal HI (neutral hydrogen atom) clouds are thus not self-gravitating. The common explanation is that there is an additional internal MHD wave pressure, besides the thermal pressure, which prevents the collapse of the cloud. The source of this MHD pressure has not been definitely identified.

2. Distortions of the Microwave Background Radiation (MBR)

The MBR can be distorted due to a gravitational redshift, a Doppler redshift, and an intrinsic variation. Let us concentrate on this third effect. In order to create galaxies, the universe can not be perfectly homogeneous. The inhomogeneities are normally assumed to be created in a primordial epoch before the epoch of the formation of the MBR. The MBR is created when the universe changes from a fully ionized plasma to a partially ionized plasma, called the decoupling epoch, $t_{\rm DEC}$. Before $t_{\rm DEC}$, the photons and the electrons (linked to the protons) are tightly coupled. A fluctuation of the baryon matter, δ_B (protons plus electrons) produces a fluctuation in the radiation density, δ_R, with $\delta_R \sim \delta_B$. Since $\delta T/T \sim (1/4)\delta_R$, we have $(\delta T/T) \sim (1/4)\delta_B$.

Let us first assume that we have a pure baryon universe (with no appreciable nonbaryonic matter (e.g., heavy neutrinos)). The growth of δ_B is proportional to the size of the universe, "a" (the scale factor). We have $a_{\rm DEC} \sim 10^{-3}$ and a_o (the present "a") ~ 1. The growth of the universe depends on the ratio of the density of the universe to the critical density of the universe, Ω. If $\Omega > 1$, the universe will eventually recollapse (i.e., a closed universe) and if $\Omega < 1$ the universe will continually expand (i.e., an open universe). For $\Omega < 1$, δ_B will grow (proportional to "a") until $a_c = [\Omega^{-1} - 1]^{-1}$. We observe today that $\delta_B \sim 1$ for $\lambda \sim 10$ Mpc. This implies that $(\delta T/T) \sim 10^{-3}[\Omega^{-1} - 1]^{1/4}$. For example, for $\Omega \sim 0.1$ (implied by local dynamical interactions of galaxies), we obtain $\delta T/T \sim 2.5 \times 10^{-3}$. However, $\lambda \sim 10$ Mpc corresponds to an angular interval on the sky of $\delta\theta \sim 7'$. Observations indicate that $\delta T/T < 1.5 \times 10^{-5}$ on this scale, implying that a pure baryonic universe does not exist.

Let us now assume that the universe is dominated by nonbaryons (e.g., heavy neutrinos). The nonbaryonic perturbation, δ_{NB}, begins to grow at the epoch of equipartition, $t_{\rm EQ}(< t_{\rm DEC})$, which occurs when the radiation density equals the matter density (baryons plus nonbaryons). Between $t_{\rm EQ}$ and $t_{\rm DEC}$, δ_{NB} grows by a factor $20\Omega h^2$, where the Hubble constant at present is $H_o = h100$ km/s-Mpc. After $t_{\rm DEC}$, the baryons fall into the gravitational potential well of the nonbaryons and $\delta_B = \delta_{NB}$. For $\Omega = 1$, $\delta_{NB}(\delta_B)$ grows by a factor $\sim 10^3$ until t_o (present). For $\delta_{NB}(\delta_B) \sim 1$ observed at present in a sphere of diameter $\lambda \sim 10$ Mpc, we obtain $\delta T/T \sim 1.2 \times 10^{-5}/h^2$, which is consistent with MBR observations on the scale $\lambda \sim 10$ Mpc ($\delta\Theta \sim 7'$).

We must emphasize that MBR is created when the plasma of the universe changes from a fully ionized plasma to a partially ionized plasma and any plasma effect occurring at this epoch, t_{DEC}, affects MBR.

Free-free emission and absorption (bremsstrahlung) and Compton scattering are particularly important processes in forming MBR. The characteristic time for the free-free process in the early universe is $t_{ff}(Z) = 0.4 \times 10^{24}(\Omega h^2)^{-2} Z^{-5/2}$ s, where the redshift $Z = a^{-1} - 1$. The characteristic time for Compton scattering is $t_C(Z) = 10^{28}(\Omega h^2)^{-1} Z^{-4}$ s. When, $t_C < t_{ff}$, a photon scatters many times before it is free-free absorbed. The effective absorption time is $\bar{t}_{ff} = (t_{ff} t_C)^{1/2} = 0.6 \times 10^{26}(\Omega h^2)^{-3/2} Z^{-13/4}$ s. Free-free processes cease to be effective in neutralizing distortions when $\bar{t}_{ff} > H^{-1}$. This occurs at a redshift $Z_{ff} \simeq 10^5(\Omega h^2)^{-6/5}$, which is much larger than Z_{DEC}.

The characteristic time for free-free processes in the above analysis assumed that all the photons have the same energy (kT). We have, however, $t_{ff}(T, \omega) \propto \omega^3[1 - \exp(-\omega/T)]^{-1}$. Thus at low frequencies, free-free processes can reestablish the Planck spectrum. The frequency, ω_B, when $t_{ff}(\omega_B) \simeq t_C$ at t_{DEC}, is $\omega_B \sim (\Omega h^2)^{1/2} T$.

The MBR, passing through hot ionized plasma clouds, becomes distorted due to the hot electrons scattering low frequency photons and making them high frequency photons. This is called the Zeldovich–Sunyaev Effect. In general, we measure the temperature of the MBR by counting the number of photons at a given low frequency. Because the Zeldovich–Sunyaev Effect removes low frequency photons, the Zeldovich–Sunyaev Effect creates a negative δT. It is approximately equal to $\delta T = (-2T_m/me)\tau_C T_r$, where τ_C is the Compton optical depth and T_m is the temperature of the hot plasma. It has been observed in various galactic clusters and is on the order of a milliKelvin.

3. The Expansion Rate of the Early Universe

Until now, in general, the primordial universe is assumed to be approximately homogeneous. At a temperature, for example of $T \sim 1$ MeV (the beginning of primordial nucleosynthesis), the universe is assumed to consist of a blackbody spectrum of photons, plus Fermi distributions of electrons (and positrons) and three families of neutrinos (and antineutrinos). (The baryon density at this temperature is $\sim 10^{-10}$ the electron density). The expansion rate of the universe is $\propto \rho^{1/2}$, where ρ is the total energy density. No plasma effects have until now been included.

According to plasma physics theory, the electromagnetic fluctuations in a plasma is predicted from the Fluctuation Dissipation Theorem, which depends on the dissipation processes in the plasma. We made a detailed analysis of the Fluctuation-Dissipation Theorem including collisionless and collisional damping processes. We obtained the electromagnetic spectrum at all frequencies and wave numbers. In particular, at high frequencies we obtained the photon black body spectrum, as expected. At low frequencies, however, we obtain a spectrum that is higher than the black body spectrum for $w \leq 4w_{pe}$, where w_{pe} is the electron plasma frequency. A higher expansion rate of the universe is thus predicted (Opher and Opher, 1997).

The spectra of fluctuations of the magnetic and electric fields, for an isotropic plasma, given by the *Fluctuation-Dissipation Theorem* (Sitenko, 1967; Akhiezer et al., 1975).

$$\frac{\langle E^2\rangle_{k\omega}}{8\pi} = \frac{\hbar}{e^{\hbar\omega/T}-1}\frac{\mathrm{Im}\,\varepsilon_L}{|\varepsilon_L|^2} + 2\frac{\hbar}{e^{\hbar\omega/T}-1}\frac{\mathrm{Im}\,\varepsilon_T}{|\varepsilon_T - (\frac{kc}{\omega})^2|^2}\,, \tag{1}$$

$$\frac{\langle B^2\rangle_{k\omega}}{8\pi} = 2\frac{\hbar}{e^{\hbar\omega/T}-1}\left(\frac{kc}{\omega}\right)^2\frac{\mathrm{Im}\,\varepsilon_T}{|\varepsilon_T - (\frac{kc}{\omega})^2|^2}\,. \tag{2}$$

The first term of Equation (1) is the longitudinal electric field fluctuations and the second term is the transverse electric field fluctuations.

In order to have a description that includes thermal effects as well as collisional effects, we use the Vlasov equation in first order (Ichimaru, 1992) with the BGK collision term (Clemmow and Dougherty, 1990). The transverse dielectric permittivity obtained is:

$$\varepsilon_T(\omega, \mathbf{k}) = 1 + \sum_\alpha \frac{\omega_{p\alpha}^2}{\omega^2}\left(\frac{\omega}{\sqrt{2}kv_\alpha}\right)Z\left(\frac{\omega + i\eta_\alpha}{\sqrt{2}kv_\alpha}\right), \tag{3}$$

where α is the label for each species of the plasma, v_α is the thermal velocity for each species, η_α is the collision frequency and $Z(z)$ is the Fried and Conte function (1961). With this transverse dielectric permittivity, we obtain the electromagnetic spectrum, that is, the sum of the magnetic and the electric field transverse spectra.

4. The Magnetic Fields and the Relativistic Electrons in Jets

Jets from active galactic nucleii are highly ionized and we observe them from the synchrotron radiation they produce. We expect that the magnetic field is frozen into the plasma. The magnetic field parallel to the axis of the jet $B_\phi \propto r^{-2}$, where r is the radius of the jet. Similarly, the poloidal magnetic field $B_\phi \propto r^{-1}V_J^{-1}$ where V_J is the jet velocity. Assuming V_J is slowly varying, we have $B_\phi \propto r^{-1}$. Thus, after a sufficient expansion of the jet (i.e., increase of r), we expect all jets to have only B_ϕ. This is not observed!

Similarly, the density of the plasma is expected to decrease as $\rho \propto r^{-2}V_J^{-1} \propto r^{-2}$. Since the plasma pressure $P_{PL} \propto \rho^\gamma \propto r^{-2\gamma}$, we have $P_{PL}/P_B \propto r^{-2\gamma}/B_\phi^2 \propto r^{-2}(\gamma - 1)$. Again, for sufficient expansion of the jet and $\gamma > 1$, $P_{PL}/P_B \longrightarrow 0$, which, again, is not observed!

The observations of the jets indicate that the magnetic field in the jet and the relativistic electrons in the jet (creating the P_{PL}) are regenerated due to the motion of the jet through the ambient medium. No satisfactory theory for this regeneration, however, exists.

5. The Collimation of Jets

Jets from active galactic nucleii appear to be highly collimated. For example, the jet of NGC 6251 extends to almost $\sim$ 1 Mpc. On the scale of tens to hundreds of kpc, it has a ramrod appearance. The jet on the order of 1–10 kpc has a knotty structure. The jet $\sim 1 pc$ is aligned with the kpc jet.

We have investigated the collimation of jets due to the creation of the Kelvin–Helmholtz instability. We showed that for a magnetic perturbation of $\Delta B/B \sim 10^{-4}$ created by the instability, a current of $I > 10^{18}$ A is produced by transit-time magnetic pumping, which is sufficient to pinch the jet (Jafelice et al., 1990).

6. The Origin of Stellar Winds

Observational data put strong limits on possible explanations of the hot solar wind. In particular, it is difficult to explain the high velocity fast solar wind $V_\infty \sim 700$ km/s as due simply to a thermal expansion from the solar surface.

We also have a problem to explain cold stellar winds in red supergiants. After a star almost terminates burning all of its hydrogen, it expands and becomes a red giant. Its mass loss is $\sim 10^7$ times larger than the solar wind. The temperature is sufficiently low that radiation pressure is insufficient to create the massive wind.

We also have a problem to explain the high-momentum stellar winds in Wolf–Rayet stars. The mass loss of Wolf–Rayet stars is enormous: $\sim 10^{10}$ times that of the solar wind. They are the most brilliant stars in the galaxy with a photospheric temperature $T \geq 30,000$ K. (For the sun, $T \sim 5500$ K).

Radiation pressure has difficulty to explain the wind source: (1) the momentum of the radiation is $\leq 1/10$ the momentum of the wind; and (2) we require an Eddington factor (radiation force/gravitational force) ≥ 0.4 at the stellar surface, which contradicts stellar evolution theory.

We note that Alfvén waves contain 1000 times more momentum per unit of energy than electromagnetic radiation.

We showed that Alfvén waves plus radiation can explain the observations of all of the above stellar winds (Jatenco-Pereira and Opher, 1989a, 1989b, 1989c; dos Santos et al., 1993a, 1993b).

7. The Origin of Filaments and Clouds Not Gravitationally Bound

We observe filaments in radio jets (e.g., Cen A), radio lobes (e.g., Cyg A) and in supernovae remnants (e.g., Crab Nebula). In particular, the radio flux is concentrated in linear structures with lengths 5–20 times their widths. Almost all of the matter is concentrated in the filaments.

We showed that these filaments can be produced by a Synchrotron-Inverse Compton Thermal Instability (SICTI). We assumed that the perturbed region is optically thick to the heating by MHD waves. We found that the condition for SICTI is:

$$q < q_{\max} \equiv \frac{[-2\nu_e + (\frac{3}{4}\tau_{sc})(2\eta - 1)]}{[(2\nu_e/2) + (2/\tau_{sc}]},$$

where $q \equiv V_A^2/V_S^2$, $V_A(V_S)$ = Alfvén (Sound) Speed, $\nu_e \equiv V_R/R$, $\tau_{sc} \equiv 3P/L_{sc}$ where P is the pressure due to relativistic electrons, L_{sc} is the total luminosity, $\eta \equiv [1 + (8\pi U_{\rm rad}/B^2)]^{-1}$ and $U_{\rm rad}$ is the radiation density (de Gouveia Dal Pino and Opher 1989a, 1989b, 1990, 1991, 1993).

Similarly, we have the problem of line-emitting clouds in quasars. Quasars were discovered over 30 years ago and their "Central Engine" is still a mystery. Their luminosities are on the order of $\sim$ 1000 times that of the Milky Way. The temperature of the clouds $\sim 10^4$ K can be compared with the temperature of the ambient medium $\sim 10^7$ K.

A possible thermal instability is naturally suggested. Equating the heating processes to the cooling processes, we obtain a curve in the log T vs log P plane in the form of a "Z". At low pressures we have only a high temperature ($\sim 10^7$ K) phase and at high pressures only a low temperature ($\sim 10^4$ K) phase. At intermediate pressures both phases can coexist (i.e., a thermal instability). The problem is that this intermediate pressure region exists at pressures below the observed pressure.

We showed that Alfvén waves, which are observed to be present, create sufficient heating to raise the thermal instability pressure region to the observed pressure (Gonçalves et al., 1993, 1996).

The short cooling times (less than a Hubble time) of the central intracluster medium (ICM) of X-ray emitting clusters of galaxies point to gas sinking towards the center of the cluster in a "cooling flow". A sizeable proportion of the cooling flows also contains extended (from $\leq$ 1 kpc to tens of kpc) optical emission line filaments around the central dominant galaxy of the cluster, which is always at the center of the flow. These extended optical filaments systems are seen only in cooling flows and are thought to be a phase of the evolution of thermal instabilities arising in the cooling flow.

One of the major problems associated with the optical filaments is their high luminosities (typically H_α luminosities are 10^{41} to 10^{42} ergs s^{-1}). Quiescently cooling gas that recombines only once at the X-ray determined mass accretion rate would not be detectable as emission line nebulae. The fact that the central regions of cooling flows often do show optical emission requires some source of energy to repeatedly reionize the gas. Several models, considering different mechanisms for ionization and heating of the gas – shocks, thermal conduction, and photoionization by soft X-rays and EUV produced in the cooling gas – have met difficulties in explaining the observed line ratios and luminosities. It is possible that some combination of photoionization and shock models could explain the observations. One example of these hybrid models involves the combination of self-absorbed irradiating mixing layers of cold clouds embedded in the hot cooling gas and emission from shocks generated in collisions between these clouds.

Jafelice and Friaça (1998) have investigated magnetic reconnection as a heating mechanism and concluded that magnetic reconnection cannot be invoked as the sole mechanism powering the optical line emission in cooling flows. However, it could be an important ingredient to explain the emission coming from low ionization lines. The models with high magnetic reconnection efficiencies exhibit a strong [OI] λ 6300 emission, and, therefore, magnetic reconnection could be an additional component together with other mechanisms producing most of the emission in H_α and higher ionization lines (as [NII] λ 6583). The existence of magnetic fields in the ICM has suggested magnetic reconnection as a heating mechanism. We investigated another heating mechanism expected to be at work in the presence of magnetic fields: Alfvén heating (AH). We explored the importance of AH not only as a dominant energy source for optical filaments in cooling flows but also as an ingredient contributing to the emission of some optical lines.

The Alfvén heating comes from the dissipation (damping) of Alfvén waves. These waves are easily generated in many cosmic plasmas, but they possess no linear damping mechanism since they are not compressive. Nonlinear damping of these waves occurs when one Alfvén wave decays into another plus a slow magnetosonic wave, or two Alfvén waves combine into one fast magnetosonic wave: the resulting magnetosonic waves can then be dissipated. The dissipation of these waves may contribute to heat the solar corona, to influence the interplanetary wave spectrum, and to determine the wave spectrum available to scatter cosmic rays. As they are not compressive, the nonlinear decay occurs due to second order effects in the wave amplitude. Interacting Alfvén waves are compressive and thus generate a new compressive wave. The resulting heating – at least in terms of nonlinear, resonance surface and turbulent damping mechanisms – has been applied to study the thermal stability in the broad line region of quasars.

Alfvén waves are created by perturbed magnetic fields. In view of this, two conditions that are fulfilled by the core of a cooling flow make it a favorable site for the generation of Alfvén waves: it contains magnetic fields, and is turbulent.

The lack of any strong hard X-ray emission due to inverse Compton from the relativistic electrons in a few clusters (generally not cooling flows) that have halo radio sources imposes a lower limit for the magnetic field of 0.1 μG over scales of up to 500 kpc. At the smaller ($\sim$ 100 kpc) scales of cooling flows, magnetic fields of typically 1–3 μG are derived corresponding to a magnetic pressure of about 1% of the thermal pressure. At the center of a cooling flow still more intense magnetic fields must be present since the field is amplified by compression as the gas flows inward. As a matter of fact, Faraday rotation and depolarization of the radio emission of extended radio sources residing in cooling flows have revealed that the magnetic field increases inward and that it can reach 20–100 μG depending on the degree of ordering of the field. In this way, at the very center of the flow, the magnetic pressure can supercede the thermal pressure.

Lines are broad throughout the filamentary region, but line widths decrease from 500–1000 km s^{-1} in or near the galactic nucleus to 100–300 km s^{-1} at the largest radii (typically 5–15 kpc). This suggests that the inner ICM is highly turbulent, with the amplitude of random velocities increasing inward. In fact, the core of a cluster is expected to be turbulent not only due to the frequent crossing of galaxies through it but also due to relics of merging of subclusters. In addition, the absence of rotation in the center of cooling flows requires rotational breaking of the gas flowing inward, and turbulent viscosity is the most likely transport process of angular momentum in a cooling flow. The emission line filaments have line widths of several hundred to $\sim$ 1000 km s^{-1}. It is possible that these velocities of the filaments are turbulent and that the hot component of the ICM is itself turbulent. Anyway, there is plenty of kinetic energy to be tapped via Alfvén waves into an energy source for the ionization of the filaments. Several damping mechanisms can be responsible for Alfvén heating. We considered the nonlinear damping and the turbulent damping for Alfvén waves.

7.1. Nonlinear Damping

An Alfvén wave is likely to dissipate because of its nonlinear interaction with either the nonuniform ambient field or another Alfvén wave. When the magnetic field is weak, (i.e., when the Alfvén velocity v_A is lower than the sound velocity, C_s), it is found that two Alfvén waves travelling in opposite directions along a magnetic field line can couple nonlinearly to produce an acoustic wave, which in turn dissipates relatively quickly. In regions of strong ($v_A > C_S$) magnetic field, one Alfvén wave can decay into another Alfvén wave and a sound wave travelling in the opposite direction. The resulting Alfvén wave has a frequency smaller than the original one and it can, in turn, decay into another lower-frequency Alfvén wave plus an acoustic wave. The cascade continues until all the Alfvenic energy has been converted to acoustic waves that dissipate rapidly. The latter possibility is considered. In our models, the onset of nonlinear heating occurs for $\beta = P_B/P_{\rm gas} = (B^2/8\pi)/nk_BT > \beta_{\rm on} > 1$, where $\beta_{\rm on}$ is the initial value of β when AH becomes important ($\beta > 1$ corresponds to $v_A/C_s > \sqrt{6/5}$ for an ideal $\gamma = 5/3$ gas). (Note that following the convention widely used in cooling flow studies, we define $\beta = P_B/P_{\rm gas}$, as where in plasma physics, $\beta_{\rm pl} = P_{\rm gas}/P_B$ denotes the "beta parameter" of the plasma.

The nonlinear damping rate is

$$\Gamma_{\rm NL} = \frac{1}{4}\sqrt{\frac{\pi}{2}}\,\xi\bar{\omega}\left(\frac{C_s}{v_A}\right)\frac{\rho\langle\delta v^2\rangle}{B^2/8\pi}\,, \tag{4}$$

where $\xi = 5$–10, C_s is the sound velocity, v_A the Alfvén velocity, $\bar{\omega}$ a characteristic Alfvén frequency and $\rho\langle\delta v^2\rangle/(B^2/8\pi)$ the ratio of the energy density of Alfvén waves to that of the magnetic field. As we are interested in AH coming from the nonlinear damping of Alfvén waves, we heuristically derive the dependences of the heating rate $H_{\rm NL}$ (erg cm^{-3} s^{-1}) with density and temperature. Thus, the nonlinear Alfvenic heating is:

$$H_{\rm NL} = \frac{\Phi_\omega \Gamma_{\rm NL}}{v_A}\,, \tag{5}$$

where Φ_ω is the wave flux, $\Gamma_{\rm NL}$ is given by Equation (1), and $v_A = B/\sqrt{4\pi\rho}$ is the Alfvén velocity. As our perturbation collapses, the component of the magnetic field perpendicular to the direction of compression is amplified. For our plane parallel geometry, $B \propto A^{-1} \propto \rho$, where A is the cross-sectional area perpendicular to the magnetic field. As a consequence, $v_A \propto \rho^{1/2}$. Let the dependence of Φ_ω on ρ be given as $\Phi_\omega \propto \rho^c$. Taking $\Phi_\omega = \rho\langle\delta v^2\rangle v_A$, we have $\rho\langle\delta v^2\rangle \propto \rho^{c-1/2}$. Since $C_s \propto T^{1/2}$, for $\bar{\omega}$ independent of the density, we obtain:

$$H_{\rm NL} \propto \rho^{2c-7/2}T^{1/2}. \tag{6}$$

For a plane-parallel geometry, $\Phi_\omega \propto A^{-1} \propto \rho$, implying $c = 1$, and Equation (6) becomes:

$$H_{\rm NL} \propto \rho^{-3/2}T^{1/2}. \tag{7}$$

7.2. Turbulent Damping

Turbulence theory indicates that the spectrum of turbulent Alfvén waves has a power law dependence in phase space

$$P(k) \propto k^{-\beta},$$

where k is the Alfvén wavenumber, β is the spectral index, and

$$\int P(k)4\pi k^2 dk$$

is the energy density of turbulent Alfvén waves. The value of the spectral index is determined by the type of the turbulence. The nonlinear interaction between outward and inward propagating waves results in an energy cascading process. A small part of the energy of lower frequency fluctuations cascades to higher frequency fluctuations until it reaches a frequency high enough for some dissipation process to occur, the most probable candidate being proton-cyclotron damping.

The observed magnetic power spectra, P_B, in the solar wind exhibits a well-developed inertial subrange resembling Kolmogorov turbulence, i.e. $P_B \propto k^{-5/3}$, over more than 3 orders of magnitude of k. This indicates a cascade to small scales. The turbulent cascade transfers wave energy from large to short scales, and, in the end, the dissipation of energy through turbulence is governed by the small scale linear absorption mechanism. Exploiting the similarity of $P_B \propto k^{-5/3}$ and Kolmogorov turbulence in ordinary fluids, the volumetric heating rate associated with the cascade process can be written as

$$H_T = \rho \langle \delta v^2 \rangle^{3/2} / L_{\rm corr}, \tag{8}$$

where ρ is the mass density, $\langle \delta v^2 \rangle$ is the velocity variance associated with the field, and $L_{\rm corr}$ is a measure of the transverse correlation length.

As in the previous subsection, we can derive the dependence of the turbulent heating rate on ρ. For a plane parallel geometry, $\langle \delta v^2 \rangle \propto \rho^{c-3/2}$ (assuming $\Phi_\omega \propto \rho^c$), and adopting $L_{\rm corr} \propto \rho^b$, we obtain:

$$H_T \propto \rho^{(3/2)c-b-5/4}. \tag{9}$$

For a plane parallel geometry, $c = 1 (\Phi_\omega \propto \rho)$, and if we assume that $L_{\rm corr} \propto B^{-1/2}$ (Hollweg, 1986, 1987), then $b = -1/2$ (since $B \propto \rho$). The resulting turbulent Alfvén wave heating is

$$H_T \propto \rho^{3/4}. \tag{10}$$

We have investigated the evolution of optical filaments since their formation out of the hot phase of the cooling flow.

The evolution of the cooling filaments is obtained by solving the hydrodynamical equations of mass, momentum and energy conservation (see Friaça, 1986, 1993; Friaça and Terlevich, 1994, 1996). The condensations are assumed to be self-gravitating. Since there is no ionization equilibrium for temperatures lower than 10^6 K, the ionization state of the gas at $T < 10^6$ K is obtained by solving the time-dependent ionization equations, for all ionic species of H, He, C, N, O, Ne, Mg, Si, S, Ar and Fe. We adopt an isochoric nonequilibrium cooling function for temperatures lower than 10^6 K, since the recombination time of important ions is longer than the cooling time at these temperatures. The cooling function and the coefficients of collisional ionization, recombination and charge exchange of the ionization equations are all calculated with the atomic database of the photoionization code AANGABA (Gruenwald and Viegas, 1992). The adopted abundances are half-solar, as derived from X-ray spectroscopy of the ICM.

We considered nonlinear and turbulent Alfvén wave heating. From Equations (7) and (10), the heating term has the form

$$H_{\rm NL} = \zeta H_o (n/n_o)^{-3/2} (T/T_o)^{1/2}, \tag{11}$$

for nonlinear heating, and

$$H_T = \zeta H_o (n/n_o)^{3/4}, \tag{12}$$

The nonlinear and the turbulent AHs represent extreme opposite dependencies of AH on density. Turbulent AH ($H_T \propto n^{3/4}$) deposits more energy at the center of the filament, where the density is higher. On the other hand, nonlinear AH ($H_{\rm NL} \propto n^{-3/2}T^{1/2}$) deposits more energy per unit volume in the outer regions of the filaments, where the density is lower. However, due to magnetic support, the central density of the cooled filament ($T \approx 10^4$ K) is only ≈ 3.3 times the initial unperturbed density, and the central decrease of the AH rate due to density variation amounts to only a factor of 6. In the end, the fact that the energy is deposited in a much cooler gas at the center than on the outside of the filament ($T \approx 10^7$ K) makes the AH somewhat more important in the colder, central gas than in the outer skin of the filament, as we can see from the timescale for heating $t_H = (3/2)nk_BT/H_{\rm NL} \propto n^{5/2}T^{1/2}$, which is 1.6 times smaller at the center than on the outside of the filament.

The calculations are intended to apply to the inner $\sim$ 10 kpc of the cooling flows, where the optical filaments are most commonly seen. The turbulent velocities of the cooling flow increase toward the center of the cluster, as inferred from the increase of the line widths near the center (Heckman et al., 1989). Since the turbulent energy of the cooling flow is the energy source of the Alfvén waves, it is only in the inner $\sim$ 10 kpc that the density of turbulent energy is high enough to make AH efficient. In this region, blobs condensating out of the cooling flow are kept warm enough by AH to produce strong line emission. Blobs cooling down in the outer regions are subject to low levels of AH and cool quiescently without noticeable optical emission. In this way, the AH mechanism is consistent with the fact that line emission does not extend to $\sim$ 100 kpc (Friaça et al., 1997).

8. The Origin of Cosmic Rays

The most popular mechanism for the acceleration of cosmic rays $\leq 10^{15}$ eV is that due to a supernova shock. In this mechanism, the shock is super-Alfvenic. Alfvén wave turbulence exists in the post-shock downstream region of the shock, and Alfvén waves also exist in the upstream pre-shock region due to high energy particles passing into the pre-shock region and creating Alfvén waves by the Cherenkov radiation condition ($V_p cos\alpha \simeq V_A$) where $V_p(\sim c)$ is the velocity of the particle and α is the angle between the velocity of the particle and the Alfvén wave.

The Alfvén waves reflect the high energy particles. With respect to the shock wave, the pre-shock Alfvén waves flow into the shock since the shock is super-Alfvenic. The Alfvén waves act as mirrors which approach each other. Each time the particle is reflected by the Alfvén waves, the particle gains energy. The process is a First-Order-Fermi Acceleration mechanism. In the linear theory the magnetic pressure (MP), the high energy particle pressure (HEPP) and the Alfvén wave pressure (AWP) are assumed to be negligible. A power law spectrum is predicted, which is observed. Also the power law index predicted is about what is observed.

However, MP, HEPP and AWP are not negligible but are comparable to the plasma pressure. When MP, HEPP and AWP are taken into account, instead of having a single discontinuity of the plasma density at the shock front, pre-shocks exist or only a smooth variation of the density exists, with no discontinuity (Medina Tanco and Opher, 1990, 1993).

We suggest that ultra high energy cosmic rays (UHECR) $> 10^{20}$ eV, are produced by rotating magnetic neutron stars with accretion discs at cosmological distances. In an intermediary step, gamma ray bursts (GRBs) are also produced by the rotating magnetic neutron stars. We assume that a current circuit (CC) with current $I \sim 10^{20}$ A is produced by the magnetic field which is twisted due to the anchoring of the field lines of the rotating neutron star in the accretion disc. The magnetic field has the same shape as that of CCs of extragalactic jets. The current flows out along the axis of rotation ($r = 0$ in cylindrical coordinates) up to a maximum distance, $z_{\max}$, and flows back at a radius $r \sim r_c$, the cocoon radius. It is assumed that $r_c \sim z_{\max} \sim R_*$, the neutron star radius. The self-inductance of the CC of length $\sim l$ is $L \sim \mu_o l$. The CC has a magnetic energy $W = 1/2LI^2 \sim 1/2\mu_o l I^2$. For $I \sim 10^{20}$ A and $l \sim R_* \sim 10^6$ cm, we obtain $W \simeq 4 \times 10^{42}$ ergs. This energy is comparable to the magnetic field energy of the neutron star.

It is assumed that the CC creates double layers (DLs). The classic double layer (Langmuir, 1929) is an electrostatic structure which can appear within a current carrying plasma and can sustain a significant potential drop, $\phi_{\rm DL}$. For a strong DL, $e\phi_{\rm DL} \gg k_B T$, the DL thickness, according to Borovsky (1983), is $d \sim 0.9\lambda_D(e\phi_{\rm DL}/k_B T)^\alpha$, where T is the temperature of the plasma, α is a constant and λ_D is the Debye length. $\lambda_D \propto (T/n)^{1/2}$. For example, for $k_B T \sim 1$ MeV and the numerical density of the plasma $n \sim 5 \times 10^{14}$ cm^{-3}, $\lambda_D \sim 3.3 \times 10^{-2}$ cm. For $e\phi/k_B T)^\alpha \sim 10^6$ with $d \sim 3 \times 10^4$ cm and $\alpha = 0.75$ as suggested by Borovsky (1983), $e\phi/k_B T)^\alpha \sim 10^8$ for $d \sim 3 \times 10^6$ cm and $\alpha = 1.0$. Since d is not accurately known, we assume that $\alpha \sim 0.75$–1. With the above parameters, $d \sim 10^6$ cm $\sim R_*$.

For d comparable to the radius of the current carrying region, the value of $\phi_{\rm DL}$ is (Peratt, 1992)

$$\phi_{\rm DL} = \left(\frac{\phi_i}{4\pi\varepsilon_0} I\right)^{1/2}, \qquad (13)$$

with $e\phi_i = m_i c^2$ and m_i the ion mass, $e\phi_i \sim 0.939 \times 10^4$ eV for protons. The power dissipated by the DL is $P_{\rm DL} \simeq \phi_{\rm DL}$. The characteristic time for the release of the energy is $\tau = W/P_{\rm DL}$. Thus $\tau = \mu_0 l I/2\phi_{\rm DL}$.

In our model, we envision that the neutron star with an accretion disc first creates the CC with a current I. The CC is then interrupted by the DL over a time τ, creating the induced voltage $\phi_{\rm DL}$ across the DL. For $I \sim 3 \times 10^{20}$ A and $l \sim 10^6$ cm, $e\phi_{\rm DL} \sim 2.8 \times 10^{15}$ eV and $\tau \sim 4.3$ s. This time is comparable to the observed time for GRBs. We suggest that GRBs are formed in the scenario outlined above: a CC of $I \sim 3 \times 10^{20}$ A is produced by a rotating neutron star with an accretion disc which then forms a DL which is active over a time $\tau = 4.3$. The potential drop across the DL is $\phi_{\mathrm{DL}} \sim 2.8 \times 10^{15}$ eV.

A very high energy density ("fireball") is formed near the DL. At this energy, the Lorentz factor for a proton is $\Gamma_p \sim 3.0 \times 10^6$ and for an electron, $\Gamma_e \sim 5.5 \times 10^9$. Many electron-positron pairs are created in the fireball which acts to cool the plasma. the fireball expands with an effective Lorentz factor $\Gamma_1 \sim 10^6$–10^{10}. The expansion of the fireball is primarily along the rotation axis due to the confining action of the electric current, $I \sim 3 \times 10^{20}$ A, which is sufficient to "pinch" a jet with a plasma energy $\leq$ magnetic energy of an axial magnetic field $\sim 10^{12}$ G. Thus, a jet with a Lorentz factor Γ_1 is formed. We assume that a DL exists in the jet plasma. For simplicity, we assume that the DL is perpendicular to the jet motion and that the effective Γ of the particles in the reference frame of the jet is $\Gamma_2' \sim \Gamma_1$. In the laboratory frame, the Γ is $\sim \Gamma_2'\Gamma$. Since $\Gamma_2' \sim \Gamma_1 \sim 3.0 \times 10^6$–$6 \times 10^9$, $\Gamma \sim 10^{13}$–10^{18}. For protons, this Γ corresponds to energies $\sim 10^{22}$–10^{27} eV, comparable to the highest energy UHECR observed, $> 10^{20}$ eV.

Acknowledgement

The author would like to thank the Brazilian financing agency CNPq for partial support.

References

Akhiezer, A.I., Akhiezer, A.I., Plovin, R.V., Sitenko, A.G. and Stepanov, K.N.: 1975, *Plasma Electrodynamics*, Pergamon Press, Oxford.

Borovsky, J.E.: 1983, *Phys. Fluids* **26**, 3273.

Clemmow, P.C. and Dougherty, J.P.: 1990, *Electrodynamics of Particles and Plasmas*, Addison-Wesley, Redwood City.

de Gouveia Dal Pino, E.M. and Opher, R.: 1989a, *Astrophys. J.* **342**, 686.

de Gouveia Dal Pino, E.M. and Opher, R.: 1989b, *Mon. Nat. R. Astr. Soc.* **240**, 573.

de Gouveia Dal Pino, E.M. and Opher, R.: 1990, *Astron. and Astrophys.* **231**, 571.

de Gouveia Dal Pino, E.M. and Opher, R.: 1991, *Astron. and Astrophys.* **242**, 319.

de Gouveia Dal Pino, E.M. and Opher, R.: 1993, *Mon. Not. R. Astr. Soc.* **263**, 687.

dos Santos, L.C., Jatenco-Pereira, V. and Opher, R.: 1993a, *Astron. and Astrophys.* **270**, 345.

dos Santos, L.C., Jatenco-Pereira, V. and Opher, R.: 1993b, *Astrophys. J.* **410**, 732.

Friaça, A.C.S.: 1986, *A&A* **164**, 1.

Friaça, A.C.S.: 1993, *A&A* **269**, 145.

Friaça, A.C.S. and Terlevich, R.J.: 1994, in: G. Tenorio-Tagle (ed.), *Violent Star Formation from 30 Douradus to QSOs*, Cambridge University Press, Cambridge, p. 424.

Friaça, A.C.S., Gonçalves, D.R., Jafelice, L.C., Jatenco-Pereira, V. and Opher, R.: 1997, *Astron. Ap.* **324**, 449.

Fried, B.D. and Conte, S.D.: 1961, *The Plasma Dispersion Function*, Academic Press, New York.

Gonçalves, D.R., Jatenco-Pereira, V. and Opher, R.: 1993, *Astrophys. J.* **414**, 57.

Gonçalves, D.R., Jatenco-Pereira, V. and Opher, R.: 1996, *Astrophys. J.* **463**, 489.

Gruenwald, R.B. and Viegas, S.M.: 1992, *ApJS* **78**, 153.

Heckman et al.: 1989, *Astrophys. J.* **338**, 48.
Ichimaru, S.: 1992, *Basic Principles of Plasma Physics*, Addison-Wesley, Redwood City.
Jafelice, L.C. and Friaça, A.C.S.: 1998, *Mon. Not. Not. R. Astr. Soc.* **324**, 449.
Jafelice, L.C., Opher, R., Assis, A.S. and Busnardo-Neto, J.: 1990, *Astrophys. J.* **348**, 61.
Jatenco-Pereira, V. and Opher, R.: 1989a, *Mon. Not. R. Astr. Soc.* **236**, 1.
Jatenco-Pereira, V. and Opher, R.: 1989b, *Astron. and Astrophys.* **209**, 327.
Jatenco-Pereira, V. and Opher, R.: 1989c, *Astrophys. J.* **344**, 513.
Langmuir, I.: 1929, *Phys. Rev.*, **33**, 954.
Medina-Tanco, G.A. and Opher, R.: 1990, *Astron. and Astrophys.* **240**, 178.
Medina-Tanco, G.A. and Opher, R.: 1993, *Astrophys. J.* **411**, 690.
Opher, M. and Opher, R.: 1997, *Phys. Rev. Lett.* **78**, 2628.
Peratt, A.L.: 1992, *Physics of the Plasma Universe*, Springer-Verlag, New York.
Sitenko, A.G.: 1967, *Electromagnetic Fluctuations in Plasma*, Academic Press, New York.

ADVANCES IN NUMERICAL MODELING OF ASTROPHYSICAL AND SPACE PLASMAS

Part II. Astrophysical Force Laws on the Large Scale

ANTHONY L. PERATT
United States Department of Energy, Washington, DC, U.S.A.; and Los Alamos National Laboratory, Los Alamos, New Mexico, U.S.A.

Abstract. Advances in the simulation of astrophysical and cosmic plasmas are the direct result of advances in computational capabilities, today consisting of new techniques such as multilevel concurrent simulation, multi-teraflop computational platforms and experimental facilities for producing and diagnosing plasmas under extreme conditions for the benchmarking of simulations. Examples of these are the treatment of mesoscalic plasma and the scaling to astrophysical and cosmic dimensions and the Accelerated Strategic Computing Initiative whose goal is to construct petaflop (10^{15} floating operations per second) computers, and pulsed power and laser inertial confinement plasmas where megajoules of energy are delivered to highly-diagnosed plasmas. This paper concentrates on the achievements to date in simulating and experimentally producing plasmas scaled to both astrophysical and cosmic plasma dimensions. A previous paper (Part I, Peratt, 1997) outlines the algorithms and computational growth.

Keywords: Numerical modeling, numerical simulation, electric space, plasma universe, cosmology, galaxies, filamentation, electrical currents, quasars, double radio galaxies

1. History of Models of Astrophysical Phenomena

While this year marks the one-hundred and second anniversary of the discovery of electric space (Peratt, 1995), i.e., the realization of the need to apply plasma physics and electromagnetic forces to the understanding of the dynamical properties, evolution, and radiation to astrophysical data was lost with the death of Kristian Birkeland (1867–1917) and not to be resurrected until two decades later by Hannes Alfvén (1908–1995). However, by that time mathematicians had devised a new model for the state of the universe, one based entirely on gravitation relying solely on the only known states of matter at that time: solids, liquids, and gases.

This paper explores the reasons why the plasma state* was so long neglected and its necessity now for explaining the dynamics of the solar system and, undoubtedly, the 'space' beyond. As space is plasma, the necessity of very large scale computational platforms are invaluable for studying its properties because of space's extreme, if not infinite, dimensions.

* The term 'plasma' was not known until coined by the Nobel chemist Irving Langmuir in 1923.

Astrophysics and Space Science **256:** 51–75, 1998.

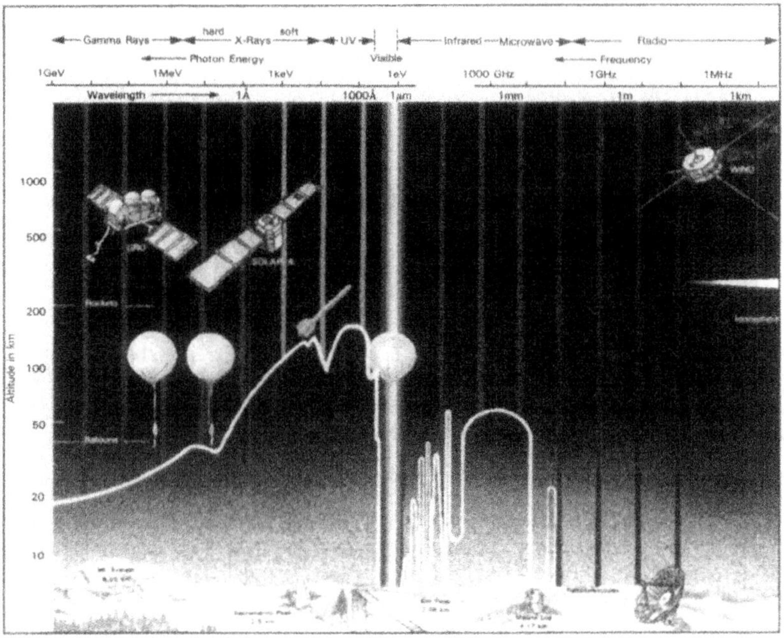

Figure 1. Altitude of penetration in the earth's atmosphere of electromagnetic radiation.

1.1. Bias of a Visual Universe

For millenia man has been privileged to observe the universe only in the visible. Thus, in the development of astronomy and astrophysics, what could be seen in a very narrow region, approximately 4000–7000 Angstroms, came to dominate the interpretation of the observed data (Figure 1). While radio astronomy became important in the 1950s, it had only minimal impact on already well-developed world models and in fact, some of the data could be claimed to support existing models. Not until the 1980s did much of the electromagnetic spectrum become known (or was even suspected to exist).*

With the exception of plant life, sight is one of the most important senses to life on earth and, neglecting phenomena such as lightning, the aurora, and the stars, favors the existence of the predominant three states of matter as found on the surface of earth: solids, liquids, and gases.

* The observation of a three degree Kelvin background temperature has, since 1965, been used to argue that the universe is finite-dimensioned. However, as early as 1896 Charles Edouard Guillaume in Meudon, France, estimated the temperature of space to be 5–6 K blackbody (Peratt, 1995). No dimension restrictions were placed on the prediction. Guillaume, together with his colleague Birkeland, advocated the dominance of electric forces in space, thus pioneering what is now called 'Electric Space'. Birkeland died as a working committee was nominating him for the Nobel prize in physics in 1917; Guillaume was awarded the Nobel prize in physics in 1920.

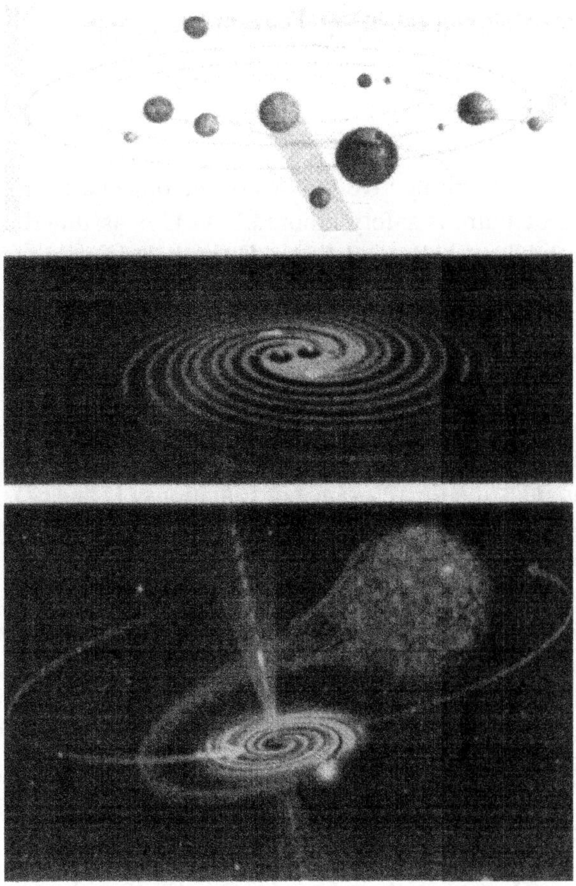

Figure 2. Newton's legacy: solid masses interacting gravitationally throughout the universe. (a) The planets and their satellites in serene and clocklike-motion that validated Newton's law of gravitation. This viewpoint, when applied to cosmic dimensions, encourages the advocacy of a mechanical clockwork universe favored since medieval times. (b) Artist's conception of astrophysical masses in gravitational interaction producing mass-density waves. (c) Artist's conception of matter at its highest density, a 'black-hole', gravitationally stripping solid and gaseous mass from a nearby star.

Thus, solids, liquids, and gases are the most identifiable forms of matter on earth and their transference to other 'masses' in the universe, such as the planets and their satellites, but also to the stars and galaxies seemed reasonable in the development of astronomical theories (Figure 2).

The merits of the artist's concepts depicted in Figure 2 are straightforward. This is a picture fully consistent with Newton's laws and the general development of astrophysics since his time. The only unknowns have to do with the properties of the solid (or gaseous or liquid) matter itself. The basic force law is

$$\mathbf{F} = m\mathbf{G} + \text{radiation forces}, \tag{1}$$

where $\mathbf{G}$ is the gravitational constant. The force between two objects separated a distance r is

$$F = -\frac{Gm_1m_2}{r^2} , \tag{2}$$

where m_1 and m_2 are the respective masses of the objects. In this scenario, space, being treated as a vacuum, is safely ignored.* As $\mathbf{G}$ is assumed constant, all interpretations of observed astrophysical forces $\mathbf{F}$ are then determined by m. Generally and especially at galactic and super-galactic dimensions, $\mathbf{F}$ is too large to be explained by the observed mass leading to the necessity of unobserved masses of various and differing magnitudes to balance Equations (1) and (2). Because the observed forces are so much greater than that predicted in this model, hypotheses such as 'black-holes' have arisen and the search for their existence has been of the highest priority in the field since their suggestion in 1969.

Today, one can question the relevance of Equations (1) and (2) to the study of the dynamical properties of astrophysical objects and the evolution of the universe.† Contrary to popular and scientific opinion of just a few decades ago, space is not an 'empty' void. Planetary and interplanetary spacecraft probes have shown space to be filled with high energy particles, magnetic fields, electric fields, and highly conducting plasma. The ability of plasmas to produce electric fields, either by instabilities brought about by plasma motion or the movement of magnetic fields, has popularized the term 'Electric Space' in recognition of the electric fields systematically discovered and measured in the solar system.

1.2. The Fourth, or 'Fundamental', State of Matter

For *plasma*, now often popularly referred to as the 'fourth state of matter',‡ the force laws are given by (Figure 3).

$$\mathbf{F} = m\mathbf{G} + e\,(\mathbf{E} + \mathbf{v} \times \mathbf{B}) + \text{radiation forces}, \tag{3}$$

where $\mathbf{E}$, $\mathbf{B}$, and $\mathbf{v}$ are the electric and magnetic field strengths, and the plasma velocity across field lines, respectively. The force between charges q is

$$F = -\frac{q^2}{4\pi\varepsilon_0 r^2} , \tag{4}$$

* Space, being the most voluminous of the cosmos, when treated as pure vacuum, gives a false sense that most of the universe is in a known state, the only unknowns being the point-like masses occupying Newton's universe. The discovery of the complexity of the planetary plasma magnetospheres proved that space is plasma with an electrodynamic complexity that exceeds that of the first three states of matter.

† At the time, the justification for the use of Equations (1) and (2) even in the near-earth environment was that any plasma ought to be electrically neutral and cancel out any electric forces.

‡ The first known use of the term '4th state of matter' occurs in Kristian Birkeland: Om Verdners Tilblivelse (On the creation of the Worlds), a lecture given in the Norwegian Academy of Science and Letters, January 31, 1913. Aars og Voss' Skoles festskrift p. 234.

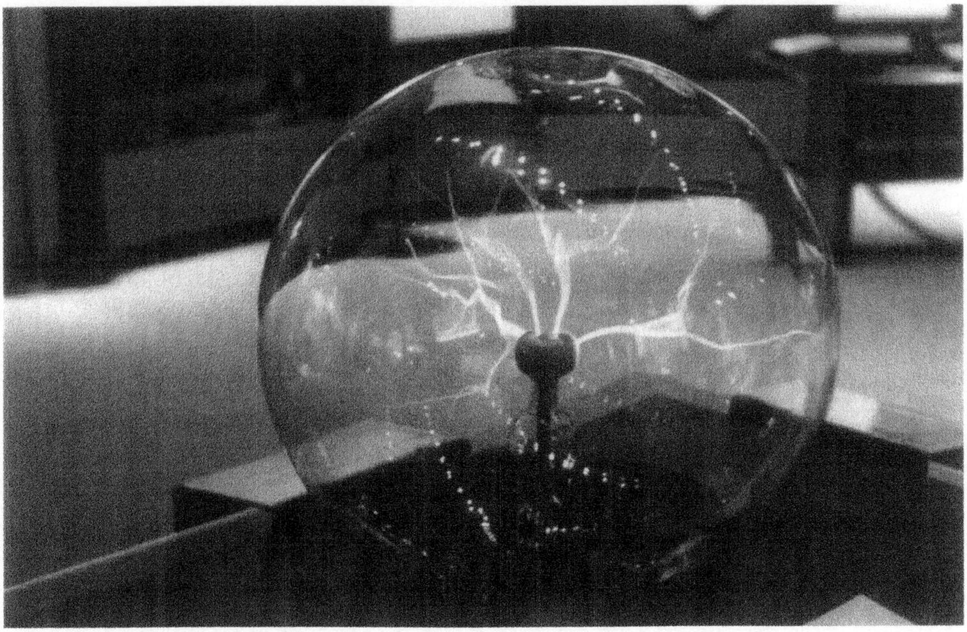

Figure 3. The fourth, or *fundamental*, state of matter. Unlike the so-called first three states of matter, the plasma state is highly energetic, filamentary, possesses a magnetic field, and is a prodigious producer of electromagnetic radiation over large regions of the electromagnetic spectrum. Courtesy of the Space Science Institute's Electric Space museum exhibition.

where ε_0 is the permittivity of free space. When the charges q are in motion with velocity v, the forces between the resulting currents $I = qv$ are

$$F = -\frac{\mu_0 I_1 I_2}{2\pi r} \,. \qquad (5)$$

It is noteworthy that Equations (4) and (5) are inherently 10^{36}–10^{39} orders of magnitude stronger than are Equations (1) and (2). It is especially noteworthy that Equation (5) follows an r^{-1} dependency instead of the r^{-2} dependencies of the gravitational and electrostatic force laws Equations (2) and (4), making it the longest range force law in the universe. In addition, Equation (5) can be positive in sign, i.e., repulsive, when the interacting currents flow in the opposite sense. This can lead to scenarios where matter distributed for hundreds of megaparsecs can appear to fall towards no observable matter (e.g., a missing 'great attractor'). The importance of electromagnetic forces on cosmic plasma cannot be overstated; even in neutral hydrogen regions ($\sim 10^{-4}$ parts ionized), the electromagnetic force to gravitational force ratio is 10^7.

But perhaps the most profound difference between plasma and the first three states of matter is that plasma is a prodigious producer of electromagnetic radiation. In high-energy pulsed power and inertial confinement fusion (ICF) experiments, plasmas are intense producers of x-rays and even gamma rays. At cosmic di-

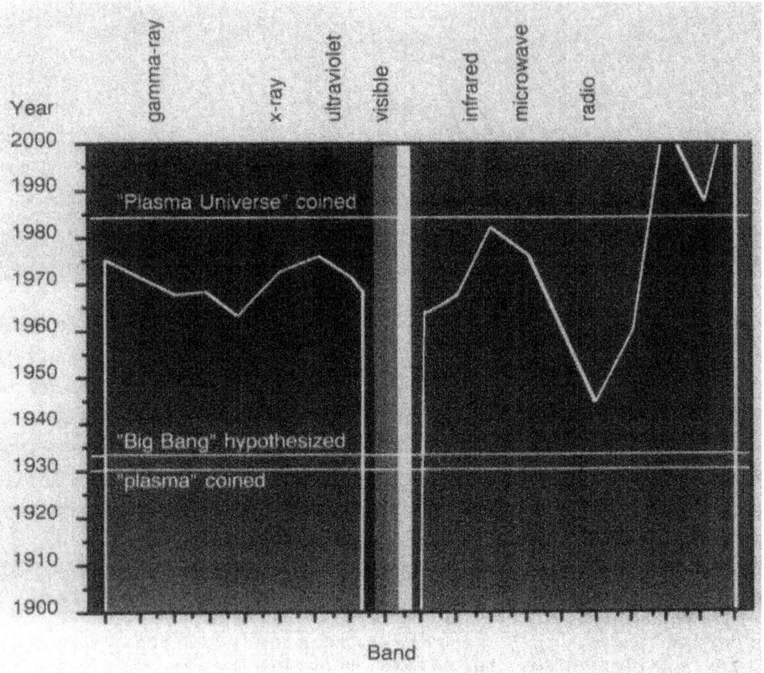

Figure 4. Measurement of the electromagnetic spectrum by wavelength by year.

mensions, charged particles accelerated by electric fields can produce energies as high as 10^{21} eV. Likewise, in the presence of magnetic fields, relativistic electrons produce synchrotron radiation, the source of much of the fluence observed by radio telescopes at radio and microwave wavelengths. Large scale plasma filaments ought also to produce very large fluences of electromagnetic radiation at extremely low frequencies, although these will be shielded by the earth's ionosphere and are yet to be observed. Figure 4 delineates the year of observational capability *versus* band that a particular portion of the entire electromagnetic spectrum became measurable. As illustrated, most of the spectrum, excluding very low frequencies, was not known prior to 1980.

Figure 5 illustrates where matter in the solid, liquid, and gaseous states is known to exist. Above the stratosphere, in the ionosphere and magnetosphere and as far beyond as we can observe, is matter in the plasma state. Below the lithosphere is again highly conducting, magnetized magma characterized by the hot plasma state of matter. Thus, the only places where the first three states of matter are know to exist with certainty in the universe are the crustal regions of the planets and their satellites in the solar system surrounding our plasma sun.*

* This observation must also be true about other stars and their surrounding systems.

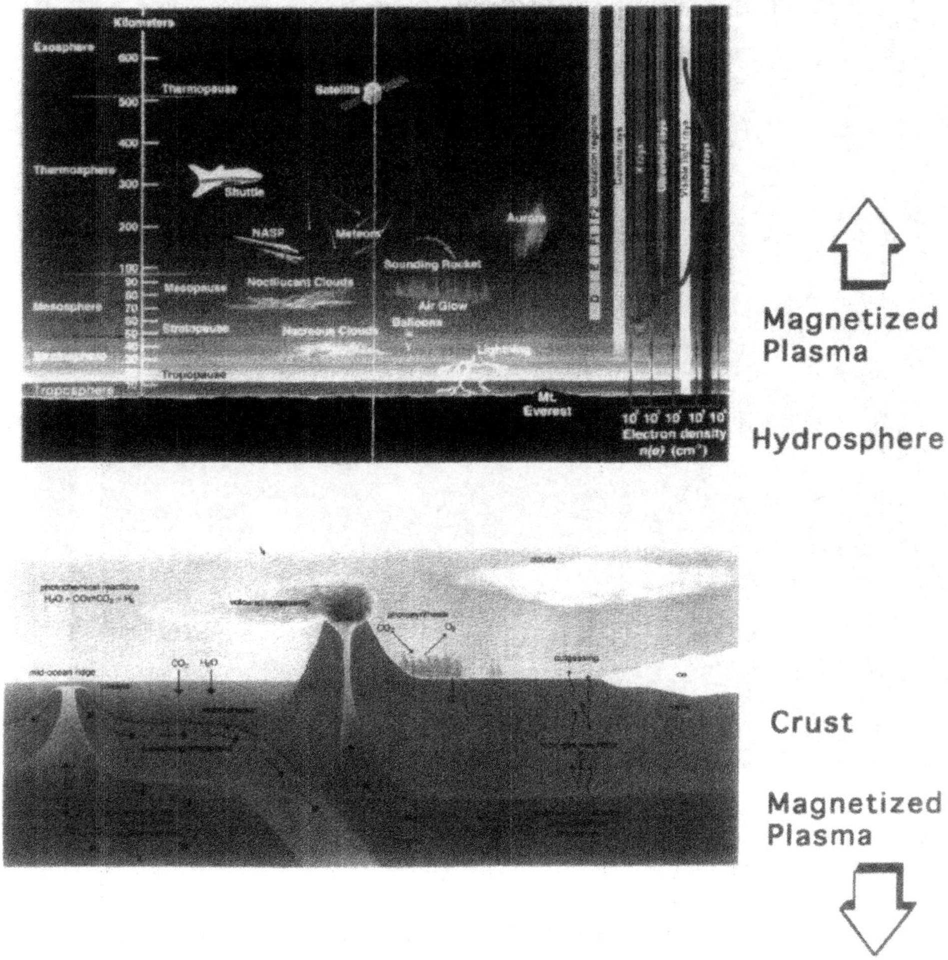

Figure 5. The planetary/satellite crusts and hydrospheres are the only known regions where a non-magnetized fluid treatment of matter is valid.

Today it is recognized that 99.999% of all observable matter in the universe is in the plasma state and plasmas are found at temperatures and densities far exceeding those that will support matter in the first three states (Figure 6).

2. Large Scale Structure of the Plasma Universe

Among the earliest predictions about the morphology of the universe that differed significantly from the common astrophysical assumption is that the universe be filamentary (Alfvén, 1950, 1981, 1990). Plasmas in electric space are energetic (because of electric fields) and they are generally inhomogeneous with constituent

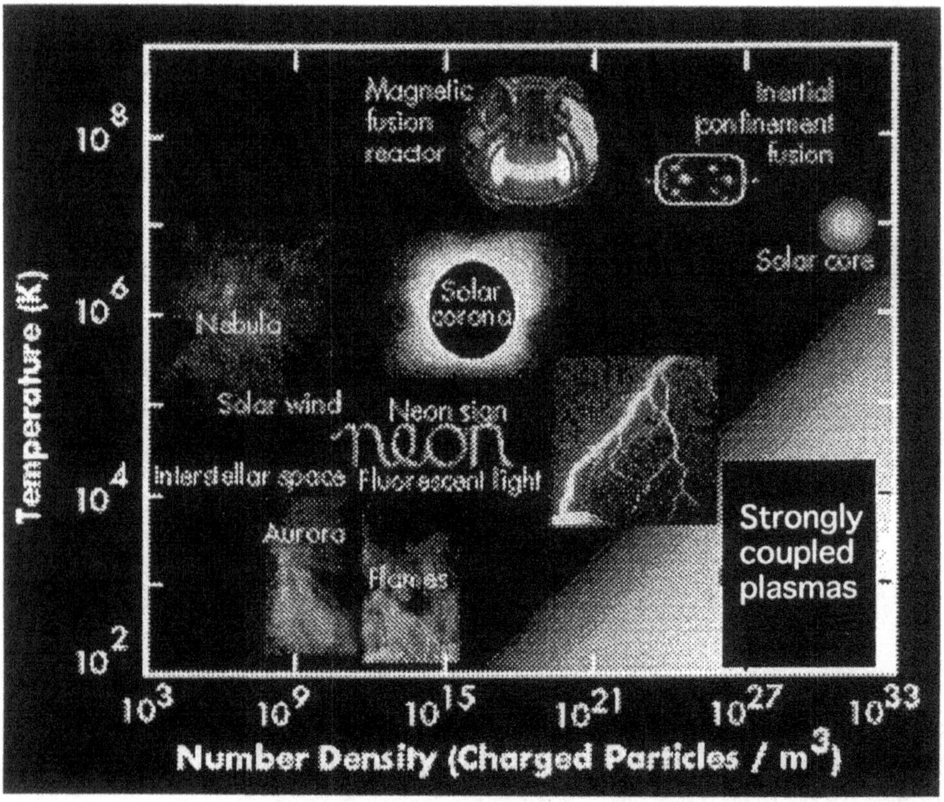

Figure 6. Remarkable range of temperatures and densities of plasmas. Courtesy of the Princeton Plasma Physics Laboratory.

parts in motion. Plasmas in relative motion are coupled by the currents they drive in each other and non-equilibrium plasma often consists of current-conducting filaments.

In the laboratory and in the Solar System, filamentary and cellular morphology is a well-known property of plasma. As the properties of the plasma state of matter is believed not to change beyond the range of our space probes, plasma with astrophysical dimensions must also be filamentary.

Additionally, transition regions have been observed that delineate the 'cells' of differing plasma types (Eastman, 1990). On an astrophysical scale, these transition regions should be observable at radio wavelengths via transition radiation signatures.

The suggestion that the universe be filamentary and cellular was generally disregarded until the 1980s, when a series of unexpected observations showed filamentary structure on the Galactic, intergalactic, and supergalactic scale. By this time, the analytical intractability of complex filamentary geometries, intense self-fields, non-linearities, and explicit time dependence had fostered the development of fully three-dimensional, fully electromagnetic, particle-in-cell simulations of plasmas having the dimensions of galaxies or systems of galaxies. It had been realized that the importance of applying electromagnetism and plasma physics to the problem of radiogalaxy and galaxy formation derived from the fact that the universe is largely a *plasma universe*.

Any imbalance in the constitutive properties of a plasma can set it in motion [if, in fact, it has not already derived from an evolving, motional state (Bohm, 1979)]. The moving plasma, i.e., charged particle flows, are currents that produce self-magnetic fields, however weak. The motion of any other plasma across weak magnetic fields produces and amplifies electromotive forces, the energy of which can be transported over large distances via currents that tend to flow along magnetic lines of force. These 'field-aligned currents,' called *Birkeland currents* in planetary magnetospheres, should also exist in cosmic plasma. The dissipation of the source energy from evolving or moving plasma in localized regions can then lead to pinches and condense states. Where double layers form in the pinches, strong electric fields can accelerate the charged particles to high energies, including gamma ray energies (Alfvén, 1981). These should then display the characteristics of relativistic charged particle beams in laboratory surroundings, for example, the production of microwaves, synchrotron radiation, and non-linear behavior such as periodicities and 'flickering'.

In contrast to the viewpoint depicted in Figure 2 is the approach by Alfvén, who reduced cosmic problems to laboratory plasma physics problems whose properties could be investigated *in situ*.* The plasma analog of Figure 2 is shown in the circuit diagram of Figure 7. The attributes of the circuit approach is that both mechanical and electrical mechanisms, including the radiation observed at non-visual wavelengths, can be studied in great detail. The circuit, which conducts currents, allows the study of the flow of charged particles in the presence of electromagnetic fields, and includes discrete particle effects such as double layers (charge separation from current instabilities) and synchrotron radiation (fast electrons in the presence of a magnetic field). When the currents are smoothed out, the fluid magnetohydrodynamic equations are obtained, which Alfvén used to great success in his study of the sun. Thus, spicules, prominences, flares, and other temporal inhomogeneities can be brought under investigation via the circuit description approach that has been highly successful in modern space science. Moreover, the circuit approach allows for the reproduction of the entire electromagnetic spectrum as well as the pinch dynamics important to the study of the formation of stars in primordial plasma

* Laboratory astrophysics, pioneered by Birkeland and used extensively by Alfvén, is a revolutionary change from classical astronomy reaching back to Ptolemy. Classical astronomy divided the study of earth and the heavens into the 'mundane' and the 'celestial', respectively, and no study of the mechanisms of the heavens could be performed on earth. The modern trappings of celestial *versus* mundane remain with us today in the form of 'black holes' and 'dark matter' whose theories are not falsifiable and therefore not open to scientific inquiry.

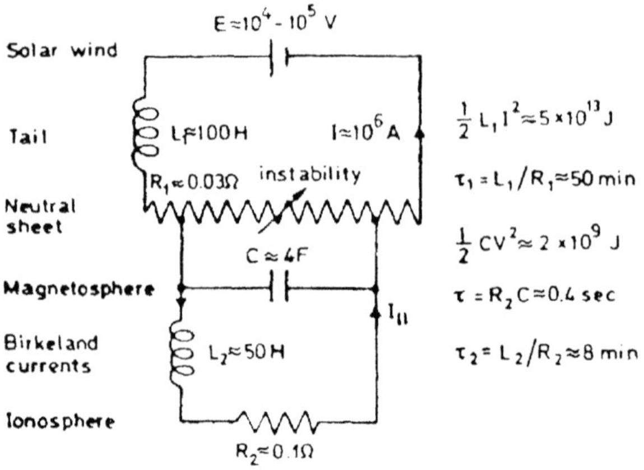

Figure 7. Generic circuit description of a space plasma problem (in this case, the flow of Birkeland currents in the Earth's magnetosphere/ionosphere.

flows. Unfortunately, the method appears unknown in classical astronomy when applied to problems of cosmic importance.

2.1. Filamentation by Birkeland Currents

An electromotive force $\int \mathbf{v} \times \mathbf{B} \cdot \mathbf{dl}$ giving rise to electrical currents in conducting media is produced wherever a relative perpendicular motion of plasma and magnetic fields exist (Akasofu, 1984; Alfvén, 1986). An example of this is the (nightside) sunward-directed magnetospheric plasma that cuts the earth's dipole field lines near the equatorial plane, thereby producing a potential supply that drives currents within the auroral circuit. The discovery of these Birkeland currents in the earth's magnetosphere in 1974 (Dessler, 1984) has resulted in the realization that the structure in auroral draperies are caused by the filamentation of Birkeland charged-particle sheets following the earth's dipole magnetic-field lines into vortex current bundles.

2.2. Galactic Dimensioned Birkeland Currents

Extrapolating the size and strength of magnetospheric currents to interstellar space leads to the suggestion that confined current flows in interstellar clouds assists in their formation (Alfvén, 1981).

As a natural extension of the size hierarchy in cosmic plasmas, the existence of galactic dimensioned Birkeland currents or filaments was hypothesized (Alfvén and Fälthammar, 1963; Peratt, 1986).

A galactic magnetic field of the order $B_G = 10^{-9}$–10^{-10} T associated with a galactic dimension of 10^{20}–10^{21} m suggests the galactic current be of the order $I_G = 10^{17}$–10^{19} A.

In the galactic dimensioned Birkeland current model, the width of a typical filament may be taken to be 35 kpc ($\approx 10^{21}$ m), separated from neighboring filaments by a similar distance. Since current filaments in laboratory plasmas generally have a width/length ratio in the range 10^{-3}–10^{-5}, a typical 35 kpc wide filament may have an overall length between 35 Mpc and 3.5 Gpc with an average length of 350 Mpc. The circuit, of course, is closed over this distance (Peratt, 1990).

2.3. Magnetic Ropes

Surface currents, delineating plasma regions of different magnetization, temperature, density, and chemical composition give space a cellular structure (Alfvén and Fälthammar, 1963). As current-carrying sheet beams collect into filaments, the morphology of the surface currents is filamentary.

For the case of tenuous cosmic plasmas, the thermokinetic pressure is often negligible and hence the magnetic field is force-free. Under the influence of the electromagnetic fields the charged particles drift with the velocity

$$v = (\mathbf{E} \times \mathbf{B}) / E^2. \tag{6}$$

The overall plasma flow is inwards and matter is accumulated in the filaments which, because of their qualitative field line pattern, are called ‘magnetic ropes’. Magnetic ropes should therefore tend to coincide with material filaments that have a higher density than the surroundings. The cosmic magnetic ropes or current filaments are not observable themselves, but the associated filaments of condensed matter can be observed by the radiation they emit and absorb.

It is because of the convection and neutralization of plasma into radiatively cooled current filaments (due to synchrotron losses) that matter in the plasma universe should often display a filamentary morphology.

3. Simulating the Astrophysical and Cosmical Plasma Phenomena

Because of the above, the modeling of astrophysical and cosmical phenomena requires a plasma treatment. This then requires a formidable computational platform: a full three-dimensional plasma treatment, with three spatial components each in electric and magnetic fields, and the inclusion of radiation packages. Because of **E** and **B**, the plasma charges tend to spiral around **B** as they accelerate along **E** or undergo other complex and unstable motion because of space-charge effects. This task of simulating electromagnetic plasma is an order of magnitude more complex than a simple gravitational simulation where gravity is the only force component on masses in simple free-fall. The remainder of this section describes the physical mechanisms and results in the simulation of cosmic plasma. A complete description of the algorithms and computational parameters have been given elsewhere (Peratt, 1996, 1997).

3.1. Synchrotron Emission from Pinched Particle Beams

One of the most important processes that limit the energies attainable in particle accelerators is the radiative loss by electrons accelerated by the magnetic field of a betatron or synchrotron. This mechanism was first brought to the attention of astronomers by Alfvén and Herlofson (1950); a remarkable suggestion at a time when plasma, magnetic fields, and laboratory physics were thought to have little, if anything, to do with a cosmos filled with isolated 'island' universes (galaxies). Synchrotron radiation is characterized by a generation of frequencies appreciably higher than the cyclotron frequency of the electrons; a continuous spectra (for a population of electrons) whose intensity decreases with frequency beyond a critical frequency (near intensity maxima); increasing beam directivity with increasing relativistic factor γ ($\gamma = (1 - \beta)^{-1/2}$); and polarized electromagnetic wave vectors.

Z-Pinches are among the most prolific radiators of synchrotron radiation known. In this regard, the Bennett-pinch, or Z-pinch, as a synchrotron source has been treated by Meierovich (1984) and Newberger (1984).

Because of their intense powers and complex morphologies, double radio galaxies and quasars have been an enigmatic problem to astronomy since their serendipitous discovery. Figure 8 illustrates the synchrotron isophotes of several of these sources.

However, if the filamentary plasma universe model is used, the isophotes of both double radio galaxies and quasars are readily explained. The radiation produced from the two nearest plasma filaments in interaction replicates both the isophotal and power spectra from double radio galaxies as shown in Figure 9. Table I delineates the basic parameters used in the interacting galactic filament simulation which are consistent with a number of previous estimates.

Because the highly relativistic electrons depicted in Figure 9 flow in direction outwards from the plane of the figure, the synchrotron radiation is also beamed in this direction (Johner, 1988).

The monochromatic power of quasars and double radio galaxies span a range of about 10^{33}–10^{39} W (Peratt, 1986b). For example, the 'prototype' double radio galaxy Cygnus A has an estimated radio luminosity of 1.6–4.4×10^{37} W. Together with the power calculated, the simulation isophotes are very close to those observed from this object (Peratt, 1986a). The left column of Figure 9 suggests that previously apparently unrelated double radio galaxies all belong to the same species but are simply seen at different times in their evolution.

3.2. Confining and Interacting Forces between Cosmic Currents

If the cosmic current is cylindrical and in a rotationless, steady-state condition, it is described by the *Carlqvist Relation*:

$$\frac{\mu_0}{8\pi} I^2(a) + \frac{1}{2} G\bar{m}^2 N^2(a) = \Delta W_{Bz} + \Delta W_k \tag{7}$$

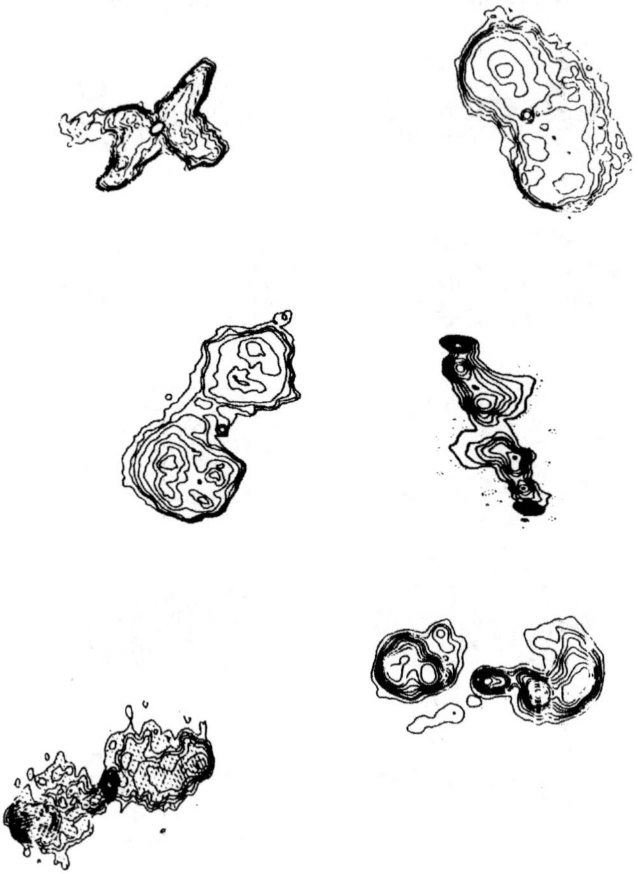

Figure 8. Synchrotron isophotes (various frequencies) of a random assortment of double radio galaxies and quasars.

for a current of radius $r = a$ where μ_0 is the permeability of free space, G is the gravitational constant, $\bar{m}$ is the mean particle mass, N is the number of particles per unit length, and ΔW_{Bz} and ΔW_k are the differential beam magnetic and kinetic energies, respectively (Peratt, 1992a).* Thus, whether or not a current or beam is gravitationally balanced, electromagnetically balanced, or force-free, depends on the magnitude of the individual terms in Equation (7).

In contrast to the gravitational and electromagnetic forces that determine the characteristic of an individual beam, interactions between beams are always dominated by electromagnetic *Biot–Savart* forces,

$$\mathbf{F}_{21} = \int \mathbf{j}_2 \times \mathbf{B}_{21} d^3 r \tag{8}$$

* When current rotation and transient phenomena are important, the *Generalized Bennett Condition* may be used in place of Equation (7) (Peratt, 1992a, chap. 2).

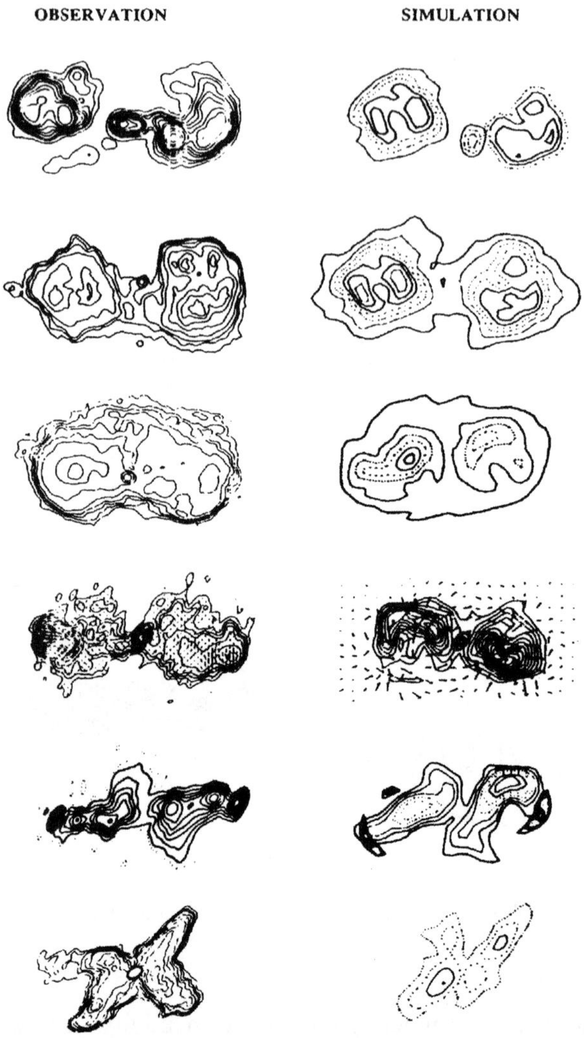

Figure 9. (a) Synchrotron isophotes (various frequencies) of double radio galaxies. (b) Simulation analogs at time 10.4 to 58.7 Myr. Time increases from top to bottom.

for all space, where $\mathbf{j_2} \times \mathbf{B_{21}}$ is the Lorentz force between the field B_{21} induced by a current I_1 on the current density j_2 at current I_2.*

Parallel axial currents within the filaments are long-range attractive, while circular (helical) currents within the filaments (as the electrons gyrate along the axial magnetic field) are short-range repulsive. If the axial currents are able to bring the

* The Biot–Savart force varies as r^{-1} and thus dominates gravitational attraction which varies as r^{-2}. 'Great Attractors', often attributed to gravitational forces between 'missing masses' display Biot–Savart, not mass attraction, characteristics.

TABLE I

Simulation derived parameters based on the radiation properties of the double radio galaxy Cygnus A.

Parameter	Simulation Value
Galactic current, I_G	2.4×10^{19} A
Galactic magnetic field, B_θ	2.5×10^{-4} G
Galactic magnetic field, B_z	2.0×10^{-4} G
Plasma temperature, T	2.0–32.0 keV
Plasma density, n_e	1.79×10^{-3} cm^{-3}
Electric field strength, E_z	62 mV/m
Synchrotron power, P_{syn}	1.16×10^{37} W
Radiation burst duration	1.28×10^{14} s
Total energy	6.3×10^{62} J

filaments close enough together so that the repulsive component of the Lorentz force becomes important, the circular currents repulse and brake, and release energy in the form of synchrotron radiation.

While a complete description of the evolution of interacting galactic currents is given elsewhere (Peratt, 1992a, 1992b), it is useful to reproduce the evolutionary sequence in this paper. Figure 10 illustrates the cross-sections of the filaments over a 10^9 yr period.

3.3. Rotation Velocities

Rotational velocities of spiral galaxies are obtained by measuring the Doppler shift of the Hα line emitted by neutral hydrogen in the spiral arms. If the galaxy is at a cant towards earth, the emission-line in the arm moving away from earth is red-shifted while the line in the arm moving towards earth is blue-shifted. The first Doppler shift measurements of the radial velocities indicated that these were flat, rather than following the Keplerian law. If galaxies were gravitationally bound systems, their outer mass should follow Kepler's laws of motion and be slower than the inner mass. The flat rotation curves of galaxies has been cited as the strongest physical evidence for the existence of dark matter. In this scenario a massive halo of dark matter has been evoked to produce the flat rotation curves. However, the rotation curves are not really flat; they show appreciable structure representative of an instability mechanism within the arms. This instability precludes the existence on any external halo of matter around galaxies that, while making the rotation curves flat, would also dampen any instability growth.

Since E_z is out of the plane of the page, the column electrons spiral downward in counter-clockwise rotation while the column ions spiral upward in clockwise

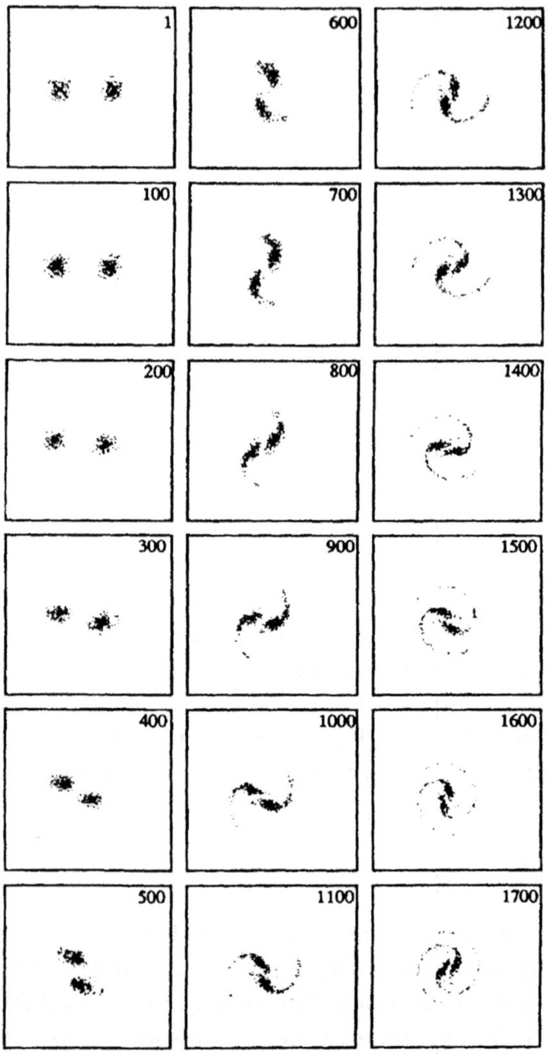

Figure 10. Single frame stills of plasma in the simulation of two adjacent Birkeland filaments: $\omega_c/\omega_p = 3.0$, $T_{e0} = T_{i0} = 32$ keV, $E_{z0} = 62$ mV/m. Total time elapsed: $\approx 10^9$ yr. The initial dimensions in frame 1 are: radius of filaments $r_{\text{filaments}} = 17.5$ kpc, distance between filaments $d_{\text{filaments}} = 80$ kpc. The length over which E_{z0} exists in the filaments is taken to be ≈ 10 kpc.

rotation. A polarization induced charge separation also occurs in each arm, which, as it thins out, produces a radial electric field across the arm. Because of this field, the arm is susceptible to the diocotron instability (Peratt, 1992). This instability appears as a wave motion in each arm and is barely discernible in the single frame photographs in Figure 10 at late times. However, the instability is readily apparent in the simulation spiral rotational velocity curve (Figure 11).

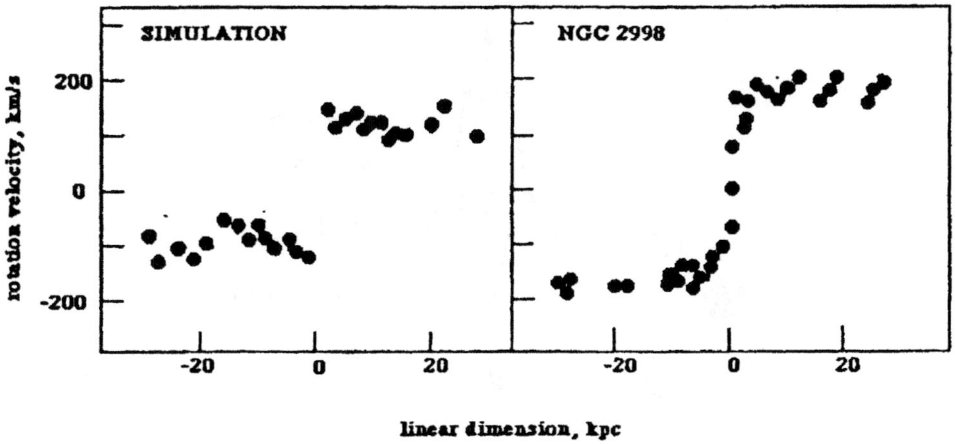

Figure 11. Spiral galaxy rotational velocity curves. Note the well–defined structure on the 'flat' portions of the curves.

The velocity consists essentially of a linearly increasing component due to a central body undergoing rigid rotation, with two 'flat' components on either side of $r = 0$ due to the trailing arms. The diocotron instability modulates the 'flat' components at the strong-magnetic-field, low-density instability wavelength $\lambda \approx 2.5\Delta r$, where Δr is the width of an arm.

3.4. The Association of Neutral Hydrogen with Galactic Magnetic Fields

Neutral hydrogen distributions are characteristic of spiral galaxies but not pre-spiral galaxy forms. Because the rotation velocities of spiral galaxies are determined by the motion of neutral hydrogen, it is desirable to know the process for neutral hydrogen accumulation in late-time galaxies.

In the plasma universe model, spiral galaxies form from the interaction of current-carrying filaments at regions where electric fields exist. The individual filaments are defined by the *Carlqvist Condition* that specifies the relationship between gravitational and electromagnetic constraining forces (Verschuur, 1995). In this model, whether or not neutral hydrogen and other neutral gases form from hydrogenic plasma depends of the efficiency of convection of plasma into the filament.

When an electric field is present in a plasma and has a component perpendicular to a magnetic field, inward convection of the charged particles occurs. Both electrons and ions drift with velocity

$$\mathbf{v} = (\mathbf{E} \times \mathbf{B})/B^2$$

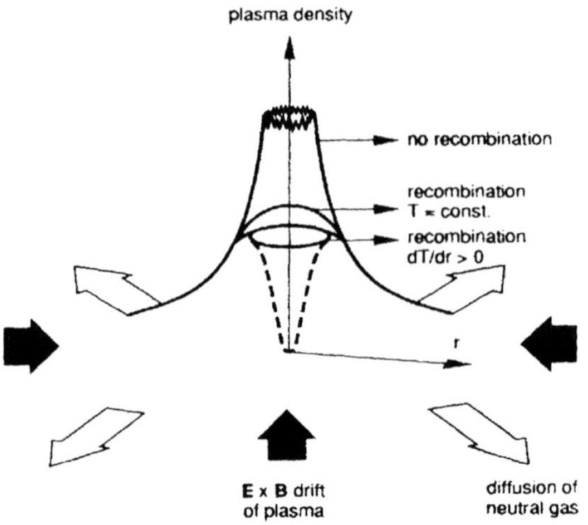

Figure 12. Plasma density profile as a function of radius shown qualitatively for three cases: No recombination, recombination with temperature T = constant, and recombination with a lower central temperature.

so that the plasma as a whole moves radially inwards. The material thus forms as magnetic ropes around magnetic flux tubes. Magnetic ropes thus contain material filaments that have a higher density than the surrounding plasma.

When a plasma is only partly ionized, the electromagnetic forces act on the non-ionized components only indirectly through the viscosity between the ionized and non-ionized constituents. For a filament, the inward radial velocity drift is

$$v_r = E_z / B_\phi$$

for the case of an axial electric field and azimuthal magnetic field (induced by the axial current I_z). Hence, at a large radial distance r, the rate of accumulation of matter into a filament is

$$\frac{dM}{dt} = 2\pi r v_r \rho_m = (2\pi r)^2 \rho_m \frac{E_z}{\mu_0 I_z} . \tag{9}$$

Marklund (1979) found a stationary state when the inward convections of ions and electrons toward the axis of a filament was matched by recombination and outward diffusion of the neutralized plasma (Figure 12). The equilibrium density of the ionized component normally has a maximum at the axis. However, because of the radiated loss of energy, the filament cools and a temperature gradient is associated with the plasma. Because of this hollow cylinders, or modifications of hollow cylinders of matter, will form about the flux tubes.

Because the radial transport depends on the ionization potential of the element, elements with the lowest ionization potentials are brought closest to axis. The most abundant elements of cosmical plasma can be divided into groups of roughly equal ionization potentials: He (24 eV); H, O, N (13 eV); C, S (11 eV); and Fe, Si,

Mg (8 eV). These elements can be expected to form hollow cylinders whose radii increase with ionization potential. Helium will make up the most widely distributed outer layer; hydrogen, oxygen, and nitrogen should form the middle layers, while iron, silicon, and magnesium will make up the inner layers. Interlap between the layers can be expected and, for the case of galaxies, the metal-to-hydrogen ratio should be maximum near the center and decrease outwardly. Both the convection process and luminosity increase with the field E_z.

For the case of a fully ionized hydrogenic plasma, the ions drift inwards until they reach a radius where the temperature is well below the ionization potential and the rate of recombination of the hydrogen plasma is considerable. Because of this 'ion pump' action, hydrogenic plasma will be evacuated from the surroundings and neutral hydrogen will be most heavily deposited in regions of strong magnetic field.

3.5. Axisymmetric and Bisymmetric Magnetic Field Patterns in Spiral Galaxies

When the ion pump-magnetic field action was discovered in the laboratory, there existed some debate as to whether galaxies possessed magnetic fields at all. Today, it is known that large-scale magnetic fields do exist in spiral galaxies. For example, the Effelsberg radio telescope has collected polarization data from about a dozen spiral galaxies at 6 to 49 cm wavelengths (Beck, 1986, 1990). Rotation measures show two different large-scale structures of the interstellar fields: Axisymmetric-spiral and bisymmetric-spiral patterns (Krause et al., 1989), as shown in Figure 13.

The orientation of the field lines is mostly along the optical spiral arms. However, the uniform field is often strongest outside the optical spiral arms. For example, in the case of galaxy IC 342, two filamentary structures were found in the map of polarized intensity (Figure 14). Their degree of polarization of $\approx 30\%$ indicates a high degree of uniformity of the magnetic field on the scale of the resolution (≈ 700 pc). These filaments extend over a length of ≈ 30 kpc and hence are the most prominent magnetic-field features detected in normal spiral galaxies so far.

A detailed analysis of the rotation measure distribution in a spiral arm southwest of the center of the Andromeda galaxy M31 (Beck et al., 1990) shows that the magnetic field and a huge HI cloud complex are anchored together (Figure 15). The magnetic field then inflates out of the plane outside the cloud. The tendency for the magnetic field to follow the HI distribution has been noted in several recent observations. Circumstantial evidence has accumulated which suggests that there is a close connection between rings of CO and Hα seen rotating in some galaxies and the magnetic fields in the nuclear regions. This is particularly apparent in observations of spiral galaxies viewed edge-on. This scenario has also been invoked for our Galaxy (Wielebinski, 1989).

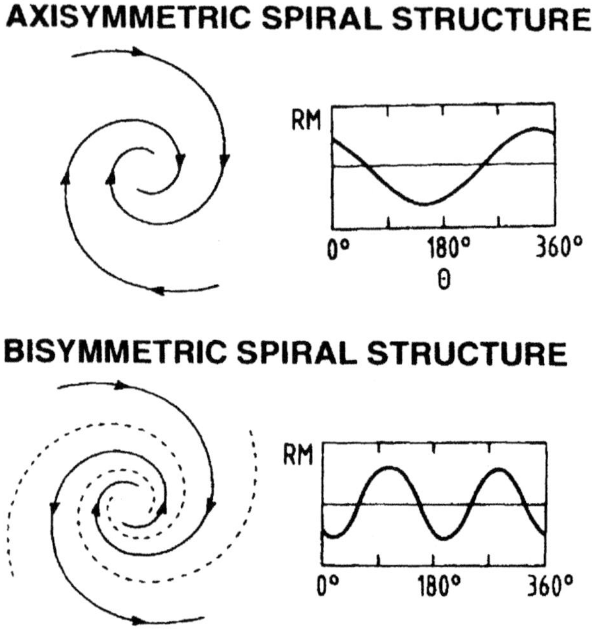

Figure 13. Definition of Axisymmetric-spiral and bisymmetric-spiral patterns.

Neutral hydrogen is detected from galaxies via the van de Hulst radio-emission line at 21.11 cm (1.420 GHz). High-resolution observation of neutral hydrogen in irregular and spiral galaxies usually reveal extended HI distributions. Contour maps of the HI typically show a relative lack of HI in the cores of spiral galaxies but high HI content in the surrounding region, usually in the shape of a 'horseshoe'. This region is not uniform but may have two or more peaks in neutral hydrogen content. Figure 16 (right) shows an example of the HI distribution in a spiral galaxy.

Figure 16 (left) shows the plasma spiral formed in this simulation overlayed on its magnetic field line (squared) isobars. The diameter of the spiral is about 50 kpc with a mass of 10^{41} kg, i.e., a size and mass of that observed from spiral galaxies. A direct comparison to observations is made by superimposing the HI distribution in NGC 4151 on its optical photograph. The observation shows two peaks in neutral hydrogen surrounding a void. The void is orientated towards one of the arms. The simulation allows the two peaks to be traced back to their origin. Both are found to be the remnants of the originally extended components, i.e., cross-sections of the original Birkeland filaments. The hydrogen deficient center is the remnant of an elliptical galaxy formed midway between the filaments, in the magnetic null.

Figure 17 shows the magnetic field orientations of the vertical (or 'parallel' to the line-of-sight) and the circumferential magnetic fields in an Sc type galaxy. Because of the acute reversals of the circumferential magnetic field, whether or not an observer sees axisymmetric or bisymmetric patterns in the synchrotron radiation

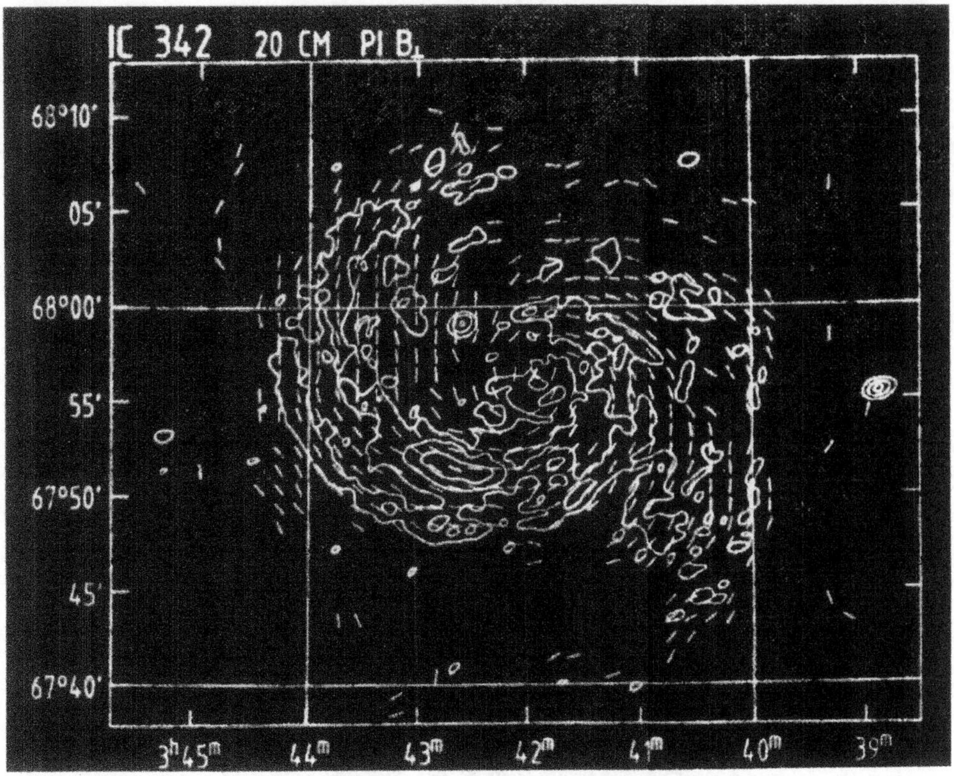

Figure 14. Contour map of the linearly polarized intensity at 1.4 GHz. The HPBW is 45″. Contour levels are 0.06, 0.12, 0.24, 0.48, 0.96 mJy/b.a. and the rms noise is 0.03 mJy/b.a. The vectors represent the orientation of the observed E-vectors and the degree of linear polarization.

depends on the location of the parallel electric fields in the radiating plasma with respect to the circumferential magnetic field.

4. Conclusions

Based on optical observations of astrophysical objects, early models of the universe were founded solely on the assumption that the mass found throughout the universe was the same as that found on the crustal regions of the planets and satellites of the solar system. Not until the 1980s was it clear that most of the observable universe as seen over large portions of the electromagnetic spectrum consisted of matter in the fourth, or fundamental state; *plasma*. However, without the computational resources to simulate astrophysical and cosmic plasmas in three dimensions and the experimental facilities to produce plasmas of similar properties with diagnostics sufficient to benchmark them to the simulations, the field languished in favor of simpler models developed in the 1920s.

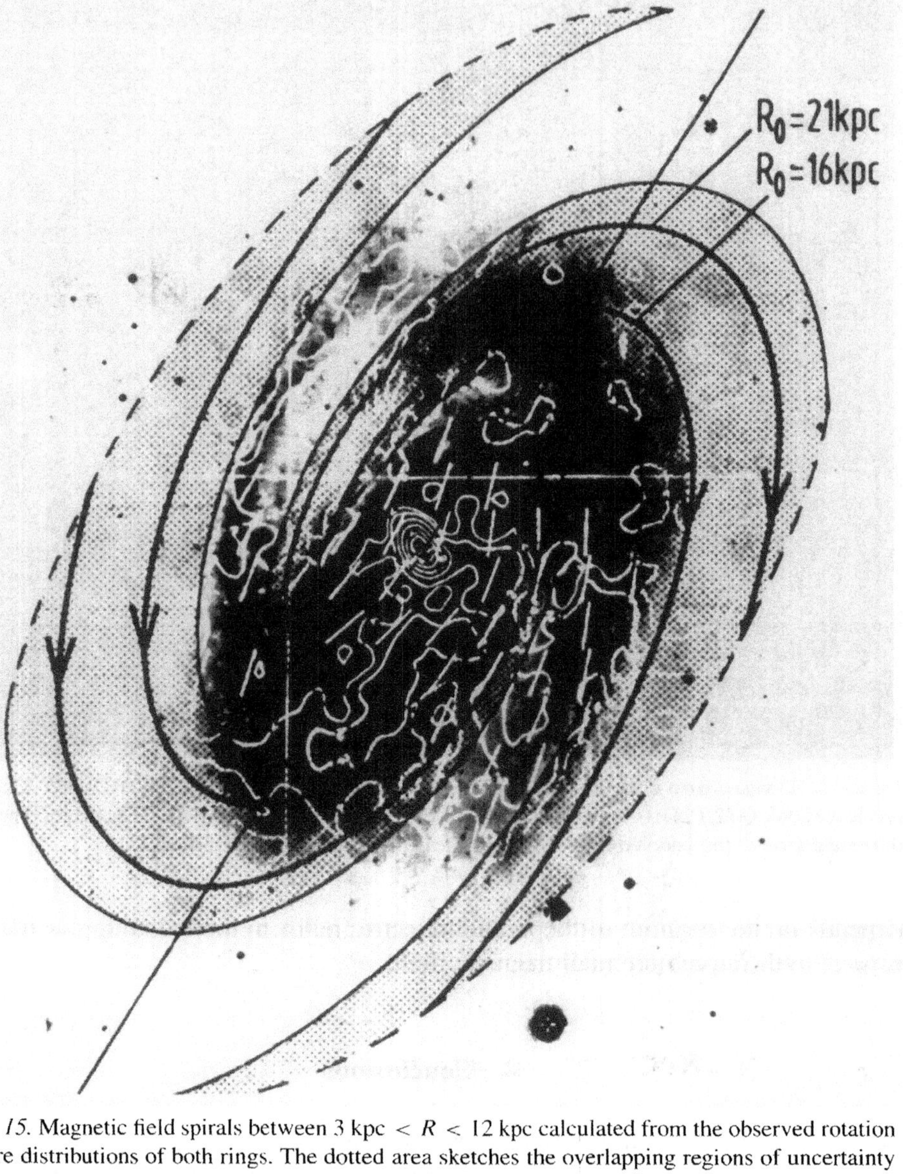

Figure 15. Magnetic field spirals between 3 kpc $< R <$ 12 kpc calculated from the observed rotation measure distributions of both rings. The dotted area sketches the overlapping regions of uncertainty concerning the locations of the spirals. The magnetic field direction is indicated by the arrows. Note that the neutral lines are located in between the dotted areas, namely in the interarm areas.

With the advent of multi-teraflop computational resources reaching towards petaflop speeds and the emergence of facilities capable of producing high-energy density plasmas that can be adequately diagnosed, the capability now exists for studying the evolution of the universe by including the most abundant state of matter observed, that of plasma. With this resource, replication of many of the fine

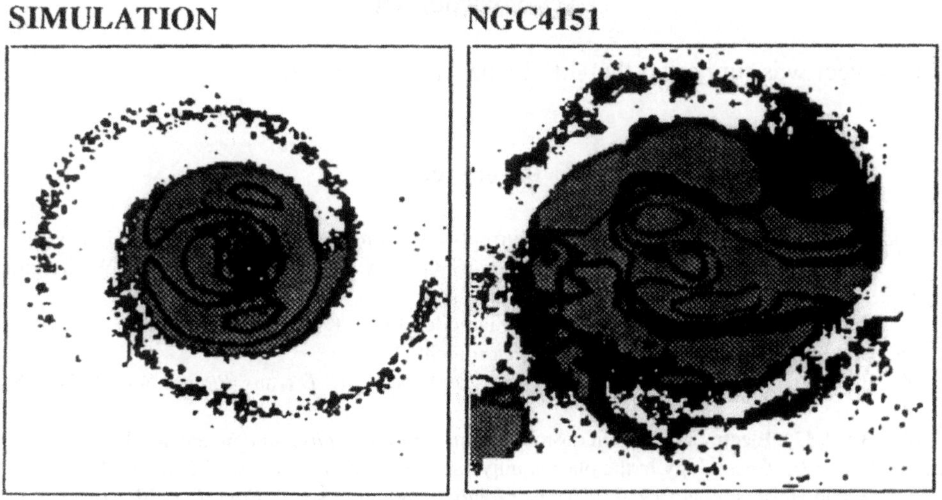

Figure 16. Left: Simulation magnetic energy density superimposed on simulation galaxy. In both cases a 'horseshoe' shaped cusp, opening towards a spiral arm, surrounds a magnetic field/HI minima core. Within the cusp, two magnetic field/HI peaks are observed. Right: HI distribution superimposed on an optical photograph of NGC 4151.

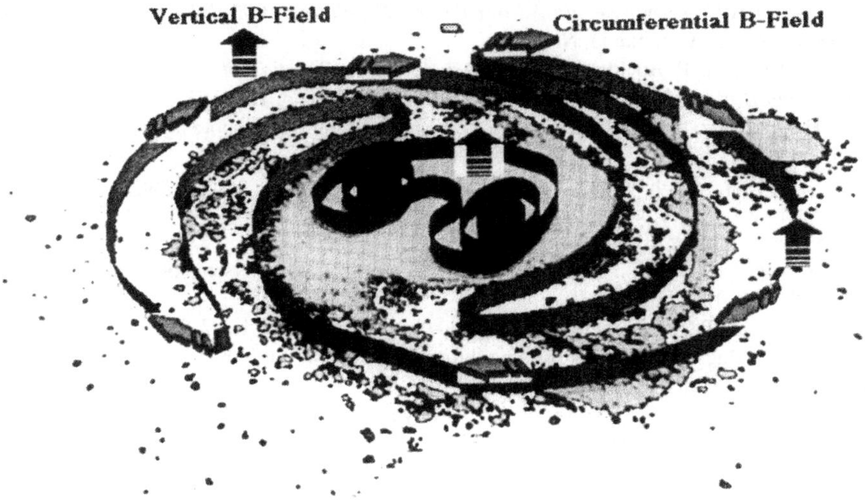

Figure 17. Simulated magnetic fields in an Sc type galaxy.

detail features observed from cosmic objects appears feasible leading to a better understanding of the nature of our infinitely evolving universe.

Acknowledgment

This project was supported the U.S. Department of Energy.

References

Akasofu, S.-I.: 1984, 'The magnetospheric currents: An introduction to magnetospheric currents', in: T.A. Potemra (ed.), *Magnetospheric Currents*, Geophysical Monograph, No. 28, Washington, DC, Amererican Geophysical Union, pp. 29–48.

Alfvén, H.: 1950, *Cosmical Electrodynamics*, Clarendon Press, Oxford/London.

Alfvén, H.: 1981, *Cosmic Plasma*, D. Reidel, Dordrecht.

Alfvén, H.: 1986, 'Double layers and circuits in astrophysics', *IEEE Trans. Plasma Sci.* **PS-14**, 779–793.

Alfvén, H.: 1987, 'Electric currents in cosmic plasma', *Rev. Geophys. and Space Sci.* **15**, 271–284.

Alfvén, H.: 1990, 'Cosmology in the plasma universe', *IEEE Trans. Plasma Sci.* **18**, 5–10.

Alfvén, H. and Herlofson, N.: 1950, 'Cosmic radiation and radio stars', *Phys. Rev.* **78**, 616.

Alfvén, H. and Fälthammar, C.-G.: 1963, *Cosmical Electrodynamics*, Oxford University Press, Oxford.

Beck, R.: 1986, 'Interstellar magnetic fields', *IEEE Trans. Plasma Sci.* **14**, 470.

Beck, R.: 1990, 'Magnetic fields in spiral galaxies', *IEEE Trans. Plasma Sci.* **18**, 33.

Bennett, W.H.: 1934, 'Magnetically self-focusing streams', *Phys. Rev.* **45**, 890–897.

Birkeland, Kr.: 1912, *The Norwegian Aurora Polaris Expedition 1902–1903, Volumes I and II*, H. Ascheough & Co., Christiania, Norway.

Bohm, D.J.: 1979, in: P. Buckley and F.D. Peat (eds.), *A Question of Physics: Conversations in Physics and Biology*, University of Toronto Press, Toronto, pp. 124–153.

Chan, C., Hershkowitz, N.: 1982, 'Transition from single to multiple double layers', *Phys. Fluids* **25**, 2135–2137.

Dessler, A.J.: 1984, 'Evolution of arguments regarding existence of field aligned currents', in: T.A. Potemra (ed.), *Magnetospheric Currents*, Geophysical Monograph, No. 28, Amererican Geophysical Union, Washington, DC, pp. 26–32.

Eastman, T.: 1990, 'Transition regions in solar system and astrophysical plasmas', *IEEE Trans. Plasma Sci.* **18**, 18–25.

Fälthammar, C.-G.: 1983, 'Magnetic-field aligned electric fields', *ESA J.* **7**, 385–401.

Fä, C.-G.: 1986, 'Magnetosphere-ionosphere interactions – Near-earth manisfestations of the plasma universe', *IEEE Trans. Plasma Sci.* **PS-14**, 616–628.

Johner, J.: 1988, 'Angular distribution of the total cyclotron radiation of a relativistic particle with parallel velocity', *Phys. Rev. A* **36**, 1498–1501.

Krause, M., Beck, R. and Hummel, E.: 1989, 'The magnetic field structure in two nearby galaxies', *Astron. Astrophys.* **217**, 4–17.

Marklund, G.T.: 1979, 'Plasma convection in force-free magnetic fields as a mechanism for chemical separation in cosmical plasmas', *Nature* **277**, 370.

Meierovich, B.E.: 1984, 'Electromagnetic collapse, problems of stability, emission of radiation and evolution of a dense pinch', *Phys. Rep.* **104**, 259–347.

Newberger, B.S., Buchwald, M.I., Karl, R.R., Moir, D.C. and Starke, T.P.: 1984, 'Synchrotron radiation from bennett beams', *Bull. Am. Phys. Soc.* **29**, 1435.

Peratt, A.L.: 1986a, 'Evolution of the plasma universe: I. Double radio galaxies, quasars, and extragalactic jets', *IEEE Trans. Plasma Sci.* **PS-14**, 639–660.

Peratt, A.L.: 1986b, 'Evolution of the plasma universe: II. The formation of systems of galaxies', *IEEE Trans. Plasma Sci.* **PS-14**, 763–778.

Peratt, A.L.: 1990, 'The evidence of electrical currents in cosmic plasma', *IEEE Trans. Plasma Sci.* **18**, 26–32.

Peratt, A.L.: 1992a, *Physics of the Plasma Universe*, Springer-Verlag, New York.

Peratt, A.L.: 1992b, 'Plasma cosmology', *Sky and Tel.*, February, 136.

Peratt, A.L.: 1995, *Plasma Astrophysics and Cosmology*, Kluwer Academic Publishers, Dordrecht.

Peratt, A.L.: 1996, *Advances in Numerical Modeling of Astrophysical and Space Plasmas*, Advanced Topics on Astrophysical and Space Plasmas, Vol. 242, Kluwer Academic Publishers, Dordrecht.

Peratt, A.L.: 1997, 'Advances in numerical modeling of astrophysical and space plasmas', *Astrophys. Space Sci.* **242**, 93–163.

Peratt, A.L. and Jones, M.E.: 1986, 'Particle-in-cell simulations of heavy ion double layers', in: A. Hirose (ed.), *IEEE Conf. Record.*, Saskatoon, Canada, p. 34.

Peter, W. and Peratt, A.L.: 1988, 'Synchrotron radiation spectrum for galactic-sized plasma filaments', *Laser and Particle Beams* **6**(3), 493–502.

Tifft, W.G.: 1995, 'Redshift quantitization', *Astrophys. Space Sci.* **227**, 25–39.

Verschuur, G.L.: 1995, 'Interstellar neutral hydrogen filaments at high galactic latitudes and the Bennett pinch', *Astrophys. Space Sci.* **227**, 187–198.

Wielebinski, R.: 1989, 'Magnetic fields in the galaxy', Max-Planck-Institut für Radioastronomie Preprint Series, No. 347.

PLASMA KINETIC THEORY I

M. PSIMOPOULOS
Plasma Group, Physics Department, Imperial College, London SW7 2BZ, U.K.

Abstract. Plasma kinetic theory is based on the Atomic Hypothesis as well as two branches of physics with different definitions of the concept of force: Gas dynamics (pressure on impact forces) and Electromagnetism (action at distance forces). A brief historical background of the above branches is given and the average force exerted on a test particle of the plasma is calculated in 1, 2 and 3 dimensions both for pressure on impact and action at distance models.

1. Introduction

Plasma kinetic theory is based on two pillars:

1. Gas dynamics.
Pressure on impact: Motion is necessary in order to have force

$$F = \langle\text{momentum transfer}\rangle/\text{time}.$$

2. Electromagnetism.
Action at distance: Force can exist without motion

$$\text{Coulomb's law} \quad \mathbf{F} = e^2 \sum_{i=1}^{N} \frac{\mathbf{r} - \mathbf{r}_i}{|\mathbf{r} - \mathbf{r}_i|^3} \,.$$

2. Atomic Hypothesis

Democritus: The world is made of atoms and the void (Russel, 1984).
Newton: The atoms interact by forces; many body problem in mechanics.

3. Historical Background

3.1. Gas Dynamics (Brush, 1983)

Torricelli, Boyle, Towneley, Pascal; concept of pressure: $p \sim V^{-1}$.
D. Bernoulli: Derivation of gas laws from hard sphere model;

Astrophysics and Space Science **256:** 77–84, 1998.

$$(1/2)\rho V^2 + p = \text{constant} .$$

Newton–Boyle: Theory of repulsive force $(1/r)$.
Herapath: Heat is nothing but molecular motion.
Carnot: Second law of thermodynamics.
Joule–Mayer: Heat-work equivalence.
Kronig, Clausius, von Helmholtz: Gas kinetic theory.
Maxwell: Velocity distribution of ideal gas.
Boltzmann: Kinetic equation of hard sphere gas at low densities; concept of entropy.
Gibbs: Notion of an ensemble of systems.
Van der Waals: Equation of state for non-ideal gases.
Einstein, von Smoluchowski, Langevin, Perrin: Brownian movement.
Development of the mathematical theory of stochastic processes.
Solution of Boltzmann's equation (Hilbert, Chapman and Cowling, Lions).
Prigogine: Solution of Liouville's equation using diagrams.
Maxwell velocity distribution for ideal gas:

$$\phi(\mathbf{v}) = \left(\frac{m}{2\pi T}\right)^{3/2} e^{-(mv^2)/(2T)} .$$

Boltzmann equation for hard sphere gas at low densities:

$$\frac{\partial\phi(\mathbf{v}_1, t)}{\partial t} = 2n_0 r_0^2 \int |(\mathbf{v}_2 - \mathbf{v}_1).\mathbf{l}|,$$

$$\{\phi(\mathbf{v}_1 + [(\mathbf{v}_2 - \mathbf{v}_1).\mathbf{l}]\mathbf{l}, t)\phi(\mathbf{v}_2 - [(\mathbf{v}_2 - \mathbf{v}_1).\mathbf{l}]\mathbf{l}, t) - \phi(\mathbf{v}_1, t)\phi(\mathbf{v}_2, t)\}d\mathbf{v}_2 d.\mathbf{l}$$

r_0: radius of sphere; $\mathbf{l}$: unit vector; $d\mathbf{l}$: surface element of unit sphere.
Note: The Maxwellian is the steady state solution of Boltzmann's equation.

3.2. Electromagnetism (Whittaker, 1951)

Electrostatics: Gilbert, Du Fay, Grey, Franklin, Aepinus, Cavendish, Coulomb, Poisson.
Electric Current: Galvani, Volta, Davy, Grothuss, De la Rive, Berzelius, Ohm.
Electromagnetism: Oersted, Ampere, Faraday, Maxwell, Hertz, Lorentz.

Basic relation between electricity-chemistry-light:
Special theory of relativity, quantum mechanics, QED.
Newton–Boscovich–Ampere world view: Action at distance.
Faraday–Maxwell–Lorentz modification: Introduction of **E**, **B** fields.

Maxwell equations

$$\nabla\mathbf{E} = 4\pi\rho,$$

$$\nabla \mathbf{B} = 0,$$

$$\frac{1}{c}\frac{\partial \mathbf{B}}{\partial t} = -\nabla \times \mathbf{E},$$

$$\frac{1}{c}\frac{\partial \mathbf{E}}{\partial t} = \nabla \times \mathbf{B} - \frac{4\pi}{c}\mathbf{j}.$$

Lorentz force:

$$\mathbf{F} = q\left(\mathbf{E} + \frac{\mathbf{v} \times \mathbf{B}}{c}\right).$$

4. Force Exerted on Test Particle

In the present lecture the average force exerted on a fixed test particle by field particles is calculated in 1, 2 and 3 dimensions (D). The theory is developed in parallel for pressure on impact and action at distance forces. In particular, it is shown how the logarithmic divergence is manifested in 3-D.

4.1. 1-D Pressure on Impact Theory of Hard Point Particles

Consider a beam n_0 (cm^{-1}), v_0 (cm/s) falling on a fixed test particle T.

Momentum transfer/particle: $dp = 2m_0v_0$
Number of colliding particles during dt: $dN = n_0v_0dt$
Total momentum transfer during dt: $dP = dNdp = 2n_0m_0v_0^2dt$
Force exerted on test particle by many beams:

$$F = 2n_0m_0 \int_{-\infty}^{+\infty} |v_0|v_0\phi(v_0)dv_0,$$

where $\phi(v_0)$ is the 1-D velocity distribution at infinity.

4.2. 1-D Action at Distance Theory

Consider a beam n_0 (cm^{-1}), v_0 (cm/s) of electrons falling on a fixed test electron T located at the origin.

Conservation of energy:

$$\frac{1}{2}m_0v^2 + \frac{e^2}{x} = \frac{1}{2}mv_0^2.$$

Speed of electrons at x:

$$|v(x)| = v_0[1 - (x_0/x)]^{1/2}.$$

$x_0 = 2e^2/m_0 v_0^2$ is the distance of minimum approach.
Continuity: $n(x)v(x) = n_0 v_0$ (in/out)
Electron number density at x:

$$n(x/v_0) = \frac{n_0 h(x - x_0)}{\sqrt{1 - \frac{x_0}{x}}},$$

where $h(x)$ is the Heaviside function.
Force exerted on test electron by the beam:

$$F = 2 \int_0^{+\infty} \frac{e^2}{x^2} n(x/v_0) dx = 2n_0 m_0 v_0^2.$$

Force exerted on test electron by many beams:

$$F = 2n_0 m_0 \int_{-\infty}^{+\infty} |v_0| v_0 \phi(v_0) dv_0.$$

Note that in 1-D the two theories give the same result.

Electron density of many beams near a test electron:

$$n(x) = \int_0^{\infty} n(x/v_0)\phi(v_0) dv_0$$

Example: $\phi(v_0) = (m_0/2\pi T_0)^{1/2} \exp(-m_0 v_0^2/2T_0)$

$n(x) = n_0 \exp(-e^2/T_0 x)$ (Boltzmann distribution).

4.3. 2-D Pressure on Impact Theory

Consider a beam n_0 (cm^{-2}), v_0 (cm/s) of hard particles (disks) falling on a fixed test particle T.

Momentum transfer/particle in the $\mathbf{v}_0/v_0$ direction: $dp = m_0 v_0(1 - \cos\chi)$; χ: deflexion angle $0 \leq \chi \leq \pi$.
Number colliding particles during dt with impact parameter within db : $dN = n_0 v_0 dt db$.
Total momentum transfer during dt in the $\mathbf{v}_0/v_0$ direction:

$dP = dNdp = m_0 v_0 (1 - \cos\chi) n_0 v_0 dt db.$
Express all quantities in terms of $\psi = (\pi - \chi)/2; 0 \le \psi \le \pi/2;$
$1 - \cos\chi = 2\cos^2\psi; b = 2r_0 \sin\psi; db = 2r_0 \cos\psi d\psi$
where r_0 is the particle radius.
$dp = 4r_0 n_0 m_0 v_0^2 \cos^3\psi d\psi dt$
Force exerted on test particle by the beam:

$$f = \frac{dP}{dt} = 8r_0 n_0 m_0 \int_0^{\pi/2} \cos^3\psi d\psi = \frac{16}{3} r_0 n_0 m_0 v_0^2.$$

Force exerted on test particle by many beams:

$$\mathbf{F} = \frac{16}{3} r_0 n_0 m_0 \int v_0 \mathbf{v}_0 \phi(\mathbf{v}_0) d\mathbf{v}_0,$$

where $\phi(\mathbf{v}_0)$ is the 2-D velocity distribution at infinity.

4.4. 2-D Action at Distance Theory

Consider a beam n_0 (cm^2), v_0 (cm/s) of electrons falling on a fixed test electron T located at the origin.

Hyperbolic orbits of field electrons:

$$r(\psi, b) = \frac{2b^2/\gamma}{(2b/\gamma)\sin\psi + \cos\psi - 1}$$

$\gamma = 2e^2/m_0 v_0^2; 0 \le b < +\infty; 0 \le \psi < 2\arctan(2b/\gamma).$

Perihelion: $\partial r/\partial\psi = 0$

$$\psi_p = 2\arctan(2b/\gamma);\ r_p = \frac{1}{2}\gamma\left[1 + \sqrt{1 + (2b/\gamma)^2}\right].$$

Envelope: $\partial r/\partial b = o$

$b_e = \gamma\tan(\psi/2); r_e = 2\gamma/(1 + \cos\psi)$ (parabola).

There are two ways in which an electron can reach a point
$M(r, \psi); r > r_e(\psi); 0 \le \psi < \pi.$

$$b_{1,2} = \frac{1}{2}\sin\psi\left(1 \pm \sqrt{1 - \frac{r_e}{r}}\right),$$

$$\partial b_{1,2}/\partial r = \frac{1}{2}\sin\psi\left(1 \pm \frac{1 - \frac{r_e}{2r}}{\sqrt{1 - \frac{r_e}{r}}}\right),$$

$\partial b_1/\partial r > 0;\ \partial b_2/\partial r < 0.$

Calculation of electron number density:

Continuity $n_0 v_0 |db| = n(r, \psi) v_\psi |dr|$
Conservation of angular momentum: $r v_\psi = b v_0$

Electron number density at (r, ψ):

$$n(r, \psi) = \frac{n_0 v_0}{v_\psi} \left| \frac{db}{dr} \right| = n_0 r \left\{ \frac{1}{b_1} \left| \frac{\partial b_1}{\partial r} \right| + \frac{1}{b_2} \left| \frac{\partial b_2}{\partial r} \right| \right\},$$

$$n(\mathbf{r}/\mathbf{v}_0) = \frac{n_0 h(r - r_e)}{\sqrt{1 - \frac{r_e}{r}}}; \quad r_e = \frac{2\gamma}{1 - \frac{(\mathbf{r}.\mathbf{v}_0)}{r v_0}},$$

where $h(x)$ is the Heaviside function.
Force exerted on test particle by the beam:

$$\mathbf{F} = -e^2 \int \frac{\mathbf{r}}{r^3} n(\mathbf{r}/\mathbf{v}_0) d\mathbf{r} = 2\pi n_0 e^2 \frac{\mathbf{v}_0}{\mathbf{v}_0}.$$

Note that in 2-D the two theories do not give same results and that the logarithmic divergence does not exists for electrons.

Electron density of many beams near a test electron:

$$n(\mathbf{r}) = \int n(\mathbf{r}/\mathbf{v}_0)\phi(\mathbf{v}_0) d\mathbf{v}_0.$$

Example: $\phi(\mathbf{v}_0) = (m_0/2\pi T_0) \exp(-m_0 v_0^2/2T_0)$

$n(\mathbf{r}) = n_0 \exp(-e^2/T_0 r)$ (Boltzmann distribution).

4.5. 3-D Pressure on Impact Theory

Consider a beam n_0 (cm^{-3}), v_0 (cm/s) of hard spherical particles falling on a fixed test particle T.

Momentum transfer/particle in the $\mathbf{v}_0/v_0$ direction:
$dp = m_0 v_0 (1 - \cos\chi)$; χ: deflection angle $0 \leq \chi \leq \pi$.
Number of colliding particles during dt with impact parameter within dB:
$dN = n_0 v_0 dt 2\pi b db$
Total momentum transfer during dt in the $\mathbf{v}_0/v_0$ direction:
$dP = dN dp = m_0 v_0 (1 - \cos\chi) n_0 v_0 dt 2\pi b db.$
Express all quantities in terms of $\psi = (\pi - \chi)/2;\ 0 \leq \psi \leq \pi/2;$

$1 - \cos\chi = 2\cos^2\psi$; $b = 2r_0 \sin\psi$; $db = 2r_0 \cos\psi d\psi$
where r_0 is the particle radius.
$dp = 16\pi r_0^2 n_0 m_0 v^2 \cos^3\psi \sin\psi d\psi dt$.

Force exerted on test particle by the beam:

$$F = \frac{dP}{dt} = 16\pi r_0^2 n_0 m_0 v_0^2 \int_0^{\pi/2} \cos^3\psi \sin\psi d\psi = 4\pi r_0^2 n_0 m_0 v_0^2.$$

Force exerted on test particle by many beams:

$$\mathbf{F} = 4\pi r_0^2 n_0 m_0 \int v_0 \mathbf{v}_0 \phi(\mathbf{v}_0) d\mathbf{v}_0,$$

where $\phi(\mathbf{v}_0)$ is the 3-D velocity distribution at infinity.

4.6. 3-D Action at Distance Theory (Psimopoulos and Rogister, 1990)

Consider a beam n_0 (cm^{-3}), v_0 (cm/s) of electrons falling on a fixed test electron T located at the origin.

$$r(\psi, b) = \frac{2b^2/\gamma}{(2b/\gamma)\sin\psi + \cos\psi - 1}.$$

$\gamma = 2e^2/m_0 v_0^2$; $0 \leq b + \infty$; $0 \leq \psi < 2\arctan(2b/\gamma)$.

Perihelion: $\partial r/\partial\psi = 0$

$$\psi_p = 2\arctan(2b/\gamma); \quad r_p = \frac{1}{2}\gamma[1 + \sqrt{1 + (2b/\gamma)^2}].$$

Envelope: $\partial r/\partial b = 0$
$b_e = \gamma \tan(\psi/2)$; $r_e = 2\gamma\ (1 + \cos\psi)$ (parabola).

There are two ways in which an electron can reach a point
$M(r, \psi); r > r_e(\psi); 0 \leq \psi < \pi$.

$$b_{1,2} = \frac{1}{2}\sin\psi \left(1 \pm \sqrt{1 - \frac{r_e}{r}}\right),$$

$$\partial b_{1,2}/\partial r = \frac{1}{2}\sin\psi \left(1 \pm \frac{1 - \frac{r_e}{2r}}{\sqrt{1 - \frac{r_e}{2r}}}\right),$$

$\partial b_1/\partial r > 0$; $\partial b_2/\partial r < 0$.

Calculation of electron number density.

Continuity: $n_0 v_0 2\pi b|db| = n(r, \psi) v_\psi 2\pi r \sin\psi |dr|$.
Conservation of angular momentum: $r v_\psi = b v_0$.

Electron number density at (r, ψ):

$$n(r, \psi) = \frac{n_0 v_0 b_0}{r v_\psi \sin\psi} \left|\frac{db}{dr}\right| = \frac{n_0}{\sin\psi}\left\{\left|\frac{\partial b_1}{\partial r}\right| + \left|\frac{\partial b_2}{\partial r}\right|\right\},$$

$$n(\mathbf{r}/\mathbf{v}_0) = n_0 h(r - r_e)\frac{1 - \frac{r_e}{2r}}{\sqrt{1 - \frac{r_e}{r}}}; \; r_e = \frac{2\gamma}{1 - \frac{(\mathbf{r}.\mathbf{v}_0)}{r v_0}},$$

where $h(x)$ is the Heaviside function.
Force exerted on test particle by the beam:

$$\mathbf{F} = -e^2 \int \frac{\mathbf{r}}{r^3} n(\mathbf{r}/\mathbf{v}_0) d\mathbf{r} = \frac{4\pi n_0 e^4}{m_0}(\ln\Lambda - 1)\frac{\mathbf{v}_0}{\mathbf{v}_0^3},$$

where

$$\Lambda = \frac{\sqrt{1 - (\gamma/\gamma_D)}}{1 - \sqrt{1 - (\gamma/\gamma_D)}}.$$

The above force is divergent at $r = +\infty$ and a cut-off was used at $r = \gamma_D$: Debye length. Clearly, this divergence is due to violation of the global neutrality conditions by neglecting e–e interactions between electrons of the beam. In a correct theory the beam should close behind the test particle and the envelope should cease to be parabolic.

Electron density of many beams near a test electron:

$$n(\mathbf{r}) = \int n(\mathbf{r}/\mathbf{v}_0)\phi(\mathbf{v}_0) d\mathbf{v}_0.$$

Example: $\phi(\mathbf{v}_0) = (m_0/2\pi T_0)^{3/2} \exp(-m_0 v_0^2/2T_0)$.
$n(\mathbf{r}) = n_0 \exp(-e^2/T_0 r)$ (Boltzmann distribution).

References

Brush, S.G.: 1983, *Statistical Physics and the Atomic Theory of Matter*, Princeton University Press.
Psimopoulos, M. and Rogister, A.: 1990, *Phys. Lett.* **149A**, 265.
Russel, B.: 1984, 'A history of western philosophy', Counterpoint.
Whittaker, E.: 1951, *A History of the Theories of Aether and Electricity*. Thomas Nelson.

PLASMA COLLECTIVE EFFECTS IN THE PRESENCE OF DUST

S.V. VLADIMIROV
Research Centre for Theoretical Astrophysics, School of Physics, The University of Sydney, New South Wales 2006, Australia

Abstract. Charged dust particles strongly affect collective processes in plasmas. Here we review recent advances in the theoretical study of collective effects in dusty plasmas. In particular, we consider processes of charging of dust by plasma currents, wave propagation and scattering, properties of Alfvén waves, and discuss collective effects in plasma-dust crystals.

1. Introduction

A great deal of interest in studying the physics of dusty plasmas, whose constituents are electrons, ions, and extremely heavy highly charged dust particles, was primarily connected with an increasing worldwide effort to model DC, RF, and microwave plasma discharge devices (where dusty particles grains appear naturally, grow in size, and play a role as natural contaminants in the processes) used in plasma-assisted materials processing for sputtering, deposition and implantation, to produce materials such as diamond films and selective surfaces for solar collectors (Selwyn et al., 1989, 1990). Another example of recent interest is the low temperature edge plasma physics in nuclear fusion devices, where dust grains emitted from walls may strongly influence anomalous transport properties (Benkadda et al., 1995; Tsypin et al., 1997).

Dust grains are usually of micron and submicron size, negatively charged, with many electrons collected on each particulate, and have a mass that is much larger than the positive ion mass. The negative charge of the dust grains could be due to different processes, such as due to bombardment by charged particles from surrounding plasmas, ultraviolet irradiation, sputtering of energetic ions (Northrop, 1992; Allen, 1992). Since the dust particles are almost immediately electrically charged being immersed in the ambient plasma, they must be coupled to each other as well as the plasma via the electric and magnetic fields (Whipple et al., 1985). The presence of chaotically moving charged dust grains has been shown to lead to considerable modification of plasma collective properties strongly affecting dispersion relations and damping of the usual plasma waves, as well as leading to specific oscillation modes associated with dust motion (Tsytovich and Havnes, 1993; Vladimirov, 1994a; Melandsø et al., 1993; Cramer and Vladimirov, 1996a, 1996b; Vladimirov and Cramer, 1996). Interaction of dust particles with

Astrophysics and Space Science **256:** 85–105, 1998.

themselves is also naturally affected by their plasma environment and therefore by plasma collective effects.

Dusty plasmas are also common in a variety of low-temperature plasmas in space environments, such as the lower ionosphere of the Earth, planetary atmospheres, asteroid zones, nebulas, and cometary tails (Goertz, 1989; Mendis and Rosenberg, 1994). Although in space physics research effects of dust grains has been studied for many years (Spitzer, 1978), only recently the role of plasma collective phenomena was recognized due to progress in experiments on comets, planetary rings, and Earth's environment (Northrop, 1992; Mendis and Rosenberg, 1994).

An exciting area of the most recent research is the plasma-dust crystal formation, theoretically predicted in (Ikezi, 1986). The success in creation of crystalline dusty structures (Chu and I, 1994; Thomas et al., 1994; Hayashi and Tachibana, 1994; Melzer et al., 1994) in laboratory radio frequency discharge plasmas makes it possible to study fundamental processes of phase transitions in a Coulomb system of charged particles (Melzer et al., 1996a; Chiang and I, 1996; Thomas and Morfill, 1996), as well as opens new possibilities for formation of new materials from dust with a big binding energy. Thus the Coulomb crystallization of dust grains in a laboratory plasma holds great potential as a model system for studies of structure, dynamics, and phase transitions in condensed matter (Melzer et al., 1996a; Chiang and I, 1996; Thomas and Morfill, 1996). This makes the topic of dusty plasmas of interest to a wide community of researchers.

The macroscopic lattices, named "dust-plasma crystals", are made of highly (negatively) charged particulates of micrometer size levitated in the sheath region above a horizontal negatively biased electrode. The grains in the lattices strongly interact via repulsive Coulomb forces, which is affected by ions and electrons of surrounding plasma. The particle system crystallizes when the interaction potential far exceeds their thermal energy (Ikezi, 1986).

Although the dust-plasma crystal systems are generally three-dimensional, in the most experiments the charged particles are located just in few layers above the horizontal electrode, where gravity is balanced by the sheath electric field. These layers act mainly as two-dimensional systems, with limited out-of-plane particle motion (Thomas and Morfill, 1996; Quinn et al., 1996). On the other hand, it was demonstrated that aligning of dust grains from different layers in the vertical plane is strongly connected with plasma collective mechanisms, namely, with the presence of super-sonic ions flowing towards the negative electrode (Vladimirov and Nambu, 1995; Vladimirov and Ishihara, 1996). It was shown (Melandsø and Goree, 1995; Ishihara and Vladimirov, 1997) that in otherwise uniform plasma, the flow leaves a polarized oscillating wake potential behind a stationary dust particle, with ions focussing to make the plasma potential positive in a local region. This positive region attracts negative particles, promoting the vertical alignment clearly observed experimentally (Melzer et al., 1996a; Chiang and I, 1996; Thomas and Morfill, 1996).

Theoretical studies of waves and instabilities in such dust-plasma crystal systems are just started (Melandsø, 1996; Vladimirov et al., 1997). At the same time, an enhanced level of plasma oscillations has been reported in experiments on dusty crystals (Chu et al., 1994; Praburam and Goree, 1996) thus allowing us to assume a significant role of plasma collective phenomena in these strongly coupled systems.

Here, recent advances in the study of collective effects in dusty plasmas are reviewed. The important point is that dust grains embedded in a plasma change its collective properties considerably and cannot always be considered just as an additional ideal plasma component (when one can directly apply the known results for multi-component plasmas), in particular, since the number of grains in the dust Debye sphere may sometimes be less than unity (Whipple et al., 1985). Moreover, the charges on dust particles are not fixed in processes of their interactions with plasma fields and other dust grains (Tsytovich and Havnes, 1993; Vladimirov, 1994a). Indeed, the floating charges are mostly determined by plasma electron and ion currents on the dust grains (Allen, 1992). The currents are naturally affected by the interactions which lead to a change of dust charges in the interaction process. This leads to qualitatively new effects (Tsytovich and Havnes, 1993; Vladimirov, 1994a, 1994b, 1994c, 1996), especially strong for low-frequency oscillations whose frequencies are comparable to the charging frequency. Below, we discuss some of this effects in more details.

Another important feature of dusty plasmas is that the presence of grains with even constant charges can strongly modify the existing wave spectra; as an example, we consider here modification of linear (Cramer and Vladimirov, 1996a) and nonlinear (Vladimirov and Cramer, 1996) Alfvén waves as well as surface Alfvén waves (Cramer and Vladimirov, 1996b) in a magnetized dusty plasma. Finally, influence of plasma collective processes on arrangements and vibrations of dust particles in the crystal-like Coulomb structures is considered. Note that the wake potential formation which leads to the vertical alignment of dust grains in the experiments, and, as a consequence, to quasi-two-dimensional features of the structures (i.e., hexagonal arrangements) when the number of dust layers is not high (the three-dimensional lattices corresponding to minimum of the potential energy are body-centered-cubic or face-centered-cubic) can also affect the specific lattice modes propagating in such systems.

2. Charging of a Dust Particle

Let us consider an unmagnetized plasma whose constituents are electrons, ions, and massive static ($m_d = \infty$) negatively charged dust grains. The latter are assumed to be point charges (i.e., their sizes are supposed to be much smaller than the effective Debye radius λ_D of the dusty plasma).

The basic charging equation is given by

$$\frac{dq}{dt} = I(q), \tag{1}$$

where q is the charge residing on a dust grain, and the current I on the dust particle is a sum of electron and ion plasma currents

$$I(q) = \sum_{\alpha} \int e_\alpha f_\alpha \sigma_\alpha(v, q) v d\mathbf{v}. \tag{2}$$

Here, the subscript $\alpha = e, i$ describes electrons or ions, e_α and f_α are the corresponding charge and distribution function of plasma particles, respectively, $e_e = -e_i \equiv -e$, $v \equiv |\mathbf{v}|$ is the absolute value of particle speed $\mathbf{v}$, σ_α is the charging cross-section (Spitzer, 1978):

$$\sigma_\alpha = \pi a^2 \left(1 - \frac{2e_\alpha q}{am_\alpha v^2}\right) \quad \text{if} \quad \frac{2e_\alpha q}{am_\alpha v^2} < 1,$$

$$\sigma_\alpha = 0 \quad \text{if} \quad \frac{2e_\alpha q}{am_\alpha v^2} \geq 1, \tag{3}$$

and, finally, m_α is the mass of the plasma electrons or ions. The last inequality in (3) gives restriction on particle charging velocities for electrons (we remind that we consider the negatively charged dust particles, i.e., $q < 0$). Thus we see that only sufficiently fast plasma electrons (which have enough energy to cross the potential barrier) can charge the dust particles.

In the state of equilibrium, we have

$$I^{\text{eq}}(Q) \equiv I_e^{\text{eq}}(Q) + I_i^{\text{eq}}(Q) = 0, \tag{4}$$

where $Q = -Z_d e$ is the equilibrium charge of the dust particulate. The equilibrium electron and ion currents are given by

$$I_e^{\text{eq}} = -e\sqrt{\frac{\pi}{2}}\, a^2 v_{Te} n_e \exp\left(-\frac{Z_d e^2}{aT_e}\right) \tag{5}$$

and

$$I_i^{\text{eq}} = e\sqrt{\frac{\pi}{2}}\, a^2 v_{Ti} n_i \left(1 + \frac{Z_d e^2}{aT_i}\right). \tag{6}$$

Here, $v_{T\alpha} = (T_\alpha / m_\alpha)^{1/2}$ is the thermal velocity, n_α is the number density, and T_α is the corresponding temperature of plasma particles. Thus the equilibrium charge can be found as a solution of the equation following from Equation (4):

$$\frac{\omega_{pe}^2}{v_{Te}} \exp\left(-\frac{Z_d e^2}{aT_e}\right) = \frac{\omega_{pi}^2}{v_{Ti}} \left(\frac{T_i}{T_e} + \frac{Z_d e^2}{aT_e}\right), \tag{7}$$

where $\omega_{p\alpha} = (4\pi n_\alpha e_\alpha^2 / m_\alpha)^{1/2}$ is the plasma frequency (electron or ion).

Furthermore, it is convenient to introduce the dimensionless variables

$$\tau \equiv \frac{T_i}{T_e}, \quad z \equiv \frac{Z_d e^2}{a T_e}. \tag{8}$$

For hydrogen plasmas, Equation (7) gives $z \simeq 2.5$ if $\tau = 1$ and $z \simeq 1.9$ if $\tau = 0.1$. Usually, Z_d is of order 10^3–10^4, but the total negative charge on the dust particles is close to (and does not exceed as a rule) the total charge of electrons. Thus, the following dimensionless parameter is of order unity

$$\mu \equiv \frac{Z_d n_d}{n_e} \sim 1, \tag{9}$$

where n_d is the dust density. However, more accurate calculations (which were done for a dust cloud in a thermal plasma, see Havnes et al., 1984) give that the following parameter is of order unity

$$P \equiv \frac{n_d}{n_e}\frac{aT_e}{e^2} = \frac{\mu}{z}. \tag{10}$$

When $z \sim 1$, these two conditions coincide.

The charging dissipative process is characterized by the charging frequency ν_d

$$\nu_d \equiv -\left.\frac{\partial I(q)}{\partial q}\right|_{q=-Z_d e} = \frac{1}{\sqrt{2\pi}}\frac{\omega_{pi}^2 a}{v_{Ti}}(1+\tau+z). \tag{11}$$

Note that the charging frequency describes the process of charging of the dust particles for small deviations from the equilibrium; it naturally contains contributions from both electrons and ions. To obtain the last expression on the right-hand side of Equation (11), Equation (7) has been used. Comparison of the rate (11) with the plasma collision frequencies (Tsypin et al., 1997; Tsytovich and Havnes, 1993) demonstrates that this process can be the most important dissipative process for dusty plasmas. For our purposes, the frequency characterizing the rate of capture of plasma electrons by dust particles ν_{ed}^{eq} is also interesting

$$\begin{aligned}\nu_{ed}^{\mathrm{eq}} &= \frac{n_d}{n_e}\int f_e \sigma_e v d\mathbf{v} = \nu_d P \frac{\tau+z}{1+\tau+z} \\ &= 2\omega_{p0}\sqrt{\frac{P n_d a^3}{2}}\sqrt{\frac{n_e}{n_{e0}}}\exp\left(-\frac{\mu_0}{P}\frac{n_{e0}}{n_e}\right),\end{aligned} \tag{12}$$

where $\omega_{p0} = (4\pi n_{e0} e^2/m_e)^{1/2}$, $\mu_0 = Z_d n_{d0}/n_{e0}$, and n_{e0} and n_{d0} are, respectively, the electron and dust densities at the initial moment. The latter expression on the right-hand side of Equation (12) is used when studying wave amplification processes in non-stationary dusty plasmas (Vladimirov, 1994c).

3. Dielectric Permittivity of a Dusty Plasma

We introduce the distribution function of dust particles as (Vladimirov, 1994a)

$$f_d = f_d(q, \mathbf{r}, t), \tag{13}$$

where the charge q is the additional independent variable. The corresponding kinetic equation is given by (we remind that the dust particles have infinite masses and, consequently, $v_d = 0$)

$$\frac{\partial f_d}{\partial t} + \frac{\partial}{\partial q} I(q) f_d = 0. \tag{14}$$

The state of equilibrium corresponds to the equilibrium distribution function $f_d^{\rm eq}$ of charges on dust particles. We have

$$Q = \frac{1}{n_d} \int q f_d^{\rm eq} dq. \tag{15}$$

The kinetics of electrons and ions are described by usual distribution functions f_α:

$$\frac{\partial f_\alpha}{\partial t} + \mathbf{v} \cdot \nabla f_\alpha + \frac{e_\alpha}{m_\alpha} \mathbf{E} \cdot \frac{\partial f_\alpha}{\partial \mathbf{v}} = - \int \sigma_\alpha v (f_d f_\alpha - f_d^{\rm eq} f_\alpha^{\rm eq}) dq. \tag{16}$$

We assume that in the equilibrium state the electron and ion capture by the dust is compensated by external sources and the equilibrium electron (ion) distribution $f_\alpha^{\rm eq}$ is isotropic. Term with magnetic field has been neglected on the left-hand side of Equation (16).

Presence of wave gives rise to small perturbations of the above distribution functions

$$f_d = f_d^{\rm eq} + \delta f_d, \quad f_\alpha = f_\alpha^{\rm eq} + \delta f_\alpha, \tag{17}$$

where $|\delta f_d| \ll |f_d|$ and $|\delta f_\alpha| \ll |f_\alpha|$. Furthermore, we linearize kinetic equations (14) and (16) with respect to these perturbations. Here, we note that, in principle, the ionization process can either be perturbed by δf_e (e.g., if the ionization is due to electron impact) or can affect dust (e.g., if the ionization is by radiation). For simplicity, we exclude such possibilities in the present investigation.

Thus, we obtain for the dust particle distribution function

$$\frac{\partial \delta f_d}{\partial t} + \frac{\partial}{\partial q} [I^{\rm eq}(q) \delta f_d] + \frac{\partial}{\partial q} [\delta I(q) f_d^{\rm eq}] = 0, \tag{18}$$

where

$$I(q) = I^{\rm eq}(q) + \delta I(q), \qquad \int I^{\rm eq}(q) f_d^{\rm eq} dq = 0,$$

$$\delta I(q) = \sum_\alpha \int d\mathbf{v} e_\alpha \sigma_\alpha v \delta f_\alpha. \tag{19}$$

The Fourier-component of the linear perturbation of the equilibrium charge on dust particles is

$$\delta Q_{\mathbf{k}\omega} = \frac{1}{n_d} \int q \delta f_d dq = \frac{i}{\omega + i\nu_d^{\mathrm{eq}}} \frac{1}{1 + G_{\mathbf{k}\omega}} \frac{1}{n_d} \int I_{\mathbf{k}\omega}^{(1)}(q) f_d^{\mathrm{eq}} dq, \tag{20}$$

and that of the linear perturbation of the electron (ion) distribution function is given by

$$\begin{aligned} \delta f_{\alpha,\mathbf{k}\omega} &= -\frac{ie_\alpha/m_\alpha}{\omega - \mathbf{k}\cdot\mathbf{v} + i\nu_{\alpha d}^{\mathrm{eq}}(v)} \mathbf{E}_{\mathbf{k}\omega} \cdot \frac{\partial f_\alpha^{\mathrm{eq}}}{\partial \mathbf{v}} \\ &+ \frac{1}{\omega + i\nu_d^{\mathrm{eq}}} \frac{1}{1 + G_{\mathbf{k}\omega}} \frac{\sigma_\alpha' v f_\alpha^{\mathrm{eq}}}{\omega - \mathbf{k}\cdot\mathbf{v} + i\nu_{\alpha d}^{\mathrm{eq}}(v)} \int I_{\mathbf{k}\omega}^{(1)}(q) f_d^{\mathrm{eq}} dq. \end{aligned} \tag{21}$$

In Equations (20) and (21), the following notations have been used:

$$I_{\mathbf{k}\omega}^{(1)}(q) = \sum_\alpha \int_{V_d} d\mathbf{v} \frac{-ie_\alpha \sigma_\alpha v}{\omega - \mathbf{k}\cdot\mathbf{v} + i\nu_{\alpha d}^{\mathrm{eq}}(v)} \left(\frac{e_\alpha}{m_\alpha} \mathbf{E}_{\mathbf{k}\omega} \cdot \frac{\partial f_\alpha^{\mathrm{eq}}}{\partial \mathbf{v}} \right), \tag{22}$$

$$G_{\mathbf{k}\omega} = \frac{-1}{\omega + i\nu_d^{\mathrm{eq}}} \sum_\alpha \int_{V_d} d\mathbf{v} \frac{\nu_{\alpha d}^{\mathrm{eq}}(v) e_\alpha \sigma_\alpha' v f_\alpha^{\mathrm{eq}}}{\omega - \mathbf{k}\cdot\mathbf{v} + i\nu_{\alpha d}^{\mathrm{eq}}(v)}, \tag{23}$$

where

$$\sigma_\alpha^{\mathrm{eq}} = \frac{1}{n_d} \int \sigma_\alpha(q) f_d^{\mathrm{eq}} dq = \sigma_\alpha(-Z_d e), \quad \sigma_\alpha' \equiv \frac{\partial \sigma_\alpha(q)}{\partial q}. \tag{24}$$

Note that σ_α' is also the step function on v, which should be taken into account by the same phase-space volume V_d of integration over $\mathbf{v}$ as in integration of expressions containing the charging cross-section (3).

To find the dielectric permittivity of longitudinal waves, we use Poisson equation. In the case of the thermal particle distributions and sufficiently high frequency $[\omega \sim \omega_{pe} \gg \max(kv_{Te}, \nu_d^{\mathrm{eq}}, \nu_{ed}^{\mathrm{eq}})]$ of the waves, we have the following approximate expression (Vladimirov, 1994a)

$$\begin{aligned} \varepsilon_{\mathbf{k}\omega}^{(l)} &\simeq 1 + \frac{4\pi}{\omega^2 k^2} \sum_\alpha \frac{-e_\alpha^2}{m_\alpha v_{T\alpha}^2} \int d\mathbf{v} \left\{ 1 - i\frac{\nu_{\alpha d}^{\mathrm{eq}}(v)}{\omega} - \frac{[\nu_{\alpha d}^{\mathrm{eq}}(v)]^2}{\omega^2} \right\} (\mathbf{k}\cdot\mathbf{v})^2 f_\alpha^{\mathrm{eq}} \\ &= 1 - \frac{\omega_{pe}^2}{\omega^2} \left[1 - i\frac{2}{3}(2 + z)\frac{\nu_{ed}^{\mathrm{eq}}}{\omega} - \sqrt{\pi z} A e^z \left(\frac{\nu_{ed}^{\mathrm{eq}}}{\omega} \right)^2 \right], \end{aligned} \tag{25}$$

where

$$A \equiv \frac{5}{4} - \frac{z}{6} + \left(\frac{5}{4} - z + \frac{z^2}{3} \right) \int_1^\infty d\tau \exp[-(\tau^2 - 1)z]. \tag{26}$$

Therefore, the charging process leads to appearing of additional imaginary and real parts in the high-frequency longitudinal dielectric permittivity. These additional terms depend on relation between the charging frequency and electron plasma frequency. For Langmuir waves, the damping due to the charging process is given by

$$\gamma_d^L \simeq -\frac{1}{3}(2+z)\nu_{ed}^{\rm eq} = -\frac{1}{3}P(2+z)\frac{\tau+z}{1+\tau+z}\nu_d^{\rm eq}. \tag{27}$$

For the transversal dielectric function we use Maxwell equations and find

$$\begin{aligned}\varepsilon_{\mathbf{k}\omega}^{(t)} &\simeq 1 + \frac{2\pi}{\omega^2 k^2}\sum_\alpha \frac{-e_\alpha^2}{m_\alpha v_{T\alpha}^2}\int d\mathbf{v}\left\{1 - i\frac{\nu_{\alpha d}^{\rm eq}(v)}{\omega} - \frac{[\nu_{\alpha d}^{\rm eq}(v)]^2}{\omega^2}\right\}(\mathbf{k}\times\mathbf{v})^2 f_\alpha^{\rm eq} \\ &= 1 - \frac{\omega_{pe}^2}{\omega^2}\left[1 - i\frac{2}{3}(2+z)\frac{\nu_{ed}^{\rm eq}}{\omega} - \sqrt{\pi}zAe^z\left(\frac{\nu_{ed}^{\rm eq}}{\omega}\right)^2\right]\end{aligned} \tag{28}$$

which exactly coincides with expression (25) for the high-frequency longitudinal dielectric permittivity. Equation (28) gives the following damping rate for electromagnetic waves due to the charging effects on dust particles

$$\gamma_d^{EM} \simeq -\frac{1}{3}\frac{\omega_{pe}^2}{\omega^2}(2+z)\nu_{ed}^{\rm eq}. \tag{29}$$

If the frequency of electromagnetic waves does not far exceed ω_{pe}, when the phase velocity is significantly influenced by the conduction current, the damping rate (29) is of the order of the electron capture rate $\nu_{ed}^{\rm eq}$. Note the real corrections to the electron plasma frequency which are also effective for electromagnetic waves.

4. Wave Scattering

The usual Thomson scattering of a wave on a test particle of charge q and mass m takes place when the incident wave causes particle oscillation in its field. The scattered power in this case is proportional to m^{-2}, and because of the very large mass of the dust particle m_d the effect decreases despite the large charge residing on the dust particle. Thus because of the large masses of dust grains comparing with masses of electrons and ions, the wave scattering on the plasma particles forming their Debye spheres can dominate over the Thomson scattering on the the test "bare" dust grain.

To find the scattering of electromagnetic waves on a dust particle, we use the procedure elaborated in (Vladimirov, 1994b). In particular, the scattering cross-section is given by

$$\sigma^{em} = \sigma^T\frac{3c^3}{16\pi}\int d\mathbf{k}\left(1+\cos^2\Theta\right)\frac{|Z_{\mathbf{k}-\mathbf{k}_0}^{\rm eff}|^2}{\omega_0^2}\delta\left(\omega_0 - \sqrt{\omega_{pe}^2 + k^2c^2}\right), \tag{30}$$

where σ^T is the cross-section of the Thomson scattering,

$$\sigma^T = \frac{8\pi}{3}\frac{e^4}{m_e^2c^3}, \tag{31}$$

c is the light speed, Θ is the angle between incident and scattered waves, subscript 0 corresponds to the incident wave with frequency $\omega_0 = (\omega_{pe}^2 + k_0^2c^2)^{1/2}$, and $Z_{\mathbf{k}}^{\mathrm{eff}} = Q_{\mathbf{k}}/e$ is Fourier component of the "effective charge" residing on the dust particle. In the linear approximation, the effective charge is given by

$$Z_{\mathbf{k}}^{\mathrm{eff}} = Z_d\frac{\varepsilon_{\mathbf{k}}^{(e)} - 1}{\varepsilon_{\mathbf{k}}}, \tag{32}$$

where $-Z_d e$, as elsewhere in this paper, is the equilibrium charge of the dust particle, and

$$\varepsilon_{\mathbf{k}} = \varepsilon_{\mathbf{k}}^{(e)} + \varepsilon_{\mathbf{k}}^{(i)} - 1 \tag{33}$$

is the complete static dielectric permittivity (i.e., $\varepsilon_{\mathbf{k}}^{(\alpha)}$ corresponds to the dielectric function of α plasma component) taking into account perturbations of the charging process.

Using the kinetic theory introduced above, we obtain in the static approximation

$$\varepsilon_{\mathbf{k},\omega=0} = 1 + \frac{1}{k^2\lambda_{De}^2}\frac{1+\tau}{\tau}\left(1 + P\frac{\tau+z}{1+\tau+z}F_{\mathbf{k}}\right), \tag{34}$$

where $\lambda_{De} = v_{Te}/\omega_{pe}$ is the electron Debye length and the formfactor $F_{\mathbf{k}\omega}$ is defined by

$$F_{\mathbf{k}} = \frac{1+\Gamma_{\mathbf{k},\omega=0}}{1+G_{\mathbf{k},\omega=0}}. \tag{35}$$

In this equation, the factor $G_{\mathbf{k}\omega}$ is given by Equation (23) and

$$\Gamma_{\mathbf{k}\omega} = \sum_\alpha \int d\mathbf{v}\frac{-ie_\alpha\sigma_\alpha' v f_\alpha^{\mathrm{eq}}}{\omega - \mathbf{k}\cdot\mathbf{v} + i\nu_{\alpha d}^{\mathrm{eq}}(v)}. \tag{36}$$

We see that the charging process leads to appearing of the additional real part in the static longitudinal dielectric permittivity. In particular, this additional term depends on the parameter P. Note also that the factor before the function $F_{\mathbf{k}}$ in the right-hand side of Equation (34) is exactly equal to $\nu_{ed}^{\mathrm{eq}}/\nu_d^{\mathrm{eq}}$, see Equation (12).

In the case of thermal particle distributions and static longitudinal perturbations $[\omega \ll \max(kv_{T\alpha}, \nu_d^{\mathrm{eq}}, \nu_{\alpha d}^{\mathrm{eq}})]$, we have the following expressions:

$$\Gamma_{\mathbf{k},\omega=0} = \frac{\pi n_d a^2 \mathrm{e}^{-z}}{Pk}\left[\frac{v_{Te}}{v_{Ti}}\frac{1}{\tau+z} + \sqrt{\frac{4z}{\pi}}\int_1^{+\infty} dx \mathrm{e}^{-x^2(z-1)}\right], \tag{37}$$

and

$$G_{\mathbf{k},\omega=0} = -\frac{\pi}{2}\frac{\tau+z}{1+\tau+z}\frac{\pi n_d a^2}{k}\left\{\int_0^{+\infty} dx\frac{x\mathrm{e}^{-x}}{x+z} + \frac{1}{1+\tau}\right.$$

$$\left.\times\left[1+\frac{2z}{\pi\tau}\int_0^{+\infty} dx\frac{\mathrm{e}^{-x}}{x}\arctan\left(\frac{k}{\pi n_d a^2}\frac{\tau x}{\tau x+z}\right)\right]\right\}. \tag{38}$$

Therefore, for the electron part of the longitudinal dielectric permittivity, which is present in Equation (32), we can easily obtain [29]

$$\varepsilon^{(e)}_{\mathbf{k},\omega=0} = 1+\frac{1}{k^2\lambda^2_{De}}\left(1+P\frac{\tau+z}{1+\tau+z}F_{\mathbf{k}}\right). \tag{39}$$

Finally, for the effective charge (32) we find

$$Z^{\mathrm{eff}}_{\mathbf{k}} = Z_d\left(1+\frac{1}{\tau}+\frac{k^2\lambda^2_{De}}{1+PF_{\mathbf{k}}(\tau+z)/(1+\tau+z)}\right)^{-1}. \tag{40}$$

Thus, we have found that the effective cross-section of scattering of electromagnetic waves on a dust particle depends not only on the scattering parameter $1/k\lambda_{De}$ but also on the parameters characterizing the charging process, in particular P.

Finally, we note that here, for the sake of simplicity, we presented only the linear theory of the scattering on the isolated dust grain. In general, nonlinear effects can be important for the process. Some of these effects (although not taking into account dust charge fluctuations) have been considered in (Tsytovich, 1992).

5. Alfvén Waves in a Magnetized Dusty Plasma

We consider here the influence on the dispersion properties of Alfvén waves of the collection of electrons and ions from the background plasma by the charged grains. This affects the equilibrium state when the electron and ion charge balance necessarily includes the charge and density of the particulates

$$-en_e + en_i - Z_d en_d = 0. \tag{41}$$

For many dusty plasmas, an appreciable proportion of the negative charge in the plasma may reside on the dust particles. In that case the dispersion properties of Alfvén in the plasma are strongly modified by the dust, because the ion Hall current is not compensated by the electron Hall current at low frequencies as well as at frequencies comparable with the ion-cyclotron frequency, so that ion cyclotron effects extend to frequencies much less than the ion-cyclotron frequency (Cramer and Vladimirov, 1996a; Vladimirov and Cramer, 1996). The waves at frequencies comparable to the ion-cyclotron frequency are also circularly polarized modes, as in the dust-free case. The Alfvén resonance (Cramer and Vladimirov, 1996a) is strongly modified in the presence of dust grains because of the imbalance of electron and ion charges.

We invoke the standard two-fluid MHD model which includes the fluid momentum equations for the plasma (singly charged cold) ions and (inertialess) electrons, the ion continuity equation, as well as Maxwell's equations ignoring the displacement current. The background magnetic field $\mathbf{B}_0$ is in the z-direction, and the dust grains have infinite masses (we thus exclude dust dynamics from consideration). Furthermore, we introduce the parameter $\delta = n_e/n_i$ which measures the charge imbalance in the plasma. The total system is supposed to be neutral with the remainder of the charge residing on the dust particles according to Equation (41).

The starting equations are Euler equations for the ion and electron velocities $\mathbf{v}_i$ and $\mathbf{v}_e$, Maxwell equations for the wave electric field $\mathbf{E}$ and magnetic field $\mathbf{B}$ (the latter also includes the background field $\mathbf{B}_0$), and continuity equations for the ion and electron densities n_i and n_e. The electron inertia is neglected. In the one-dimensional case (being interested in evolution parallel to the external magnetic field), we introduce the right(left)-circularly polarized magnetic field components $B_\pm = B_x \pm iB_y$ and find

$$\frac{\partial \rho}{\partial t} + \frac{\partial(\rho v_z)}{\partial z} = 0, \tag{42}$$

$$\frac{\partial B_\pm}{\partial t} = -\frac{\partial}{\partial z}\left[\frac{1}{\delta}\left(v_z B_\pm - v_\pm\right)\right] \mp \frac{i v_A^2}{\Omega_i}\frac{\partial}{\partial z}\left(\frac{1}{\rho\delta}\frac{\partial B_\pm}{\partial z}\right), \tag{43}$$

$$\frac{dv_\pm}{dt} = \mp i\Omega_i \frac{1-\delta}{\delta}\left(v_z B_\pm - v_\pm\right) + \frac{v_A^2}{\rho\delta}\frac{\partial B_\pm}{\partial z}, \tag{44}$$

$$\frac{dv_z}{dt} = \mp \frac{i\Omega_i}{2}\frac{1-\delta}{\delta}\left(v_\pm B_\pm^* - v_\pm^* B_\pm\right) - \frac{v_A^2}{2\rho\delta}\frac{\partial}{\partial z}\left(|B_\pm|^2\right), \tag{45}$$

where the right(left)-circularly polarized velocity component has been introduced analogously to the magnetic field component: $v_\pm = v_x \pm i v_y$. In Equations (42–45), we have introduced the dimensionless magnetic field amplitude normalized as $\mathbf{B} = \mathbf{B}/B_0$, where $B_0 = |\mathbf{B}_0|$, and omitted the subscript i for the ion velocity. Furthermore, $\rho = m_i n_i/\rho_0$ is the normalized density of the ion component of the plasma, where $\rho_0 = m_i n_{i0}$ and n_{i0} corresponds to the equilibrium ion concentration, $\Omega_i = eB_0/m_i$ is the ion gyrofrequency, and $v_A = (B_0^2/\mu_0\rho_0)^{1/2}$ is the Alfvén velocity.

Linearizing Equations (42–45) with respect to the wave magnetic field, we find

$$\left(\frac{\partial^2}{\partial t^2} \mp i\Omega_i \frac{1-\delta_0}{\delta_0}\frac{\partial}{\partial t} - \frac{v_A^2}{\delta_0}\frac{\partial^2}{\partial z^2} \pm i\frac{v_A^2}{\Omega_i\delta_0}\frac{\partial^3}{\partial t\,\partial z^2}\right) B_\pm = 0, \tag{46}$$

where $\delta_0 = n_{e0}/n_{i0}$ measures the unperturbed electron and ion number density imbalance. First, we consider the case

$$kv_A \ll \Omega_i \frac{1-\delta_0}{\sqrt{2(1+\delta_0)}} < \Omega_i, \tag{47}$$

(region A) and find the following approximate solutions:

$$\omega_{-1} \equiv \omega_{HF} \simeq \Omega_m \left[1 + \frac{k^2 v_A^2}{(1-\delta_0)^2 \Omega_i^2}\right], \quad \omega_{+1} \equiv \omega_{LF} \simeq \frac{k^2 v_A^2}{(1-\delta_0)\Omega_i}, \tag{48}$$

where

$$\Omega_m = \Omega_i \frac{1-\delta_0}{\delta_0}. \tag{49}$$

If we consider the higher wavenumber range defined by

$$\Omega_i \frac{1-\delta_0}{\sqrt{2(1+\delta_0)}} \ll kv_A \ll \Omega_i, \tag{50}$$

(region B) then we obtain

$$\omega_{HF,LF} \simeq kv_A \sqrt{\frac{1+\delta_0}{2}} \pm \frac{\Omega_i}{2}\left(1 - \delta_0 + \frac{k^2 v_A^2}{\Omega_i^2}\right). \tag{51}$$

i.e. an almost linear dispersion relation, with a displacement in frequency due to the charge imbalance and ion cyclotron dispersion, of opposite sign for the two modes.

The linear electromagnetic modes do not involve perturbations of ρ or v_z. Furthermore, the fluctuations in ρ and v_z are assumed to be of small amplitude [of order $|B_\pm|^2$ from Equation (45)], so we can linearize Equations (42–45) in these quantities and find

$$\begin{aligned} &\frac{\partial}{\partial t}\left[\frac{\partial B_\pm}{\partial t} \mp i\Omega_m B_\pm \pm \frac{iv_A^2}{\delta_0 \Omega_i}\frac{\partial^2 B_\pm}{\partial z^2}\right. \\ &\left. + \frac{\partial}{\partial z}\left(\mp \frac{iv_A^2 \Delta\rho}{\delta_0 \Omega_i}\frac{\partial B_\pm}{\partial z} + \frac{v_z B_\pm}{\delta_0} + \frac{v_\pm \Delta\rho}{\delta_0^2}\right)\right] \\ &= \frac{1}{\delta_0}\frac{\partial^2}{\partial z^2}\left(v_A^2 B_\pm - v_z v_\pm\right). \end{aligned} \tag{52}$$

The higher frequency mode is assumed to propagate at a frequency close to Ω_m, so that we introduce the envelope amplitudes of the wave magnetic field and ion velocity perturbations as $B_{-1} = b_1 \exp(-i\Omega_m t)$ and $v_{-1} = v_1 \exp(-i\Omega_m t)$. Retaining the largest dispersion and nonlinear terms, we thus find

$$i\frac{\partial b_1}{\partial t} + \frac{v_A^2}{\delta_0^2 \Omega_m}\frac{\partial^2 b_1}{\partial z^2} + i\frac{1+\delta_0}{\delta_0}\frac{\partial(v_z b_1)}{\partial z} + \Omega_m \Delta\rho b_1 = 0. \tag{53}$$

For the parallel component of the ion velocity perturbation, we have

$$\frac{\partial v_z}{\partial t} = -\frac{i\Omega_m}{2}\left(v_1^* b_1 - v_1 b_1^*\right) - \frac{v_A^2}{2\delta_0}\frac{\partial(|b_1|^2)}{\partial z}\,. \tag{54}$$

Equations (53) and (54), together with the linearized continuity equation (42), form the set of equations describing evolution of the higher frequency mode amplitude.

For the lower frequency mode, we find that the nonlinear contribution in the approximation used is negligible containing high orders of field derivatives. We also should mention that for the lower-frequency mode, the dust dynamics can be of large importance, so that the starting equations should be modified to include dust dynamic terms. For the range of wavenumbers corresponding to the region B, defined by Equation (50), the same method of stretched coordinates as used for the dust-free case can be invoked. As a result, there are no qualitatively new features comparing with the dust-free case, although in the resulting DNLS equation the dispersion and nonlinear coefficients are affected by the presence of dust.

We now seek wave solutions of the weakly nonlinear equations for the higher frequency mode of region A. We have the linear solution of Equations (53) and (54) in the form $b_1 = b_{10}\exp(-i\Omega t + iKz)$ and $v_1 = v_{10}\exp(-i\Omega t + iKz)$, where b_{10} and v_{10} are constants, and $\Omega \simeq K^2 v_A^2/\delta_0^2\Omega_m$ is the dispersion correction to the frequency Ω_m. Furthermore, we allow for a slow dependence of the amplitudes b_{10} and v_{10}, as well as their phase Θ, on t and z and consider a propagating solution only, where all functions (including phases) depend on $Z = z - v_0 t$. In this case, for substitution in Equation (54) we have $v_1(Z) = -\delta_0 v_0 b_1(Z)$. This is an important simplification which holds only under the assumption of the dependence of both amplitudes and phase on Z.

Assuming that all perturbations vanish at $Z \to \pm\infty$, we can write Equation (53) in the form of a mixed DNLS–NLS equation

$$-iv_0\frac{db_1}{dZ} + D\frac{d^2 b_1}{dZ^2} + i\mu\frac{d}{dZ}\left(|b_1|^2 b_1\right) + \nu|b_1|^2 b_1 = 0, \tag{55}$$

where the coefficients are $D = v_A^2/\delta_0^2\Omega_m$, $\mu = (1+\delta_0)v_A^2/2\delta_0 v_0$, and $\nu = \Omega_m v_A^2/2\delta_0^2 v_0^2$. Furthermore, separating the imaginary part we find $K(Z) = (v_0/2D) - (3\mu b_{10}^2/4D)$, and $\Omega(Z) = K(Z)v_0$. The equation for the real part of (55) can be written as

$$\frac{d^2 b_{10}}{dZ^2} = -\frac{\partial U(b_{10})}{\partial b_{10}}\,, \tag{56}$$

where the nonlinear potential is

$$U(b) = \frac{v_0^2}{8D^2}b^2 + \left(\frac{\nu}{4D} - \frac{\mu v_0}{8D^2}\right)b^4 + \frac{\mu^2}{32D^2}b^6. \tag{57}$$

The potential (57) describes nonlinear conoidal waves; the corresponding solutions can be written in terms of elliptic functions. Note that when $\nu < \mu v_0/2D$ or $v_A^2 <$

$(1+\delta_0)\delta_0^2 v_0^2$, a metastable equilibrium state with $b_{10} \neq 0$ appears. A separatrix dividing the quasiperiodic oscillations in this case corresponds to a localized wave packet. Also, we find that no solutions in the form of standard soliton solutions occur if the conditions opposite to the above are applicable.

6. Formation of Dust Crystals

The possibility of formation of Coulomb quasilattices (Ikezi, 1986) involving the micro-meter sized highly charged dust particulates was first time demonstrated experimentally in works (Chu and I, 1994; Thomas et al., 1994; Hayashi and Tachibana, 1994; Metzer et al., 1994). In the experiments, the dust-crystal structure is observed in the sheath region of a radio-frequency discharge plasma where there is balance between the gravitational and electrostatic forces. The distance between the dust grains is of order Debye length λ_D.

It is well known that in the sheath region strong ion flow is established; according to the Bohm criterion the speed of the flow exceeds the ion acoustic velocity (Vladimirov and Nambu, 1995; Vladimirov and Ishihara, 1996). The drag force on a test particle in a plasma with finite flows necessarily includes collective effects. It was shown that if the speed of the flow exceeds the velocity of ion oscillations in the flow, an oscillating stationary wake is formed behind the static test particle (Vladimirov and Nambu, 1995; Vladimirov and Ishihara, 1996; Melandsø and Goree, 1995; Ishihara and Vladimirov, 1997). The effect is similar to the Cooper pairing of electrons in superconductors, and was earlier studied, e.g., for two-component electron-ion plasmas (Nambu and Akama, 1985). We note that the collective mechanism can be responsible for the oscillatory potential in the direction parallel (Vladimirov and Nambu, 1995; Ishihara and Vladimirov, 1997) as well as perpendicular (Vladimirov and Ishihara, 1996; Ishihara and Vladimirov, 1997) to the flow, and the attraction due to the wake potential can overcome the static Coulomb repulsion. The characteristic spacing in this case is of order Debye length λ_D in agreement with the experiments.

We consider the cylindrical geometry (ρ, φ, z); the plasma ion flow is in the $-z$-direction with velocity v_{i0}. The test dust particle of the charge Q is placed on the position $(0, 0)$. Furthermore, we calculate the potential behind the test particle downstream the flow within the wake cone: $|z| > \rho\sqrt{M^2 - 1}$, where $M = v_{i0}/v_s$ is the Mach number and $v_s = (T_e n_i / m_i n_e)^{1/2}$ is the sound velocity. Note that because we are interested in static dust particulates placed in the ion flow, the Cherenkov angle depends only on the speed of the flow. The static Coulomb repulsion between the dust particles is strong if the distance between them is less than the Debye length. Here, we note that the main contribution to the effective Debye length λ_D in a typical discharge conditions Chu and I, 1994; Thomas et al., 1994; Hayashi and Tachibana, 1994; Melzer et al., 1994, 1996a; Chiang and I, 1996) is due to the plasma electrons: $\lambda_D = \lambda_{De} \equiv (T_e/4\pi n_e e^2)^{1/2}$. Below, we are interested in the case where dust particles are subject to the Debye screening potential as well as the wake potential which arises when the ion flow velocity exceeds the critical speed.

We write the electrostatic potential of the static dust particle as

$$\Phi(\mathbf{r}) = Q \int \frac{d\mathbf{k}}{2\pi^2 k^2} \frac{\exp(i\mathbf{k}\cdot\mathbf{r})}{\varepsilon(\mathbf{k}, -k_z v_{i0})}, \tag{58}$$

where $\mathbf{k} = (\mathbf{k}_\perp, k_z)$. The dielectric response function of the plasma in the presence of the finite ion flow with the speed v_{i0} ($v_s < v_{i0} \ll v_{Te}$, where $v_{Te} = (T_e/m_e)^{1/2}$ is the electron thermal velocity and m_e is the electron mass) is calculated under condition

$$k v_{Ti} \ll |k_z v_{i0}| \ll k v_{Te}, \tag{59}$$

where $v_{Ti} = (T_i/m_i)^{1/2}$ is the ion thermal velocity, T_i is the ion temperature, and, as we stated already, the flow is in $-z$-direction. For the plasma dielectric response function, we have

$$\varepsilon(\mathbf{k}, -k_z v_{i0}) = 1 + \frac{1}{k^2\lambda_D^2} - \frac{\omega_{pi}^2}{(-k_z v_{i0} + i0)^2}, \tag{60}$$

where $\omega_{pi} = (4\pi n_i e^2/m_i)^{1/2}$ is the ion plasma frequency. Note the imaginary part $+i0$ (appearing due to causality) which is important when integrating over k_z.

Furthermore, we separate in Equation (58) (using the inverse of function (60)) the static Debye and oscillating wake potentials and obtain outside the Mach cone

$$\Phi(\mathbf{r}) = \Phi_D(\mathbf{r}), \tag{61}$$

where

$$\Phi_D(\mathbf{r}) = \frac{Q}{|\mathbf{r}|} \exp(-|\mathbf{r}|/\lambda_D), \tag{62}$$

and inside the Mach cone

$$\Phi(\mathbf{r}) = \Phi_W(\mathbf{r}), \tag{63}$$

where $\Phi_W(\mathbf{r})$ involves the collective effects caused by the oscillations in the ion flow:

$$\Phi_W(\mathbf{r}) = Q \int \frac{d\mathbf{k}}{2\pi^2 k^2} \frac{k^2\lambda_D^2\omega_{\mathbf{k}}^2 \exp(i\mathbf{k}\cdot\mathbf{r})}{(1 + k^2\lambda_D^2)[(-k_z v_{i0} + i0)^2 - \omega_{\mathbf{k}}^2]}. \tag{64}$$

Here, $\omega_{\mathbf{k}} = k v_s/(1 + k^2\lambda_D^2)^{1/2}$ is the characteristic frequency of the oscillations in the flow; it naturally appears when we equal to zero the plasma response function (60). The potential (64) describes the strong resonant interaction between the oscillations in the ion flow and the test particulate when $|k_z v_{i0}|$ is close to $\omega_{\mathbf{k}}$.

In dimensionless units normalized by the inverse of the Debye length, we can write Equation (64) as

$$\Phi_W(\mathbf{r}) = \frac{Q}{\lambda_D M^2} \int \frac{dk_z d\mathbf{k}_\perp}{2\pi^2} \frac{k^2}{1+k^2} \frac{\exp(i\mathbf{k}\cdot\mathbf{r}/\lambda_D)}{[(k_z - i0)^2 + k_0^2][(k_z - i0)^2 - k_1^2]}, \quad (65)$$

where $k^2 = k_z^2 + k_\perp^2$ and $k_{0,1}^2 = \pm(1 - M^{-2} + k_\perp^2)/2 + [k_\perp^2 M^{-2} + (1 - M^{-2} + k_\perp^2)^2/4]^{1/2}$. Note that the contribution from the poles at $k_z = \pm ik_0$ provides the non-oscillating part which modifies the static Debye shielding scale (62) in plasmas with finite ion flows.

After integration over k_z in (65) we find

$$\Phi_W(\mathbf{r}) = -\frac{Q}{\lambda_D M^2} \int \frac{d\mathbf{k}_\perp}{2\pi |k_1|} \frac{k_\perp^2 + k_1^2}{k_0^2 + k_1^2} \frac{\exp(ik_\perp \rho \cos\varphi/\lambda_D)}{1 + k_\perp^2 + k_1^2} \sin(|k_1 z|/\lambda_D). \quad (66)$$

Taking into account the cylindrical symmetry of the wake cone, we obtain

$$\Phi_W(\rho, z) = -\frac{Q}{\lambda_D M^2} \int_0^{+\infty} \frac{k_\perp dk_\perp}{|k_1|} \frac{k_\perp^2 + k_1^2}{k_0^2 + k_1^2} \frac{J_0(k_\perp \rho/\lambda_D)}{1 + k_\perp^2 + k_1^2} \sin(|k_1 z|/\lambda_D), \quad (67)$$

where J_0 is the Bessel function of zero order.

It is difficult to find analytically the mathematically exact expression for the potential (67). Thus, below we present an estimation to evaluate the asymptotic behavior of the function. The main contribution to the integral (67) comes from $k_\perp \sim 1$. Then for the distance small in the perpendicular direction, $k_\perp \rho \ll 1$, and for $|z| > \lambda_D \sqrt{M^2 - 1}$, the main contribution to the stationary wake potential is given by

$$\Phi_W(\rho = 0, z) \approx \frac{q_t}{|z|} \frac{2\cos(|z|/L_s)}{1 - M^{-2}}, \quad (68)$$

where $L_s = \lambda_D \sqrt{M^2 - 1}$ is the effective length.

From Equation (68), we can conclude that the wake potential is attractive for $\cos(|z|/L_s) < 0$. On the other hand, for the distance $\rho > \lambda_D$ and $|z| > \lambda_D \sqrt{M^2 - 1}$ we find the following approximate expression for the wake potential:

$$\Phi_W(\rho, z) \simeq \frac{Q}{1 - M^{-2}} \sqrt{\frac{\lambda_D}{2\pi\rho}} \times \left\{ \frac{\cos[(\pi/4) + (z_-/\lambda_D\sqrt{M^2 - 1})]}{z_-} + \frac{\cos[(\pi/4) - (z_+/\lambda_D\sqrt{M^2 - 1})]}{z_+} \right\}, \quad (69)$$

where $z_\pm \equiv |z| \pm \rho\sqrt{M^2 - 1} > 0$ (we remind that the oscillating potential exists only in the wake of the test particle, i.e., for $z < 0$ and $|z| > \rho\sqrt{M^2 - 1}$). Obviously, this function oscillates as we change ρ or z. Because oscillating potential

(69) is proportional to the same dust particle charge Q as the static Debye potential (62), and contains no screening exponential, there are regions in space which correspond to the change of the effective potential sign and, hence, to the most probable positions of the particulates. We stress that these regions are not only on the line $\rho = 0$. The effective periodic spacing in the plane perpendicular to the flow is of order Debye length; note that this can be seen not only from approximate expression (69), but also from more exact formula (67).

Thus, we have demonstrated that collective effects can provide the oscillating potential on the line as well as in the plane perpendicular to the direction of the ion flow downstream the dust particle. Since in the experiments the most probable effective spacing is of order Debye length (the estimation is $\sim 0.2 \div 1.7\lambda_D$), we see that the polygon structure can appear in plasma crystal formation. Indeed, because the wake potential cannot change the sign of the effective potential at the distances less than the λ_D, the dust particulates are not expected to be arranged with the distances less than the Debye length. At the same time, the characteristic spacing of the polar radius-vector in the plane perpendicular to the flow is also of order λ_D. Therefore, we can expect the particulates on the equal distances λ_D on the periphery of the circle of the radius of order λ_D. This may correspond to polygon of order not higher than hexagon. Note that hexagonal structures were observed practically in all the experiments on dust crystallization (Chu and I, 1994; Thomas et al., 1994; Hayashi and Tachibana, 1994; Melzer et al., 1994, 1996a; Chiang and I, 1996).

The proposed mechanism can provide a qualitative explanation of the arrangement of particles downstream the test dust particulate. For quantitative picture, it is necessary to calculate the potential of ensemble of dust particulates as well as contribution of other factors such us gravitation, plasma inhomogeneity, etc. Moreover, the regular arrangement of the particles in the first (with regard to the ion flow) layer of the plasma crystal needs more investigations; we mention only that if the regular quasicrystal structure is established in downstream layers, this may also affect the arrangement of particles in the first layer.

7. Vibrational Lattice Modes in the Dust Crystals

The considered above dust-crystal structures are observed in the sheath region where there is balance between the gravitational and electrostatic forces, and in many experiments it consists of just a few layers of dust particles levitating above the horizontal negatively biased electrode (Chu and I, 1994; Thomas et al., 1994; Hayashi and Tachibana, 1994; Melzer et al., 1994, 1996a, 1996b; Chiang and I, 1996; Peters et al., 1996).

Recently, it was demonstrated (Melandsø, 1996; Vladimirov et al., 1997) that lattice waves can propagate in the one-dimensional chain of strongly coupled dust particles. The waves include motion of dust particles in horizontal direction (which is the simplest situation) (Melandsø, 1996) as well as in the vertical direction (when the grain oscillations depend also on the shape of the external potential of the electrode) (Vladimirov et al., 1997). Note that motion of the dust grains in the vertical direction can provide a useful tool for determining the grain charge (Melzer et al., 1996b; Peters et al., 1996).

We demonstrate here that oscillations of the dust grains in the vertical plane can lead to a novel low-frequency mode. We note that excitation of this mode can also be responsible for phase transitions in the system which are the subject of particular recent interest, see, e.g., (Thomas and Morfill, 1996; Quinn et al., 1996). The mode is characterized by an optic-mode-like inverse dispersion (i.e., its frequency decreases with the growing wave number) if $kr_0 \ll 1$ where k is the wave number, r_0 is the interparticle distance, and only nearest-neighbour interactions are taken into account.

Consider vibrations of a one-dimensional horizontal chain of particulates of equal mass M separated by the distance r_0. We assume that the interaction potential between particles can be approximated by the Debye law

$$\Phi = \frac{Q}{r}\exp\left(-\frac{r}{\lambda_D}\right). \tag{70}$$

In addition to the electrostatic Debye shielded force, the gravitational force Mg and the sheath electrostatic force $F_e = QE(z)$ act on the dust grains in the vertical direction z. The balance of forces in the linear approximation with respect to small perturbations δz of the equilibrium at $z = 0$ gives the equation for vertical oscillations

$$M\frac{d^2\delta z_n}{dt^2} = \frac{Q^2}{r_0^3}\,\mathrm{e}^{-r_0/\lambda_D}(1 + r_0/\lambda_D)(2\delta z_n - \delta z_{n-1} - \delta z_{n+1}) - Mg + F_e. \tag{71}$$

Here

$$F_e - Mg = -\gamma\delta z_n, \tag{72}$$

where γ is a constant assuming linear variation of the sheath electric field, and δz_n is the vertical deviation of a particle number n from its equilibrium position. We note that although in general particles oscillate in the vertical as well as in horizontal direction, in the linear approximation their transverse vibrations (which are the subject of this study) and longitudinal vibrations are not coupled. Substituting $\delta z_n = A\exp(-i\omega t + iknr_o)$ into Equation (71) gives the dispersion of the vertical oscillations

$$\omega^2 = \frac{\gamma}{M} - \frac{4Q^2}{Mr_0^3}\,\mathrm{e}^{-r_0/\lambda_D}(1 + r_0/\lambda_D)\sin^2\frac{kr_0}{2}. \tag{73}$$

We see that for $k = 0$ the characteristic frequency is given by $\omega^2 = \gamma/M$, decreasing with growing wave number when $kr_0 \ll 1$.

To estimate the effective width of the potential well γ, we consider the standard model of the sheath (Vladimirov et al., 1997), which considers Boltzmann distributed electrons. For simplicity, we ignore the influence the dust grains may cause on the field distribution in the sheath region. The ion continuity equation gives the ion density n_i in terms of the density n_0 in the plasma bulk

$$n_i(z) = n_0 \left[1 - \frac{2e\phi(z)}{m_i v_0^2}\right]^{-1/2}, \tag{74}$$

where v_0 is the speed of the ion flow towards the negatively charged electrode, m_i is the ion mass, and $\phi(z)$ is the sheath potential. From Poisson's equation, we then obtain

$$\frac{d^2\phi(z)}{dz^2} = 4\pi e n_0 \left[\exp\left(\frac{e\phi(z)}{T_e}\right) - \left(1 - \frac{2e\phi(z)}{m_i v_0^2}\right)^{-1/2}\right]. \tag{75}$$

This equation can be integrated once to give [applying the boundary conditions $E(\infty) = \phi(\infty) = 0$]

$$E^2(z) = 8\pi n_0 T_e \left\{\exp\left(\frac{e\phi(z)}{T_e}\right) - 1 + \frac{v_0^2}{v_s^2}\left[\left(1 - \frac{2e\phi(z)}{T_e}\frac{v_s^2}{v_0^2}\right)^{1/2} - 1\right]\right\}, \tag{76}$$

where $v_s^2 = T_e/m_i$ is the squared sound speed.

Linearizing Equation (76) with respect to small potential and field variations, we find

$$\delta E \approx \frac{4\pi n_0 T_e}{E_0}\left[\exp\left(\frac{e\phi_0}{T_e}\right) - \left(1 - \frac{2e\phi_0}{T_e}\frac{v_s^2}{v_0^2}\right)^{-1/2}\right]\frac{e\delta\phi}{T_e}, \tag{77}$$

where the electric field E_0 is at $z = 0$, and for the dust grains it is balanced by the gravity

$$|QE_0| = Mg. \tag{78}$$

Now, we assume that the sheath electric field near the position of the dust grains can be linearly approximated. Thus we find

$$\delta E \approx -4\pi n_0 e \left[\exp\left(\frac{e\phi_0}{T_e}\right) - \left(1 - \frac{2e\phi_0}{T_e}\frac{v_s^2}{v_0^2}\right)^{-1/2}\right]\delta z, \tag{79}$$

and the effective width of the potential well is given by

$$\gamma = 4\pi e|Q|n_0 \left[\exp\left(\frac{e\phi_0}{T_e}\right) - \left(1 - \frac{2e\phi_0}{T_e}\frac{v_s^2}{v_0^2}\right)^{-1/2}\right]. \tag{80}$$

Equations (76) and (78) for the sheath potential ϕ_0 in the equilibrium position can be solved numerically. When $e\phi_0 \ll T_e$, the characteristic frequency is approximately given by

$$f_0 = \frac{1}{2\pi}\sqrt{\frac{\gamma}{M}} \approx \frac{1}{2\pi}\sqrt{\frac{g(1 - v_s^2/v_0^2)}{\lambda_D}} \simeq 20\ \mathrm{Hz}, \tag{81}$$

where we assumed $\lambda_D \approx (T_e/4\pi n_0 e^2)^{1/2} \sim 2 \times 10^{-2}$ cm and $v_0^2/v_s^2 \sim 1.5$.

Thus, the vertical oscillations of a one-dimensional chain of dust grains levitating in the sheath field of a horizontal negatively biased electrode can give rise to a specific low-frequency mode which is characterized by inverse optic-mode-like dispersion when the wave lengths far exceed the intergrain distance. Excitation of the mode may stimulate phase transitions in the system. We note that considering the one dimensional chain we ignored the ion drag force which is not important in this situation. However, for several layers of dust grains in the vertical direction, the ion drag force which in a plasma with finite flows necessarily includes such collective effects as generation of the wake fields studied in the preceding section must be taken into account.

Acknowledgment

This work has been supported by the Australian Research Council.

References

Allen, J.: 1992, *Phys. Scripta* **T45**, 497.
Benkadda, S., Tsytovich, V.N. and Verga, A.: 1995, *Comm. Plasma Phys. Contr. Fusion* **16**, 321.
Chiang, C. and I, L.: 1996, *Phys. Rev. Lett.* **77**, 647.
Chu, J.H. and I, L.: 1994, *Phys. Rev. Lett.* **72**, 4009.
Chu, J.H., Du, J.-B. and I, L.: 1994, *J. Phys. D.* **27**, 296.
Cramer, N.F. and Vladimirov, S.V.: 1996a, *Phys. Scripta* **53**, 586.
Cramer, N.F. and Vladimirov, S.V.: 1996b, *Phys. Plasmas* **3**, 4740.
Goertz, C.K.: 1989, *Rev. Geophys.* **27**, 271.
Havnes, O., Morfill, G.E. and Goertz, C.K.: 1984, *J. Geophys. Res.* **89**, 10999.
Hayashi, Y. and Tachibana, K.: 1994, *Jpn. J. Appl. Phys.* **33**, L804.
Ikezi, H.: 1986, *Phys. Fluids* **29**, 1764.
Ishihara, O. and Vladimirov, S.V.: 1997, *Phys. Plasmas* **4**, in press.
Melandsø, F.: 1996, *Phys. Plasmas* **3**, 3890.
Melandsø, F. and Goree, J.: 1995, *Phys. Rev. E* **52**, 5312.
Melandsø, F., Aslaksen, T.K. and Havnes, O.: 1993, *Planet. Space Sci.* **41**, 321.
Melzer, A., Trottenberg, T. and Piel, A.: 1994, *Phys. Lett. A* **191**, 301.
Melzer, A., Homann, A. and Piel, A.: 1996a, *Phys. Rev. E* **53**, 2757.
Melzer, A. et al.: 1996b, *Phys. Rev. E* **54**, R46.
Mendis, D.A. and Rosenberg, M.: 1994, *Ann. Rev. Astron. Astrophys.* **32**, 419.

Nambu, M. and Akama, H.: 1985, *Phys. Fluids* **28**, 2300.
Northrop, T.G.: 1992, *Phys. Scripta* **T45**, 475.
Peters, A. et al.: 1996, *Phys. Lett. A* **223**, 389.
Praburam, G. and Goree, J.: 1996, *Phys. Plasmas* **3**, 1212.
Quinn, R.A. et al.: 1996, *Phys. Rev. E* **53**, R2049.
Selwyn, G.S. et al.: 1989, *J. Vac. Technol. A* **7**, 2758.
Selwyn, G.S. et al.: 1990, *Appl. Phys. Lett.* **57**, 1876.
Spitzer, L., Jr.: 1978, *Physical Processes in the Interstellar Medium*, Wiley, New York.
Thomas, H.M. and Morfill, G.E.: 1996, *Nature* **379**, 806.
Thomas, H. et al.: 1994, *Phys. Rev. Lett.* **73**, 652.
Tsypin, V.S., Vladimirov, S.V., Elfimov, A.G., Tendler, M., de Assis, A.S. and de Azevedo, C.A.: 1997, *Phys. Plasmas*, in press.
Tsytovich, V.N.: 1992, *Phys. Scripta* **45**, 521.
Tsytovich V.N. and Havnes, O.: 1993, *Comm. Plasma Phys. Contr. Fusion* **15**, 267.
Vladimirov, S.V.: 1994a, *Phys. Plasmas* **1**, 2762.
Vladimirov, S.V.: 1994b, *Phys. Rev. E* **50**, 1422.
Vladimirov, S.V.: 1994c, *Phys. Rev. E* **49**, R997.
Vladimirov, S.V.: 1996, in: T. Hada and H. Matsumoto (eds.), *Nonlinear Waves and Chaos in Space Plasmas*, Terra Scientific, Tokyo, Chapter 7.
Vladimirov, S.V. and Cramer, N.F.: 1996, *Phys. Rev. E* **54**, 6762.
Vladimirov, S.V. and Ishihara, O.: 1996, *Phys. Plasmas* **3**, 444.
Vladimirov, S.V. and Nambu, M.: 1995, *Phys. Rev. E* **52**, R2172.
Vladimirov, S.V., Shevchenko, P.V. and Cramer, N.F.: 1997, *Phys. Rev. E*, in press.
Whipple, E.C., Northrop, T.G. and Mendis, D.A.: 1985, *J. Geophys. Res.* **90**, 7405.

SHOCK STRUCTURES IN PLASMAS CONTAINING VARIABLE-CHARGE MACRO PARTICLES

S.I. POPEL
Institute for Dynamics of Geospheres, Leninsky pr. 38, Bldg. 6, Moscow 117979, Russia

V.N. TSYTOVICH
General Physics Institute, Vavilova 38, Moscow 117942, Russia

M.Y. YU
Theoretical Physics I, Ruhr University Bochum, D-44780 Bochum, Germany

Abstract. The results on shock waves in plasmas containing variable-charge dust grains are reviewed. The problems where the shock waves in dusty plasmas can be important are formulated. The dusty plasma is presented as an open, highly dissipative system where there is a tendency to self-organization. An anomalous dissipation originating from the charging processes results in a possibility of existence of a new kind of shock waves. The shock waves related to this dissipation can be important for description of star formation, shocks in supernova explosions, particle acceleration in shocks, formation of dusty crystals, etc.

1. Introduction

The study of plasmas containing variable-charge macro particles (dust particles) is important (Whipple et al., 1985; Notni and Tiersch, 1987; Lipschultz, 1987; Goertz, 1989; Sheehan et al., 1990; Carlile et al., 1991; Jellum et al., 1991) for the understanding of many astrophysical phenomena, such as planetary rings, cometary tails, asteroids, and interstellar clouds, as well as the near-Earth environment, such as the ionosphere, noctilucent clouds and auroras. It is also relevant to many laboratory and technological plasmas, such as low-temperature RF and DC glow discharges, RF plasma etching and the wall region of hot plasmas, as well as to the understanding of many novel materials, such as some semi-conductors, optical fibers and dust crystals (Thomas et al., 1994; Chu and Lin, 1994).

The characteristic features of dusty plasmas are the following (Tsytovich, 1997):

(1) Dust particles can survive for a long time in a plasma, and even grow, over a wide range of electron temperatures – from several electronvolts up to several tens of electronvolts. This temperature interval corresponds, in most experiments, to low-temperature plasmas.

(2) The characteristic size of dust particles is of the order of 1 μm. Often it is possible to refer to particles with larger sizes as the dust ones if their size is less than

Astrophysics and Space Science **256:** 107–123, 1998.

the Debye length λ_D. For example, plasmas of planetary rings are characterized by $\lambda_D \sim$ 10–30 m and the distribution of particles over the sizes a is

$$n_d(a) = \frac{2a_0^2}{a^3} n_d, \quad n_d = \int_{a_0}^{\sim\lambda_D} n_d(a) da,$$

where n_d is the density of the dust particles.

(3) Dust particles acquire big charges during very short periods of time. The charging process can be due to field emission, ultraviolet irradiation, microscopic plasma currents flowing into the dust grains, etc.

(4) If the charging is due to the microscopic currents then the dust particles which appear in a plasma (by injection or in situ growth) soon acquire very large charge corresponding to the floating potential. Dust charge increases until the thermal electron flux on a dust particle becomes equivalent to the thermal ion flux. This means that the equilibrium charge of dust grains is *negative* and $Z_d e^2/a \approx \beta T_e$, where T_e is the electron temperature in energy units, $-e$ is the electron charge, a is the size of the dust particle, and Z_d is the charge of the dust particle in the units of electron charge. The numerical factor β appears in many circumstances to be close to 2 and depends on the plasma density logarithmically, on the ratio of electron to ion temperatures, as well as on certain other plasma parameters. For micron-sized dust particles, electron temperature $T_e \approx 3$ eV, and, for hydrogen plasmas, $Z_d \approx 10^4$.

(5) The charges of dust grains are *variable* because the magnitude of the charge depends on plasma parameters and their perturbations.

(6) Plasma-dust systems are *always* open systems, as the plasma particle fluxes recombining on the dust particles, as well as the energy fluxes absorbed by the dust particles, need to be supported by external source of particles and energy. The rate of dissipation is high with a tendency for self-organization and for the formation of long lived localized and dissipative structures.

(7) Real processes on a dust grain surface are complicated. In addition to the absorption of plasma particles (electrons and ions), reflection of the plasma particles, surface recombination, secondary emission of electrons, etc. can be important. The approximation when the ions and electrons entering the dust grain recombine into neutral atoms, which then re-enter the plasma and re-ionize, thus preserving the number of ions and electrons, can often be used.

Self-organization on the macroscopic level in dusty plasmas can result in formation of macroscopic structures (dusty crystals, dust clouds, etc.). It can also make a contribution to the process of formation of planetary rings. The parameter which determines the rate of possible self-organization process on the macroscopic level is the rate of dissipation, as in open systems it characterizes the ability of the system to form self-organized structures. The rate of dissipation, $\Sigma_{\rm pl}$, in the presence of volume ionization, can, to a certain approximation, be written as the ratio of the dissipated energy to the energy per unit volume (Tsytovich, 1997):

$$\Sigma_{\rm pl} = \frac{\psi^E n_d}{E} \approx \omega_{pi} P R \frac{a}{\lambda_{Di}} f(\tau, z), \quad P = \frac{n_d Z_d}{n_e}, \quad \tau = \frac{T_i}{T_e}, \quad z = \frac{Z_d e^2}{a T_e},$$

$$R = \frac{n_i T_i}{n_n T_n + n_e T_e + n_i T_i + n_d T_d}, \quad f(\tau, z) = \frac{z^2 + 2z + 2\tau z + 2\tau + 2\tau^2}{z\tau},$$

ω_{pi} is the ion plasma frequency, λ_{Di} is the ion Debye length; the subscripts e, i, n, and d in the designations for the density n and the temperature T are related to electrons, ions, neutrals, and dust particles, respectively. The value of $\Sigma_{\rm pl}^{-1}$ determines the characteristic time $T_{\rm so}$ of the development of self-organization. Estimates give $T_{\rm so} \sim 10^8$ years in interstellar clouds, $T_{\rm so} \sim 10^4$ s in planetary rings, and $T_{\rm so} \sim 10^2$ s in typical laboratory experiments on dusty crystal formation and on plasma etching.

Here we consider self-organization on microscopic level. In this case self-organization is related to the development of nonlinear plasma processes which can result in formation of nonlinear (coherent and dissipative) structures, such as solitons, collapsing cavities, shock waves, etc. (Popel and Yu, 1994; Popel and Yu, 1995; Popel et al., 1996; Popel et al., 1997). We are interested in the problem of shocks in dusty plasmas. Although this problem is of great importance, its investigation is still in the initial stage. Nevertheless, the basic results, which demonstrate the new physics in formation of the shocks in dusty plasma, are obtained (Popel et al., 1996; Popel et al., 1997; Tsytovich and Popel, 1997) and we review here these results.

Shock waves often arise in nature because of a balance between wave breaking nonlinear and wave damping dissipative forces (Zel'dovich and Raizer, 1967). Collisional and collisionless shock waves can appear because of friction between the particles and wave-particle interaction (Sagdeev, 1966; Chen, 1974), respectively. The presence in the plasma of the macro particles with variable charges results in the new mechanism of dissipation. An anomalous dissipation appears which originates from the charging processes. This dissipation can, in particular, give rise to shock waves. Here we demonstrate the possibility of existence of such shock waves (Popel et al., 1996; Popel et al., 1997; Tsytovich and Popel, 1997). They are collisionless in the sense that they do not involve electron-ion collisions. However, in contrast to the classical collisionless shock waves, the dissipation due to dust charging involves interaction of the electrons and ions with the dust grains in the form of microscopic grain currents.

Here we discuss also the ways of description of shock waves in space dusty plasmas (Tsytovich and Popel, 1997). Their importance is associated with different astrophysical applications (Spitzer, 1978). For example, according to modern concepts (Kaplan and Pikel'ner, 1982; Kaplan and Pikel'ner, 1974), the formation of stars occurs mainly in interstellar dust-molecular clouds after compression shock waves propagate through them, creating the initial density condensations for further gravitational contraction. The presence of the dust in the interstellar clouds can influence significantly even the magnitude of the sound velocity, not to mention the shock wave propagation. Moreover, the processes like dust agglomeration and the neutral initiated attraction of the dust in the interstellar clouds are possible (Tsytovich, 1997). They can change the threshold of self-contraction presented in (Kaplan and Pikel'ner, 1982; Kaplan and Pikel'ner, 1974). Several other examples of astrophysical situations where shocks in dusty plasmas can be important are given below, in the section, where the problems of shock waves in dusty plasmas are formulated. We note also that the existence of shock waves can be important to laboratory dusty plasmas. In particular, shock waves have been observed in the experiment which had a purpose of plasma-dust crystal formation and was carried out by the group of G.E. Morfill in the state of artificial weightlessness (Morfill, 1996).

2. The Exact Solutions

Dusty plasma is usually a complicated system where it is rather difficult to carry out clear analytical investigation of nonlinear processes. Nevertheless, there is a situation when one can show analytically (Popel et al., 1996; Popel et al., 1997) the new characteristic features in shock wave propagation. This situation corresponds to the following assumptions:

(1) The plasma can be considered as the uniform unmagnetized collisionless one. Its constituents are electrons, ions, and dust particles.

(2) The time scale corresponds to ion acoustic wave propagation.

(3) The dust-grain charge variation is solely due to the microscopic electron and ion grain currents originating from the potential difference between the plasma and the grain surface.

(4) The dust particles are massive ($m_i Z_d \ll m_d$, where $m_{i,d}$ are the ion and dust masses). Then the dust can be considered as stationary and the density n_d is constant in the ion acoustic time scale.

(5) In the absence of perturbations the quasineutrality condition $n_{i0} = n_{e0} + Z_d n_d$, where the subscript 0 denotes unperturbed quantities, holds.

(6) The orbit-limited probe model (Chen, 1965; Barnes et al., 1992) is valid.

(7) The ions and electrons entering the dust grain recombine into neutral atoms, which then re-enter the plasma and re-ionize, thus preserving the number of ions and electrons.

(8) Nonlinear waves propagate along the axis x.

According to the orbit-limited probe model, we have for equilibrium electrons and kinetic ions (Barnes et al., 1992; Tsytovich and Havnes, 1993; Whipple, 1981; Goree, 1994):

$$I_e \approx -\pi a^2 e \left(\frac{8T_e}{\pi m_e}\right)^{1/2} n_e \exp\left(\frac{eq_d}{aT_e}\right), \tag{1}$$

and

$$I_i = \sqrt{\frac{\pi}{2}} a^2 v_{Ti} e n_i \left[2 \exp\left(-\frac{v_i^2}{2v_{Ti}^2} \right) + \sqrt{2\pi} \frac{v_{Ti}}{v_i} \left(1 + \frac{v_i^2}{v_{Ti}^2} - \frac{2eq_d}{am_i v_{Ti}^2} \right) \operatorname{erf}\left(\frac{v_i}{\sqrt{2} v_{Ti}} \right) \right], \tag{2}$$

where $q_d = -Z_d e$ is the dust grain charge, m_e is the electron mass, v_{Tj} is the thermal speed of the species j $(= i, e)$, $v_j \mathbf{e}_x$ is the fluid velocity of the species j, and $\operatorname{erf}(x)$ is the error function. Here, the complete formula (Whipple, 1981) for I_i is used in order to account for the slow ions, which should be included for physical consistency in the low φ region of the shock, where φ is the electrostatic potential. Indeed, the cold ion limit (when the ion temperature $T_i = 0$) used for the ion acoustic waves cannot be applied in considering the charging process. In fact, for small v_j/v_{Tj} the unperturbed charge of the dust grain as determined from the current balance condition $I_e + I_i = 0$ is given (Tsytovich and Havnes, 1993) by $\omega_{pe}^2 \exp(-z_0)/v_{Te} = \omega_{pi}^2 (T_i/T_e + z_0)/v_{Ti}$, where $z_0 \equiv Z_{d0} e^2/aT_e$ is the dimensionless dust particle charge and ω_{pe} is the electron plasma frequency. Clearly, there is no solution for $T_j \to 0$.

The average charge of the stationary dusts is governed by the equation (Tsytovich and Havnes, 1993; Vladimirov, 1994; Ma and Yu, 1994; Popeland and Yu, 1994, 1995)

$$\partial_t q_d = I_e(q_d) + I_i(q_d), \tag{3}$$

for charge conservation.

We consider a quasi-stationary structure moving with a speed V in the x direction, satisfying the conditions $v_{Ti} < V \ll v_{Te}$ and $T_i \ll T_e$ for ion acoustic wave propagation. Thus, all parameters depend only on $\xi = x - Vt$. The electron density is taken to be Boltzmann $[n_e = n_{e0} \exp(e\varphi/T_e)]$ with constant T_e. We shall also assume that in the moving frame $v_i \gg v_{Ti}$, which imposes a lower limit on the magnitude of the electrostatic potential φ of the nonlinear waves. From the ion conservation equations, one obtains

$$n_i = M n_{i0} \left(M^2 - 2\varphi \right)^{-1/2}, \tag{4}$$

$$v_i = c_s (M - \sqrt{M^2 - 2\varphi}), \tag{5}$$

where we have used the normalization $e\varphi/T_e \Longrightarrow \varphi$, $V/c_s \Longrightarrow M$, and $\xi/\lambda_{De} \Longrightarrow \xi$. Here, $c_s = (T_e/m_i)^{1/2}$ is the ion acoustic speed, λ_{De} is the electron Debye length.

From the Poisson equation we can obtain

$$d_\xi^2 \varphi = \exp(\varphi) + \left(1 + \frac{\delta z}{z_0} \right) Z_{d0} d - \frac{M(1 + Z_{d0} d)}{\sqrt{M^2 - 2\varphi}}, \tag{6}$$

where $d = n_{d0}/n_{e0}$. The normalized charge perturbation $\delta z = -e\delta q_d/aT_e$ is governed by

$$d_\xi \delta z = j \equiv \frac{A}{\sqrt{M^2 - 2\varphi}} \times \left[\sqrt{\frac{2}{\pi}} \exp\left(-\gamma^2\right)\left(\frac{1+2\gamma^2}{\sqrt{2}\gamma} + \frac{\sqrt{2}(z_0 + \delta z)T_e}{\gamma T_i}\right)\operatorname{erf}(\gamma)\right] - \sqrt{\frac{8}{\pi}}\frac{A}{M}\left(1 + z_0\frac{T_e}{T_i}\right)\exp\left(\varphi - \delta z\right), \tag{7}$$

where $A = a[(1 + Z_{d0}d)/4\lambda_{De}](T_i/T_e)^{1/2}$, $\gamma = [M - (M^2 - 2\varphi)^{1/2}](T_e/2T_i)^{1/2}$, and we have used Equations (1–3). Note that j is the normalized perturbed total current density. Clearly, all solutions must satisfy $\varphi \leq M^2/2$. In fact, it can be shown that for $v_i \gg v_{Ti}$ (or $|\varphi| \gg M\sqrt{T_i/T_e}$), the dust-charge variation is always important.

Equations (6) and (7) can be numerically integrated provided that the existence conditions for shock solutions are known. For the latter to exist, there must be two distinct asymptotic values for φ (say at $\xi \to \pm\infty$), and the derivatives of the perturbed quantities must vanish there. In this limit, we obtain from Equations (6) and (7)

$$\delta z = f_1(\varphi) \equiv -z_0 + \frac{z_0}{Z_{d0}d}\left[\frac{M(1 + Z_{d0}d)}{\sqrt{M^2 - 2\varphi}} - \exp(\varphi)\right], \tag{8}$$

as well as the current balance condition $j = 0$.

The solution of $j = 0$ can be implicitly written as $\delta z = f_2(\varphi)$. It has the properties that for $\varphi \ll 1$, $f_2(\varphi) \approx \varphi\left(1 - 1/M^2\right)/\left[1 + (z_0 + T_i/T_e)^{-1}\right]$, and for $\varphi \geq 1$, $f_2(\varphi)$ is approximated by

$$\delta z \approx \varphi - \ln\left[\sqrt{\frac{\pi T_e}{8T_i}}\frac{M\beta\gamma_0}{\sqrt{M^2 - 2\varphi}}\left(1 + 2\frac{z_0 + \varphi}{\gamma_0^2}\right)\right], \tag{9}$$

where $\beta = (1 + T_e z_0/T_i)^{-1}$ and $\gamma_0 = M - (M^2 - 2\varphi)^{1/2}$. Near $\varphi = M^2/2$, f_2 is a decreasing function of φ with the minimum value $f_2(\varphi = M^2/2) = -z_0 - M^2/2$. Furthermore, if

$$M^2 > M_0^2 \equiv \left(1 + \frac{1 + Z_{d0}d}{Z_{d0}d}z_0 G\right)\left(\frac{z_0 G}{1 + Z_{d0}d}\right)^{-1}, \tag{10}$$

where $G = (1 + z_0 T_e/T_i)/[1 + (z_0 + 1)T_e/T_i]$, then the equation $f_1(\varphi) = f_2(\varphi)$ (so that Equation (8) and $j = 0$ are satisfied simultaneously) has the solutions $\varphi = 0$ and $\varphi = \varphi_A$. We note also that M_0 satisfies $1 < M_0^2 < 1 + Z_{d0}d$. As an example, for the typical parameters $Z_{d0}d = 2$, $T_e/T_i = 10$, $a/\lambda_D = 0.01$, and $M = 1.7$,

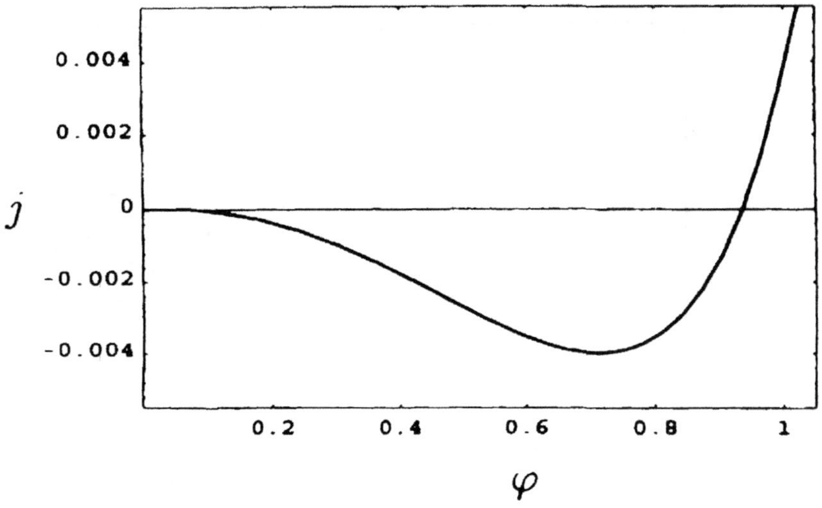

Figure 1. The dependence of j on φ for $Z_{d0}d = 2$, $T_e/T_i = 10$, $a/\lambda_D = 0.01$, and $M = 1.7$. The solutions of the equation $j = 0$ are $\varphi = 0$ and $\varphi = \varphi_A \approx 0.935$.

one finds $z_0 \approx 1.23$ and $\varphi_A \approx 0.935$. On the other hand, if inequality (10) is not satisfied, the only solution of $f_1(\varphi) = f_2(\varphi)$ is $\varphi = 0$, so that a shock wave solution is not possible. Thus, the inequality (10) is necessary for the existence of two asymptotic solutions, namely $\varphi = 0$ and $\varphi = \varphi_A$ (say at $\xi \to +\infty$ and $-\infty$, respectively).

We note that $\varphi = 0$ and $\varphi = \varphi_A$ are also exact solutions of Equations (6) and (7), since the latter differential equations are homogeneous. Furthermore, both solutions correspond to $\delta E \equiv -\nabla\varphi = -d_\xi\varphi = 0$. A solution bridging these two asymptotic ones can be obtained by making the Ansatz that the second derivative in Equation (6) is negligible. In this case we obtain from Equations (6) and (8)

$$d_\varphi \xi = \frac{1}{j} d_\varphi \delta z = \frac{z_0}{Z_{d0} j d} \left[\frac{(1 + Z_{d0}d)M}{(M^2 - 2\varphi)^{3/2}} - \exp(\varphi) \right], \tag{11}$$

which can be integrated. The zeros of the denominator on the right-hand side of Equation (11) correspond to $j = 0$, or the asymptotic values of $\varphi(\xi)$. Within the range $0 < \varphi < \varphi_A$ the function j is negative (see Figure 1).

A solution for φ can now be obtained. In Figure 2 the functions $f_1(\varphi)$ and $f_2(\varphi)$ for (a) $M^2 \leq 1 + Z_{d0}d$, and (b) $M^2 > 1 + Z_{d0}d$, are presented schematically. For case (a), the minimum value (for $\varphi \geq 0$) of $f_1(\varphi)$ is attained (see Figure 2a) at $\varphi = 0$, while for case (b), it is reached (see Figure 2b) at $0 < \varphi = \varphi_1 < \varphi_A$. Thus for the case (a) the factor in the square brackets on the right-hand side of Equation (11) is positive for $0 < \varphi < \varphi_A$, so that $d_\varphi \xi \leq 0$. This means that $\delta z = \delta z_A$ (corresponding to $\varphi = \varphi_A$) is positive. In fact, for the parameters in our example we obtain $\delta z_A \approx 0.31$. It follows that for case (a) there exist solutions of Equation (11) in the form of a decreasing function with $\varphi \to \varphi_A > 0$ for

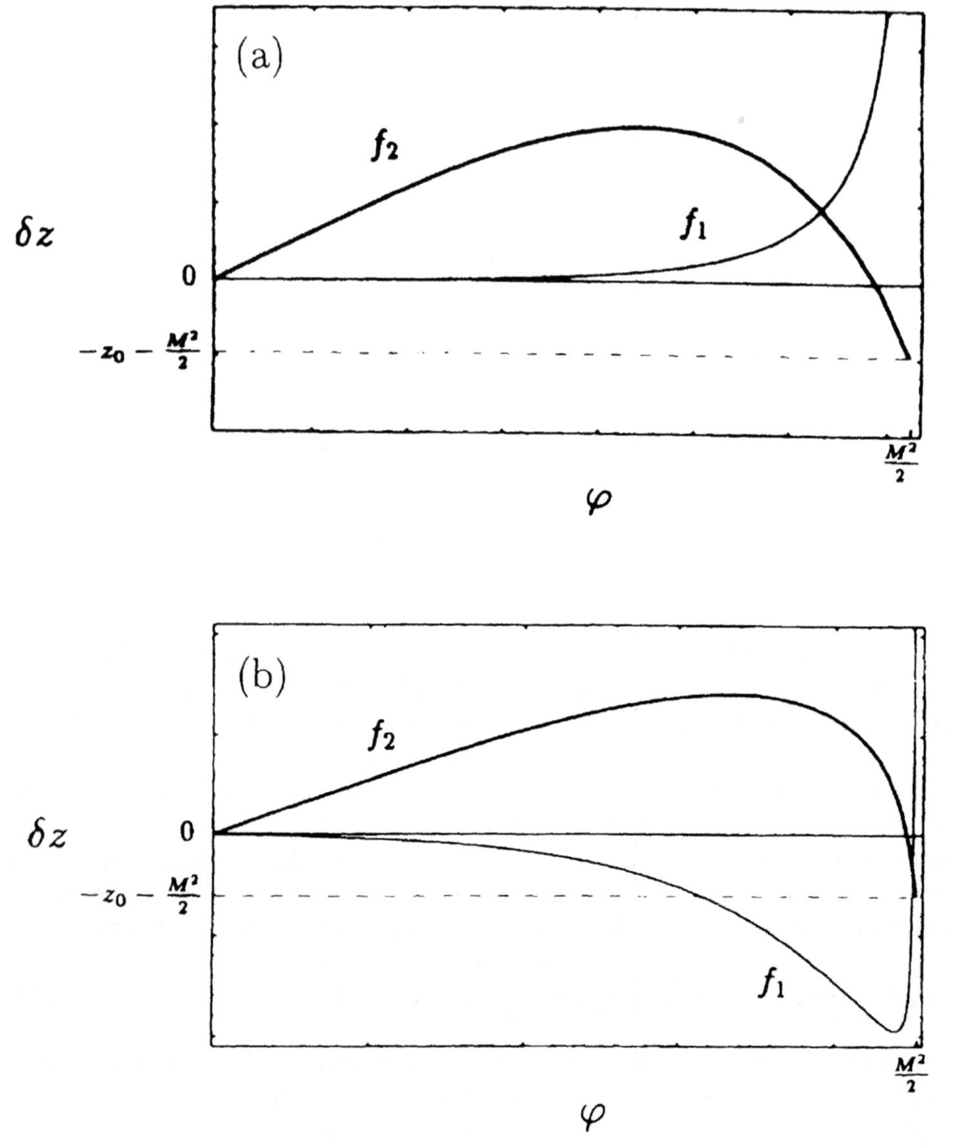

Figure 2. Schematic representations of the functions $f_1(\varphi)$ and $f_2(\varphi)$ for (a) $M^2 \leq 1 + Z_{d0}d$, and (b) $M^2 > 1 + Z_{d0}d$, with the inequality (9) satisfied. The dashed lines represent $\delta z = -z_0 - M^2/2$.

$\xi \to -\infty$, and $\varphi \to 0$ for $\xi \to +\infty$. In Figure 3 the profile of the potential $\varphi(\xi)$ satisfying Equation (11) is shown. A similar profile for the normalized charge perturbation δz can be deduced from Equation (7) by recalling that j is always negative in the region of interest. On the other hand, for case (b) we have $d_\varphi \xi < 0$ for $0 \leq \varphi < \varphi_1$, and $d_\varphi \xi > 0$ for $\varphi_1 < \varphi \leq \varphi_A$. Furthermore, $\xi \to +\infty$ for both

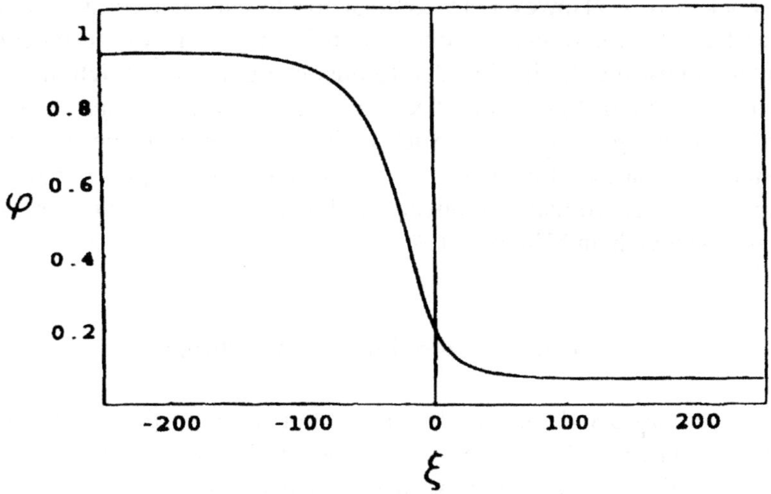

Figure 3. The $\varphi(\xi)$ profile for the nonlinear wave structure. The parameters are the same as those of Figure 1.

$\varphi \to 0$ and $\varphi \to \varphi_A$. Thus, there are no shock-like solutions of Equation (11) in this case since $\varphi(\xi)$ is determined only for $\xi > \xi_0$, where ξ_0 is a finite constant.

In the analysis we have neglected the second derivative d_ξ^2 in Equation (6), equivalent to assuming quasineutrality in the wave dynamics. For waves with amplitude φ_A near or exceeding unity, this Ansatz is valid if the width of the shock front $\Delta\xi$ greatly exceeds unity. For the shock wave in Figure 3, we have $\Delta\xi \sim 100 \gg 1$, so that the Ansatz is well justified. Indeed, one can formally express the applicability of the quasineutrality condition in terms of the plasma parameters and the wave amplitude as $4\lambda_{De}/a \gg \exp[(z_0/Z_{d0}d)\exp(\varphi_A)]$, which is satisfied for sufficiently intense [but $\varphi_A < (1 + Z_{d0}d)/2$] waves. The analysis of very weak waves ($\varphi_A/M^2 \leq a/\lambda_{De}$) would require the consideration of all terms in Equation (6).

Thus, electrostatic shock solutions with potentials of the order of T_e/e can exist in a uniform dusty plasma with Boltzmann electrons, inertial ions, and immobile but *variable charge* dust grains. The shock speed (Mach number) must satisfy both Equation (10) and the condition (a). For most dusty plasmas, where $n_{e0} \approx n_d$, this speed is near the ion sound speed $(T_e/m_i)^{1/2}$. In principle, such shocks could be easily generated and/or detected. The shock amplitude φ_A is an increasing function of M. The maximum value of the amplitude corresponds to $M^2 = 1 + Z_{d0}d$, and is given by $\varphi_A|_{\max} \sim (1 + Z_{d0}d)/2$. The increase of the charging frequency (Popel et al., 1997), which characterizes the dissipative process arising from the charge variation of the dust particles, results in a decrease of the shock front width.

The shock wave discussed here is collisionless in the sense that it does not involve electron-ion collisions (the source of dissipation for the collisional ion acoustic shock waves). However, in contrast to the classical collisionless shock waves (Sagdeev, 1966), where the dissipation is only from the turbulent wave-particle interactions, the dissipation due to dust charging (Tsytovich and Havnes, 1993) involve the interaction of the electrons and ions with the dust grains in the form of microscopic grain currents. Such a situation can occur because for most dusty plasmas the charging frequency $\nu_q = \omega_{pi}^2 a(1 + z_0 + T_i/T_e)/\sqrt{2\pi} v_{Ti}$ is much larger than the ion-ion collision frequency and also exceeds the electron collision frequencies (Tsytovich and Havnes, 1993).

3. Shock Waves in Space Dusty Plasmas

Above we have considered the simplified situation when in the plasma there are only electrons, ions, and dust grains. In astrophysical situations the dusty plasma contains dust particles, electrons, ions, and *neutrals*. Interaction of dust particles with ions and neutrals results in the possibility of the so-called bombardment force (Tsytovich et al., 1996) resulting in attraction between the dust particles which, in turn, means the new, more complicated physics of the shock waves. It is the presence of neutrals in dusty plasmas that should be taken into account in consideration of the problem of star formation. In this section, following the paper (Tsytovich and Popel, 1997), we present the way how one can develop the hydrodynamics for description of stationary shock waves in a dusty plasma containing electrons, ions, neutrals, and dust grains.

The shocks in stationary media can be described by conservation laws for particle momentum and energy fluxes through the surface of the shock. In a dusty plasma the situation is more complicated because the dusty plasma is *an open* system. Therefore, we restrict our consideration only to the case when at the front of the shock wave the dusty plasma is stationary and not moving. This means that there should be some external ionization which supports the density of the plasma particles (in the absence of the source of electrons and ions the plasma is recombined very rapidly on dust (Tsytovich, 1997)). The ionization can be produced by electromagnetic radiation of burning stars, by cosmic rays, by electron beams, etc. We will not determine the conditions behind the shock front, because they should appear as a result of the disturbances produced by the shock. However, we assume that behind the shock front there is a source of the plasma particles with the same rate of particle generation as at the front. The dissipation which gives rise to shock waves is again related to the plasma recombination on the grains, i.e. to the dust charging process. The momentum, the energy, and the number of electrons, ions, and neutrals are not conserved at the shock front, but the total momentum, the total mass density, and the total energy of the particles (electrons, ions, neutrals, dust grains) are supposed to be conserved. Unfortunately, the latter restricts significantly the possibility for the finding the shock front structure. Indeed, the characteristic width of the front is determined by the process of electron and ion absorption by the dust grains. Thus the investigation of the shock front structure requires, in particular, solution of the momentum and energy equations for electrons and ions, and dust particles, separately.

The basic equations for the investigation of the shock waves in the dusty plasma are modified as compared to the case of the conventional gas dynamics. We assume the one-dimensional propagation of the shock waves. We use the continuity equation for the ion mass density where the source of ions q_i is introduced:

$$\frac{\partial(n_i m_i)}{\partial t} + \frac{\partial(n_i u_i m_i)}{\partial x} = -\bar{\nu}_i n_i m_i + q_i. \tag{12}$$

Here $\bar{\nu}_i$ is the effective frequency (Tsytovich, 1997) of the disappearance (recombination) of ions on dust particles, u_i is the directed velocity of ions.

Usually the mass density of electrons is much less than that of ions, and can be neglected. The recombination of the ions on dust particles conserves the mass density, since the attachment of the ions to the dust particles increases the mass of the latter. The continuity equation for the dust particle mass density can be written in the form corresponding to the conservation of the total dust and ion mass densities in the absence of the source q_i:

$$\frac{\partial(n_d m_d)}{\partial t} + \frac{\partial(n_d u_d m_d)}{\partial x} = \bar{\nu}_i n_i m_i. \tag{13}$$

The source q_i in Equation (12) can be expressed in terms of the frequency of unperturbed quantities (characterized by the subscript 0) before the shock front because we assume that it compensates the losses of electrons and ions due to the recombination on dust particles:

$$q_i = \bar{\nu}_{i,0} n_{i,0} m_i. \tag{14}$$

The value $\bar{\nu}_i$ is a function of n_d, n_i, Z_d, and, in general case, of the relative velocity between the ions and the dust. The latter in the region before the front of the shock should be equal to zero because of its relaxation due to the ion-dust collisions. Thus $\bar{\nu}_{i,0}$ depends on $n_{d,0}$, $n_{i,0}$, z_0 (where z_0 is the dimensionless dust particle charge before the shock front).

Equations (12) and (13) allow us to find the equation describing the evolution of the total mass density:

$$\frac{\partial(n_d m_d + n_i m_i)}{\partial t} + \frac{\partial(n_d m_d u_d + n_i m_i u_i)}{\partial x} = \bar{\nu}_{i,0} n_{i,0} m_i. \tag{15}$$

The change in dust masses is a very slow process, so that during the time of particle passage through the shock front structure (in the frame of reference of the shock) one can neglect the dust particle mass increase, i.e. the term on the right-hand side of Equation (15) can be omitted. Thus one can obtain the mass density conservation law in the following form:

$$(n_{d,1} m_d + n_{i,1} m_i) u_1 = (n_{d,0} m_d + n_{i,0} m_i) u_0, \tag{16}$$

where the subscript 1 denotes the quantities behind the front.

The difference between the similar relation in the conventional gas dynamics and Equation (16) is that the latter contains the mass densities of both the ions and the dust. We have taken into account in Equation (16) that due to the relaxation process the dust and the ions have the same velocities in the regions behind the shock front and at the shock front. We note that in dust-molecular interstellar clouds the mass densities of ions and dust are comparable usually. Furthermore, there are the situations when the value $n_d m_d$ is approximately one order of magnitude larger than $n_i m_i$. These estimates can be obtained using the accepted data of dust-molecular interstellar clouds (Tsytovich, 1997). However, these estimates can be modified because in reality there are some uncertainties in the data. On the other hand, the dust clouds can accumulate the ions. Consequently, the ion mass density in the cloud can be larger than in the surrounding space.

From the condition of stationarity of the plasma in the presence of the source, one can find from Equation (12) and the condition of the absence of the spatial dependence of the ion mass density both behind and before the shock front:

$$\bar{\nu}_{i,1} n_{i,1} = \bar{\nu}_{i,0} n_{i,0}. \tag{17}$$

This equation is inherent in *only* the gas dynamics of the *dusty* plasma.

We note that an appearance of an electron due to the ionization of a neutral means automatically an appearance of an ion. Thus the quasineutrality condition should be fulfilled behind and before the shock front. This implies that the validity of the steady-state condition (14) for ions results in the validity of the steady-state condition for electrons. This means that the relation for electrons analogous to Equation (17) (where the subscripts i should be substituted by the subscripts e and $\bar{\nu}_e$ is the effective frequency (Tsytovich, 1997) of the recombination of electrons on dust particles) should be fulfilled if Equation (17) is valid.

From Equation (12) we find in the frame of reference of the shock:

$$\frac{\partial (n_i u_i)}{\partial x} = -(\bar{\nu}_i n_i - \bar{\nu}_{i,0} n_{i,0}). \tag{18}$$

The similar equation is valid for electrons with a substitution in (18) of the subscript e for that i. Although for the equilibrium values we have $\bar{\nu}_{i,0} n_{i,0} = \bar{\nu}_{e,0} n_{e,0}$, the corresponding nonequilibrium values are not equal to each other. Thus subtracting from Equation (18) the analogous equation for electrons one can obtain the equation expressing the fact of charge conservation in the system:

$$\frac{\partial Z_d}{\partial t} + \frac{\partial (Z_d u_d)}{\partial x} = \bar{\nu}_e n_e - \bar{\nu}_i n_i. \tag{19}$$

In the frame of reference associated with the shock front the first term of Equation (19) should be omitted.

As for the neutrals, here we neglect the binary particle collisions between the neutrals and electrons as well as the neutrals and ions. We note that usually the neutrals are attached to the dust particles for a finite residence time. Thereafter they leave the dust grains acquiring the dust surface temperature. Thus their number is conserved

$$\frac{\partial n_n}{\partial t} + \frac{\partial (u_n n_n)}{\partial x} = 0, \tag{20}$$

but their temperature becomes close to the dust temperature. The latter condition is used below in the equations for the momentum conservation.

The conservation equation for the ion momentum at the front of the shock is:

$$n_i \frac{\partial (m_i u_i)}{\partial t} + n_i m_i u_i \frac{\partial u_i}{\partial x} + \frac{\partial (n_i T_i)}{\partial x} = -e n_i \frac{\partial \varphi}{\partial x} - \tilde{\nu}_i m_i (u_i - u_d) n_i, \tag{21}$$

where the momentum exchange between the ions and the neutrals is not taken into account; φ is again the electrostatic potential, $\tilde{\nu}_i$ is given by expression (3.68) in (Tsytovich, 1997). The similar equation (with $\tilde{\nu}_e$ determined by (3.67) in (Tsytovich, 1997) can be written for electrons.

Assuming that the temperature of the neutrals is equal to the dust temperature T_d, one obtains for the neutrals:

$$n_n \frac{\partial (m_n u_n)}{\partial t} + n_n m_n u_n \frac{\partial u_n}{\partial x} + \frac{\partial (n_n T_d)}{\partial x} = -\nu_n m_n (u_n - u_d) n_n, \tag{22}$$

where $\nu_n m_n (u_n - u_d) n_n$ is the force of friction between neutrals and dust (Tsytovich, 1997).

For the dust particles one has to take into account the momentum transfer in collisions with ions and neutrals (the momentum transfer in collisions with electrons is negligibly small):

$$n_d \frac{\partial (m_d u_d)}{\partial t} + n_d m_d u_d \frac{\partial u_d}{\partial x} + \frac{\partial (n_d T_d)}{\partial x}$$
$$= Z_d n_d \frac{\partial \varphi}{\partial x} - \tilde{\nu}_i m_i (u_d - u_i) n_i - \tilde{\nu}_e m_e (u_d - u_e) n_i - \nu_n m_n (u_d - u_n) n_n. \tag{23}$$

Equations (21–23) are satisfied in the regions before the shock front and behind it because of the absence of the spatial variations, stationarity of all parameters, and equality of the dust and ion velocities. Using Equations (15), (21), (22), and the quasineutrality condition, one can find the following relationship for the balance of the energy density:

$$(m_i n_{i,0} + m_d n_{d,0} + m_n n_{n,0}) \frac{u_0^2}{2} + n_{i,0} T_i + n_{e,0} T_{e,0} + (n_{d,0} + n_{n,0}) T_{d,0}$$
$$= (m_i n_{i,1} + m_d n_{d,1} + m_n n_{n,1}) \frac{u_1^2}{2}$$
$$+ n_{i,1} T_{i,1} + n_{e,1} T_{e,1} + (n_{d,1} + n_{n,0}) T_{d,1}, \tag{24}$$

which is different from the corresponding one in the conventional gas dynamics by the presence of the terms related to the dust.

Since the system is the open one, there exists the enegry absorbed by the dust grains which should be in a balance with plasma heating and cooling to provide the conditions for the existence of a stationary shock wave. In reality, such a balance can be established because the ionization source heats the plasma usually and there also exists plasma cooling due to the thermal radiation of the dust particles. Below the external heating sources are denoted by q_i^E, q_e^E, and q_d^E. The energy transfer to electrons and ions by the external sources q_e^E and q_i^E, respectively, under the stationary conditions is equal to the energy transfer to dust particles. The Coulomb collisions of electrons and ions with dust particles are elastic and do not contribute to the energy transfer to dust particles. Thus only the charging collisions contribute to the energy transfer to them. Let us denote the rates of the energy transfer to the dust particles (contributed by electrons and ions, respectively) as $\hat{\nu}_e n_e T_e$ and $\hat{\nu}_i n_i T_i$, where $\hat{\nu}_e$ and $\hat{\nu}_i$ are the frequencies of the transfer (determined by the charging frequency ν_q). The stationarity conditions before and behind the shock front can be expressed in the following form:

$$\hat{\nu}_{e,0(1)} n_{e,0(1)} T_{e,0(1)} = q_e^E, \quad \hat{\nu}_{i,0(1)} n_{i,0(1)} T_{i,0(1)} = q_i^E. \tag{25}$$

Thus, one can relate the magnitudes of the ion (electron) temperatures behind and before the shock front:

$$\hat{\nu}_{i(e),0} n_{i(e),0} T_{i(e),0} = \hat{\nu}_{i(e),1} n_{i(e),1} T_{i(e),1}. \tag{26}$$

Furthermore, the stationarity means the equality of the heating and cooling processes. In space plasmas the cooling process is usually due to the heat radiation of the dust grains. In accordance with Stefan–Boltzmann's law of radiation, the rate of radiation losses in optically thin plasmas is proportional to $T_d^4 n_d$. Thus we have:

$$n_{d,0} T_{d,0}^4 = n_{d,1} T_{d,1}^4. \tag{27}$$

We emphasize that Equations (26) and (27) are valid in the case when the sources of heating do note depend on plasma parameters including the dust density and the dust temperature. Below we assume the validity of this assumption.

If the radiation is dominated in the energy transfer then the equation

$$n_d T_d^4 = n_{d,0} T_{d,0}^4 \tag{28}$$

can be used instead of the energy flux conservation equation even for the parameters of the dust in the vicinity of the shock front.

The equations relating the parameters before and behind the shock front can be used for definition of the parameters behind the shock front if those before the shock front are known. The generalized Hugoniot relations for the shock waves in dusty plasmas are obtained in (Tsytovich and Popel, 1997) under the following assumptions:

(1) Equality of the heating and cooling processes in the shock.

(2) The radiation is dominated in the energy transfer. Equation (28) can be used instead of the energy flux conservation equation.
(3) The dust pressure is small as compared to the neutral, electron and ion pressures.
(4) The dust and the neutrals dominate in the mass density.
(5) The temperatures of the neutrals and the dust particles are equal to each other.

These generalized Hugoniot relations are much more complicated than the Hugoniot relations appeared in the conventional gas dynamics. The new relationships appear. For example, in order to find the relationship between the dust charge behind the shock front and the Mach number of the shock wave it is necessary to solve a set of equations (including the transcendential ones). This problem can be solved only numerically.

Here we mention only the width of the shock front for the case when the Mach number of the shock wave $M \gg 1$. In the case considered in this section the width is determined (Tsytovich and Popel, 1997) by the mean free path of electrons and ions in collisions with the dust:

$$\Delta\xi \sim \lambda_{D0}^2 n_{e0}/a n_{d0} Z_{d0}, \tag{29}$$

where

$$\frac{1}{\lambda_D^2} = \frac{1}{\lambda_{De}^2} + \frac{1}{\lambda_{Di}^2} \,. \tag{30}$$

The magnitude of $\Delta\xi$ is much less than the particle mean free path for binary (electron-ion, electron-electron, ion-ion) collisions. We emphasize again that the process of the collisions of electrons and ions with the dust particles results in the charging of the latter. Thus the dissipation which gives rise to shock waves is again related to the dust charging process and the waves considered in this and in the previous sections have the same nature. The fact that the width of the shock fronts in dusty plasmas is rather narrow (as compared to the classical collisionless shock waves) is related directly to the general properties of dusty plasma as a highly dissipative system and the system, where collisions of electrons and ions with the dust particles can dominate very often.

4. Problems of Shock Waves in Dusty Plasmas

The examples we have given make clear the nature of shock waves in dusty plasmas. However, they serve mainly as illustrations of how the shock waves in dusty plasmas can be described. In reality, many other processes can be important and a more comprehensive analysis of the shock waves should be used. Here we outline the problems which can be important for applications in astrophysical and laboratory plasmas:

(1) *Star formation in shocks.* One beleives that most of new stars is formed in

dust-molecular clouds, shock waves initiating this process. They create the increase in density sufficient for the gravitational self-compression (Jeans instability). The observations show that the presence of dust is well correlated with star formation. In particular, there are direct observations of star formation in dust clouds. The problem is to determine which process is more important: condensation due to the dust attraction or the gravitational self-compression in the shock wave.

(2) *Shocks in supernova explosions.* The layer of dust behind the supernova shock is observed usually. The problem is to verify whether the layer of dust is related to the process of dust condensation behind the shock wave front.

(3) *Particle acceleration in shocks.* The problem is to investigate whether shocks in dusty plasmas can result in the process of Fermi acceleration of electrons and ions. Indeed, the dust in the regions before the front of the shock and behind it can serve as scattering centers to return the fast particle to the shock front and to perform a subsequent acceleration. These dust centers of scattering could be much more effective than the turbulent scattering by MHD waves excited by fast particles.

(4) *Possible formation of dusty crystals behind the shock.* The shocks, which near the front are in gaseous state and behind the front are in solid (crystal) state, are completely new self-organized structures. The possibility of formation of such self-organized structures would allow one to introduce the notion of phase transition in shock wave.

We note also the problems of the influence of the dust attraction, the dissipation in atomic collisions, the agglomeration of the dust, as well as the high power radiation on the shock wave propagation.

5. Conclusions

Thus we have reviewed briefly the main investigations and results on the shock waves in dusty plasmas, discussed their possible applications, and formulated the problems where the shock waves in dusty plasmas can be important. The main conclusions are the following:

(1) The dusty plasma is an open system where there is a tendency to self-organization. The consideration of nonlinear processes is important in description of formation of coherent and dissipative structures.

(2) An anomalous dissipation originating from the charging processes results in a possibility of existence of a new kind of shock waves. They can exist in the presence of collisionless electrons and ions but cannot be considered as collisionless in the exact sense.

(3) The shock waves related to the dissipation originating from the dust particle charging processes can be important for description of star formation in shocks, shocks in supernova explosions, particle acceleration in shocks, formation of dusty crystals, etc.

Acknowledgements

This work was supported by the Russian Foundation for Basic Research (Grant No. 96-02-16456-a) and by the Sonderforschungsbereich 191 Niedertemperatur Plasmen. One of the authors (S.I.P.) would like to thank the Organizing Committee of VII LAWPP 97 for supporting his participation in the Workshop.

References

Barnes, M.S., Keller, J.H., Forster, J.C., O'Neill, J.A. and Coultas, D.K.: 1992, *Phys. Rev. Lett.* **68**, 313.
Carlile, R.N., Geha, S., O'Hanlon, J.F. and Stewart, J.C.: 1991, *Appl. Phys. Lett.* **59**, 1167.
Chen, F.F.: 1965, in: R.H. Huddlestone and S.L. Leonard (eds.), *Plasma Diagnostic Techniques*, Academic, New York, Chap. 4.
Chen, F.F.: 1974, *Introduction to Plasma Physics*, Plenum, New York, Chap. 8.
Chu, J. and Lin, I.: 1994, *Physics A* **205**, 183.
Goertz, C.K.: 1989, *Rev. Geophys.* **27**, 271.
Goree, J.: 1994, *Plasma Sources Sci. Technol.* **3**, 400.
Jellum, G.M., Daugherty, D.E. and Graves, D.B.: 1991, *J. Appl. Phys.* **69**, 6923.
Kaplan, S.A. and Pikel'ner, S.B.: 1974, *Ann. Rev. Astron. Astrophys.* **12**, 113.
Kaplan, S.A. and Pikel'ner, S.B.: 1982, *Interstellar Medium*, Harvard University Press, Cambridge, MA.
Lipschultz, B.: 1987, *J. Nucl. Mater.* **145**, 15.
Ma, J.X. and Yu, M.Y.: 1994, *Phys. Plasmas* **1**, 3520.
Morfill, G.E.: 1996, Private communication.
Notni, P. and Tiersch, H.: 1987, *Astron. Astrophys.* **187**, 796.
Popel, S.I. and Yu, M.Y.: 1994, *Phys. Rev. E* **50**, 3060.
Popel, S.I. and Yu, M.Y.: 1995, *Contrib. Plasma Phys.* **35**, 103.
Popel, S.I., Yu, M.Y. and Tsytovich, V.N.: 1996, *Phys. Plasmas* **3**, 4313.
Popel, S.I., Yu, M.Y. and Tsytovich, V.N.: 1997, *Contrib. Plasma Phys.* **37**, 3.
Sagdeev, R.Z.: 1966, in: M.A. Leontovich (ed.), *Reviews of Plasma Physics*, Vol. 4, Consultants Bureau, New York, p. 23.
Sheehan, D.P., Carlillo, M. and Heidbrink, W.: 1990, *Rev. Sci. Instrum.* **61**, 3871.
Spitzer, L.J.: 1978, *Physical Processes in the Interstellar Medium*, John Wiley, New York.
Thomas, H., Morfill, G., Demmel, V. and Goree, J.: 1994, *Phys. Rev. Lett.* **73**, 652.
Tsytovich, V.N.: 1997, *Physics-Uspekhi* **40**, 53 [*Uspekhi Fizicheskikh Nauk* **167**, 57 (1997)].
Tsytovich, V.N. and Havnes, O.: 1993, *Comments Plasma Phys. Controlled Fusion* **15**, 267.
Tsytovich, V.N. and Popel, S.I.: 1997, 'Shock waves in space dusty plasmas', in preparation.
Tsytovich, V.N., Khodataev, Ya.K., Bingham, R. and Resendes, D.: 1996, *Comments Plasma Phys. Controlled Fusion* **17**, 287.
Vladimirov, S.V.: 1994, Phys. Plasmas **1**, 2762.
Whipple, E.C.: 1981, *Rep. Prog. Phys.* **44**, 1198.
Whipple, E.C., Northrop, T.G. and Mendis, D.A.: 1985, *J. Geophys. Res.* **90**, 7405.
Zel'dovich, Ya.B. and Raizer, Yu.P.: 1967, *Physics of Shock Waves and High-Temperature Hydrodynamic Phenomena*, Academic, New York.

REVIEW OF INERTIAL CONFINEMENT FUSION

M.G. HAINES
Blackett Laboratory, Imperial College, London SW7 2BZ, U.K.

Abstract. The physics of inertial confinement fusion is reviewed. The trend to short-wavelength lasers is argued, and the distinction between direct and indirect (soft X-ray) drive is made. Key present issues include the non-linear growth of Rayleigh–Taylor (R–T) instabilities, the seeding of this instability by the initial laser imprint, the relevance of self-generated magnetic fields, and the importance of parametric instabilities (stimulated Brillouin and Raman scattering) in gas-filled hohlraums. Experiments are reviewed which explore the R–T instability in both planar and converging geometry. The employment of various optical smoothing techniques is contrasted with the overcoating of the capsule by gold coated plastic foams to reduce considerably the imprint problem. The role of spontaneously generated magnetic fields in non-symmetric plasmas is discussed. Recent hohlraum compression results are presented together with gas bag targets which replicate the long-scale-length low density plasmas expected in NIF gas filled hohlraums. The onset of first Brillouin and then Raman scattering is observed. The fast ignitor scheme is a proposal to use an intense short pulse laser to drill a hole through the coronal plasma and then, with laser excited fast electrons, create a propagating thermonuclear spark in a dense, relatively cold laser-compressed target. Some preliminary results of laser hole drilling and 2-D and 3-D PIC simulations of this and the $> 10^8$ Gauss self-generated magnetic fields are presented. The proposed National Ignition Facility (NIF) is described.

1. Introduction

Over the past 25 years laser-driven inertial confinement fusion has evolved from 1 kJ lasers to 100 kJ lasers with 1MJ lasers now in the planning, with which it is hoped that ignition, i.e., a thermonuclear burn, will be triggered.

At the same time it was discovered that going to high values of $I\lambda^2$ (I = laser intensity, λ = laser wavelength) led to the energy being deposited in a small fraction of electrons which would have energy in the 20 keV to 1 MeV range. Thus for obtaining good hydrodynamic coupling and also preventing preheat it was preferable to go to short-wavelength and an intensity of such a value that collisional inverse bremsstrahlung is the principal absorption mechanism rather than resonance absorption. Early experiments with CO_2 lasers ($\lambda = 10.6\ \mu$m) and neodymium glass lasers at 1.06 μm have given way to experiments at $\lambda = 0.35\ \mu$m, the third harmonic of neodymium-glass and KrF lasers ($\lambda = 0.26\ \mu$m) at intensities of up to 10^{15} W/cm^2.

It was also realised that a spatially uniform value of the ablation pressure cannot be achieved using thermal conductive smoothing of laser non-uniformities, but the laser beams themselves must be smoothed by some scheme: random phase plates

Astrophysics and Space Science **256:** 125–139, 1998.

(RPP) (Kato and Mima, 1984), induced spatial incoherence (ISI) (Lehmberg and Obenschain, 1983) or smoothing by spectral dispersion (SSD) (Skupsky et al., 1989). However, there could be a residual problem in that the early laser beam which still has spatial non-uniformities irradiates the target when it is cold and has a steep intensity gradient such that the critical and ablation surfaces are close to each other. This early imprinting could lead to seeding of the Rayleigh–Taylor instability. To overcome this a foam overlay with an additional thin ($\sim$ 100) layer of gold has been employed with very promising results (Desselberger et al., 1995; Dunne et al., 1995), so that the so-called laser direct-drive scheme for fusion is a strong contender for ICF.

A few years ago the indirect scheme (Figure 1) was declassified. Here the spherical capsule containing the D-T fuel is suspended in a hohlraum, i.e. a hollow cavity, which is energised by lasers (or particle beams or a Z-pinch) to a radiation temperature greater than 250 eV, i.e. soft X-rays. This scheme overcomes the problem of the spatial non-uniformities of the lasers and has another advantage of obtaining higher ablation velocities through the absorption of the soft X-rays at higher densities. The price paid for this is the inefficiency of the conversion of laser energy to soft X-rays, and the losses of soft X-rays through the holes needed to send in the laser beam (see Figure 1). Recent work at Sandia has however demonstrated a 15% wall-plug conversion efficiency in a collapsing wire array Z-pinch. Both direct and indirect schemes also have to take care to have beam balance for uniform irradiation. In scaling up to hohlraums for the 1.8MJ National Ignition Facility (NIF) it is found that not only does one have to fill the hohlraum with gas to reduce the ablative expansion of the gold wall which would otherwise fill the cavity, but also the spatial location of the beam spots has to be adjusted during the pulse in order to optimise the uniformity of soft X-ray deposition on the capsule. However the introduction of gas into the hohlraum can lead to the triggering of parametric instabilities such as stimulated Brillouin scattering, stimulated Raman scattering and filamentation. Gas bag experiments that mimic the low density long scale length plasmas expected in NIF hohlraums will be reported.

Next we return to high $I\lambda^2$ interactions, but now in an ultra-intense short pulse laser irradiation, relevant to the ignitor concept (Tabak et al., 1994). The idea behind this is to compress the fuel to the necessary density but at a much lower temperature. This requires less energy. Then a short pulse intense laser is focussed onto the target, and, because of the relativistic ponderomotive pressure, will drill a hole through the coronal plasma to the dense fuel; a further increase in intensity will give a burst of relativistic electrons which will collisionally heat the dense plasma locally, causing a thermonuclear spark which should propagate through the compressed fuel. Experiments and modelling of hole drilling by short pulse lasers the intensity of which is such that the quiver velocity of the electrons is relativistic will be reported.

A novel feature of this laser channel formation is the spontaneous generation of magnetic fields associated with the lack of cancellation of the net inward flow of

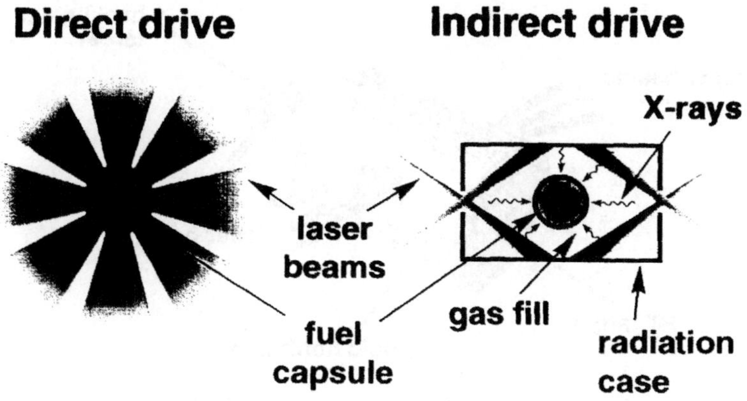

Figure 1. Concepts of direct and indirect drive schemes in laser fusion.

relativistic electrons by the cold return flow. In simulations (Wilks et al., 1992) a field of 250 MG has been found, and in 3-D simulations (Pukhov and Meyer ter Vehn, 1996) this might be the cause of pinching of the channel. Magnetic fields can also be generated in the coronal plasma by the $\nabla T \times \nabla n$ effect, but these are of opposite sign.

Lastly, a brief mention will be made of the proposed NIF laser and of the recent Z-pinch hohlraum results found at Sandia National Laboratory.

2. The Rayleigh–Taylor Instability

2.1. Introduction

The Rayleigh–Taylor instability can occur in ICF during both the implosion phase and during the deceleration of the shell by the compressed and burning fuel at the centre. This instability arises in its simplest form when a heavier fluid is lying on top of a lighter fluid (with gravity downwards). An upward acceleration $\underline{a}$ is entirely equivalent to a downward gravity $\underline{g}$. Hence it follows that a system is unstable if $\underline{a} \cdot \nabla\rho > 0$. In ICF the acceleration is essentially given by $\underline{a} = -\nabla p/\rho$, and thus it is the layer of plasma just inside the ablation surface, satisfying $\nabla p.\nabla\rho < 0$ which leads to the instability. The instability also arises during the deceleration phase of the shell when the compressed fuel at the centre is igniting. During the ablation phase the presence of the ablation velocity v_A leads to a reduction in the growth rate γ to give, approximately (Takabe et al., 1983)

$$\gamma = (kg)^{1/2} - 3kv_A, \tag{1}$$

where k is the wavenumber of the perturbation. After a linear phase characterised by exponential growth in time t, an initially sinusoidal spatial perturbation takes

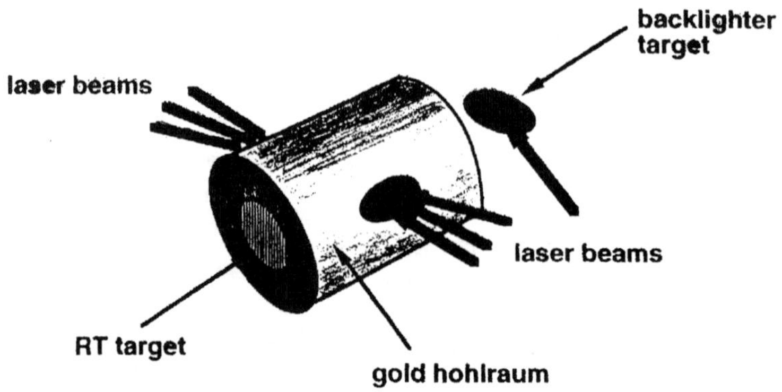

Figure 2. Arrangement for the planar acceleration of a foil in an AWE design laser driven hohlraum.

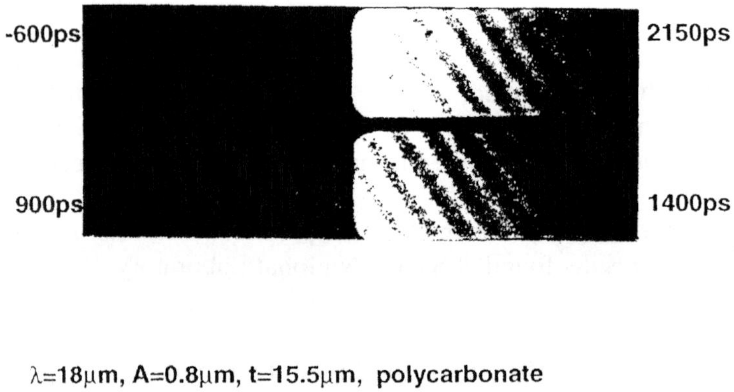

Figure 3. A sequence of four soft X-ray radiographs showing Rayleigh–Taylor growth for short wavelength modulated targets accelerated by X-rays in a hohlraum.

the form of bubbles rising linearly with t and spikes of heavy fluid falling under gravity with a displacement proportional to t^2.

2.2. Radiation Driven Acceleration

In a recent experiment (Taylor et al., 1996a) using a laser energised hohlraum a planar foil with a series of perturbing grooves was accelerated, as shown in Figure 2. A probing back-lighter X-ray beam, generated by focussing another laser beam on a high Z target, was used to perform radiographic measurements of the growth of the R–T instability.

Four radiographic X-ray images at different times are presented in Figure 3, showing growth in amplitude of the spatial modulations in line density.

With two initial wavelengths of perturbation of 10 and 15 μm the non-linear coupling led to a Fourier spectrum, shown in Figure 4.

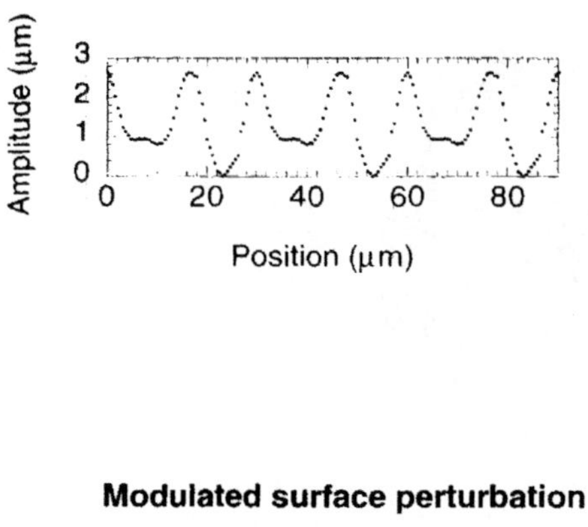

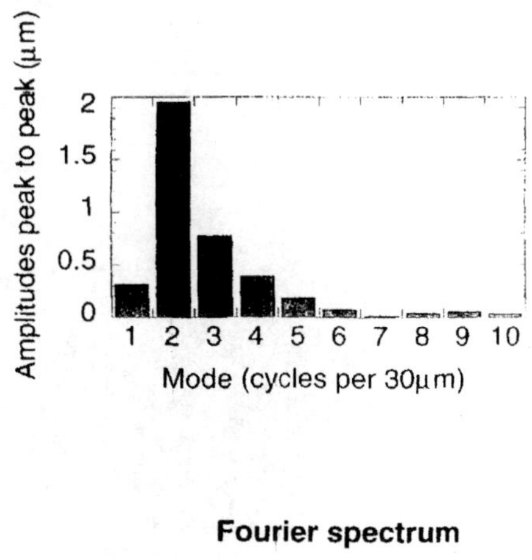

Figure 4. The modulated surface perturbation and the Fourier spectrum from a two mode target with 10 and 15 μm initial wavelength perturbation.

2.3. Cylindrical Convergence

In work carried out by Los Alamos staff on NOVA at Livermore, a hollow cylinder with a $m = 12$ dudecagon 9 μm perturbation was placed in a laser driven hohlraum (Willi, private communication). The sequence of transmission images along the cylinder shows the penetration of the ablatively driven R–T instability into the inside surface, leading to a large departure from cylindrical symmetry of the converging shell. A larger effect would be expected in the case of spherical convergence.

2.4. Indirectly Driven Spherical Implosions

Recently up to 20 kJ of 351 nm laser energy has been used to make a two ring irradiation pattern on the inside wall of a hohlraum similar to proposed NIF targets, using the OMEGA laser at LLE Rochester (McCrory, private communication). The imploded spherical core was clearly observed, and is shown in Figure 5.

3. Imprinting and the Use of Foam Overlayer

Regardless of the laser smoothing schemes employed there is a real problem caused by the initial laser imprint. This is because it takes several coherence times for the present smoothing schemes to be effective. Experimental observations of laser imprinting over 2000 ps have been modelled in a 3-D hydrocode (in a joint IC/NRL collaboration) (Taylor et al., 1996b), and as seen in Figure 6, there is a close agreement in their development.

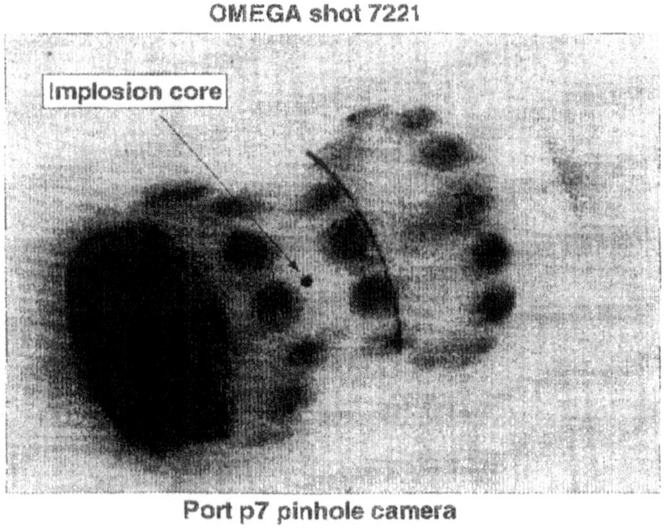

Figure 5. The implosion core is clearly seen in X-ray emission from a hohlraum implosion on the OMEGA laser system.

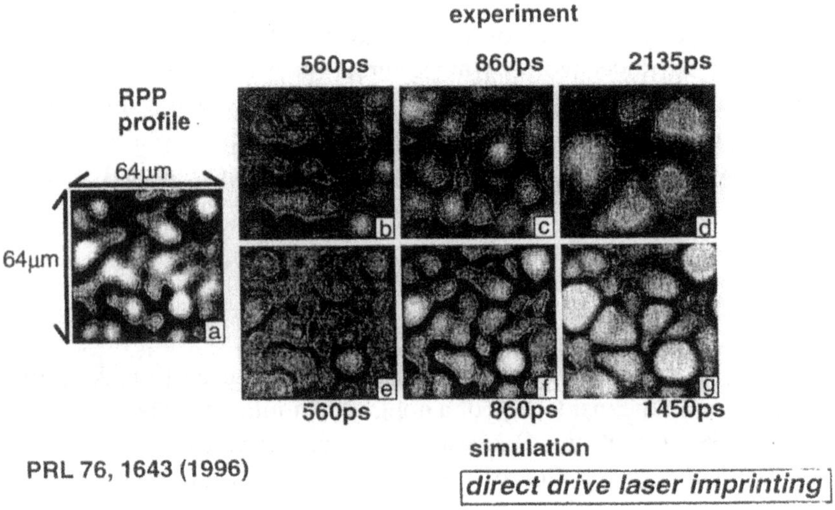

Figure 6. Observations of direct drive laser imprinting are compared to 3-D hydrocode simulations.

If a preformed plasma is created through, for example, an initial X-ray drive (the indirect-direct scheme) it is found that the density scalelength is difficult to adjust and also a shock is generated in the solid by the soft X-ray pulse.

A more effective scheme to remove the effect of the initial imprint is to place a foam overlay on the surface of the solid target. This can be 50 to 250 μm in thickness and coated with some 100 Å of high Z material such as gold. The non-

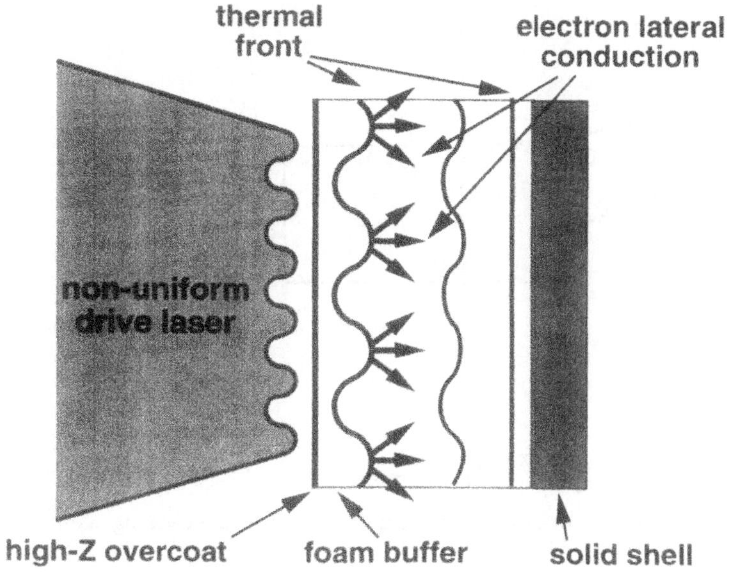

Figure 7. The use of gold coated foam buffer overlay for smoothing of the non-uniform laser energy deposition by diffusive electron conduction.

uniform laser irradiation will initially be absorbed by the gold which will radiate soft X-rays. These X-rays will drive a supersonic ionising thermal front through the low density foam. Over the 50 to 250 microns of the propagation through the foam there is adequate time for lateral electron thermal smoothing of the non-uniform drive as shown schematically in Figure 7. An experiment performed at Limeil by the Imperial College group used a soft X-ray radiographic technique to follow in time the propagation of the supersonic ionising front driven by the X-rays emitted by the thin gold overlayer. Figure 8 shows the layout of the experiment and Figure 9 the streaked X-ray radiograph (Willi, 1996). The same technique was used to follow the subsequent acceleration of the plastic foil. By comparing the rear side trajectories of foils accelerated by direct laser ablation or by using a foam overlay it was found that the acceleration of the foil was not decreased by the addition of the foam. In this case the foil was 12 μm thick mylar, and the foam was 50 μm thick with 50 mg/cc density with a 250 Å gold layer. The importance of the gold layer in reducing the effect of laser imprinting is shown in Figure 10, where the standard deviation of optical depth modulations growing in time is shown for five cases: direct drive, gold coated foam direct drive (FDD), uncoated foam-foil, tight focus FDD, and a direct drive with ISI smoothing. Thus, gold coated foam overlays appear to allow direct laser drive with a minimum of imprinting which in turn minimises the seeding of Rayleigh–Taylor instabilities. This is confirmed by taking X-ray radiographs normal to an initially smooth target surface; a coherent laser beam leads to many imprint spots that grow in amplitude in time while no structures

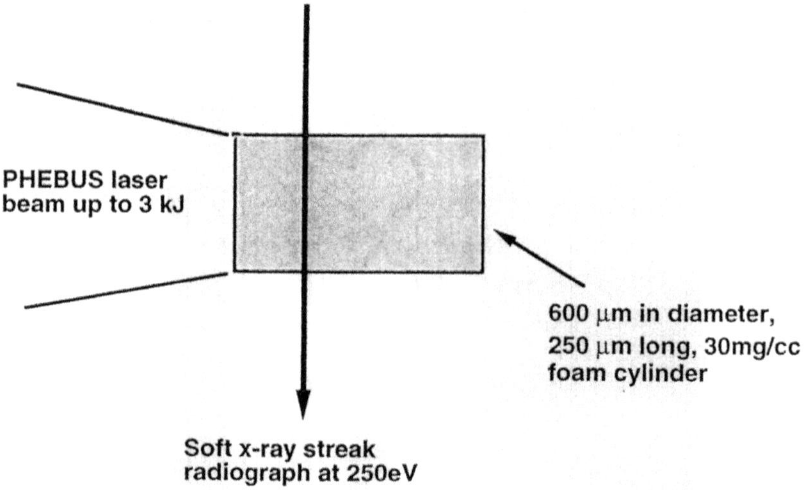

Figure 8. Schematic drawing of experimental target for supersonic preheating of directly driven foam targets.

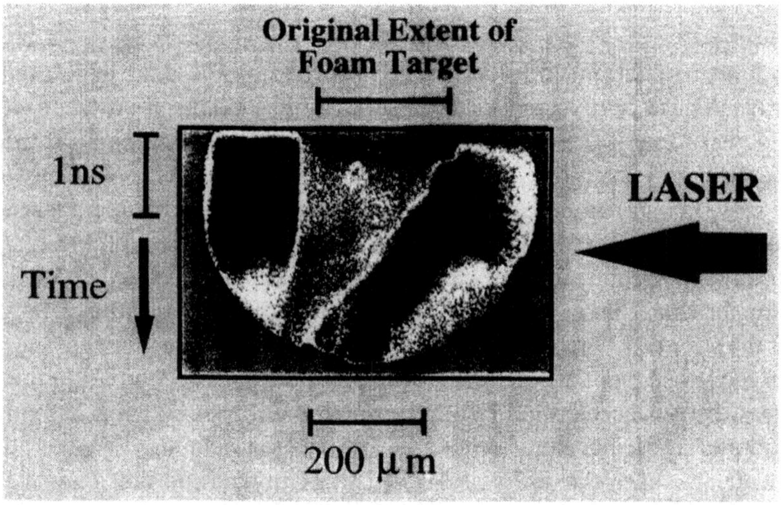

Figure 9. A streaked soft X-ray radiograph showing the X-ray preforming wave from the gold overlay travelling supersonically into the low density foam.

appear in the case of a gold coated foam-foil. Such techniques can give micron spatial resolution if the transmitted image is focussed via a spherical multi-layered mirror.

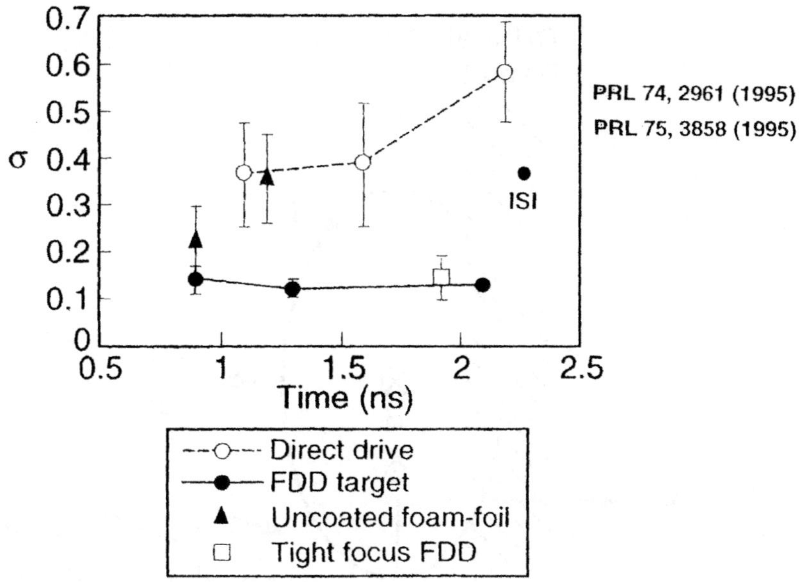

Figure 10. A plot of the standard deviation of optical depth modulations due to laser imprinting for a foil, uncoated foam-foil and a gold coated foam direct drive (FDD) target.

4. Self-Generated Magnetic Fields

In a perfectly spherically symmetric target there will be an electrostatic field $\underline{E}$ which is equal to

$$-\frac{\nabla p_e}{n_e e} - \frac{\beta}{e}\nabla T_e$$

where β is the thermoelectric coefficient, p_e, n_e and T_e are the electron pressure, temperature and density, respectively. The simplest mechanism for the spontaneous generation of magnetic fields occurs when, due to laser irradiation non-uniformities the temperature at the critical surface and coronal plasma has transverse variation. Taking Faraday's law and the form of E above gives

$$\frac{\partial B}{\partial t} = -\nabla \times \underline{E} = \nabla \times \left(\frac{\nabla p_e}{n_e e}\right) = \frac{\nabla T_e \times \nabla n_e}{n_e e} \tag{2}$$

which shows that if gradients of temperature and density are not parallel a magnetic field is generated. Usually, there is a high intensity laser spot, and then the magnetic field is an azimuthal field surrounding the axis of this spot (Stamper et al., 1971), as shown in Figure 11. In magnitude if ∇T_e is $(1\ \text{keV})/(10\ \mu\text{m})$ and $\nabla n_e/n_e$ is $1/(10\ \mu\text{m})$ then $\underline{B}$ will rise to a megagauss in 10 ps.

There are two main effects of this field: one is the reduction in transverse electron thermal conduction if $\omega_e \tau_{ei} > 1$ (ω_e being the electron cyclotron frequency and τ_{ei} the electron-ion collision time), and the other is a $\underline{J} \times \underline{B}$ force or a magnetic

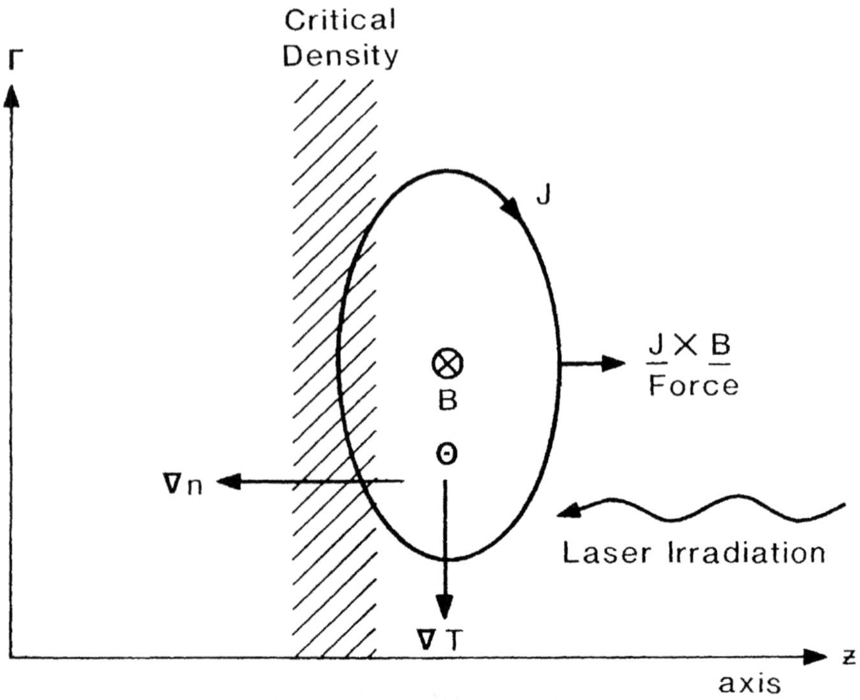

Figure 11. Schematic drawing of current and magnetic field generation for a focused laser beam on to a plane target.

pressure effect that can cause pinching on the axis and fast ion blow-off (Craxton and Haines, 1978).

There are other sources of the magnetic field: for example, if the radiation pressure is non-uniform or if the ponderomotive stress associated with the quiver velocity of the electrons in the laser is non-uniform in the transverse direction, then the magnetic field can be convected, not only by the plasma flow velocity but also by the hotter more collisionless electrons which provide the heat flux (Nishiguchi et al., 1985) (called the Nernst effect). The latter can cause amplification of the field near the ablation surface (Nishiguchi et al., 1985). A review of the sources of magnetic field, its convection, amplification and diffusion is given in (Haines, 1986).

It is found that with short intense and tightly focussed laser pulses, relativistic energies are given to the electrons in the focal spot region. When they stream out and return along different trajectories in the low density corona a magnetic field similar to that described in Equation (2) is produced. But those electrons that stream inward and penetrate the target behave in a manner more similar to the problem of relativistic electron beam-plasma interactions. A cold return current is induced, but around the cylinder enclosing the inward relativistic electron beam

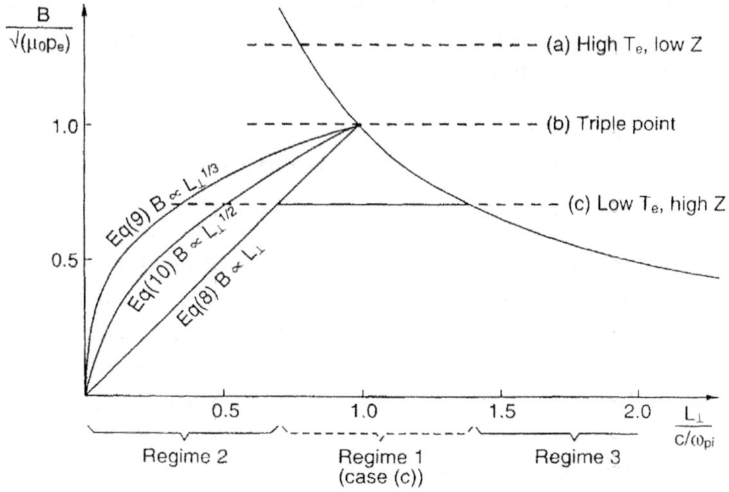

Figure 12. Theoretical model showing the saturated value of the magnetic field *versus* the transverse temperature gradient scalelength in laser produced plasmas.

there is a skin depth associated with the gradual fall-off of return current. This return current has to be driven by an electric field, but it will have a curl at the edge of the beam and hence there will be a magnetic field generated. In PIC simulations (Wilks et al., 1992) fields of 250 MG are generated. Because of the absence of collisions the skin depth and hence the magnitude of the magnetic field will be determined by the collisionless skin depth c/ω_{pe}, or, for a hot plasma, the anomalous skin effect will apply. This magnetic field can develop to such a magnitude that it can cause pinching of the channel carrying the laser beam and fast electrons (Pukhov and Meyer ter Vehn, 1996).

A recent paper (Haines, 1997) has explored the saturation mechanisms that determine the magnitude of the magnetic field near the critical surface. Figure 12 shows how the magnetic field (normalised to $\sqrt{(\mu_0 p_e)}$) varies with the transverse scale length $L_\perp$ (defined by $T_e/\nabla_\perp T_e$). At high $L_\perp$ ablative convection dominates giving $B \sim L_\perp^{-1}$ while at low $L_\perp$ anomalous resistive diffusion associated with the lower hybrid turbulence will lead to B increasing with some (fractional) power of $L_\perp$. The maximum normalised B occurs when $L_\perp$ is equal to the ion collisionless skin depth, c/ω_{pi}. But if the plasma is resistive (low T_e, high Z) this maximum will be reduced.

Magnetic fields can also be generated in a spherically symmetric plasma through instabilities. These instabilities can arise from $\nabla T_e \times \nabla n_e$ effects in Rayleigh–Taylor instabilities, or as a Weibel instability through an anisotropic distribution function arising for example from the heat flux (Epperlein and Bell, 1987), or from a thermal instability in the cold return current (Haines, 1981).

The importance of magnetic fields in, say, the laser imprinting problem is still an area of research, but it is more certain that in intense, short pulse laser-plasma interactions magnetic field can play a dominant role.

5. The Gas-Filled Hohlraum

In the next generation of experiments on indirect drive it will be necessary to include a gas in the hohlraum in order to reduce significantly the speed of ablation of the gold wall of the hohlraum. A gas such as methane is considered, and at a pressure such that its electron density when ionised by the laser beams passing through it will be 10% of critical density. Under the action of laser beams at 0.35 μm wavelength and 10^{15} Wcm^{-2} intensity the electron temperature will rise to 3 to 5 keV.

Experiments have been carried out at Lawrence Livermore National Laboratory using gas bags and the NOVA laser to simulate the long scale length gas filled regions of a NIF target. One of the problems that might occur is that stimulated Brillouin scattering (SBS) will occur and that this will lead to a back reflection of the laser beams. In these experiments (MacGowan, 1996) nine heater beams are employed at low intensity and then an interaction beam is launched through the preformed plasma. Indeed, as seen in Figure 13, SBS is generated by the interaction beam, and some reflection occurs, but then stimulated Raman scattering builds up and appears to quench the SBS. The mechanism for this is still not understood. With the SRS another problem can arise through the acceleration of very hot electrons that might cause unwanted preheat of the capsule in the hohlraum.

6. Fast Ignitor

The fast ignitor (Tabak et al., 1994) is a scheme for obtaining a much higher gain in thermonuclear yield for less total laser energy. It is achieved by separating out the two requirements of high density, i.e. high ρR (where ρ is the final density and R the final fuel radius) and ignition temperature ($>$ 5 keV). A smaller and slow driver is first employed to obtain a dense, but relatively cold, compressed fuel with the correct ρR value. Then an intense short laser pulse is employed to drill a hole through the coronal and ablator plasma to the fuel. Finally, an intense short laser pulse is delivered which energises electrons to relativistic energy in a volume of the fuel sufficient to heat the ions to fusion temperatures and start a thermonuclear burn wave. Clearly, the linear dimensions of this volume should be greater than the mean-free-path of the alpha particles in order that a burn wave is achieved. Figure 14 illustrates the target gain *versus* laser energy expected with the fast ignitor scheme.

The hole drilling relies on the ponderomotive force on the electrons oscillating relativistically in the laser field. There are three parts to this time-averaged force:

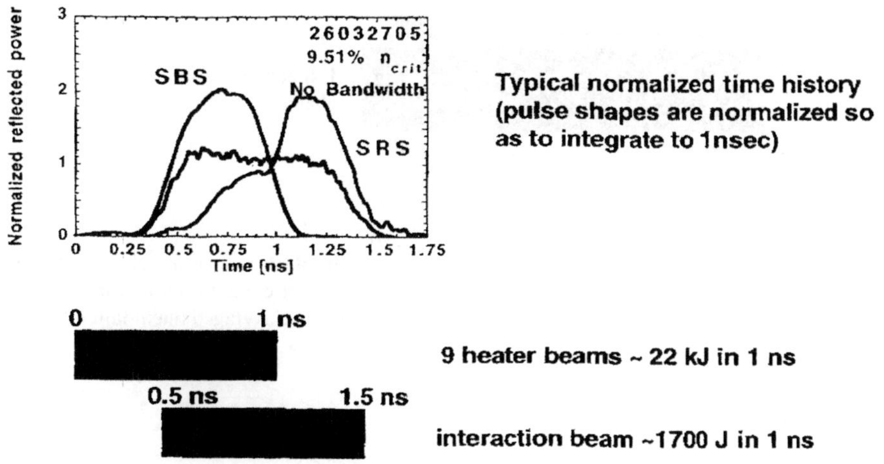

Figure 13. Gas bag experiments at LLNL using nine heater laser beams (22 kJ in 1 ns) followed by a single interaction beam (1.7 kJ in 1 ns) showing stimulated Brillouin scattering and stimulated Raman scattering. The plasma is heated for the first nanosec, then cools during the final 500 ps when probe is left on.

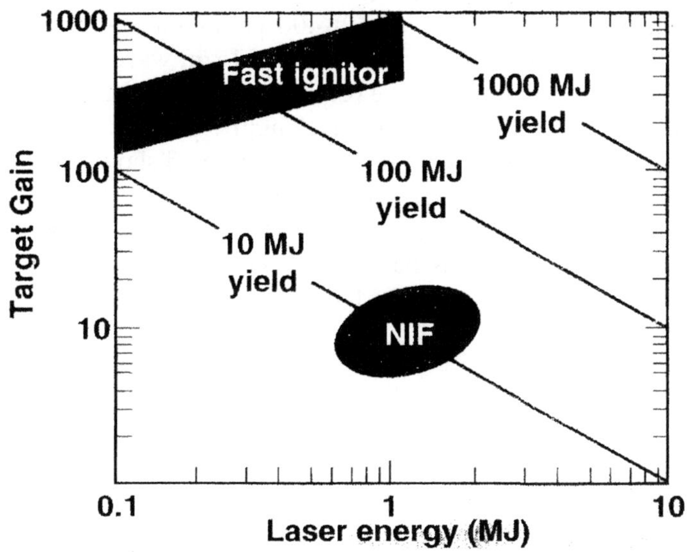

Figure 14. Target gain *versus* laser energy showing the fast ignitor scheme compared to NIF.

the $\langle \underline{\tilde{J}} \times \underline{\tilde{B}} \rangle$ force associated with the oscillatory electron current $\underline{\tilde{J}}$ and the laser magnetic field $\underline{\tilde{B}}$, the divergence of the stress tensor associated with the electron quiver velocity, and the electric field force ρE where ρ is the charge density. The last can act as a restoring force on the electrons, but it is the primary force on the ions that over a longer timescale removes them from the region. Through this effect a hollow channel can be created.

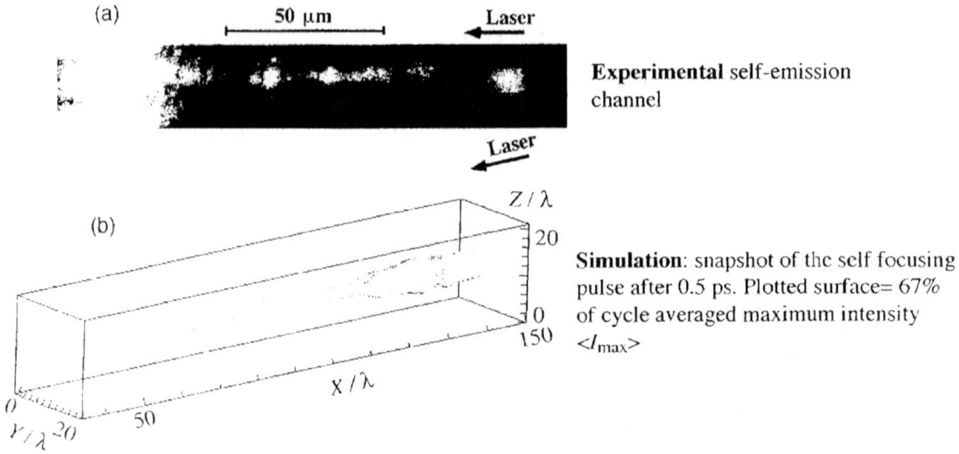

Figure 15. Self-emission measured in a short intense laser pulse channelling experiment is compared to 3-D PIC code simulations.

An experiment has been carried out at the Rutherford Appleton Laboratory (Borghesi et al., 1997) in which a thin Formvar foil of 0.1 to 0.5 μm thickness was heated by a 0.5 μm wavelength 400 ps laser beam at an intensity of 10^{13} W/cm^2. Then a 1 ps interaction beam of 1.054 μm wavelength and intensity of 5×10^{18} W/cm^2 was focused onto the preformed underdense plasma. Self emission at $2\omega_0$ and using a probing shorter wavelength 1 ps laser, interferometry and schlieren images of the channel were recorded. Figure 15 shows the self-emission channel in the experiment together with a 3-D PIC code (Pukhov and Meyer ter Vehn, 1996) simulation of these experimental conditions. Good agreement is found. Megagauss magnetic fields have also been detected through polarimetric measurements. Future experiments are planned for hole-drilling with the petawatt lasers now under construction.

7. The Future

With the building of the US National Ignition Facility (NIF) commencing this year, the subject of ICF will be moving from laboratory scale experiments to equipment approaching the anticipated scale of a reactor. Indeed ignition and thermonuclear burn are computed to occur in the 1.8 MJ NIF hohlraum target energised by 192 laser beams. At the same time, as mentioned above, sub-picosecond laser pulses of up to 10^{15} W (petawatt) are also being developed which can test much of the physics of the fast ignitor concept.

In the neighbouring field of dense Z-pinches, Sandia National Laboratory have imploded multiple wire arrays using both Saturn and PBFA-Z generators. In the latter case up to 20 MA of current in 240 tungsten wires has yielded 1.8 MJ of soft X-rays in a 4 ns pulse (Spielman, 1997). Larger volume ($\sim$ 10 cm^3) hohlraums can be produced by this technique but at a lower temperature of 100 eV. This relatively inexpensive work can be considered as complementary, but is probably capable of scaling up as advances in pulsed power technology are made.

In the more immediate future there will be much interest in improving the techniques for direct drive laser compression. Here there is much scope for the smaller research groups to develop ideas such as the foam overlays to capsules which are currently having much success.

References

Borghesi, M., Mackinnon, A.J., Barringer, L., Gaillard, R. Gizzi, L.A., Meyer, C., Willi, O., Pukhov, A. and Meyer-ter-Vehn, J.: 1997, *Phys. Rev. Lett.* **78**, 879.

Craxton, R.S. and Haines, M.G.: 1978, *Plasma Phys.* **20**, 487.

Desselberger, M., Jones, M.W., Edwards, J., Dunne, M. and Willi, O.: 1995, *Phys. Rev. Lett.* **74**, 2961.

Dunne, M., Borghesi, M., Iwase, A., Jones, M., Taylor, R., Willi, O., Gibson, R., Goldman, S.R., Mack, J. and Watt, R.G.: 1995, *Phys. Rev. Lett.* **75**, 3858.

Epperlein, E.M. and Bell, A.R.: 1987, *Plasma Phys. Controlled Fusion* **29**, 85.

Haines, M.G.: 1981, *Phys. Rev. Lett.* **47**, 917.

Haines, M.G.: 1986, *Can. J. Phys.* **64**, 912.

Haines, M.G.: 1997, *Phys. Rev. Lett.* **78**, 254.

Kato, Y. and Mima, K.: 1984, *Phys. Rev. Lett.* **53**, 1057.

Lehmberg, R.H. and Obenschain, S.P.: 1983, *Opt. Commun.* **46**, 27.

MacGowan, B.J.: 1996, Private communication. See also IAEA Int. Conf. 1996 on Plasma Phys. and Controlled Fusion. *Fusion Energy* **3**, 181 (IAEA, Vienna).

McCrory, R.L.: 1997, Private communication.

Nishiguchi, A., Yabe, T. and Haines, M.G.: 1985, *Phys. Fluids* **28**, 3683.

Pukhov, A. and Meyer ter Vehn, J.: 1996, *Phys. Rev. Lett.* **76**, 3975.

Skupsky, S., Short, R.W., Kessler, T., Craxton, R.S., Letzring, S. and Soures, J.M.: 1989, *J. Appl. Phys.* **66**, 3456.

Spielman, R.: 1997, Private communication. See also Dense Z-Pinches, AIP Conf. Proc. **409**, 101, 1997.

Stamper, J.A., Papadopoulos, K., Sudan, R.N., Dean, S.O., McLean, E.A. and Dawson, J.M.: 1971, *Phys. Rev. Lett.* **26**, 1012.

Tabak, M. et al.: 1994, *Phys. Plasmas* **1**, 1626.

Takabe, H., Montierth, L. and Morse, R.L.: 1983, *Phys. Fluids* **26**, 2299.

Taylor, R.J., Nuruzzaman, S., Meyer, C. and Willi, O.: 1996a, *Annual Report to the Central Laser Facility 1996*, RAL-TR-96-066, 30.

Taylor, R., Dahlburg, J., Iwase, A., Fyfe, D.E., Gardner, J.H. and Willi, O.: 1996b, *Phys. Rev. Lett.* **76**, 1643.

Wilks, S.C., Kruer, W.L., Tabak, M. and Langdon, A.B.: 1992, *Phys. Rev. Lett.* **69**, 1383.

Willi, O.: 1996, Private communication. See also IAEA Int. Conf. 1996 on Plasma Phys. and Controlled Fusion. *Fusion Energy* **3**, 69 (IAEA, Vienna).

Willi, O.: 1997, Private communication.

A NEW APPROACH TO ICF

J.G. LINHART
Dipartimento di Fisica, Università degli Studi di Ferrara, via Paradiso, 44100 Ferrara, Italy

Abstract. After a brief outline of the history of inertial confinement in nuclear fusion research, a new approach is described in which a micro Z-pinch is used as a spark, first igniting a D-T transition channel and finally a cylinder of advanced fuel. A fusion reactor based on this concept will operate with a repetition rate of around 1 shot/minute and a yield of 10 tons TNT/shot.

At the 1958 Geneva conference on peaceful applications of nuclear energy Jim Tuck of Los Alamos laboratory considered, for the first time, a possibility of unconfined DT plasma as a source of energy on a reactor level (Tuck, 1958). His conclusion was negative, mainly because at that time nobody was audacious enough to consider plasma densities equal or superior to the solid state density – a realm monopolized then by the H-bomb people.

The idea of confining a plasma inertially, i.e. by putting a heavy tamper around it, was first formulated in 1961 at the Salzburg conference – the concept and the term "inertial confinement" was presented on that occasion (Linhart et al., 1962).

At first, in order to deal with plasmas of density inferior to 10^{20} ions/cm^3, conceivable at that time, one had to assume that the plasma was thermally insulated from the tamper by a magnetic field. This implied some kind of axially symmetric plasma (Maisonnier et al., 1966; Alikhanov and Konkashbaev, 1974; Linhart, 1982) and obviously magnetic fields in the MGauss range (Euratom, 1966).

It was not until 1969, when during a Summer school at Varenna, Kidder (1969) from LLNL introduced the idea of a DT sphere compressed by ablation produced by a laser. Since then most of the ICF work centered on laser (and later ion beams) induced ablation of spherical pellets.

It soon became evident that in order to produce a spherically symmetric compression, the laser must illuminate the pellet uniformly – the permitted nonuniformity being at most a few percent. The way to smooth the incident light is to interpose a uniformizing element. Thus around 1980 and later several forms of hohlraum, surrounding the DT pellet, were proposed (indirect compression).

It was also realized that only a central spark capable of igniting a highly compressed DT mantle could give a sufficiently high gain, necessary to compensate the many inefficiencies of the driver and of the ablation drive.

Astrophysics and Space Science **256:** 141–149, 1998.

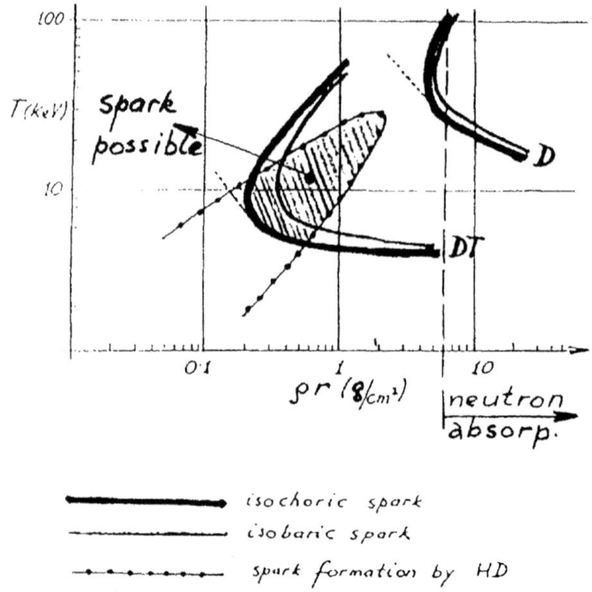

Figure 1.

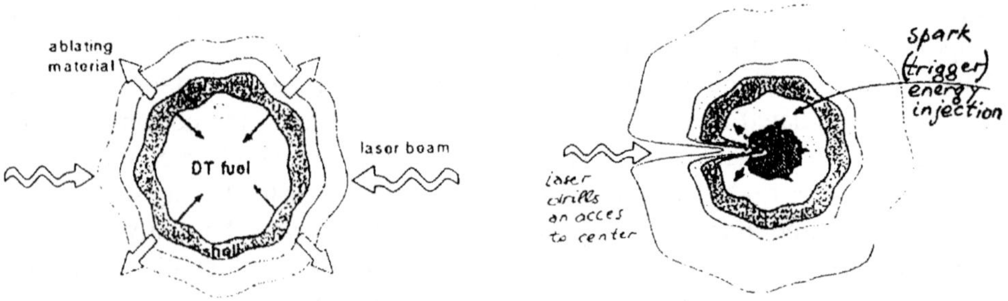

Figure 2.

More recently it was shown (Basko, 1990) that the central spark generated by hydrodynamic process (converging shocks) is capable to ignite only a DT fuel (Figure 1). Even this may prove difficult because it requires a fine control of the ablation process – a feat made even more difficult by the interposition of a hohlraum (or hotraum).

It is for this reason that the people working with spherical pellets started recently to shift to schemes where, even though the DT is still compressed spherically – the spark (hot spot, trigger) is generated by some other mechanism – either on the periphery of the pellet or, drilling a hole in the compressed material, somewhere inside it (Tabak et al., 1994; Caruso and Pais, 1996; fast ignitor, Figure 2).

This could result in considerable saving in the driver energy, unfortunately, the technical difficulties appear to be formidable. With the fast ignitor scheme the

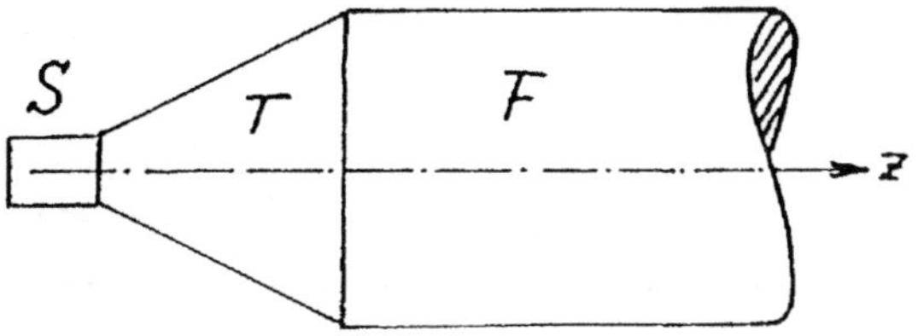

S – spark
T – transition section
F – the main fuel

Figure 3.

original concept of the spherical pellet and spark is thus approaching the concept of axially symmetric spark and target (Figure 3).

However, the axial concept of detonator (spark), transition *T* and the main thermonuclear charge has been for a long time the leit-motif in most H-bombs.

With the discovery of new techniques capable of compressing the detonator to more than 5 times its solid density (n_s), both the mass of the trigger and of the thermonuclear charge became so small that a class of explosive devices described as tactical weapons (originally known as neutron bombs) has been developed, capable of explosions as small as 1 kton.

As a result of this development of mini-bombs several workers at LLNL (Call and Moir, 1990; Szoke and Moir, 1991) have tried to resurrect the old dream of E. Teller – the Ploughshare, which at the level of 1 kton becomes more attractive. The project is called PACER (Figure 4).

Unfortunately, the containment of such explosions seems still rather problematic and moreover the detonator remains Pu, be it only a fraction of a kilogram per shot. The efforts to achieve still smaller explosions imply the ability to compress Pu to well over 10 times n_s – this calls for very sophisticated detonator mechanisms and for driver energies in excess of 10 MJ (Linhart, 1992). At the moment the progress in this direction appears too complicated.

It seems that a possible way of disposing of the Pu-trigger is to start the spark directly in a DT plasma. This obviously presents an additional problem to that of compressing Pu. The DT plasma must be heated to an ignition temperature of at least a few Kev, requiring very short times for the compression and eventual inertial confinement of the spark. Also the energy output of the spark is not directly capable of starting a detonation wave in deuterium or in LiD. We shall, therefore, need a transition *T* between the DT spark and a D channel, capable of amplifying the spark output.

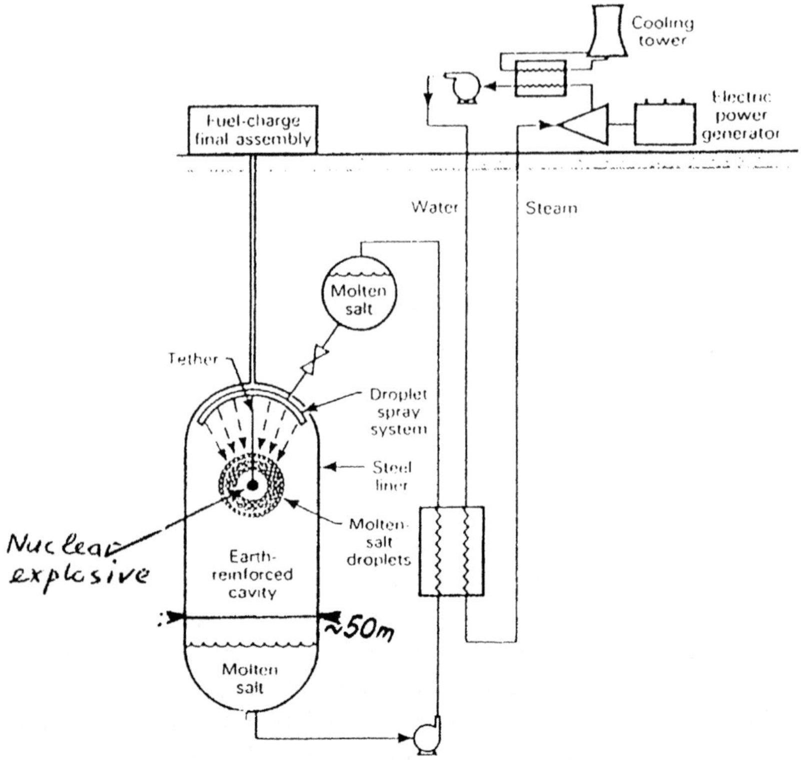

Figure 4.

This can be achieved if the medium in the transition is still D-T. The mechanism proposed here is a conical equivalent to that described by Miley (1981) for spherical pellets (project AFLINT, Figure 5).

Taking a conical slice of the AFLINT target one has the advantage of the DT spark becoming accessible and thus it can be compressed and heated independently of the T and F sections. Also the total energies of the conical-cylindrical slice are much smaller than those required in AFLINT (Linhart, 1970, 1993).

We suggest that the easiest way to realize such a small, hot and highly compressed spark and T section is using a minute Z-pinch (Bortolotti et al., 1993; Linhart, 1994).

1.5 and 2-dimensional numerical simulations suggest that a micro Z-pinch is indeed capable of starting a detonation wave in a larger D-T channel – particularly if this channel is compressed and inertially confined by a heavy liner (Figure 6) (Vichrev et al., 1993, 1996; Bilbao and Linhart, 1996, 1998).

An attractive feature of a DT-pinch is, that provided its length L is not too large, the fusion α's can be shot out of the spark and be directly responsible for the ignition of the detonation wave in the T sector.

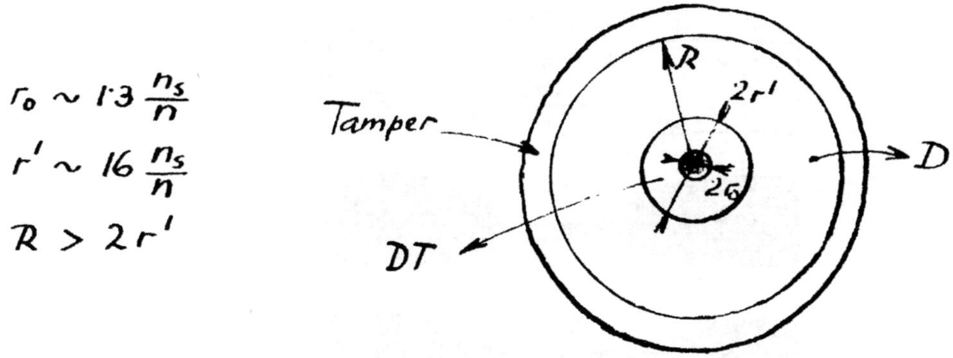

Figure 5.

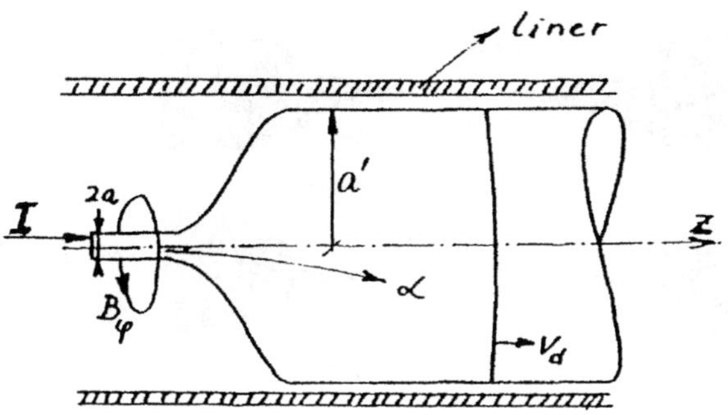

Figure 6.

It can be shown (Bilbao and Linhart, 1996, 1998) that, in order for the α-ignition to be dominant, the pinch half-length

$$L = g \cdot 10^{12} \frac{T^{3/2}}{n \ln \Lambda}, \quad g < 1,$$

where T is the temperature of the spark and $\ln \Lambda$ the Coulomb logarithm. The current I required to flow in S and T is (Bilbao and Linhart, 1998):

$$I \geq 0.12 \cdot 10^{-8} \left(\frac{\ln \Lambda}{g(G+1) \ln \frac{a_0}{a}} \right)^{1/5} N^{1/2} \left(\frac{a'}{a} \right)^{2/5} \ldots (\mathrm{MA}),$$

where G is the α-fusion gain in S, a_0 the initial radius of S, N the ion line density and a' the radius of T. Taking values of these parameters that could be realized in a micro Z-pinch it follows that I is of the order of 10 MA. However, given the small radius and length of the spark (a_0 and L will be of the order of 100 μm) it is clear that, were the pinch collapse to occur within a few Alfvén transit times, the current rise times must be only a few nanoseconds (Bilbao and Linhart, 1996).

Let us now present a typical result of 2-dimensional simulation of a pinch whose $a_0 = 100\ \mu$, $L = 200\ \mu$m, $a_0' = 200\ \mu$m and the current rises as $I = I_0 + I_1(t/t_0)^2$ and $I_0 = 1/2$ MA, $I_1 = 2.5$ MA, $t_0 = 2.5$ nsec.

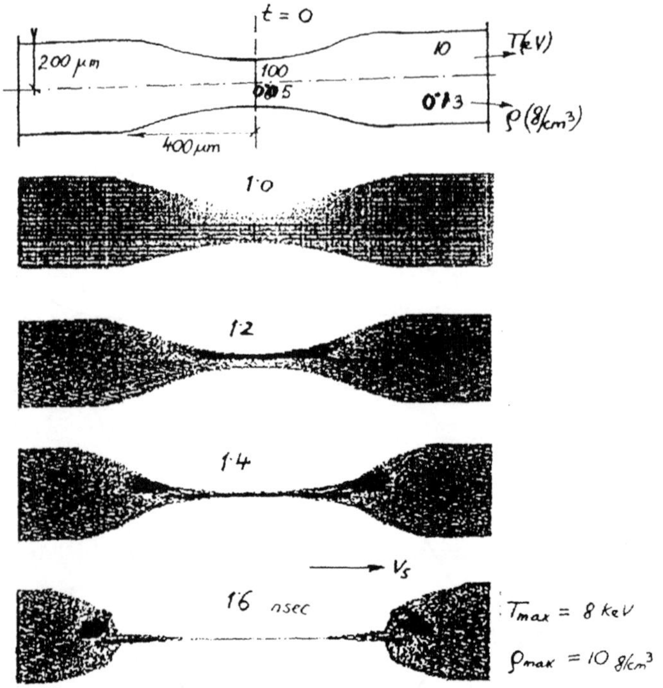

Figure 7.

The initial situation is that of a Z-pinch in Bennett equilibrium with a nearly surface current (Figure 7). After $t = t_0$ the current remains constant $I = 3$ (MA). Initially, $T_0 = 100$ eV in the S section and 10 eV in the a' channel.

Such a rapid current rise, as assumed here, can be achieved only by a fast liner imploding on a B_ϕ field (Linhart, 1995). The speed V_L of the liner must be at least 10 cm/μsec. The driver of such a liner could be

1. fast Z-pinch impacting on the liner (Linhart, 1979; Hartman et al., 1994),
2. ablation drive (laser of ion beam or black body radiation).

The choice of the driver will be determined by economic considerations. In Figure 8 are shown graphs representing the economically permissible cost C of the device to be exploded based on the cost E of a kWh of electricity (Gilinski and Hubbard, 1968; Moir, 1996) and on the fraction $1 = c/Q$, where Q is the total cost per shot.

Assuming e.g. $C = 100$ \$/shot, $q = 0.1$, $E = 5$ cents, the energy liberated per shot must be about 40 GJ (i.e. 8 tons TNT). Such a blast can be contained in a steel or steel-lined cavity whose radius is 20 meters. In order to generate 1 GW thermal the reprate must be about 1 shot/40 secs (Figure 9).

The amount of tritium in the spark and in the transition region T will be negligible (less than 10 g/day) and can be supplied by an external tritium source, although the tritium recovery is advisable and practicable. The neutrons emitted

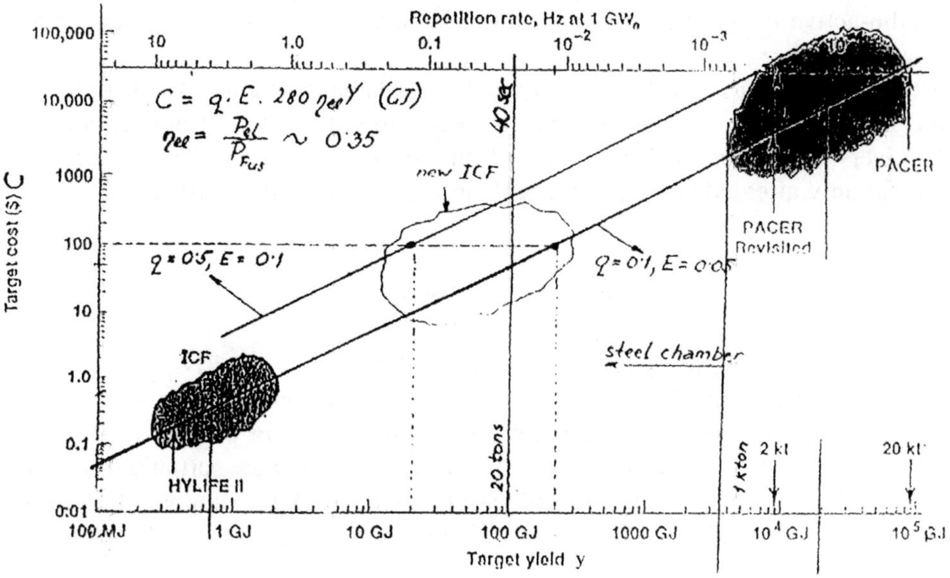

Figure 8.

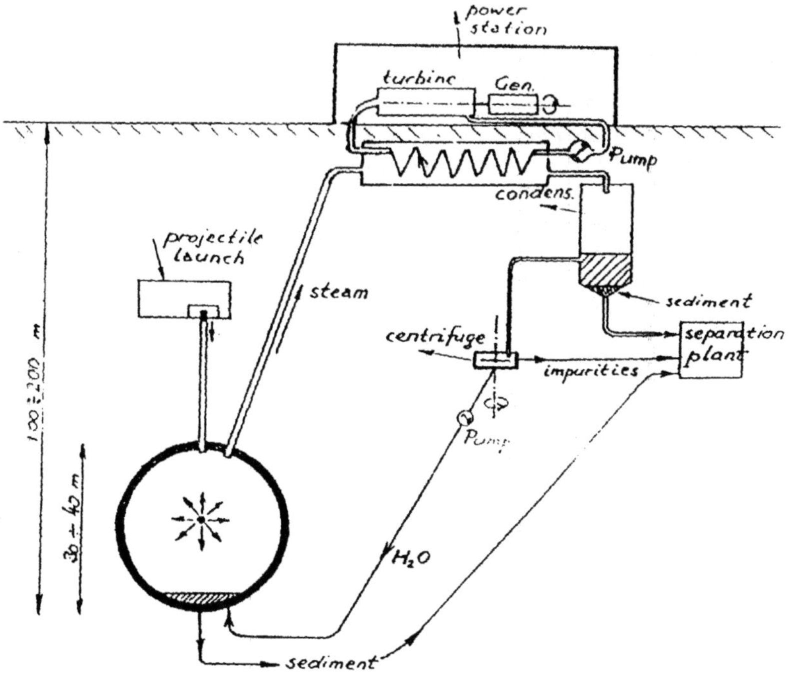

Figure 9.

and not absorbed in the fuel, must be absorbed in a way which produces only non-radio-active materials. This is feasible in a large cavity filled with steam and additives.

The considerations of a large cavity and the related problems of transparency of the steam, etc., to laser or ion beams poise limitations to the application of such drivers. Consequently, the design of an explosive device to be used can be attempted only after exhaustive numerical simulations and pilot experiments.

Acknowledgements

Our recognition is due to Drs. R. Moir and C.W. Hartman of LLNL for sending us reprints of their papers related to this subject and to many participants of the workshops on 2-dimensional simulation of fusion detonation processes (Prague, 1993, Ferrara, 1995, London, 1996). Funds for this research were partially provided by Ferrara Ricerche and University of Ferrara with the kind support of Professor P. Dalpiaz. My colleague Professor L. Bilbao has provided the 2-dimensional calculations in Figure 8.

References

Alikhanov, S.G. and Konkashbaev, I.K.: 1974, in: *Proc. 5th Int. Conf. Plasma Phys. and Contr. Nucl. Fusion Rel.*, Vol. 2, IAEA-CN-33, Vienna, p. 577.
Basko, M.M.: 1990, *Nucl. Fus.* **30**, 2443.
Bilbao, L. and Linhart, J.G.: 1996, *Plasma Phys. Rep.* **22**, 457.
Bilbao, L. and Linhart, J.G.: 1998, in: *Proc. 1996 ICENES Conf.*, in press.
Bortolotti, A. et al.: 1993, in: *Proc. 3rd Int. Conf. Dense Z-Pinches*, AIPress, p. 372.
Call, Ch.J. and Moir, R.W.: 1990, *Nucl. Sci & Eng.* **104**, 364.
Caruso, A. and Pais, V.A.: 1996, *Nucl. Fus.* **36**, 745.
Gilinski, V. and Hubbard, H.: 1968, *Nucl. Fus.* **8**, 69.
Hartman, C.W. et al.: 1994, *MGauss Mag. Field Gen. and Pulsed Power Appl.*, Nova Sci. Publs., p. 873.
Kidder, R.E.: 1969, in: *Proc. Int. School of Phys. E. Fermi*, course 48, pp. 315–327.
Knoepfel, H. and Herlach, F. (eds): 1966, *Proc. 1st MGauss Conference*, Frascati.
Linhart, J.G.: 1970, *Nucl. Fus.* **10**, 211.
Linhart, J.G.: 1979, *Nucl. Fus.* **19**, 264.
Linhart, J.G.: 1982, *Unconventional Approaches to Fusion*, Plenum Press, New York, pp. 233 and 247.
Linhart, J.G.: 1992, *Czech. J. Phys.* **42**, 985.
Linhart, J.G.: 1993, in: *Proc. Int. Workshop Plasma Phys.*, Pichl, p. 284.
Linhart, J.G.: 1994, *Nuovo Cimento* **16D**, 281.
Linhart, J.G.: 1995, in: *2nd Nat. Symp. Plasma 95*, Warsaw, p. 71.
Linhart, J.G. et al.: 1962, *Nucl. Fus. Suppl.* **2**, 739.
Maisonnier, Ch. et al.: 1966, in: *Proc. Int. Conf. Culham*, 1965, IAEA, Vienna, p. 345.
Miley, G.H.: 1981, *Laser Interaction and Related Phenomena*, Plenum Press, New York.
Moir, R.W.: 1996, private communication.

Szoke, A. and Moir, R.W.: 1991, *Technol. Rev.* **94**, 20.
Tabak, M. et al.: 1994, *Phys. of Plasmas* **I**, 1626.
Tuck, J.: 1958, in: *Proc. Int. Conf. on Peaceful Uses of Atomic Energy*, Geneva, Vol. 32, p. 3.
Vichrev, V.V. et al.: 1993, *Fiz. Plasmy* **19**, 79.
Vichrev, V.V. et al.: 1996, *Plasma Phys. Rep.* **22**, 97.

HOHLRAUM TARGETS DRIVEN BY CLUSTER ION BEAMS FOR INERTIAL CONFINEMENT FUSION

C. DEUTSCH and N.A. TAHIR
Laboratoire de Physique des Gaz et Plasmas, Université Paris IX, Bât. 212, 91405 Orsay, France

O. GEB and J.A. MARUHN
Institut für Theoretische Physik, Universität Frankfurt, Postfach 111932, 60054 Frankfurt, Germany

Abstract. Correlated ion stopping of charged debris resulting from the fragmentation of energetic cluster ions focussed on Au foam converters is considered for driving indirectly a thermonuclear capsule. The 1D simulation of converter time evolution demonstrates a very high conversion efficiency of projectile energy into hard X-ray photons building up a very hot ($T_r > 300$ eV) thermal bath. Intense and energetic cluster ion beams thus demonstrate considerable potential as a novel driver for inertial confinement fusion.

Presently, in the highly active field of inertial confinement fusion (ICF) driven by intense heavy ion beam (HIB) the so-called indirect drive approach is given a lot of attention (Lindl, 1995; Lindl et al., 1986; Ho et al., 1994; Murakami and Meyer ter Vehn, 1991). It essentially implies a three steps process. The driver energy is first focussed on purposely laterally located converters on a two-sided hohlraum cavity enclosing a smaller and spherical capsule with the thermonuclear fuel deuterium + tritium (DT) in it (see Figure 1). Then the converters get heated up after a few nsec irradiation, up to a few hundreds of eV temperature. They fill the hohlraum cavity with hard X rays for 10–12 nsec.

The corresponding photons then gradually thermalize with a radiative temperature ~ 300 eV, through repeated bouncings on the casing inner surface enclosing the hohlraum and the capsule outer surface. The given heat bath is then able to provide a highly isentropic compression of the inner capsule. To optimize these processes one has to focus kiloamps of heavy ions ($V \sim c/3$) on a 3 mm radius spot area on converters, in order to get a huge power deposition Pd. The crux of this well established scenario is that one should transport the ion of masse M_i in a lowest charge state Z_i, in order to minimize space charge effects as well as the triggering of electromagnetic instabilities. On the other hand, the much higher projectile charge Z_i' in plasma target secures a beam-target coupling $\sim Z_i'^2$ through a standard stopping process in the hot converters. Those rather contradictory re-

Astrophysics and Space Science **256:** 151–159, 1998.

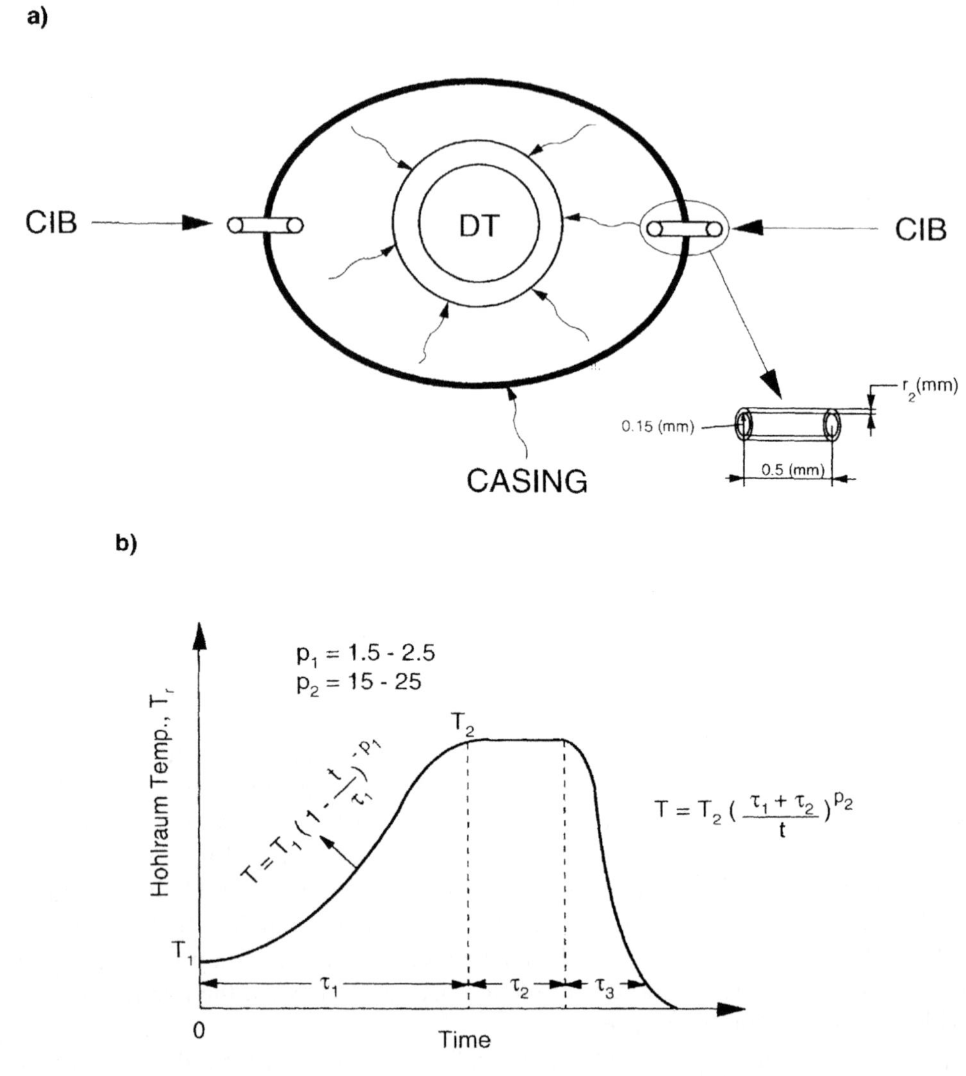

Figure 1. Hohlraum design. (a) Target structure with cylindrical Au foam (1 g/cc) convertors surrounded by a corona (thickness r_2 of solid Au. (b) Pulse shape profile (Table I).

quirements on the projectile ion charge could be strongly relaxed or even resolved by considering the focalization of intense cluster ion beams (CIB). The production, characterization, linear acceleration and even circular storage (Kroto, 1996; Larsson, 1995) of supermolecular arrangements of basic atomic entities is now progressing at a very swift pace (Kroto, 1996; Larsson, 1995; Scheier and Märk, 1994). For instance, C_{60}^{+} is one of the best known and more robust of them. In contradistinction to atomic ions, there is no M_i limitation for cluster ions. So, on can use them on a much larger $Z_i/Mi \ll 1$ scale, allowing $Z_i > 2$, provided the

TABLE I

Pulse parameters.

Prepulse temperature	$T_1 = 100$ eV
Prepulse power	$P_1 = 10$ TW/cm^2
Maximum temperature	$T_2 = 300$ eV
Maximum power	$P_2 = 830$ TW/cm^2
Prepulse duration	$\tau_1 = 22$ nsec
Main pulse duration	$\tau_2 = 8$ nsec
Shape factor	$P_1 = 2.0$
Shape factor	$P_2 = 20.0$

given structure can sustain linear acceleration or storage (Kroto, 1996; Larsson, 1995). This is indeed the case for C_{60}^{n+} with $-2 \leq n \leq 7$ (Scheier and Märk, 1994).

The CIB-converter coupling $\sim (\sum_i Z_i')^2$ features a strongly enhanced and correlated stopping (ECS) arising from inflight and dynamical electrostatic interferences between the ion debris (charge Z_i') resulting from the initial fragmentation of a cluster ion impacting the target dense electron fluid (Deutsch and Tahir, 1992; Eliezer et al., 1995; Tahir et al., 1994). The sum over i runs over all those fragments. This coupling is advantageously contrasted to the usual HIB coupling $\sim (\sum_i Z_i')$ corresponding to uncorrelated ion projectiles, in otherwise identical beam-target conditions.

As a result one expects a much shorter penetration indepth. Those specific CIB features have already been used to design highly innovative direct drive scenarios (Deutsch and Tahir, 1992; Eliezer et al., 1995), giving credence to the momentum rich beam concept due to Maschke (1984).

Here we intend to unravel their powerful potentialities for the indirect drive approach to ICF.

The much shorter cluster ion ranges make it possible to envision very thin convertor depths accommodating the corresponding very short CIB ranges. A few typical range-energy relationships of related concern are given in Figure 2 for Au target at various cold matter density. So, in contradistinction to hohlraums (Lindl, 1995; Lindl et al., 1986; Ho et al., 1994; Murakami and Meyer ter Vehn, 1991) currently designed for heavy ion drivers, a much larger volume is left unoccupied for further isotropization of the thermal photon bath around capsule and converters.

The stopping data displayed in Figure 2 are then introduced into a 1D hydrodynamical simulation (Ramis et al., 1988) based on the code MULTI to investigating the time evolution of an Au foam cylindrical convertor (Figure 3) 0.15 mm in radius and 0.5 mm long wrapped in a corona (r_2 thick) in solid gold. The latter is expected altogether with a 0.01 g/cc D_2 gas fill in hohlraum to restrict the transverse radial expansion to a few mm only, 10 nsec after the peak CIB illumination has begun

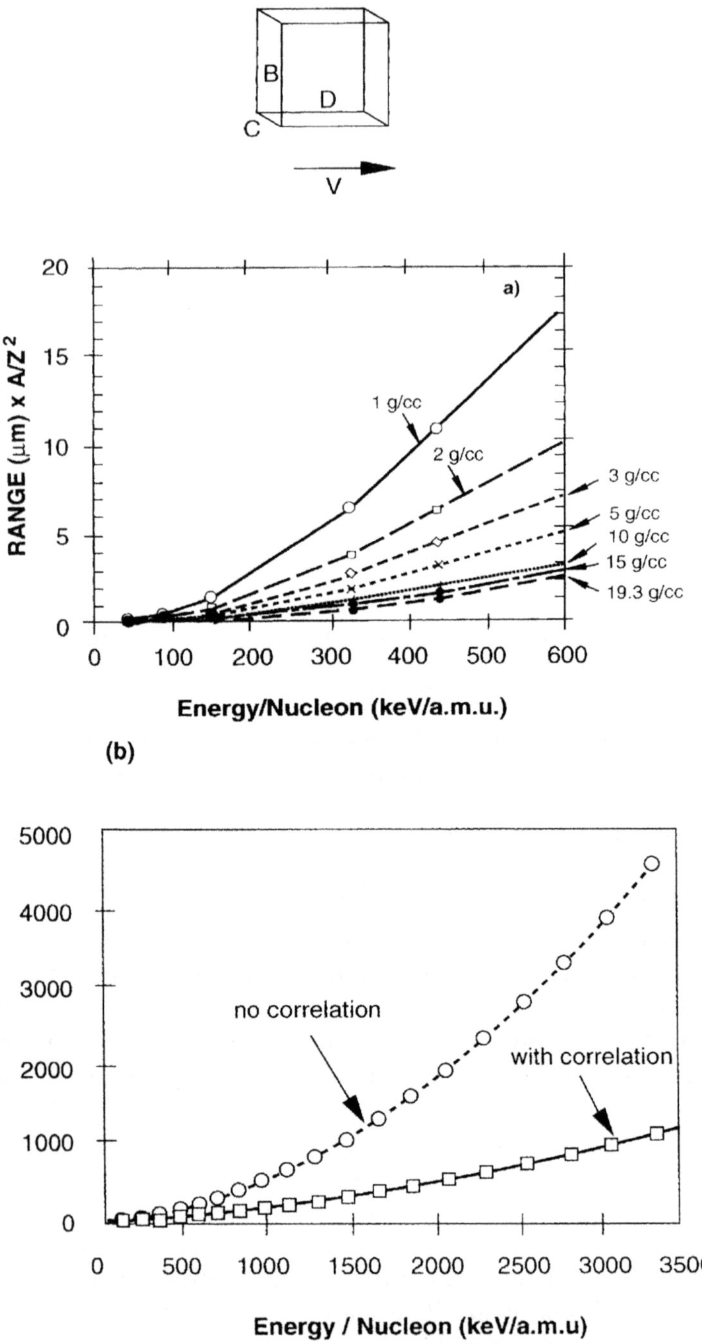

Figure 2. (a) Range-energy relationships for correlated stopping of eight ion debris (charge Z, atomic mass A) on vertices of a two atomic unit (a.u.) edge cubic box flowing in foam Au target at several densities with one edge/overall drift velocity. (b) As in (a) for $Z = 1$ and gold foam at 1/10 solid density.

a)

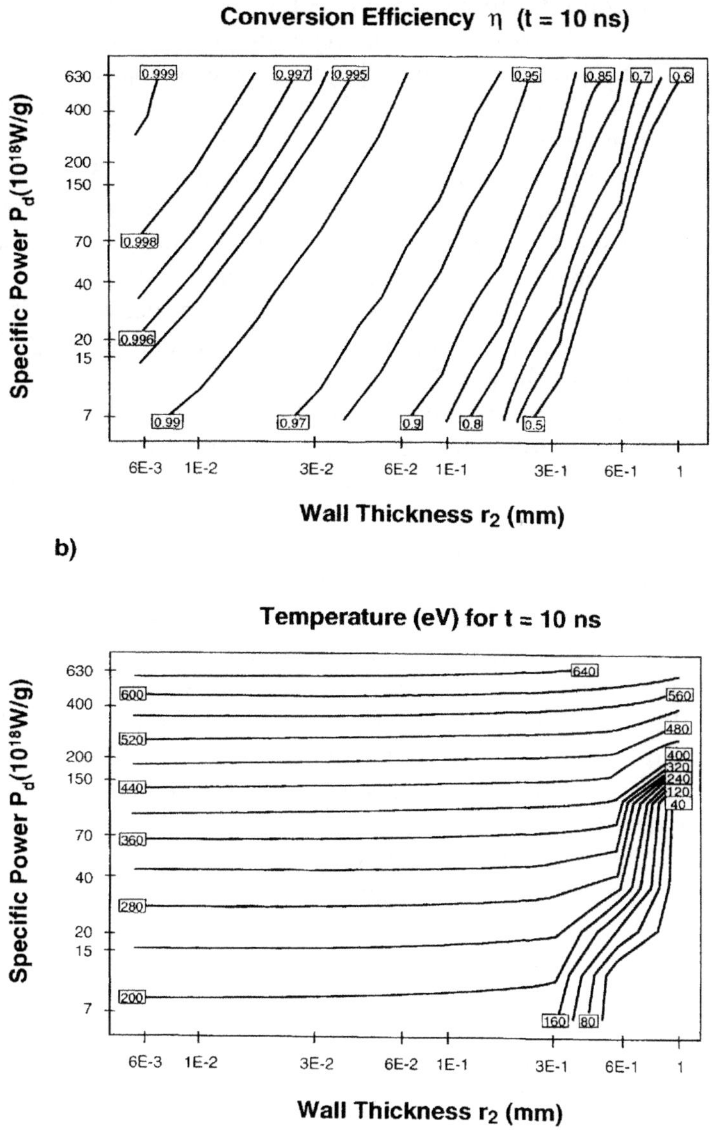

Figure 3. 1D simulation (MULTI) of convectors evolution 10 nsec after the beginning of peak illumination. (a) Conversion efficiency η of CIB kinetic energy into X-rays. (b) Hohlraum radiative temperature T_r.

according to the pulse shaping featured by Figure 1b and Table I. The specific deposited power Pd is considered in the 10^{18-20} W/g range. Figures 3 exhibit Pd variations in terms of the corona thickness r_2 parametrized respectively by η, CIB energy conversion efficiency into hard photons and radiative hohlraum temperature T_r. The given convertor illumination conditions thus correspond to an energy transfer in the (1–100) MJ range. Envisioning a fullerene (C_{60}) CIB producing ion fragments eightfold correlated in target (Figure 2), which appears as a reasonable estimate for the fragment distributions available to the presently considered target scenario fixed for instance at 10 MJ irradiation level demands 10 MA of C_{60} on each convertor for 8 nsec with a projectile kinetic energy $\cong$ 1.75 Mev/a.m.u.

The required linear accelerating structures seem to lie within the realm of present technology making use of induction linac facilities provided sufficiently intense cluster ion sources may be developed (Bangerter, 1993).

Figure 3a demonstrates that for $r_2 \geq 0.1$ mm, η ranges between 0.85 and unity. On the other hand, T_r remains remarkably r_2-independent for $T \gtrsim 300$ eV. It increases rapidly with Pd up to values > 600 eV.

These results look promising enough to qualify unambiguously CIB as a serious candidate to the indirect drive approach for particle driven fusion.

They also secure a swift thermalization of the produced hard photons in the hohlraum.

The issues of conversion and symmetrization have also been considered at length. It has been shown that in a simple two converters configuration it is not possible to achieve the high level of 98% symmetry in the radiation field. However, more sophisticated configurations including radiation shields placed at appropriate places in the hohlraum, make it possible to reach this required symmetry level (Kang et al., 1993). Then the problem of capsule implosion can be reduced to a 1D problem handled with a three-temperature computer code (Tahir et al., 1982, 1986) providing simulation results of compression, ignition and thermonuclear burn of a typical reactor-size target.

The target initial conditions are shown in Figure 4a while the target initial parameters are presented in Table II. The target consists of a fusion capsule that contains 5 mg solid DT shell whose inner radius is 3.02 mm and is filled with a low density DT gas. The fuel shell is followed by a radiation shield made of a low-, material doped with high-Z atoms. The advantages of using such a radiation shield instead of a pure high-Z radiation shield have already been presented (Tahir et al., 1982). The radiation shield is followed by a carbon ablator shell. The capsule is enclosed in a solid gold casing and the radius of this hohlraum casing is three times the outer capsule radius. The cavity is filled with a black body radiation characterized with a radiation temperature T_R which varies in time as shown in Figure 1b. This shaped input pulse is essential to achieve a high target gain (Kidder, 1979; Bodner, 1981). The precise pulse parameters used in this set of calculations are given in Table I. The heat bath radiation ablates the material from the capsule surface that generates an ablation pressure which is responsible for driving the implosion.

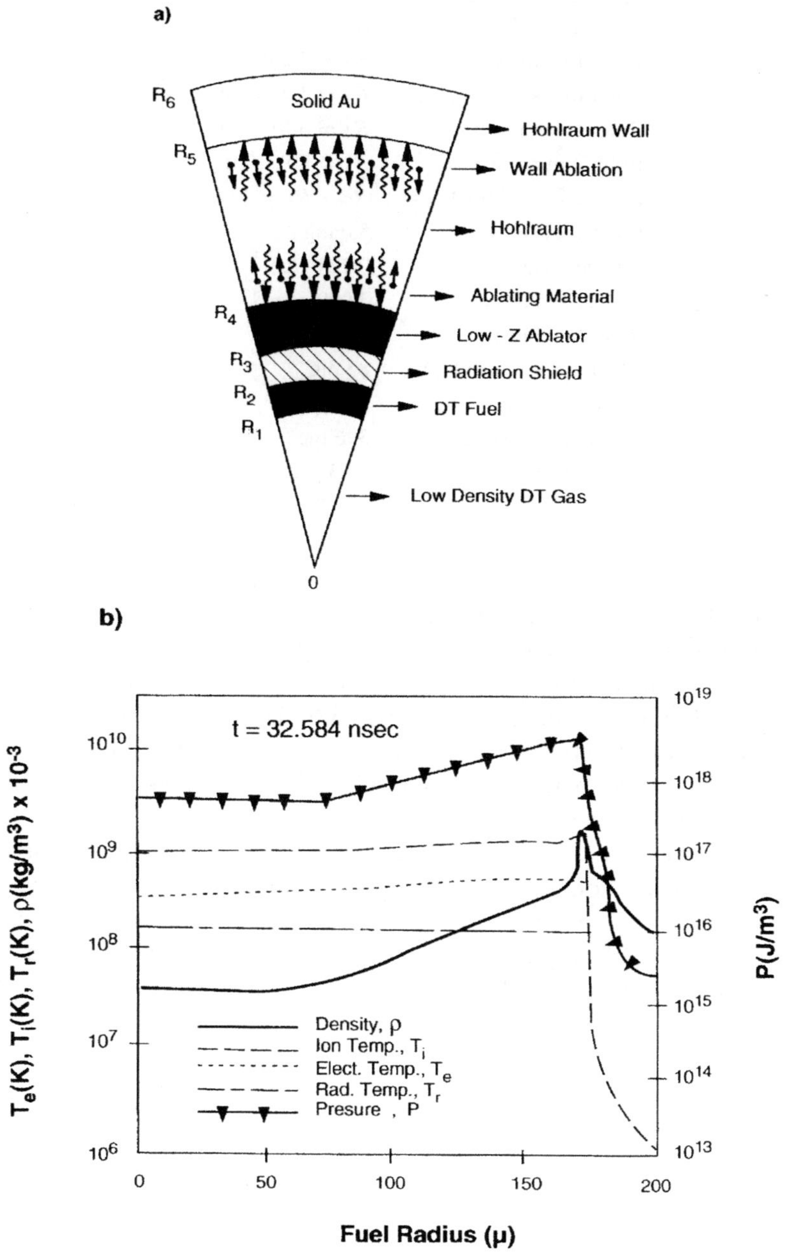

Figure 4. 1D simulation of inner capsule compression. (a) Capsule design (Table II). (b) Final compression stage. The ignited hot spot spreads all over the compressed fuel.

TABLE II
Capsule and target parameters.

Inner fuel radius	$R_1 = 3.020$ mm
Outer fuel radius	$R_2 = 3.206$ mm
Outer rad. shield radius	$R_3 = 3.248$ mm
Outer ablator radius	$R_4 = 3.516$ mm
Hohlraum radius	$R_5 = 10.510$ mm
Hohlraum wall thickness	$R_6 - R_5 = 15.0\ \mu$m
DT fuel mass	5.0 mg
DT fuel density	0.224 g/cm^3
Rad. shield density	2.000 g/cm^3
Rad. shield mass	11 mg
Ablator mass (solid C)	85 mg
Ablator density	2.20 g/cm^3
Casing mass (solid Au)	396 mg
Casing density	19.3 g/cm^3

In Figure 4b we plot the electron temperature, T_e, the ion temperature, T_i, the radiation temperature, T_r, the density, ρ and the pressure P vs. the capsule radius at $t = 32.584$ ns, when burn has spread into the entire DT fuel. The latter has then absorbed 3.8 MJ radiation energy while about 5.6 MJ radiation energy is lost to the casing, when the target gets 9.4MJ of CIB kinetic energy. The implosion is shown to yield 965 MJ output energy with a capsule gain $G_c \sim 254$. The overall target gain drops to a still handsome $G_r \cong 100$.

In summary, we have conducted the first investigation of indirectly driven ICF through intense cluster ion beams. Making use of much shorter ion ranges in uniformly irradiated convertors, we have been able to demonstrate em unusually high conversion efficiency of projectile kinetic energy into hard X-rays altogether with a radiation temperature in the 300–600 eV for a driver energy spanning the [1–100] MJ interval. We expect those encouraging results to foster more investigation on the production and acceleration of intense cluster ion beams.

References

Bangerter, R.O.: 1993, *Nuo. Cim.* **106A**, 1445.
Bodner, S.J.: 1981, *J. Fusion Energy* **1**, 219.
Deutsch, C. and Fromy, P.: 1995, *Phys. Rev. E* **51**, 632.
Deutsch, C. and Tahir, N.A.: 1992, *Phys. Fluids B* **4**, 3735.
Eliezer, S., Martinez-Val, J.M. and Deutsch, C.: 1995, *Laser Part. Beam* **13**, 43.
Ho, D.-M., Lindl, J.D. and Tabak, M.: 1994, *Nucl. Fusion* **34**(8), 1081.

Kang, K.H., Lutz, K.-J., Tahir, N.A. and Maruhn, J.A.: 1993, *Nuo. Cim.* **106A**, 1865.
Kidder, R.E.: 1979, *Nucl. Fusion* **19**, 223.
Kroto, H.: 1996, *Science* **271**, 877.
Larsson, M.: 1995, *Rep. Prog. Phys.* **58**, 1267.
Lindl, J.D.: 1995, *Phys. Plasmas* **2**, 3933.
Lindl, J.D., Bangerter, R.O., Mark, J.W.-K. and Pan, Y.L.: 1986, in: M. Reiser, T. Godlove and R. Bangerter (eds.), *AIP Conf. Proc. 152: Heavy Ion Inertial Fusion*, American Institute of Physics, New York, p. 89.
Maschke, A.W.: 1984, in: Hirao, Y. (ed.), *Proceedings of HIF 84 International Symposium on Heavy Ion Fusion*, Institute for Nuclear Study, University of Tokyo, p. 168.
Murakami, M. and Meyer ter Vehn, J.: 1991, *Nucl. Fus.* **31**, 1315.
Ramis, R., Schmalz, R. and Meyer-ter-Vehn, J.: 1988, *Comput. Phys. Comm.* **49**, 475.
Scheier, P. and Märk, T.D.: 1994, *Phys. Rev. Lett.* **73**, 54.
Tahir, N.A. et al.: 1982, *Nucl. Fusion* **32**, 581.
Tahir, N.A., Long K.A. and Laing, E.W.: 1986, *J. Appl. Phys.* **60**, 898.
Tahir, N.A., Hoffmann, D.H.H., Maruhn, J.A. and Deutsch, C.: 1994, *Nucl. Instr. Meth. B* **88**, 127.

THE INTERACTION PHYSICS OF THE FAST IGNITOR CONCEPT

C. DEUTSCH, H. FURUKAWA, K. MIMA, M. MURAKAMI and K. NISHIHARA
Institute for Laser Engineering, Osaka University, Suita 565, Osaka, Japan

Abstract. The interaction of relativistic electrons produced by ultrafast lasers focussing them on strongly precompressed thermonuclearfuel is analytically modelled. Energy loss to target electrons is treated through binary collisions and Langmuir wave excitation. The overall penetration depth is determined by quasielastic and multiple scattering on target ions. It thus appears possible to ignite efficient hot spots in a target with density larger than 300 g/cc.

Since its inception (Tabak et al., 1994) in 1994, the so-called Fast Ignitor Scenario (FIS) proposed to ease the indirect drive approach to inertial confinement fusion (Lindl, 1995) of hollow pellet containing the thermonuclear fuel deuterium + tritium (DT) has been the object ofmany intensive investigations.

Most of them, conducted through numerical simulations (Bonnaud and Lefebvre, 1995; Busquet, 1995a; Pukhov and Meyer-Ter-Vehn, 1996), have already addressed basic issues concerning the capability of ultrafast lasers with irradiance $I \geq I_{18} \equiv 10^{18}$ W/cm^2, to bore a channel in the corona of the precompressed target fuel. The latter is expected to be prepared through powerful lasers (Lindl, 1995) or intense heavy ion beams (Deutsch, 1986) suitably synchronized. This novel and time segmented scenario (Tabak et al., 1994), has to be appreciated as the latest sophistication elaborating upon the already overexploited disparity between the cheap compression cost ($\sim n^{2/3} \sim 1.1 \times 10^7$ J/g) of strongly degenerate Fermi electrons and the high toll requested for plasma heating ($\sim 6 \times 10^8$ J/g). In these regards, a very significant improvement previously introduced is the hot spot ignition of a small fraction (a few percent only) of the cold compressed fuel (Lindl, 1995).

Our main concern in the present work is to investigate analytically within a simple model the coupling (Wilks, 1992) to a precompressed target of the relativistic electron beams (REB) in the MeV energy produced by the ultrafast lasers currently under development with a supercompressed DT fuel (Sakabe, 1995). More specifically, we intend to critically investigate the REB capabilities to igniting hot spots well localized within the overal spherically, supercompressed DT.

Following recent numerical simulations (Bonnaud and Lefebvre, 1995; Busquet, 1995a; Pukhov and Meyer-Ter-Vehn, 1996), we take it for granted that the REB propagation may be approximated in a cylindrical geometry on an acceleration dis-

Astrophysics and Space Science **256:** 161–168, 1998.

tance $\sim$ 200 μm through a very steep density gradient (Busquet, 1995b). Nonetheless, we shall work our model in a homogeneous approximation with a fixed beam density n_b and a target plasma density n_p. Here we are striving for proof of principle arguments rather than for quantitative accuracy.

The REB interaction physics contemplated here looks like a remake of a very similar one evisionned in the mid-seventies (Yonas, 1979) for a REB direct driven compression. However, the present targets are at least 500 times more dense than the former ones. In order to coordinate their proposal, Tabak et al. (1994) advocate a 6 μm range for 1 MeV electrons fully stopped in 300 g/cc DT fuel. Such a figure looks reasonable for igniting a hot spot much smaller than the surrounding dense compressed DT core. However, the uncertainties still plaguing the REB direct drive approach (Yonas, 1979; Sudan, 1988; Nardi and Zinamon, 1978), through a solid density target, make it compulsory to pay a thorough attention to similar issues encountered in the present scenario. The REB energy range considered is fixed by the laser irradiance through the relationship ($\lambda_\mu = 1$ μm) (Wilks, 1993)

$$T \text{ (MeV)} = 0.51 \left[\left(1 + 0.7 \left(\frac{I}{I_{18}} \right) \lambda_\mu^2 \right)^{1/2} - 1 \right] \tag{1}$$

with $1 \leq (I/I_{18}) \leq 20$ so $0.152 \leq T$ (MeV) ≤ 1.44.

FIS prescriptions (Tabak et al., 1994) recommend a 3 kJ REB energy at 1 MeV for hot spot ignition. This amounts to a current $\sim 3 \times 10^8$ Amp. Considering a compressed core radius $\sigma \cong 50$ μm a plausible channel radius $a = \sigma/4$ yields an average beam density $n_b \sim 1.3 \times 10^{22}$ e-cm^{-3}. It should also be noticed that a 300 g/cc DT core at the usual 5 keV temperature (Lindl, 1995) appears as a fully ionized and weakly coupled ($\Lambda \sim 5 \times 10^{-3}$) hydrogenic plasma with $n_p \sim 10^{26}$ e-cm^{-3}.

As far as the REB propagation is concerned, we consider a monochromatic beam. Any transient collective modes excited in the channel is expected to be instantaneously collisionally damped according to (T_p = target temperature) (Nardi and Zinamon, 1978).

$$\frac{n_p \text{ (cm}^{-3})}{T_p^3 \text{ (eV)}} \geq 10^{11}, \tag{2}$$

easily fulfilled in the present situation.

A noticeable exception might be afforded by the electromagnetic filamentation (Weibel) through transverse modes, one thus encounters a situation quite germane to the filamentation of intense ion beams (Deutsch, 1986). However, the presently produced REB within compressed core do not have to satisfy the same stringent focussing conditions. Moreover, very reccent 2D numerical simulations (Adam and Guérin, 1996) makes it clear that the REB velocity is mostly axial for the first 20–25 femtoseconds. Then the beam starts also a lateral expansion, which is likely to stabilize the filamentation process (Deutsch, 1986).

Moreover, some insight into the REB axial velocity distribution may be gained from 2D simulations (Adam and Guérin, 1996) showing it as a superposition of a low velocity plateau flanked with a high and narrow velocity component. So, the first part might be used for uniform channel heating while the second one is essentially modelled by the monochromatic beam considered here which penetrates the dense compressed core.

This huge current is expected to pinch the REB through an azimutal magnetic field $B_0 \sim 4.8 \times 10^{10}$ G. Nonetheless, the linear beam density is still able to secure a very high space charge with Budker parameter $\geq 10^3$ so the standard Alfvén–Lawson limit is easily overcome. Such enormous B_0 values could a priori be a concern for the REB-target interaction itself.

Fortunately, the resulting target electron Larmor radius $\sim 3.8 \times 10^{-8}$ cm still remains much larger than the corresponding Debye length $\sim 5 \times 10^{-9}$ cm and the mean particle interdistances $\sim 1.35 \times 10^{-19}$ cm in a DT compressed at 300 g/cc and 5 keV temperature.

As a consequence, it is the target density which takes responsibility for shaping the energy loss and multiple scattering processes which we consider now.

The ratio $n_b/n_p \leq 10^{-4}$ demonstrates that in spite of its huge current the REB should be taken dilute in the overcompressed target with a mean electron interdistance larger by a good order of magnitude compared to that in the target. The REB-target interaction is then reducible to that of a linear superposition of isolated charges. Focussing attention on the most significant stopping mechanisms, we include binary electron-electron collisions through a plasma-adapted Möller expression (Val'chuk et al., 1995)

$$-\frac{dE}{dx} = \frac{2\pi n_p e^4}{m_e \beta^2 c^2} \times \left[\ell n \frac{1}{2\tau_{\min}} + 0.125 \left(\frac{\tau}{\tau+1} \right) - \frac{(2\tau+1)}{(\tau+1)^2} \ell n 2 + 1 - \ell n 2 \right], \quad (3)$$

with $\tau_{\min}$ = ratio of projectile electron wavelength λ_e to target Debye length λ_D, $\tau = \gamma - 1$ in terms of the usual Lorentz parameteres $\beta = V/c, \gamma = (1-\beta^2)^{-1/2}$, and the excitation of Langmuir collective modes (Bohm and Pines, 1952)

$$-\frac{dE}{dx} = \frac{2\pi n_p e^4}{m_e \beta^2 c^2} \times \ell n \left[\frac{V}{\omega_p \lambda_D} \times \left(\frac{2}{3} \right)^{1/2} \right]^2, \quad (4)$$

in terms of the target electron plasma frequency ω_p.

We are entitled to restrict to a continuous slowing down approximation because large and sudden energy losses are likely to happen very rarely. This point is well documented by the fact that a bremsstrahlung contribution comparable to the above ones would request a beam energy $\sim 800/Z + 1$ MeV much larger ($Z = 1$) than those considered here. Also, electron-position pair production remains totally negligible below 10 MeV (Shearer et al., 1973). However, in contradistinction to the simpler ion stopping (Deutsch et al., 1989) case we have to give up the straigh line approximation for the projectile trajectory. Due attention now has to be paid to the quasielastic and highly erratic motion of the relativistic electrons experiencing multiples scattering on target ions. Such a process is essentially quantified by the square average deflection per unit path length ($Z = 1$, $A = 2$) (Hubbel and Birkoff, 1982; Nigam et al., 1959)

$$\lambda^{-1}\ (\text{cm}^{-1}) = 8\pi \left(\frac{e^2}{m_e c^2}\right)^2 \times \frac{Z(Z+1)}{A\beta^4}(1-\beta^2)$$
$$\times \left[\ell n\left(\frac{137\beta}{Z^{1/3}(1-\beta^2)}\right) + \ell n(1.76) - \left(1 + \frac{\beta^2}{4}\right)\right]. \quad (5)$$

Putting together the stopping contributions (3) and (4) allows us to compute a continuous range for a 90% energy loss of 1 MeV electron, as

$$R = \frac{m_e c^2}{4\pi n_p c^4} \times \int_{0.3025}^{0.8836} \frac{dV}{(1-V)^{3/2}} \times D(V)^{-1} = 34.66\ \mu\text{m}, \quad (6)$$

with

$$D(V) = \ell n(68.53V) + \ell n(68.026V^{1/2})$$
$$+ \frac{(1-\sqrt{1-V})^2}{8} - (2\sqrt{1-V} + V - 1) + 1 - \ell n 2$$

for a stopping target with 300 g/cc and 5 keV. It should be noticed that the range (6) in a hot target with classical electrons is larger than the corresponding one at $T = 0$ with fully degenerate ones. The latter being notoriously less responsive to the projectile electric field (Deutsch et al., 1984).

This R value is comparable with the core extension σ. So, we really need an efficient packing mechanism to wind the projectile trajectories within a much smaller domain in the compressed core. This winding process is easily quantified by the maximum penetration depth ℓ_0 of the given REB.

The precise calculation of this parameter has for many years been the hard core of that subfield of nuclear and particle physics devoted to particle detection (Rossi, 1963). In the present context, we find it very useful to use a classical result due to Hemmer and Farquar (1968) which is based on stochastic arguments summarizing a great deal of previous efforts. So, considering a slab thickness containing every projectile trajectory whatever their orientation with respect to beam axis is, one gets the simple relationship

$$R = \ell_0 + \frac{1}{2}\frac{\ell_0^2}{\lambda} + \frac{1}{2}\frac{\ell_0^3}{\lambda^2}, \quad (7)$$

between the continuous winded range R and the maximum penetration depth ℓ_0.

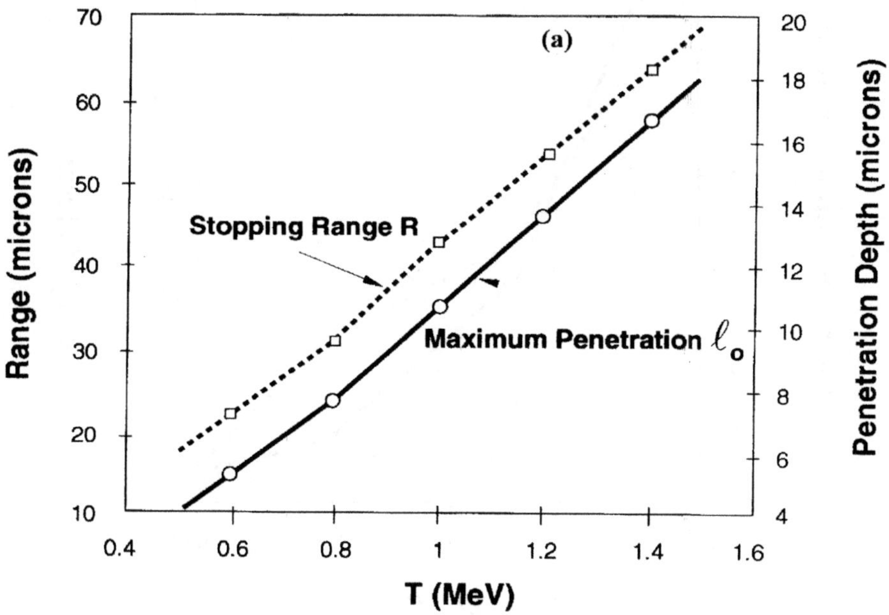

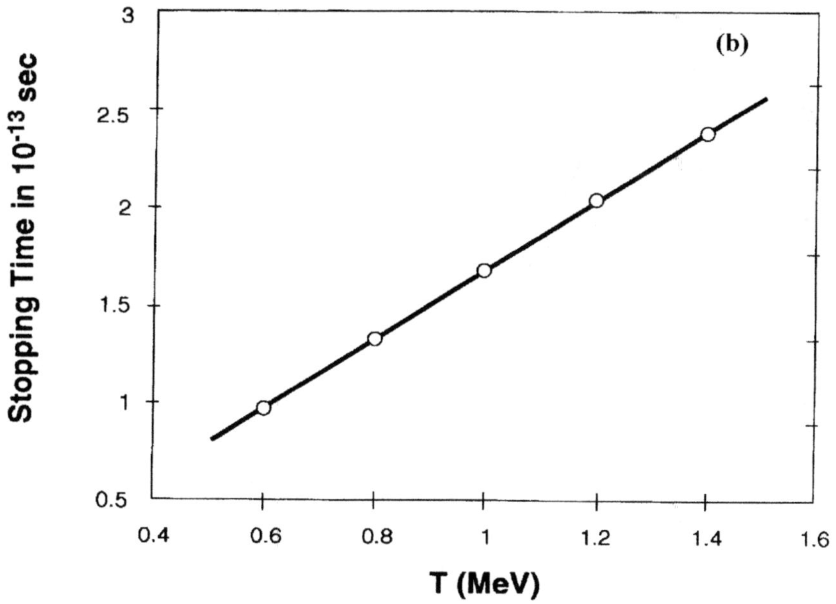

Figure 1. (a) REB range R (μm) and maximum penetration depth ℓ_0 (cm) in a 300 g/cc DT target at 5 keV and $0.5 \leq T$ (MeV) ≤ 1.5. (b) Corresponding stopping time t_{stop}.

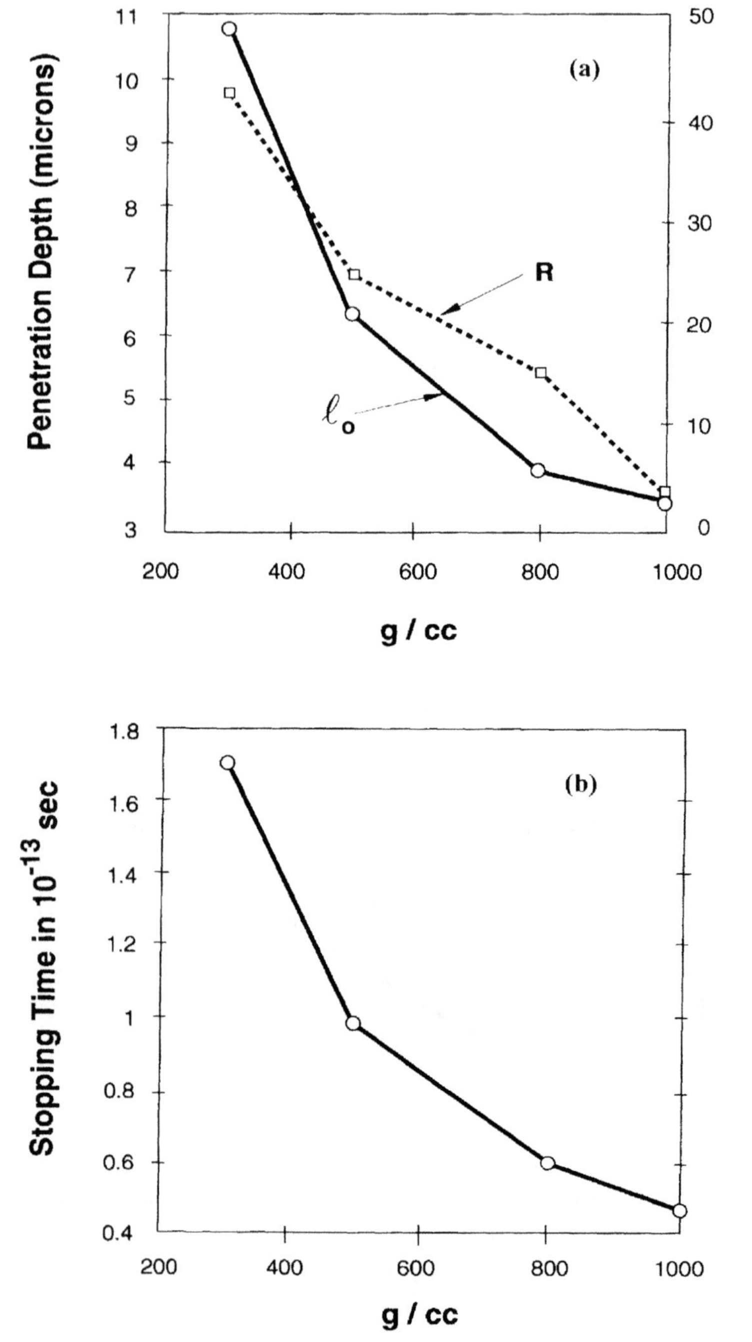

Figure 2. (a) R and ℓ_0 for $T = 1$ MeV, and target density ranging from 300 up to 1000 g/cc, with 5 keV temperature. (b) Corresponding stopping time t_{stop}.

Figure 1a features simultaneously R and ℓ_0 for the energy range $0.5 \leq T$ (MeV) ≤ 15. In order to qualify the FIS as a coherent ignition scenario, we have also to request a sufficiently short stopping time

$$t_{\mathrm{stop}} = -\frac{1}{c} \times \int_{E_{\max}}^{E_{\max}} \frac{(1 + \frac{E}{m_e c^2})}{\left[\frac{E}{m_e c^2}\left(\frac{E}{m_e c^2} + 2\right)\right]^{1/2}} \times \frac{dE}{dE/dx} \,. \tag{8}$$

The corresponding one $\sim 10^{-13}$ sec exhibited in Figure 1b seems compatible with further equilibration time between hot electrons and thermonuclear ions in ignited target, which allows for an adequate hot spot extension highlighted by $4 \leq \ell_0$ (μm) ≤ 18.

Those results are indeed putting the interaction physics of the FIS on serious grounds. However, other related issues such as the conversion efficiency of the laserlight into energetic REB also have to be positively addressed. Such concerns are motivating some authors (Wilks et al., 1992; Gibbon and Bell, 1992) to advocate an even more precompressed DT core. This explains that we investigate in Figure 2, the target density density dependence of the above results for a 1 MeV REB and a target density between 300 and 1000 g/cc. Then, we really witness a drastic reduction of R, ℓ_0 and t_{stop}.

On the other hand, we also proceeded to a systematic variation of the target temperature between 1 and 5 MeV. The above results are then left practically unchanged. Large T variations remains below 2.5%.

In summary, we have demonstrated through a simple but efficient analytical modelling of the laser produce REB in interaction with the supercompressed DT core, that the fast ignitor scenario is actually able to yield thermonuclear ignition. Additional insight should be gained by incorporating realistic boundary conditions to the REB propagation scheme.

Acknowledgements

It is a great pleasure to thank many experts who have significantly contributed to the present work through pleasant discussions. Amongst them, we owe special thanks to G. Bonnaud, M. Busquet, M. Descroisette, Y. Kato, E. Lefebvre, J. Lindl, J. Meyer-ter-Vehn, P. Mulser, M. Tabak and Y. Yamanaka.

References

Adam, J.C. and Guérin, S.: 1996, Private Communication.
Bohm, D. and Pines, D.: 1952, *Phys. Rev.* **85**, 338.
Bonnaud, G. and Lefebvre, E.: 1995, *Phys. Rev. Lett.* **74**, 200.
Busquet, M.: 1995a, *Chocs* **13**, 57.

Busquet, M.: 1995b, Private Communication.

Deutsch, C.: 1986, *Ann. Phys. Fr.* **11**, 1.

Deutsch, C., Maynard, G. and Minoo, H.: 1984, *Las. Int. Rel. Plasma Phom.* **6**, 1010.

Deutsch, C., Maynard, G., Bimbot, R., Gardès, D., Servajean, A., Fleurier, C., Hoffmann, D. and Weyrich, K.: 1989, *Nucl. Instr. Meth.* **A278**, 38.

Gibbon, P. and Bell, A.R.: 1992, *Phys. Rev. Lett.* **68**, 1535.

Hemmer, P.C. and Farquar, I.E.: 1968, *Phys. Rev.* **168**, 294.

Hubbel, H.H. and Birkoff, R.D.: 1982, *Phys. Rev.* **A26**, 2460.

Lindl, J.: 1995, *Phys. Plasmas* **2**, 3933.

Nardi, E. and Zinamon, Z.: 1978, *Phys. Rev.* **A18**, 1246.

Nigam, P., Sundaresan, M.K. and Ta-You-Wu: 1959, *Phys. Rev.* **115**, 491.

Pukhov, A. and Meyer-Ter-Vehn, J.: 1996, *Phys. Rev. Lett.* **76**, 3995.

Rossi, B.: 1963, *Rev. Mod. Phys.* **35**, 23.

Sakabe, S.: 1995, Private Communication.

Shearer, J.W., Garrison, J., Wong, J. and Swain, J.E.: 1973, *Phys. Rev.* **A8**, 1582.

Sudan, R.N.: 1988, in: A.A. Galeev and R.N. Sudan (eds.), *Basic Plasma Physics II*, N.H.P.C., Amsterdam, Chap. 3, p. 337.

Tabak, M., Hammer, I., Glinsky, M.E., Kruer, W.L., Wilks, S.C., Woodworth, J., Campbell, E.M., Perry, M.D. and Mason, R.J.: 1994, *Phys. Plasmas* **1**, 1626.

Val'chuck, V.V., Volkov, N.B. and Yalovets, A.P.: 1995, *Plasma Phys. Rep.* **21**, 159.

Wilks, S.C.: 1993, *Phys. Fluids* **B5**, 2603.

Wilks, S.C., Kruer, W.L., Tabak, M. and Langdon, A.B.: 1992, *Phys. Rev. Lett.* **69**, 1383.

Yonas, G.: 1979, *IEEE Trans. Nucl. Sci.* **NS-26**, 610.

CHARGE COMPENSATED ION BEAM PROPAGATION IN A REACTOR SIZED CHAMBER

J.L. VAY and C. DEUTSCH
LPGP, Université Paris XI, Bât. 212, 91405 Orsay, France

Abstract. Final ion propagation in the space charge compensated scheme (ITEP) is contrasted to Hylife II scenario. For the latter, we identify a partial beam neutralization only through electron background. The former displays an acceptable focalization at pellet. The background electron temperature has a significant influence on beam minimum radius. Transverse emittance is given a specific attention.

Energetic and intense nonrelativistic ($V_b \sim c/3$) heavy ion beams focussed on tiny and hollow pellets with a thermonuclear deuterium + tritium (DT) fuel are now routinely considered for producing an inertially driven compression (Barnard et al., 1996; Reiser, 1994; Deutsch, 1996). Up to now, a lot of attention has therefore been given to ion-target interaction (Barnard et al., 1996; Reiser, 1994; Deutsch, 1996) and beam dynamics as well (Reiser, 1994; Deutsch, 1996; Friedman et al., 1992). However, in order to evaluate carefully the characteristics and merits of the whole concept elaborating on the scheme of heavy ion driven inertial fusion one also needs a complete end to end and multidimensional simulation of the ion beams successive manipulation all along their way, from ion sources to pellet. In particular, we intend to pay here thorough attention to the ion beam propagation in target vicinity, within reactor chamber, where the thermonuclear fusion energy is mostly converted into energetic neutrons and hard X-rays.

Outstanding engineering and system studies have already been devoted to the reaction chamber part of the whole accelerator scenario (Barnard et al., 1996; Moir, 1991, 1992). However, the extreme technical performances imposed upon the materials involved as well as the very severe time constraints requested by an implosion duration in the several tens of nanosecond range demand that the relevant interaction physics be carefully cleared up, at a basic level. Specifically, one has to check that the inflight stability with no emittance growth can be secured for kiloamperes of intense ion beams propagation under the influence of their own space charge while interacting with a residual pressure of Flibe (BeF_2 + LiF) gas around pellet.

In particular, due attention has to be paid to the interaction between ion beam and the ion debris resulting from the Flibe molecules fragmentation. The basic

Astrophysics and Space Science **256:** 169–175, 1998.

TABLE I

Ion beam parameters used for contrasting Hylife II (positive ions) and ITEP (charge compensated) heavy ion beams in reaction chamber.

$M_{/\text{ion}}$	200 amu
$E_{/\text{ion}}$	10 GeV
N_{gas}	5×10^{13} cm^{-3}
I_{average}	3.125 kA
Longitudinal profile	parabolic
$R_{\text{init max}}$	6 cm
V_{th}	1×10^{-4} c
$\varepsilon_{x\ \text{rms}}$	5 mm.mrad
Focalization distance	3.14 m

issues underlying a clear understanding of the beam dynamics demand a first principle definition of the beam stability in its own reference frame.

Toward this goal, we have implemented a fully electromagnetic particle in cell (PIC) code in three dimensions (Vay and Deutsch, 1996, 1998) with careful treatment of boundary conditions.

We consider intense and monokinetic Pb^+ ion beams propagating typically at one-third speed light ($V_b = c/3$), in a converging geometry, to secure focalization on a few millimeters spot size on target. Fiducial beam parameters of current interest for heavy ion fusion (Barnard et al., 1996; Reiser, 1994; Deutsch, 1996; Friedman et al., 1992) are given in Table I.

Beam-Flibe gas interaction is essentially monitored by Pb^+ stripping and BeF_2 ionization. Total gas ionization cross-section is as $\sigma_{\text{ion}}^{\text{gas}} = 0.9 \times \sigma_{\text{ion}}^{\text{BeF}_2} + 0.1 \times \sigma_{\text{ion}}^{\text{LiF}}$ out of the linear approximations $\sigma_{\text{ion}}^{\text{BeF}_2} = \sigma_{\text{ion}}^{\text{Be}} + 2\sigma_{\text{ion}}^{\text{F}}$ and $\sigma_{\text{ion}}^{\text{LiF}} = \sigma_{\text{ion}}^{\text{Li}} + \sigma_{\text{ion}}^{\text{F}}$.

In order to release the strong and always pervading space charge constraints on the final focussing of the intense ion beams on target, one is tempted to reduce it drastically by considering a fully dynamical charge compensation. A significant step in this direction has recently been taken by Koshkarev et al. at ITEP (Moscow) (1993) by proposing to accelerate simultaneously positive and negative ions of same species and opposite ionicity at same energy. The corresponding hardware, outlined in Figure 1, is built on two identical and parallel low energy linac sections. Their highest energy ($\sim$ 1 MeV/u.m.a.) extremities are then bent toward each other.

Then successive funnelling of positive and negative ions can occur within a third and central linear track parallel to linacs.

Eight ion sources of eight of platinum ions ($Pt^{\pm}_{192}$, $Pt^{\pm}_{194}$, $Pt^{\pm}_{196}$, $Pt^{\pm}_{198}$) are divided into two symmetrical arrays according to the sign of ion charge and mass. Sources

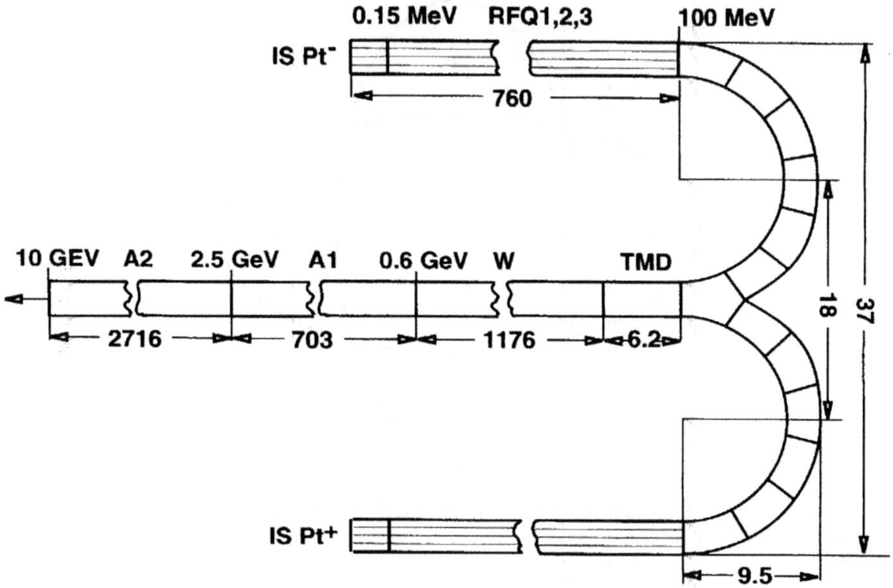

Figure 1. The total scheme of linac for charge compensated acceleration: IS – ion sources; RFQ1,2,3 – injector channels; TMD – transverse matching device; W – Wideroe channel; A1 and A2 – Alvarez channels; all lengths are measured in meters (Koshkarev et al., 1996).

product identical parallel beams: 0.15 MeV energy, 0.2 cm radius, transverse emittance of 24π cm-rad in each direction. The distance between arrays (measured from the middle axes) is 36 m. The sources of each array are positioned symmetrically with respect to the common axis of symmetry. The heaviest isotope source is located at largest distance from that axis.

The given negative ions are stable enough to sustain acceleration. Ion losses have to be kept below one percent during acceleration, accumulation and beam transport. In this context, it is a fortunate occurrence that heavy atoms with atomic mass > 120 display an enhanced electron affinity. As far as the manipulation of intense ion beams is concerned, advantages appear substantial.

The compensation of space charge forces through the simultaneous acceleration of negative ions permits the transport of very intense beams ($P \geq 100$ TW) in a unique channel.

Putting altogether ions with slightly different mass and same quantity of motion allows for a rather large velocity dispersion ($\pm$ 2%) in final focussing system while the dispersion of the quantity of motion remains below 0.4%.

– In contradistinction, to standard accelerator scenarios (Barkhudaryan et al., 1993), beam funnelling may be operated for several charge to mass ratios q/A. This induces a much smaller dilution in longitudinal phase space.

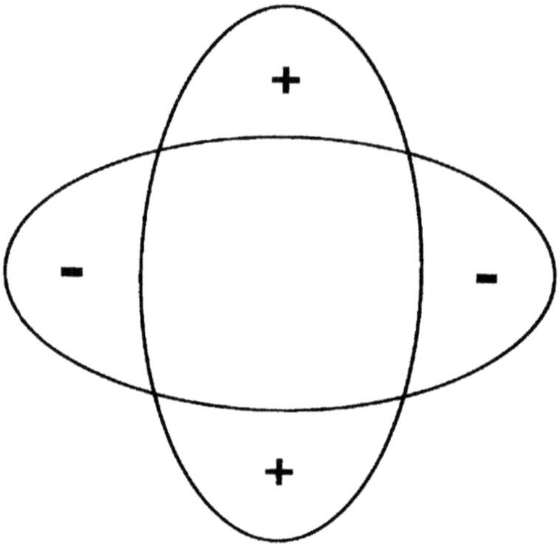

Figure 2. Transverse cross-section of charge compensated ion beam built on positive ($q = 1$) and negative ($q = -1$) ions with same mass flowing in crossed geometry.

- Compression factor can get significantly enhanced (≥ 200) altogether with a reduction of the radiofrequency (RF) electric field in view of the small phase space volume and compensation of the longitudinal space charge force.

Transverse cross-section of the presently considered space charge compensated beam is given in Figure 2. To get closer to the actually available acceleration linear geometries, we take it as a strong but not full overlap of two elliptical beams for each charge state.

Maximum ion temperature (200 keV) is three times smaller than the noncompensated one (cf. Figure 3). Apparent dissymmetry between positive and negative is only an artefact due to selecting a given beam cut in Figure 2. Perveance K parametrizes the beam space charge effects (Reiser, 1994). Electron temperature stay much lower than the ion's. Presently, the beam self-fields are much weaker in similar situations with no charge compensation. The plateau is now about 2–3 keV in lieu of 15 keV with positive ion beam only.

Figure 4 exhibits $\varepsilon_{x\ \mathrm{rms}}$ emittance for ion beams of either charge and the former three propagation scenarios. As far as practial applications to beam-target interactions are considered, it appears highly gratifying that the time dependence remains rather weak.

Finally, it appears now appropriate to extend the diagram beam radius-propagation penetration in reaction chamber in the laboratory system to the configurations of present concern. In Figure 5, we thus contrast the usual heavy ion chamber scenario (Hylife II) (Moir, 1991, 1992) to the space charge neutralized (Koshkarev, 1993; Koshkarev et al., 1993) for the same beam and chamber parameters (see Table I).

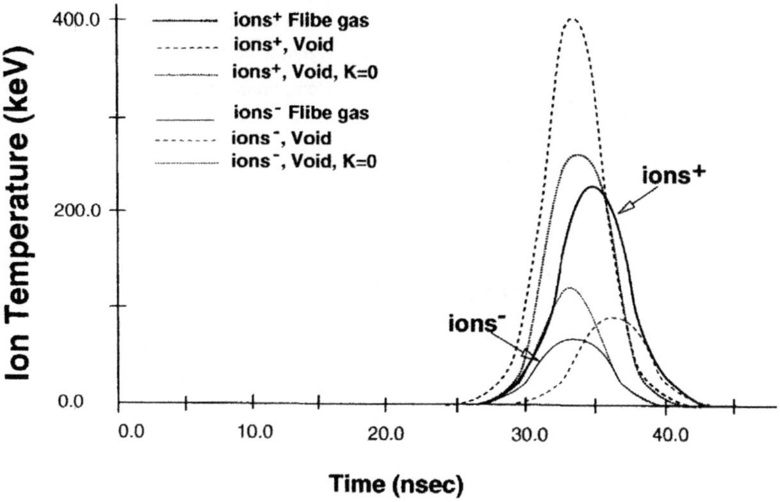

Figure 3. Time evolution of ion temperatures (over entire beam) for positive and negative ion beams in several distinct propagation schemes.

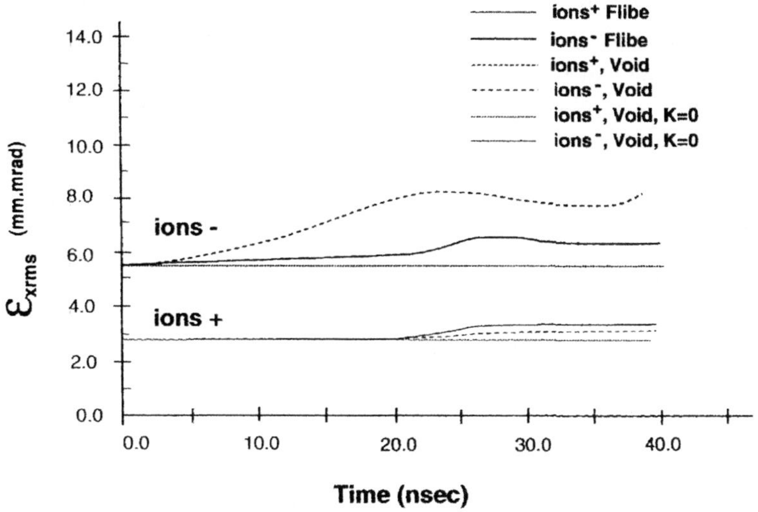

Figure 4. Time dependence of ε_{xrms} emittance (averaged over entire beam) for the propagation of a space charge compensated ion beam in Flibe, void with $K \neq 0$ and $K = 0$, respectively.

It appears highly rewarding that the space charge compensated beam converges to a much smaller radius ($\sim$ 2.5 mm) than the noncompensated Hylife II one ($\sim$ 5.7 mm). Provisionally, it may be stated that charge compensation might be of great help in securing adequate beam focalization on target.

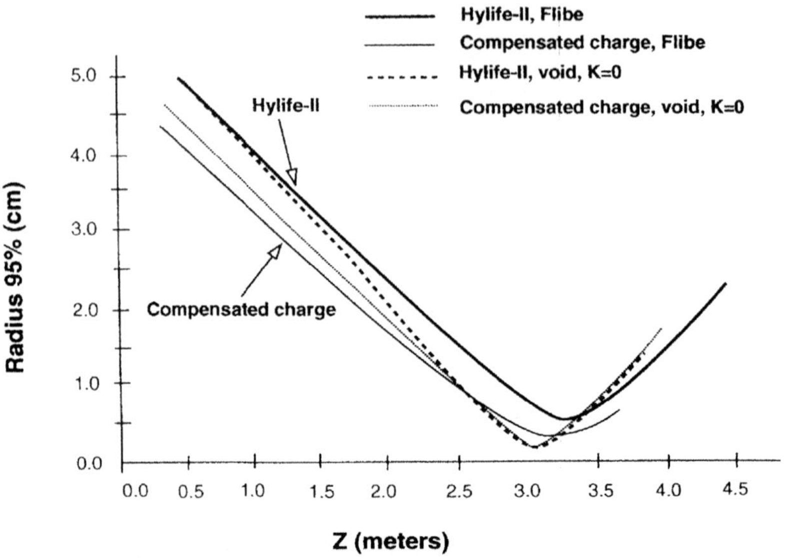

Figure 5. Beam radius in terms of penetration distance.

Electromagnetic simulation codes have been developed for investigating the final propagation of intense ion beams in a reaction chamber envisioned for driving inertially a thermonuclear pellet. They make use of a novel but simple wave absorption algorithm at walls of simulation grid. They provide access to the dependence of minimum focalization radius with respect to ion beam and Flibe gas stripping cross-sections. Those codes have unambiguously highlighted the significance of the background electron temperature for the focalization radius at pellet. We have also been able to proceed for the very first time to a detailed chamber analysis of a novel accelerator scheme advocating space charge compensation. For this case, it has been shown that a 2.5 mm beam spot size on pellet is achievable for ballistic transport with parameters of significance for heavy ion driven inertial fusion.

Acknowledgements

We have derived great benefits from fruitful discussions with J.C. Adam, R. Bangerter, N. Barboza, D.A. Callahan, A. Friedman, W. Herrmannsfeldt, D.G. Koshkarev, B. Langdon, R. Moir and M. Reiser.

References

Barkhudaryan, A.V., Koshkarev, D.G. and Zenkevich, P.P.: 1993, *Nuo. Cim.* **106A**, 1751.

Barnard, J.J., Fessenden, T.J. and Lee, E.P. (eds.): 1996, *Proc. of the International Symposium on Heavy Ion Inertial Fusion*, 6–9 Sept. 1995, Princeton, U.S.A., *Fus. Eng. Des.* **32–33**, 1.

Deutsch, C.: 1986, *Ann. Phys. Fr.* **11**, 1.
Deutsch, C.: 1996, *Fusion Technology* **29**, 306.
Friedman, A., Grote, D.P. and Haber, I.: 1992, *Phys. Fluids* **B4**, 2203.
Koshkarev, D.G.: 1993, *Nuo. Cim.* **106A**, 1567.
Koshkarev, D.G. et al.: 1996, *Fus. Engrg. Des.* **32–33**, 355.
Moir, R.W.: 1992, *Part. Accel.* **37–38**, 459.
Moir, R.W.: 1991, *Fusion Technology* **19**, 617.
Reiser, M.: 1994, *Theory and Design of Charged Particle Beams*, John Wiley and Sons, New York.
Vay, J.L. and Deutsch, C.: 1996, *Fus. Engrg. Des.* **32–33**, 467.
Vay, J.L. and Deutsch, C.: 1998, to be published.

SAWTOOTH INSTABILITY IN TOKAMAK PLASMAS

R. J. HASTIE
Imperial College, London, UK

UKAEA, Culham Science Centre, Abingdon, Oxon, UK, (UKAEA/Euratom Fusion Association)

Abstract. The Sawtooth instability is a familiar feature in Tokamak plasmas. It appears as a regularly recurring reorganisation of the core plasma. A brief survey of the experimental observations on many Tokamaks is presented. A qualitative description of the relevant theoretical ideas and of how they have evolved from early magnetic reconnection models in response to increasingly detailed experimental data, is also presented.

1. Introduction

The Tokamak is a toroidal plasma confinement device in which a combination of toroidal and poloidal magnetic fields is employed to confine hot plasma (Figures 1(a–d)). A strong toroidal magnetic field is created by external coils while the poloidal magnetic field is generated by currents in the plasma in the toroidal direction. This current is induced by transformer action or can be driven by launching electromagnetic waves (at ion cyclotron, lower hybrid or electron cyclotron frequencies) into the plasma. Injection of high energy beams of neutral atoms (NBI), with the primary objective of heating the plasma, also drives toroidal current. Typical plasma density and temperature in the plasma core (near the magnetic axis in Figure 1) are $n_e \approx 10^{20}$ m^{-3} and $T \approx 1 - 10$ keV, while the energetic ions created by neutral beam injection have energies up to 100 keV and those ions heated by resonance with ion cyclotron waves have energies up to, and in excess of, 1 MeV. Typically the density, $n_e(r)$, temperatures $T_e(r)$, $T_i(r)$ and toroidal current density $j_\phi(r)$, are all peaked functions of the minor radius r, falling to small values at the plasma boundary ($r = a$), where the edge plasma is in direct contact with a set of limiting structures (limiter plasmas) or is defined by a magnetic separatrix (divertor plasmas). The different types of plasma boundary will be of little concern to the topic of this review which is devoted to events in the 'plasma core', a region which we can roughly define as that volume for which $r/a \lesssim 1/2$, (Figure 1(a)).

Generally speaking Tokamak plasmas are predicted, and observed, to suffer from a variety of instabilities; i.e. they do not exist in a totally quiescent state and their transport properties are not those predicted by collisional transport theory. It appears that fine scale instabilities (known as micro-instabilities) are usually

Astrophysics and Space Science **256:** 177–204, 1998.

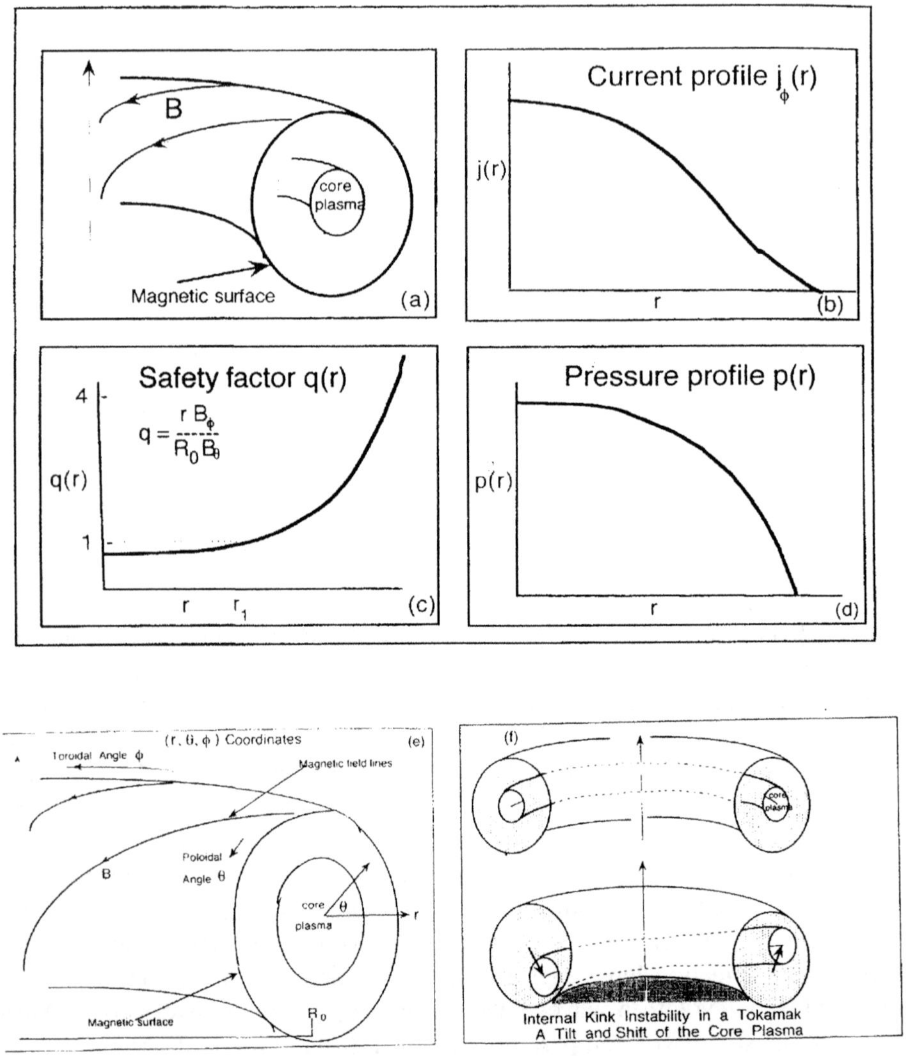

Figure 1. (a) Geometry of a Tokamak. (b), (c), (d), typical profiles for the toroidal plasma current $j_\phi(r)$, the field line winding number, or 'safety factor', $q(r)$ and the pressure $p(r)$. (e) The magnetic coordinates r, θ, ϕ. (f) Geometry of the internal kink instability in a torus.

present and determine local transport properties. It is also evident that certain large scale instabilities can occur and that, when they do, complete loss of plasma confinement often results. The stability boundaries for such gross instabilities therefore determine the normal operating regime of Tokamaks.

The Sawtooth instability is, however, an interesting exception to the above statement. It is a macroscopic instability which affects a significant volume of the plasma (the plasma core), but which is benign in the sense that termination of the discharge does not follow. The core plasma quickly recovers and the instability repeatedly occurs in a regular manner.

This review is concerned with the current understanding of the Sawtooth instability in Tokamak plasmas.

2. The Sawtooth Instability

The Sawtooth instability manifests itself as a regular periodic reorganisation of the core plasma surrounding the magnetic axis in Tokamaks. First reported by von Goeler et al. [1] in 1974, it is evident on many diagnostics in all Tokamaks, both large and small. The complete cycle of events can be subdivided into three regimes: (i) the Sawtooth ramp phase; a quiescent period during which the plasma density and temperature increase approximately linearly with time (see Figure 2) in the plasma core; (ii) the precursor oscillation phase, during which a helical magnetic perturbation grows and, due to toroidal rotation of the plasma, the soft X-ray (SXR) emission, electron cyclotron emission, ECE, (measuring $T_e(r)$) and interferometric data (measuring $n_e(r)$) show growing oscillatory behaviour; (iii) the collapse phase, in which $n_e(r)$, $T_e(r)$ and SXR intensity fall rapidly to lower values. In ohmically heated Tokamaks the resulting modulations of $n_e(r, t)$ and $T_e(r, t)$ are less than 10% and the period of the complete cycle, τ_s, increases with the size of the device. In TEXTOR (major radius $R_0 = 1.75$ m) $\tau_s \simeq 20$ msec and in JET ($R_0 \simeq 3$ m) $\tau_s \simeq 100$ msec.

This review adopts a historical perspective to indicate how theoretical ideas and interpretation have developed in response to the increasingly detailed experimental data over the past two decades.

Initially, 1974, the only relevant theoretical calculations [2, 3] concerned the linear instability [2] and saturated amplitude [3] of the ideal MHD internal kink mode, with $m = n = 1$ (where m is the poloidal mode number and the toroidal mode number n is represented by $k_z R_0$ in a periodic cylinder model of a Tokamak), which occurs when the value of the safety factor, $q(r) = rB_\phi/R_0 B_\theta$, falls below unity on the magnetic axis; i.e. whenever the current profile, $j_\phi(r)$, becomes sufficiently peaked that the axial current density $j_\phi(0)$ exceeds the critical value $2B_\phi(0)/R_0$. The ideal internal kink instability in a toroidal plasma takes the form of a tilt and shift of the core plasma, (Figure 1(f)), in the region where the 'safety factor', q, is less than 1. The safety factor is a measure of the rate at which magnetic field lines wind around the symmetry axis as they pass around the magnetic axis ($r = 0$ in Figure 1(e)) i.e. field line tractories are defined by $d\phi/d\theta = q(r)$. Thus, on the magnetic surface where $q = 1$ all field lines close on themselves after just one toroidal (or poloidal) circuit (see Figures 1(c,e) for a characteristic $q(r)$ profile in a Tokamak and for a description of the magnetic coordinates (r, θ, ϕ)).

Using the theoretical results and understanding established in References 2 and 3, Kadomtsev [4, 5] proposed a mechanism for the complete sawtooth cycle. He

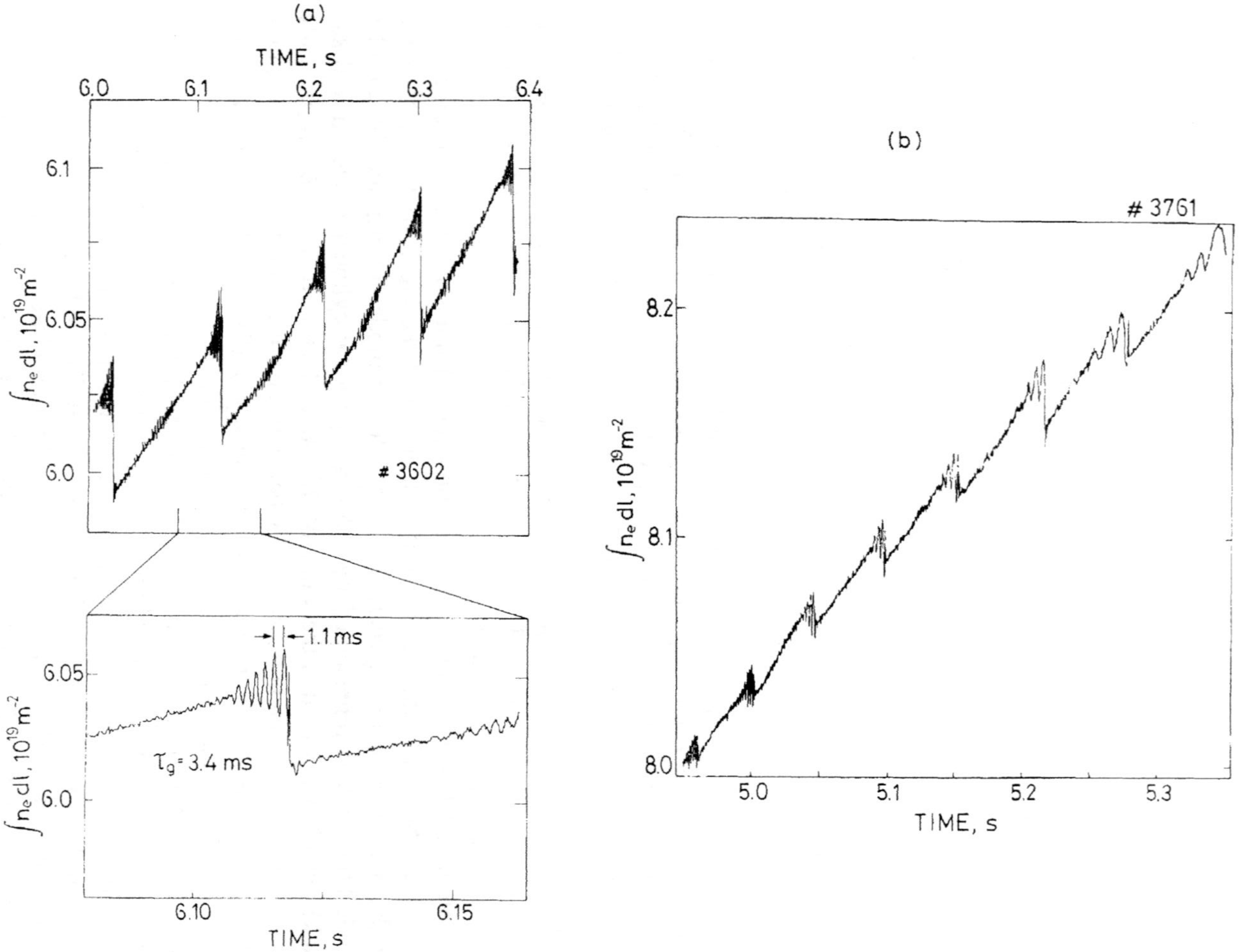

[illegible] discharges on the JET Tokamak, showing the ramp, precursor oscillation and collapse phases of the Sawtooth

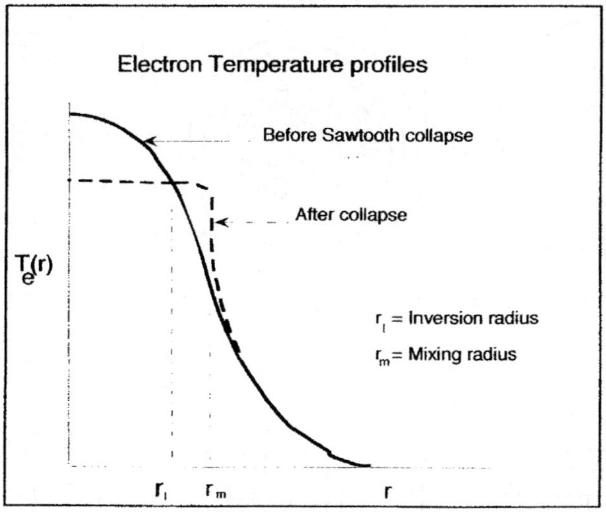

Figure 3. Typical electron temperatures $T_e(r)$, before and after the collapse phase of the Sawtooth instability. The mixing radius r_m defines the radial limit of the sudden change in $T_e(r)$, while the inversion radius r_I is defined as the point at which $\delta T = (T_{final} - T_{initial}) \simeq 0$.

visualised that the kink mode would drive magnetic reconnection of ψ^*, the flux through a helical ribbon bounded by the magnetic axis and a closed field line on the $q(r) = 1$ magnetic surface. Kadomtsev suggested that this reconnection process would continue until all $\psi^*(r) = \int_0^r r dr B_\theta(1-q) > 0$ was annihilated and a new, cylindrically symmetric, equilibrium was established with $q(0) > 1$. The final equilibrium state was calculated precisely in terms of the initial (at onset of the kink instability) state, and the time for the reconnection event to run to completion was estimated as $\tau_k \approx (\tau_\eta \tau_A^*)^{1/2}$ with $\tau_\eta = \mu_0 r_1^2/\eta$, $\tau_A^* = r_1(\mu_0\rho)^{1/2}/B_\theta(r_1)(1 - q_0)$, where $q(r_1) \equiv 1$. A consequence of this magnetic reconnection is that hot plasma near the magnetic axis ($r < r_1$) mixes with cooler plasma beyond the $q = 1$ surface ($r > r_1$), resulting in a flattened temperature (and pressure) profile (Figure 3) and a drop in $T(r = 0)$. It was then visualised that, during the succeeding ramp phase, ohmic heating in the new core plasma would cause temperature, and hence current, peaking so that $q(0, t)$ would fall below unity again and the cycle would repeat. One attraction of this elegant explanation is that it involves only a single mechanism; the weakness that, since there is no linear stability threshold for the $m = n = 1$ kink mode in cylindrical geometry if $q_0 < 1$, the quiescent ramp phase of the Sawtooth cycle is difficult to explain, and its duration is outside the scope of the model.

Following this proposed mechanism, numerical simulations [6, 7] (see Figure 4) were published and confirmed that a resistive MHD fluid in cylindrical equilibrium with $q(0) < 1$ did reconnect in such a fashion. Careful analysis of the soft X-ray emission from the TFR Tokamak during the precursor oscillation and collapse,

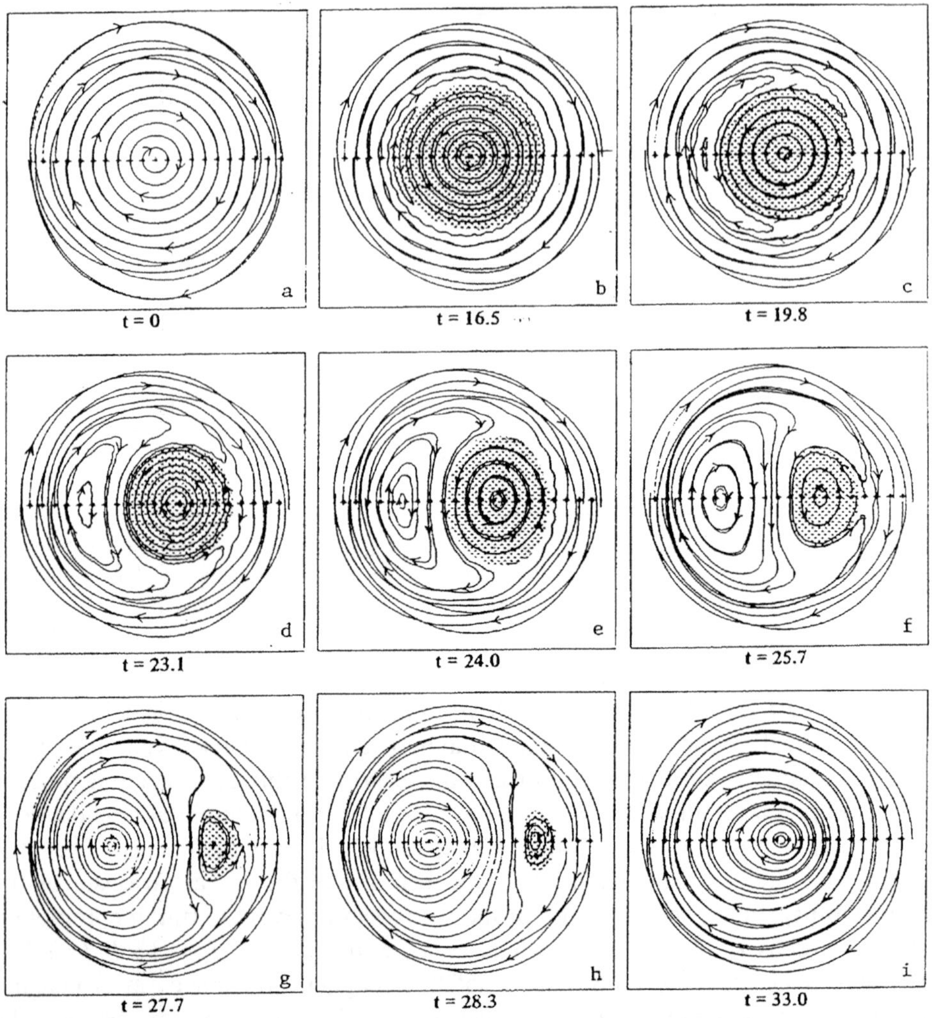

Figure 4. Approximate structure of magnetic surfaces during the reconnection process, showing growth of the, crescent-shaped, magnetic island, shrinking of the hot core plasma, until the island takes its place and a new cylindrically symmetric equilibrium is established (from Reference 6).

however, led Dubois et al. [8, 9, 10] to the conclusion that, although a reconnection process involving a change of magnetic topology, with $m = n = 1$ island growth, does commence as Kadomtsev suggested, this process appears to be interrupted by the sudden thermal collapse in the core when the magnetic island is still relatively small. These authors, therefore, proposed that a rapidly growing secondary instability is responsible for the observed thermal collapse, and concluded that it is uncertain whether or not the reconnection process proceeds to completion ($q(0) > 1$).

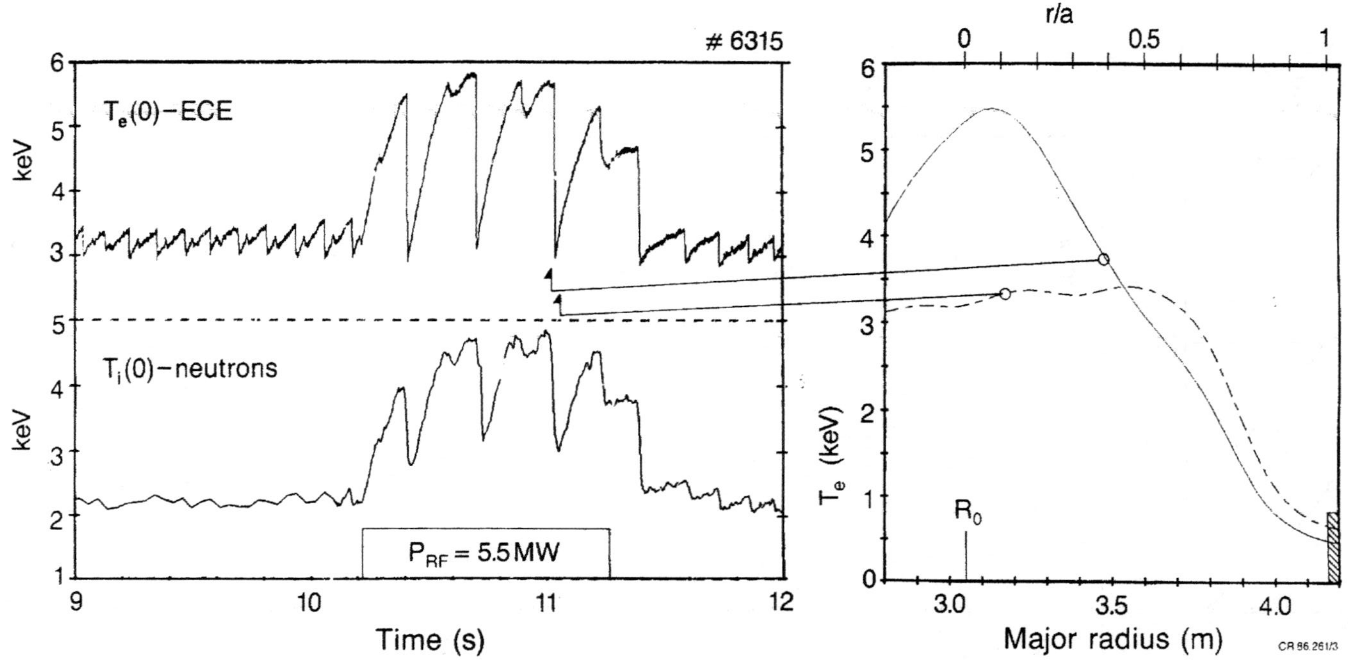

Figure 5. Longer Sawtooth periods after auxiliary heating is switched on suggest that the kink mode is not driven unstable by plasma pressure. Also note the dramatic collapse (50%) of the central electron temperature T_e in this JET discharge with auxiliary heating (ICRH).

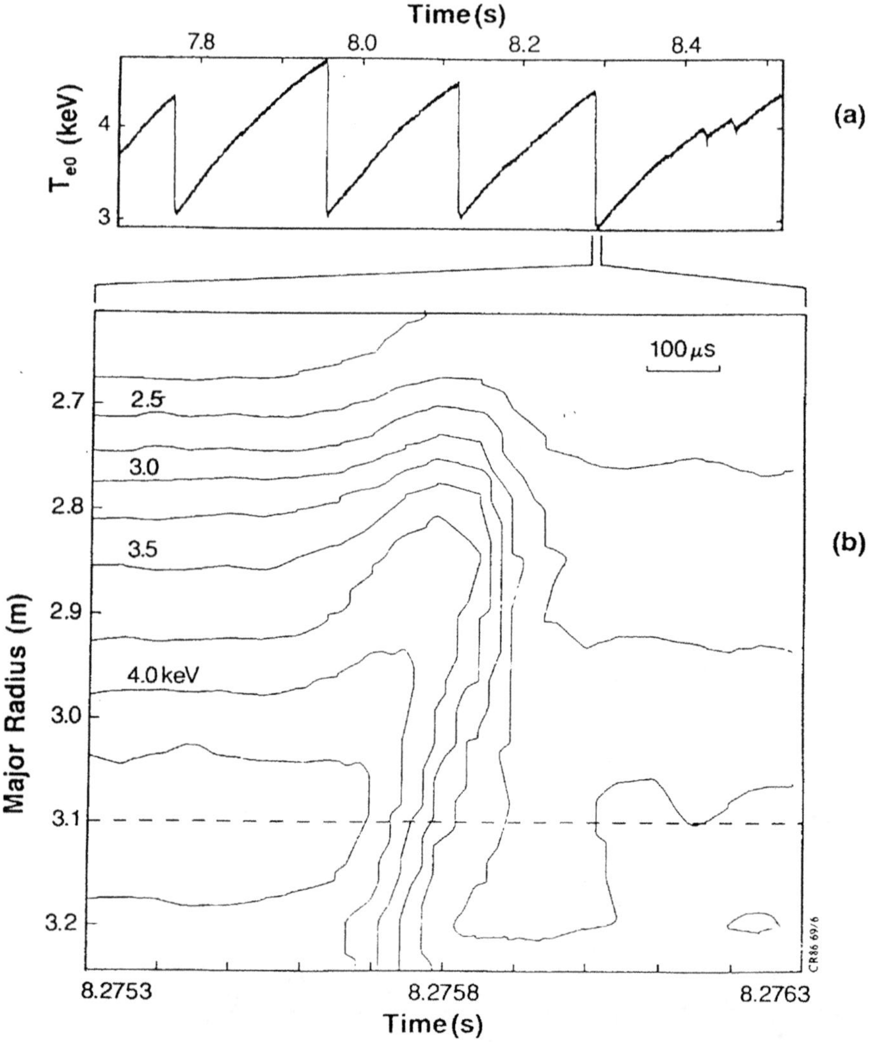

Figure 6. Fast Sawtooth collapse events without a precursor oscillation. The kink displacement of the plasma core is still a feature (Figure 6b), but occurs on the same 100μsec timescale as the thermal collapse. Figure 6b shows the electron temperature as a function of major radius R, and time t.

During the 1980's a new generation of large Tokamaks began operating and revealed new variations in Sawtooth behaviour which stimulated further theoretical activity. Partial, or compound, Sawteeth [11] were reported on the DIII Tokamak where, although the repetition period, τ_s, remained the same, normal Sawteeth collapse events alternated with events in which the thermal redistribution was localised to an off axis region, and the magnetic axis was largely unaffected. Parail and Pereversev [12] had suggested a modification of the Kadomtsev model in-

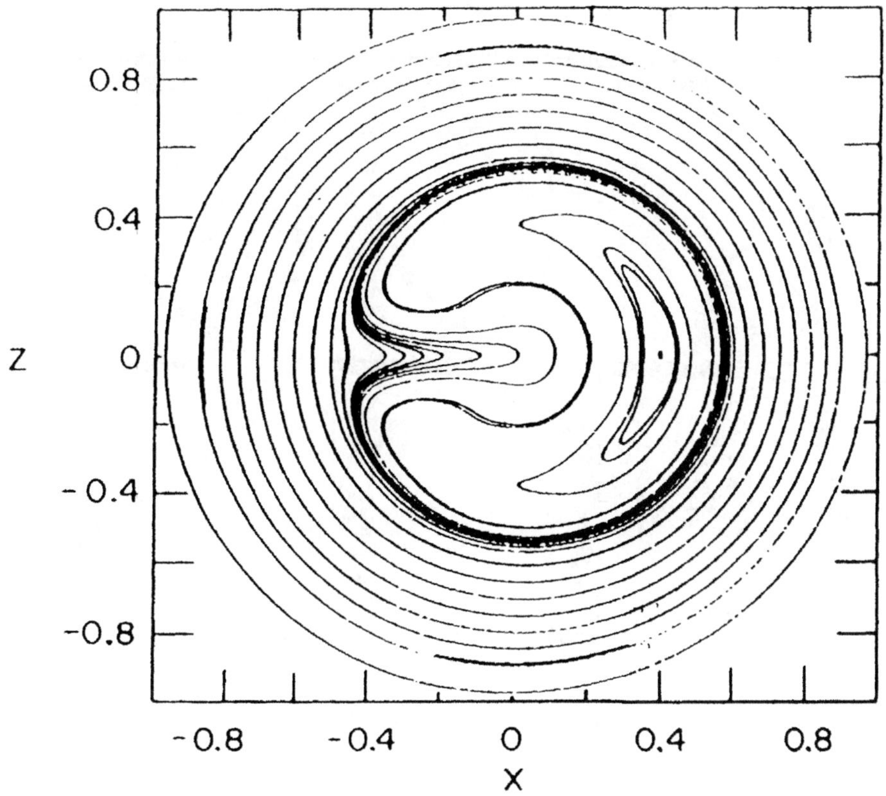

Figure 7. Magnetic surface shapes in the non-linear phase of the $m = n = 1$, Ideal MHD kink instability, when $q \sim 1$ in the plasma core. There is no reconnection. The hot core plasma distorts to a crescent shape and is pushed aside; a bubble of cooler plasma is drawn into the concave side of the crescent, in a 'quasi-interchange' motion.

volving a non-monotonic profile of $q(r)$ with two $q(r) = 1$ surfaces and Pfeiffer [11] extended this in an attempt to explain the new observations. In JET [13] auxiliary heating (both neutral beam injection, NBI, and ion cyclotron resonance heating, ICRH) resulted in a more rapid rise of temperature and pressure during the Sawtooth ramp, but longer Sawtooth repetition times. This suggests that the destabilisation of the MHD kink mode is not caused by plasma pressure ($\beta = 2\mu_0 p/B^2$), but by a 'magnetic trigger' (Wesson [14]), i.e. by some property of the current profile $j_\phi(r)$ or of the $q(r)$ profile. JET also reported Sawteeth in which the slowly growing precursor oscillation is absent [15] (Figure 5) so that the $m = n = 1$ kink instability appears suddenly with rapidly increasing growth rate γ ($\gamma^{-1} \approx 100$ μsec, at maximum growth rate, and rising to that value in 100 μsec). These observations in the mid 1980's led Wesson [15] to suggest that the MHD kink instability seen at the collapse phase might be of ideal MHD character (hence the fast growth). However, to make such instability possible in an ideal MHD model

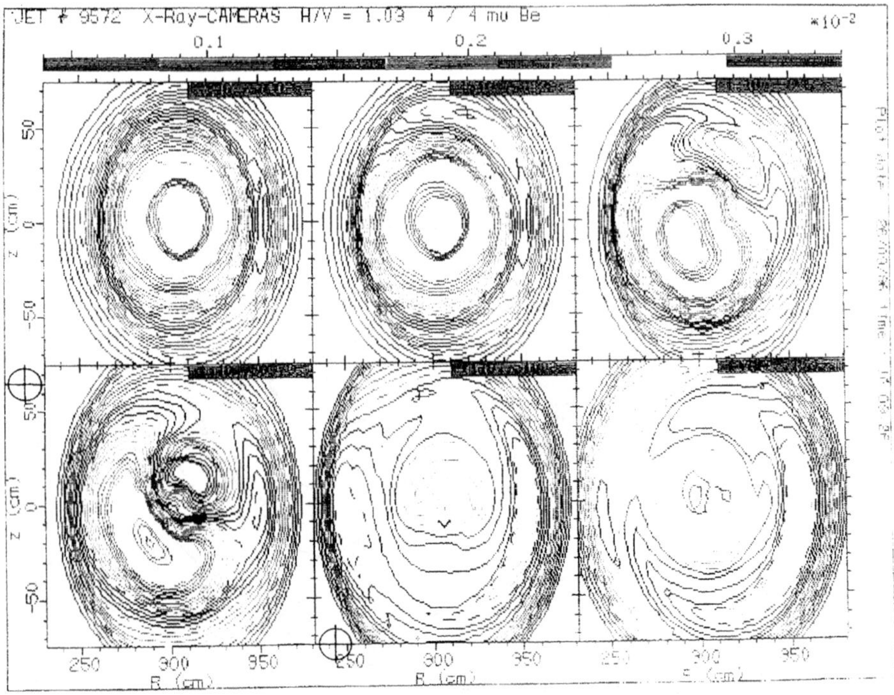

Figure 8. Contours of constant SXR emission during a Sawtooth collapse in the JET Tokamak, displaying the 'hot-crescent, cold-bubble' or 'quasi-interchange' structure.

at relatively low β, it is necessary that $q(r) \simeq 1$ in the entire core region to a high accuracy. In a region where the safety factor $q(r) \equiv 1$, the helical pitch of the magnetic field coincides exactly with the pitch of an $m = n = 1$ kink displacement and no magnetic field line bending occurs. Even a small deviation of $q(r)$ from unity requires significant field line bending energy in a kink displacement and consequent MHD stability. Thus Wesson was led to postulate that $q(r)$ is flat in the core and $q(r) = 1$. Non-linear MHD simulations (see Figure 7) show that the nature of the plasma displacement in such an ideal MHD instability differs considerably from that in a $q(0) < 1$ reconnecting instability. Magnetic surfaces are strongly distorted so that the hot plasma core becomes crescent shaped and a 'bubble' of cold plasma is drawn into the concave crescent. This 'hot crescent, cold bubble' structure, which Wesson called a 'quasi-interchange', contrasts markedly with the cool island (crescent shaped), hot core (circular shape) of the Sykes-Wesson [6] reconnection simulation (Figure 4). During the 1980's, analysis of the SXR emission from Tokamaks became more sophisticated and tomographic methods using up to three SXR cameras and many chordal measurements were employed. Reconstruction of the contours of constant SXR emission on JET (Figure 8) [16] revealed the 'hot crescent, cold bubble' structure predicted by Wesson for kink instability in a $q(r) \simeq 1$ equilibrium.

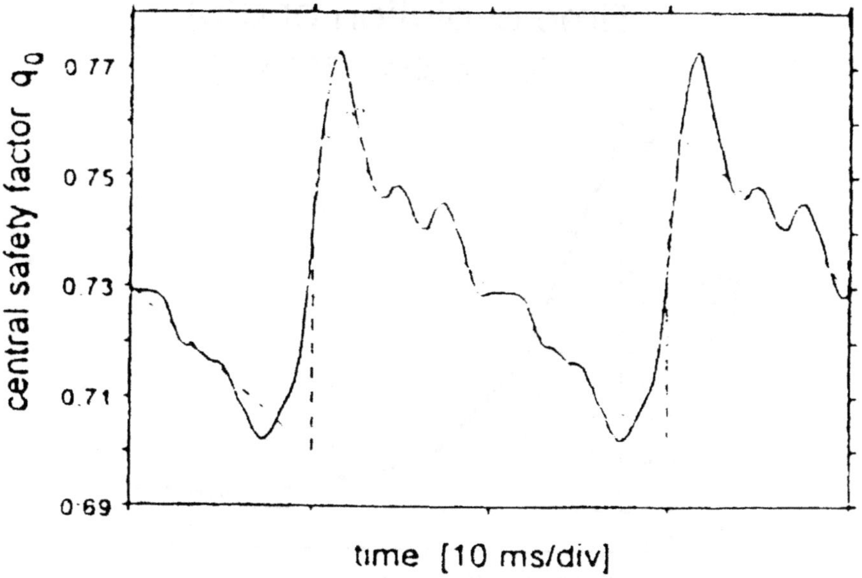

Figure 9. Behaviour of $q(0)$, during a Sawtooth on TEXTOR. The axial value of q never rises above 0.8. Box-car averaging techniques were used to sum the signals from many similar Sawteeth, and enable the $\sim 5\%$ sudden changes in q_0 to be monitored.

Thus by 1986 three sawtooth mechanisms had been proposed.

1. Magnetic reconnection driven by the $m = n = 1$ kink instability.
2. Interrupted reconnection, with secondary instability causing the thermal collapse.
3. Quasi-interchange instability in a $q(r) \simeq 1$ core region.

Meanwhile, a new generation of numerical simulations was being published. First, in cylindrical geometry, Denton, Drake and Kleva [17] had convincingly produced many complete sawtooth cycles and shown that strongly anisotropic thermal conductivity is necessary for this. Aydemir et al. [18] and Park et al. [19] extended the studies to toroidal equilibria. In all cases complete reconnection occurred, with $q(0) > 1$ immediately after each thermal collapse. The reconnection process proved to be rather slow in these simulations, broadly agreeing with the Kadomtsev estimate, τ_k, which is an order of magnitude longer than that observed in the high temperature plasmas of large auxiliary heated Tokamaks like JET, TFTR and DIII-D. Many theorists now turned their attention to investigating physical effects leading to faster reconnection. Electron inertia (linear [20] and non-linear [21]), Hall terms [22, 23], anomalous electron viscosity [24] and neo-classical parallel viscosity [25] have all been added to the conventional resistive Ohm's law, and all these effects speed up reconnection.

The 1980's, however, also brought the first reported attempts to measure the poloidal magnetic field in Tokamaks and hence directly measure the profile of the

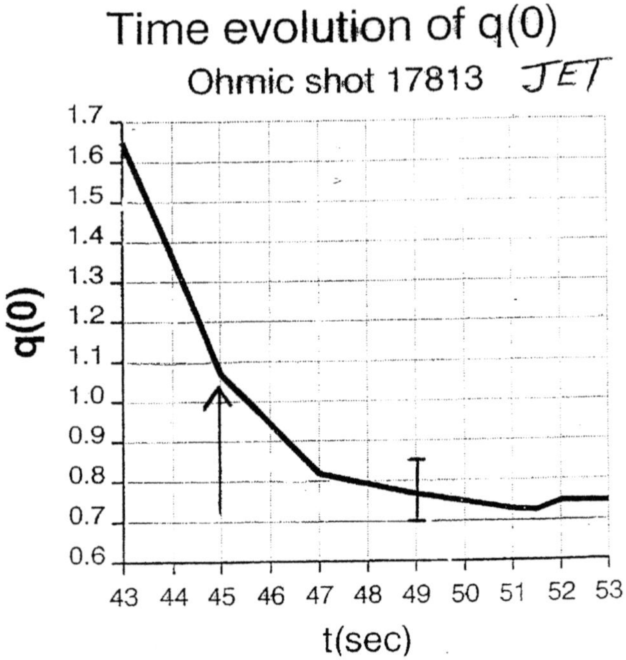

Figure 10. Evaluation of $q(0)$ in a JET ohmic discharge. $q(0)$ continues to decrease after Sawtoothing commences at $t = 45$. Averaging smoothes out the small modulations which occur at each Sawtooth event, which are consequently not visible (from Reference 28).

safety factor $q(r)$. A pioneering measurement by electron scattering of laser light in the DITE Tokamak was reported by Forrest *et al.* [26] in 1978, and indicated that $q(0)$ might fall significantly below unity. The first routinely accurate measurements were carried out on TEXTOR by Soltwisch et al. [27] by measuring the Faraday rotation of a plane polarized FIR micro-wave beam. Blum et al. [28] employed similar methods on JET, while the motional Stark effect (MSE) was used [29, 30] to determine $q(r)$ in DIII-D and TFTR. Figures (9–12) show results of those measurements on TEXTOR [31], JET [28] and TFTR [32]. In each case $q(0) \simeq 0.75$, and varies by $\simeq 5\%$ at each collapse event before slowly declining during the succeeding ramp period. These surprising observations indicate that at no time during a Sawtooth cycle does the axial value of the safety factor, q_0, rise above unity. This appears to rule out both the Kadomtsev and Wesson models for the Sawtooth and raises many new questions, some of which will be examined in the remainder of this review.

'The Trigger problem'

The polarimity or MSE measurements of $q(r)$, together with measurements of $n_e(r,t)$, $T_e(r,t)$, $T_i(r,t)$ make it possible to construct a detailed picture of the Tokamak equilibrium and its evolution during Sawtoothing. In particular one can

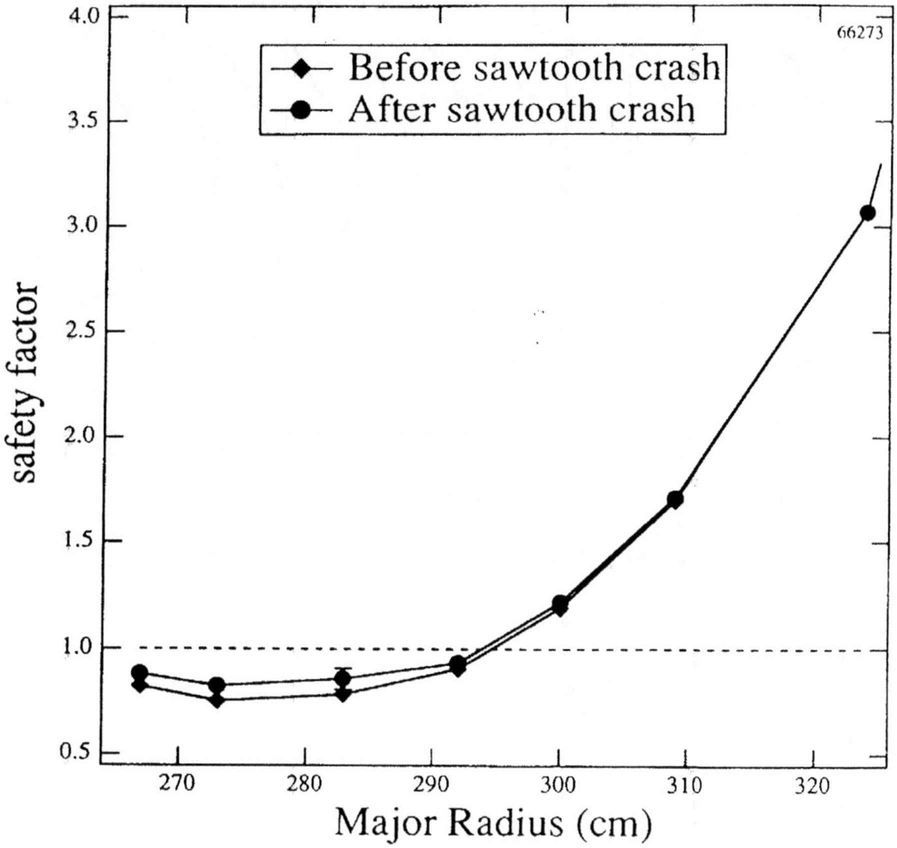

Figure 11. $q(r)$ profiles just before, and just after, a Sawtooth collapse on the TFTR Tokamak, measured using the Motional Stark Effect (from Reference 32).

seek to correlate equilibrium properties at the onset of the $m = 1$ instability (which initiates the collapse phase of Sawteeth) with the crossing of a theoretical stability boundary for the internal kink mode. The theoretical boundary differs, of course, as additional physical effects are introduced into the linearized 'MHD' equations. In increasing complexity, one can assess: (1) Ideal, single fluid, MHD, (2) Resistive single fluid MHD, (3) Two fluid, resistive equations incorporating diamagnetic effects (i.e. $\omega \lesssim \omega_*$ with $\omega_* \approx \rho_i v_i / a^2$ being the diamagnetic frequency), (4) Two fluid plus kinetic treatment of a collisionless high energy ($E \gg kT_i$) population of ions.

The results of such comparisons have been disappointing.

1. The ideal MHD theory of the internal kink mode yields the stability criterion

$$\delta W_{\text{mhd}} > 0 \tag{1}$$

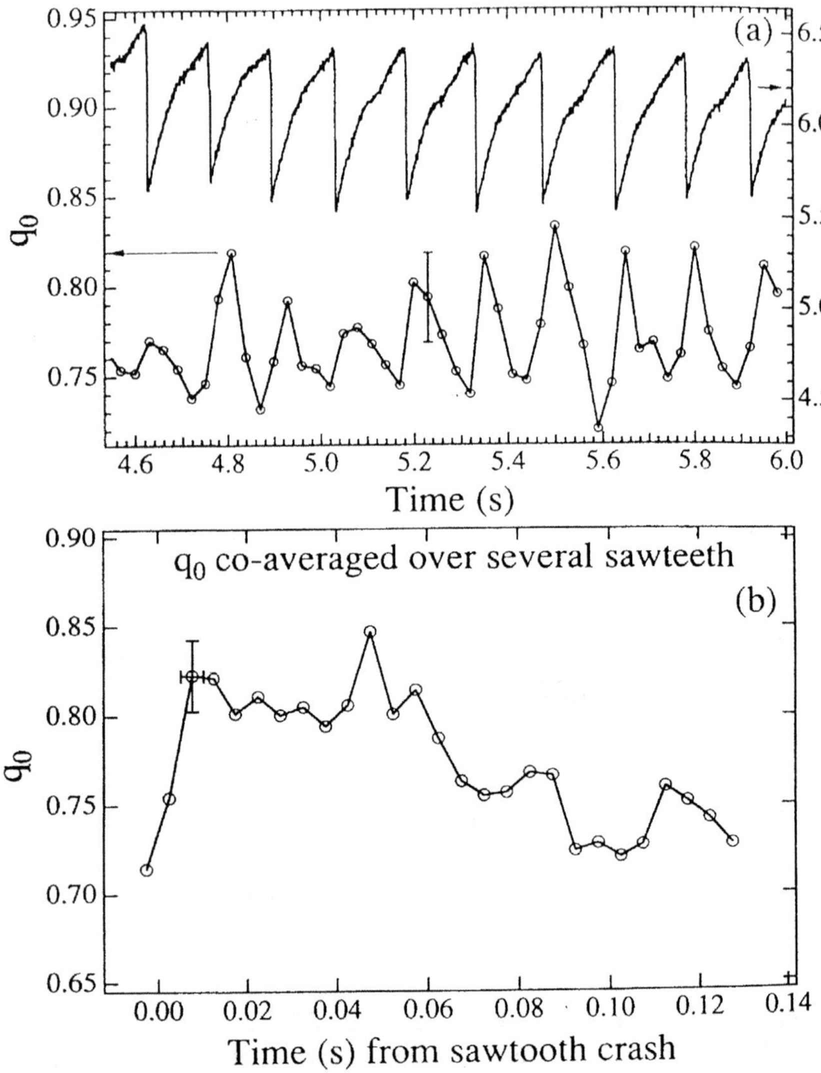

Figure 12. (a) q_0 measurements using MSE during Sawtoothing on the TFTR Tokamak. (b) q_0 evolution during a Sawtooth period, obtained by averaging 9 consecutive Sawtooth events (from Reference 32).

where δW_{mhd} is the potential energy required to perturb the plasma by an $m/n = 1/1$ displacement, with coupling to 0/1, 2/1 and more remote poloidal harmonics induced by the 2-D nature of a toroidal Tokamak equilibrium. In an elegant short paper Bussac et al. [33] calculated the analytic form for δW_{mhd}

for a toroidal equilibrium in the limit of large aspect ratio, $R/a \gg 1$, obtaining the result,

$$\delta W_{\mathrm{mhd}} = 2\pi^2 \varepsilon_1^4 \xi_0^2 R_0 B_0^2 \delta \hat{W}_{\mathrm{mhd}}, \tag{2}$$

with ξ_0 being the amplitude of the kink displacement and

$$\delta \hat{W}_{\mathrm{mhd}} = \delta W_0 - \beta_p \delta W_1 - \beta_p^2 \delta W_2, \tag{3}$$

where

$$\beta_p = -\frac{2}{B_\theta^2(r_1)} \int_0^{r_1} \frac{r^2}{r_1^2} \frac{dp}{dr} dr$$

is a measure of the poloidal beta confined within the magnetic surface where $q(r) = 1$, $\varepsilon_1 = r_1/R$ with $q(r_1) \equiv 1$ and the 'poloidal' beta is the ratio of the plasma pressure p to the magnetic pressure of the poloidal component of the field (Figure 1(e)). In particular Bussac et al. showed that the ideal kink energy of a toroidal plasma is crucially dependent on the toroidal curvature of the equilibrium and is quite different from that of a straight (cylindrical) plasma with the same $p(r)$, $q(r)$ and $j(r)$ profiles.

The coefficients δW_n in Equation (3) depend on the global current profile (or equivalently the $q(r)$ profile). The calculation showed that the δW_n are positive, implying stability ($\delta \hat{W}_{\mathrm{mhd}} > 0$) for $\beta_p \to 0$, and that for typical current profiles in Tokamaks the critical value of β_p above which instability should appear is $\sim 0.2 - 0.3$. The 2-D linear stability computer codes which were developed during the following decade have revealed that the Bussac et al. [33] calculation, although asymptotically correct for $R/a \to \infty$, is rather inaccurate when applied at values of the aspect ratio ($R/a \sim 3$ or 4) which are typical of most Tokamaks. Nevertheless, such computer codes still predict [34] ideal mhd stability at values of β_p below $0.15 \sim 0.2$ for realistic Tokamak equilibria. In conflict with this prediction, Tokamak discharges, in which the only plasma heating is ohmic heating, regularly display Sawtoothing activity at much smaller values of β_p (~ 0.05), implying that the 1/1 kink instability observed at each Sawtooth collapse is not of ideal MHD character.

The single fluid resistive MHD model [35, 36] is also unsuccessful in accounting for Sawteeth. This model predicts linear instability at any value of β_p, when $q(0) < 1$. There are two forms of reconnecting instability in an Ideal mhd stable plasma, ($\delta \hat{W}_{\mathrm{mhd}} > 0$), when $q(0) < 1$. One is known as the resistive kink mode, with growth rate,

$$\gamma \tau_A = s_1^{2/3} S^{-1/3}, \tag{4}$$

where $s_1 = r_1 q'(r_1)$ is the shear of the magnetic field lines at the $q = 1$ surface, and $S = \tau_\eta / \tau_A$ is the Lundquist number, with $\tau_A = \sqrt{3} R_0 / V_A$ and $V_A = B/\sqrt{\mu_0 \rho}$ is the Alfven speed. The other is the tearing instability with growth rate,

$$\gamma \tau_A = 0.55 s_1^2 (\varepsilon_1^2 \delta \hat{W}_{\mathrm{mhd}})^{-4/5} S^{-3/5} \tag{5}$$

The tearing instability scaling, $\gamma \tau_A \propto S^{-3/5}$, applies if $\varepsilon_1^2 \delta \hat{W}_{\mathrm{mhd}} > s_1^{5/3} / S^{1/2}$ and the growth rate increases as Ideal marginal stability is approached,

($\delta\hat{W}_{\rm mhd} \rightarrow 0+$). The resistive kink scaling applies when $\delta\hat{W}_{\rm mhd}$ is small. Larger, Alfvenic, growth is predicted when $\delta\hat{W}_{\rm mhd} < 0$. Since the resistive fluid model predicts an unstable plasma during the ramp phase when $q(0) < 1$, with linear growth time of order 5–10 msec, it cannot explain the quiescent ramp phase of the Sawtooth cycle, during which no mhd activity is observed and which can have a duration of many seconds (many hundreds of e-folding times of the resistive instability) in large Tokamaks with auxiliary heating. One possible explanation of the stable ramp phase in this model involves subtle distortions of the current profile close to the $q = 1$ radius. Such equilibria have been investigated by several authors [37–39] and the existence of resistively stable equilibria, with $q_0 < 1$, has been demonstrated. However, such equilibria have locally non-monotonic current densities, $j_\phi(r)$, and transport simulations of Tokamak discharges show that resistive diffusion ought to render them short lived, if they exist at any time.

Two fluid plasma models have also been explored [40, 22] and show more promise. The physics of the reconnecting layer at r_1 is now complicated by Hall terms in Ohm's law and gyro-viscous effects in the ion stress tensor, which appear as diamagnetic frequency terms, $\omega_* \propto \rho_i v_i / r L_n$ (ρ_i = ion Larmor radius, v_i = ion thermal speed, $L_n^{-1} = d\ln n_e/dr$) which greatly exceed τ_η^{-1} and strongly modify the single fluid resistive dispersion relation. Asymptotic matching of layer solutions to those of the outer ideal region leads to a dispersion relation of the (symbolic) form [40],

$$\delta\hat{W}_{\rm mhd} = \delta\hat{W}_{\rm layer}(s_1, \omega_*\tau_\eta, \ldots) \tag{6}$$

where $\delta\hat{W}_{\rm layer}$ also has complicated dependence on the density and temperature gradients at r_1. Analysis of this dispersion relation [40] has revealed stable regions of parameter space which might account for quiescence during the Sawtooth ramp. When $\delta\hat{W}_{\rm mhd} \approx 0$ (i.e. for β_p near to its critical value for ideal instability) a relatively simple stability criterion for reconnecting instabilities may be derived from (6), namely [41, 42]

$$s_1 < 1.4(\beta^2 \frac{R^3}{L_n^2 L_p})^{\frac{1}{3}} \tag{7}$$

for stability, where $L_p^{-1} = d\ln p/dr$. Studies of Sawteeth on the TFTR Tokamak [42] comparing two different types of discharge (known as 'supershots' and 'L-modes') found that in discharges where this criterion was satisfied no Sawtoothing occurred, while in those which violated the inequality (7), (the L-modes in fact), normal Sawtoothing was present. Remarkably, however, the supershot discharges, in which no Sawteeth were seen, had very high values of β_p ($\approx 1 - 2$), greatly in excess of the critical value for ideal instability ($\beta_p \sim 0.2$). This led Levinton et al. [42] to conjecture that ideal unstable equilibria may behave as though they are marginally stable and that Sawtoothing is controlled by the two-fluid dynamics of the reconnecting layer under these conditions (i.e. by inequality (7)). This idea is very plausible. In essence it

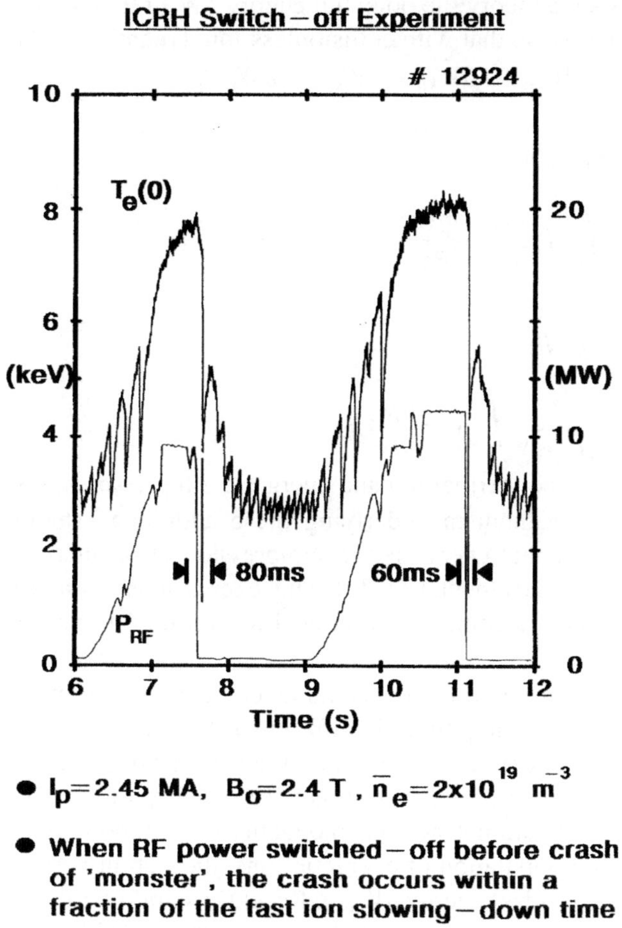

Figure 13. Stabilising effect of energetic ions demonstrated by switch-off of ICRH. This triggered a Sawtooth collapse in 60 - 80 msec, a delay comparable to the slowing down time of the trapped energetic ions generated by the ICRH.

suggests that an ideal unstable equilibrium, with $\delta \hat{W}_{\rm mhd} < 0$, merely adjusts by kinking slightly in the core, $r < r_1$, region. The resulting field line bending energy in the new, slightly distorted but topologically unchanged, equilibrium returns $\delta \hat{W}_{\rm mhd}$ to zero. A precise calculation of this process in an ideal mhd plasma [3] showed that the amplitude of the kink displacement is, in practice, extremely small.

2. In experiments [43] with auxiliary heating, long Sawtooth periods were terminated by a Sawtooth collapse that followed switch off of ICRH with a time lag of $\sim 60 - 80$ ms – comparable to the slowing down time of the energetic ions (Figure 13). This suggests that a collisionless kinetic theory may be re-

quired. In such a theory the potential energy $\delta\hat{W}$ in Equation (6) acquires two additional terms, so that with collisionless ion dynamics (6) becomes

$$\delta\hat{W}_{\text{total}} \equiv \delta\hat{W}_{\text{mhd}} + \delta\hat{W}_{ki} + \delta\hat{W}_{kE} = \delta\hat{W}_{\text{layer}} \tag{8}$$

where [44]

$$\delta\hat{W}_{ki} \simeq \varepsilon_1^{-3/2}\beta_i \tag{9}$$

with

$$\beta_i = \frac{5}{2}\int_0^1 dx x^{3/2}\frac{2p_i(x)}{B_0^2}, \qquad x \equiv r/r_1$$

and [45]

$$\delta\hat{W}_{kE} \simeq \varepsilon_1^{-\frac{1}{2}}\beta_E^* \tag{10}$$

with

$$\beta_E^* = -\frac{2}{B_\theta^2(r_1)}\int_0^1 x^{3/2}\frac{dp_E}{dx}dx$$

and p_i, p_E are the thermal ion and energetic ion pressures, respectively. The physical mechanism underlying these additional, stabilising, contributions to the potential energy is the compression of magnetically trapped ions by the kink displacement.The differing expressions (9) and (10) arise from different limits of a common dynamical response of thermal and high energy ions.

The complex dispersion relation (8) contains elements which may account for the wide range of Sawtooth ramp times in differing Tokamaks. According to the Levinton [42] conjecture, energetic ions will be irrelevant if $\delta\hat{W}_{\text{total}} < 0$ (ideal instability even with energetic ions present). In this case stability of the Sawtooth is determined by the plasma parameters at r_1 (inequality (7)). If, however, $\delta\hat{W}_{\text{total}} > 0$ due to the presence of a population of energetic ions, the stability criterion is modified [40] by the (positive) ideal energy $\delta\hat{W}_{\text{total}}$. Recently stability criteria based on these ideas and results have been incorporated in a transport code in an attempt to predict the Sawtooth repetition time τ_s for a D-T plasma in the proposed ITER device. It was found [46] that the alpha particles produced by D-T fusion had a significant effect in lengthening τ_s. The electron and ion dynamics of the reconnecting layer are, however, even more complex than has been assumed in the derivation of (7). In current Tokamak experiments a long mean-free-path analysis, $\lambda_{mfp} \gg R_0$, incorporating magnetic trapping effects is required. Work on this topic has recently been reported [47–49].

Any linear stability analysis, however complex, will predict the emergence of a slowly growing mode – a precursor oscillation – as a linear stability boundary is crossed by a slowly evolving Tokamak equilibrium. The experimental observation of precursorless Sawteeth, in which the initial kink perturbation has a large growth rate, appears to require a non-linear instability mechanism to explain it. Work on this topic is also under development in the form of 'small island' stability theory. This seeks to describe a non-linear regime in which some magnetic reconnection has occurred and the resulting magnetic island at the $q = 1$ surface has become wider than the narrow layer of linear theory, in which non-ideal effects, such as resistivity, control plasma dynamics.

At this point, the predictions of linear and non-linear calculations differ because the change of magnetic topology, implicit in the non-linear island, permits rapid longitudinal particle motion (in kinetic theory) or longitudinal transport (in fluid theory) from one side of the 'layer' to the other. In linear theory this role can only be performed by cross field drifts, (or cross field transport processes in fluid theory).

Now, in high temperature Tokamaks the linear layer width is predicted to be of the order of a few mm, and magnetic perturbations associated with islands of this width are difficult to detect. Thus, the first detectable kink perturbation seen during a precursorless Sawtooth event may already be in a non-linear phase of its dynamics. Consequently one possible explanation of the sudden growth of the $m/n = 1/1$ kink in precursorless Sawteeth, may lie in the differing stability boundaries of 'linear' and 'small-island' stability theory. In the event that the linear stability boundary is higher (e.g. a larger value for the critical β_p) than the 'small island' stability boundary, a plasma evolving slowly, from a linearly stable to linearly unstable state, could pass to a strongly unstable non-linear state before any perturbations were detected experimenally. In such a scenario, it is the linear stability boundary which determines the 'trigger' for the Sawtooth event, but the subsequent dynamic evolution is controlled by non-linear effects. Zakharov et al. [41] have investigated such a situation using the two fluid equations and found that, in contrast to single fluid theory, rapid exponential growth is predicted in the non-linear phase with

$$\gamma \tau_A \propto s_1 \frac{\rho_i}{r_1} \tag{11}$$

Before turning to other interesting experimental observations, it may be useful to summarise this section.

Since the Sawtooth collapse is *invariably* accompanied by a 1/1 kink displacement, much work has been carried out on the linear stability of this mode. The calculations involve asymptotic matching at a critical layer where $q = 1$. Relatively simple calculations, neglecting inertia and dissipation, are used to calculate a measure of the potential energy, $\delta \hat{W}_{\text{total}}$, available in the outer regions of the plasma (away from the narrow reconnecting layer at $q = 1$) to drive instability. The appropriate model for this relies on collisionless ion dynamics, and treats energetic ions in a different limit from thermal ions when calculating the compressional energy.

The complementary calculation, within the layer, in the simplest possible limit, involves only inertia. Realistically, however, the layer calculations need to take account of long mean-free-path transport processes [47–49], of toroidal geometry, of finite Larmor radius (FLR) averaging effects and of the diamagnetic frequency effects which Hall terms and gyro-viscosity generate in two fluid calculations. Reconnection is predicted by single fluid theory, but the diamagnetic corrections of two fluid theory suppress reconnection under certain conditions (inequality (7)).

Linear stability calculations may correctly determine the marginal stability boundary, and hence the Sawtooth 'Trigger' condition, but they fail to account for the sudden appearance of 1/1 displacements with large (100 μsec^{-1}) growth rates. For this 'small island' stability theory [41] is required. This topic is not yet highly developed in the context of Sawteeth, but it offers the possibility of an explanation for sudden growth.

SOME MISCELLANEOUS OBSERVATIONS

A number of other experimental observations throw light on the sawtooth phenomenon. A few are briefly outlined below.

'Snakes' Pellet injection in JET quite frequently results in the creation of a high density closed tube of plasma (the Snake) [50] on the $q = 1$ surface when a pellet of sufficient size reaches the core. Snakes are long lived phenomena aligned to the closed magnetic field lines of the $q = 1$ surface. They survive many sawtooth collapses, with the thermal and density redistribution taking place around them. At each Sawtooth event, the radial position of the Snake suddenly decreases slightly, to increase slowly during the subsequent ramp phase of the Sawtooth. The Snake appears to be a reliable 'marker' for the $q = 1$ surface so that these observations have been taken as further evidence that a Kadomtsev reconnection is not taking place at each collapse, and that $q(r)$ experiences relatively small changes.

Magnetic shear at $q = 1$. Pellet injection in JET also furnishes a novel and elegant means of estimating the magnetic shear $s = rq'/q$, at the $q = 1$ surface [51]. Light emitted by the ablating pellet, shows a pronounced dip as the pellet passes through $q = 1$ where field lines close on themselves after just one circuit of the Tokamak. At this point the source of electrons impinging on the pellet is much reduced (since the pellet effectively passes through its own 'shadow') resulting in reduced emission. Close to the $q = 1$ surface this mechanism remains partially operative and, by measuring the width (in time) of the reduced emission, Gill et al. [51] were able to calculate the shear s_1. The method yields a rather indirect measurement of s_1 and one might suspect that other physics (ablation instabilities) might produce similar effects. However, accepting the geometric interpretation of the reduced emission, the results are surprising. Gill et al. deduced $s_1 \simeq 1{-}2 \times 10^{-2}$ with no correlation between the measured value and the time within the Sawtooth ramp (early, mid, or late) that the pellet was injected (Figure 14). For comparison, a simple parabolic $q(r)$ with $r_1/a \sim 1/3$ yields a value of $s_1 \approx 0.16$ (an order of magnitude larger).

Sawtooth suppression. Over the years a number of methods of suppressing Sawteeth in Tokamaks have been discovered experimentally, and these may throw some light on aspects of the sawtooth, and certainly on the 'Trigger mechanism'. The following are three examples:

1. By localised ECRH/ECCD (Electron cyclotron resonance heating and current drive) tuned to resonance at $r = r_1 + \delta$, $(\delta/r_1 \ll 1)$ i.e. just outside the

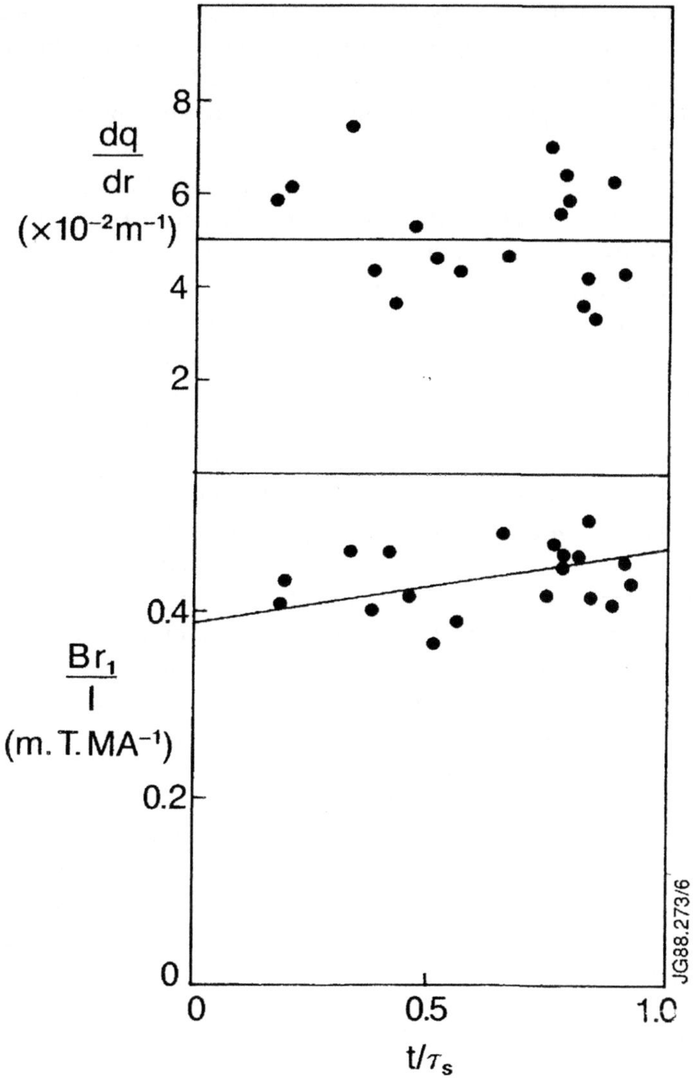

Figure 14. Magnetic shear, $s_1 = r_1 \frac{dq}{dr}(r_1)$ at the $q = 1$ surface as inferred from light emitted as a pellet passes through $q = 1$. Note the lack of correlation between values of $q'(r_1)$ and the time elapsed since the last Sawtooth collapse (from Reference 51).

$q = 1$ surface [52]. Evidently the mechanism involves linear stabilisation of the $m = n = 1$ internal kink mode (ideal or resistive) by current profile control. Calculations showing the stabilising effect of local perturbations in the current prolile have been published [37–39].

2. By generation of energetic trapped ions. A combination of NBI and ICRH appears to be most effective, and Sawtooth periods in excess of 8 secs have

been obtained on JET [53] by this means. In this case, linear stability theory shows that an energetic trapped ion population can stabilise the $m/n = 1/1$ ideal internal kink at moderate values of $1 - q(0)$, but whether this mechanism could be responsible for complete Sawtooth suppression, or merely lengthening of the Sawtooth period is not clear.

3. By application of externally generated resonant magnetic perturbations (RMPs) with $m = 2$, $n = 1$. The RMP is generated by coils external to the plasma, and creates a magnetic island at the $q = 2$ surface which exerts a drag on the plasma at that radius. A sufficiently large RMP brings the plasma to rest at $q = 2$ (the toroidal rotation of the plasma is suppressed) and when this occurs Sawteeth are suppressed [54]. Reduction of the amplitude of the RMP (reducing the drag) results in spin up of the plasma and resumption of normal Sawtoothing. The way in which this mechanism suppresses the kink mode is not understood but it points to the importance of (sheared) plasma rotation in the stability of the kink mode: a neglected subject.

Impurity transport during a Sawtooth collapse. One explanation for the rapid transport during a Sawtooth collapse has received considerable attention [55, 56, 57] and deserves a special mention. In this model it is proposed that the $m = n = 1$ helical magnetic perturbation, interacting with the axisymmetric equilibrium fields of the Tokamak, causes stochastic break up of the regularly nested magnetic surfaces in the plasma core. Magnetic field lines 'wander' radially as they circuit the torus and, it is argued, rapid longitudinal transport (parallel to B) is responsible for the sudden thermal collapse of the core plasma.

A recent experiment [58] on the JET device has a bearing on this suggestion. In the experiment a small amount of heavy impurity (Nickel) was introduced at the plasma edge. The Nickel ionized and was transported toward the core but, for reasons that are not understood, the Nickel ions did not penetrate beyond a radius of about $r/a \approx 1/3$, comparable to the radius of the $q = 1$ surface. However, at the next Sawtooth event the Nickel impurity was observed to suddenly invade the core region on the same 50 μsec timescale as the thermal collapse. Wesson et al. [58] point out that on this timescale, whereas electrons with an energy of around 5 keV travel about 1500 metres (equivalent to nearly 100 transits of the JET torus), the massive Nickel ions cover a longitudinal distance of only 5 metres (less than one transit of the machine). It follows that the sudden radial influx of Nickel ions from $r/a > 1/3$ into the axis cannot be explained in terms of magnetic stochasticity. Instead, Nickel ions must be drifting *across* the magnetic field.

A possible explanation, which still involves magnetic stochasticity, runs as follows. At sufficiently large amplitude of the $m = n = 1$ magnetic perturbations, magnetic stochasticity may indeed become significant. This will permit rapid (on $\sim$ 100 μsec timescale) exchange of passing electrons (up to 50% of the electrons are magnetically trapped and cannot follow wandering field lines to different radii) between the axial region $r \ll r_1$ and the surrounding, $r \sim r_1$ region. Unequal electron densities and/or trapping fractions in these two regions may cause the

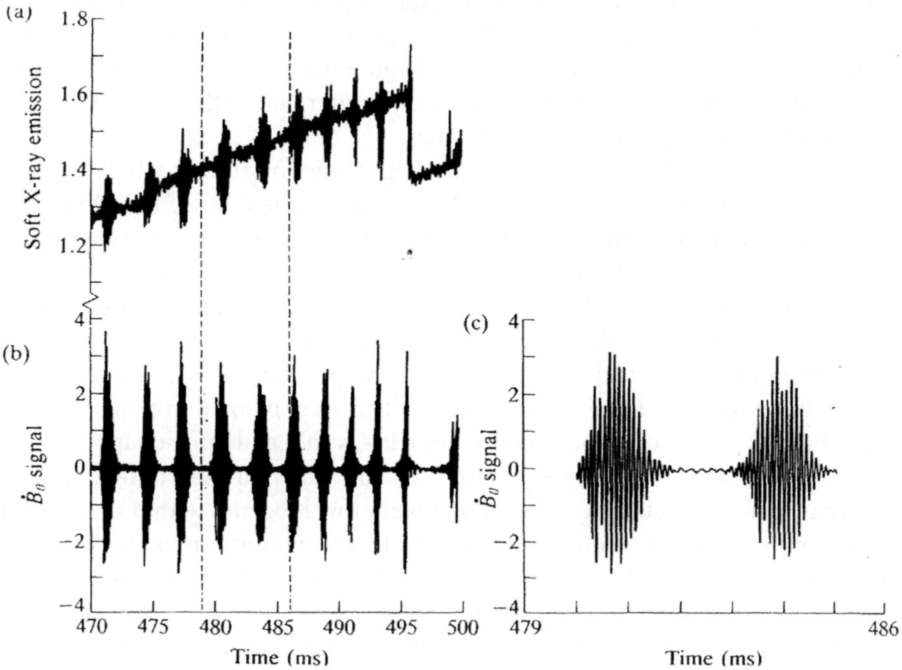

Figure 15. (a,b) Regular bursts of fishbone activity on PDX as seen on soft X-ray emission and magnetic fluctuations. (c) Characteristic signature of the magnetic fluctuation signal for the Fishbone instability on the PDX Tokamak (from Reference 60).

build up of electrostatic potentials with $e\phi/T_e \lesssim 0(1)$. Under the influence of such electrostatic potential, ions, impurities and magnetically trapped electrons will $\mathbf{E} \times \mathbf{B}$ drift together across the magnetic field in a large scale convective motion. Such a detachment of plasma from the magnetic field might explain the 'quasi-interchange' structure of the SXR emission contours noted earlier [25] and would not require flat, $q(r) = 1$, equilibria in the Tokamak core. Magnetic stochasticity remains an essential element in this scenario, with the other element being the different longitudinal mobility of electrons and ions.

However other observations on JET [59] cast doubt on this suggestion also. Duperrex et al. [59] made detailed measurements of the magnetic perturbations outside the plasma and found that their evolution matched very closely that of the distortions in the SXR emission from the plasma core. They concluded that the core displacements are of a purely electromagnetic character, rather than of an electrostatic character.

The Fishbone instability. In 1983 a new form of $m = n = 1$ internal kink instability was observed on the PDX experiment [60]. Like the precursor oscillation of the Sawtooth instability, the Fishbone instability appears as a slowly growing, oscillating, signal on the SXR emission and other diagnostics, and can be attributed to a growing $m = n = 1$ kink displacement of the core plasma, rotating in the toroidal direction. Unlike the Sawtooth instability, no rapid thermal collapse in the core is observed. Instead, the oscillatory signal first grows and then decays in amplitude, giving the characteristic signal (Figure 15) which led to the terminology 'Fishbone instability'. Periods of quiescence alternate with bursts of Fishbone activity (Figure 15) in a fairly regular manner.

A good qualitative theoretical understanding of the instability, which has stood the test of time, was quickly provided [61]. The unique feature of the PDX device which caused it to exhibit the instability so clearly, lay in the fact that the angle of neutral beam injection (NBI) was nearly perpendicular to the main toroidal field. Consequently, most of the high energy ions became magnetically trapped. Like the mirror trapped particles of the Earth's Magnetosphere, which precess around the earth in the equatorial direction, magnetically trapped ions in a Tokamak precess in the toroidal direction. The Fishbone instability was explained in terms of a magnetic perturbation whose toroidal phase velocity is in Landau resonance with the precessional velocity of the energetic NBI ions and which draws on their energy to grow. Because the relevant frequency is relatively low it occurs within the spectrum of the Alfvén continuum, so that Alfvén continuum damping must be overcome. Because of this there must be a sufficiently high pressure of the trapped energetic ions, and previous experiments with NBI did not have sufficient auxiliary heating power, or the necessary perpendicular injection geometry, to exceed the threshold for instability.

Much theoretical work quickly followed. Some of this focused on the new Fishbone instability, and in particular on the possibility that the mode is a discrete eigenmode, weakly damped in the absence of energetic ions, within a gap in the Alfvén continuum [62]. However, since it had become clear that a relatively low pressure of trapped energetic ions could be responsible for the new kink instability, attention now also focused on the effect of such ions on the MHD (ideal or resistive) internal kink mode which had been thought to be responsible for the Sawtooth instability. As a result it is now accepted that quantitative predictions of Sawtooth characteristics must necessarily take account of the collisionless dynamics of energetic ion populations. Where these have been created by external heating schemes, such as NBI or ICRH, the relevant ion distribution functions are neither Maxwellian in energy nor isotropic in pitch angle ($\theta = \cos^{-1}(v_\parallel/v)$). In such cases the resulting stability calculations are complicated. Simpler calculations, and possibly more robust predictions, are possible when the energetic ions are the α-particles produced by D-T fusion. Here the distribution is isotropic and the calculations more tractable [63, 64]. A comprehensive review of the theoretical work on energetic ion effects in Sawteeth and Fishbones has been given by Porcelli [65].

The absence of sudden thermal collapse events during Fishbone activity provides another clue to the mechanism involved in the thermal collapse events in Sawteeth. According to the theory of Fishbones the frequency of such modes is such that they should cause little or no magnetic reconnection. This suggests that a change in magnetic topology may be an essential ingredient in the fast thermal collapses of Sawteeth, where a transition from a mildly unstable linear phase to a more strongly unstable, small island, phase may be crucial. In contrast, non-linear effects in the Fishbone 1/1 kink tend to be stabilising since the mode expels the energetic ions which provide the drive.

Summary and Conclusions

The understanding of the Sawtooth instability in Tokamak plasmas is improving but is nevertheless still incomplete. Early explanations in terms of a single magnetic reconnection phenomenon now appear to be untenable. Instead the Sawtooth collapse phase appears to require at least two ingredients: the initial growth of an MHD internal kink instability causing some magnetic reconnection in the vicinity of the $q = 1$ surface, together with a secondary event which is responsible for the very rapid collapse of the core density $n_e(r = 0)$ and temperatures $T_j(r = 0)$. The precise nature of this secondary event remains unclear. It may be a pressure driven instability as has been suggested by, e.g., Bussac et al. [66], Gimblett et al. [67], or it may be sudden stochastization of the magnetic field structure at a critical amplitude of the kink perturbation as suggested by, e.g., Mercier [55], Lichtenberg et al. [56, 57].

Observations on TORE-SUPRA [68] showing a sudden burst of increased density fluctuations (in the frequency range 500 kHz–1 MHz) during the thermal collapse give strong support for the former explanation. It has also been suggested [69, 70] that Sawteeth may be triggered by a critical level of turbulence unconnected with the $m = n = 1$ kink mode (whose appearance, in this model, is a consequence rather than a cause of the collapse event).

In the case of the, so-called, precursorless Sawteeth, the very rapid onset of the $m = n = 1$ kink perturbation cannot be reconciled with slow evolution of an equilibrium through marginal stability but might be explained by a transition to a more unstable, 'small island', non-linear regime.

Some progress has been made in correlating the linear stability boundary for the $m = n = 1$ kink mode with experimental observations of the equilibrium at onset of the kink, but this requires a two fluid treatment of the inertial layer at the $q(r) = 1$ surface, and a calculation of the collisionless dynamics of any high energy ions (with $E \gg kT_e, kT_i$) which are present in the core plasma.

The measurements, using polarimetry or the motional Stark effect, which indicate that the axial value of the safety factor, $q(r = 0)$, never rises to, or above, unity at any time during the Sawtooth cycle, cannot be reconciled with the non-linear fluid MHD simulations in which the initial Kadomtsev reconnection process invariably continues to completion, resulting in a $q(0) > 1$, post collapse, state. However, if these measurements indicating that $q(0) < 1$ could be discarded most of the other observations become mutually compatible and compatible with non-linear simulations of discharges in which $q(r)$ is flat and close to unity. In addition, it is a fact that, on certain Tokamak devices, the measurements of $q(r)$ have suggested values which are indeed very close to unity. McCormick et al. [71] used Zeeman splitting of radiation from injected Lithium atoms to measure $q(0) \approx 1$ in the ASDEX Tokamak and recent MSE measurements on DIII-D [72] gave $q(0) = 0.96$ before and $q(0) = 1.02$ after a Sawtooth collapse. In such cases a hybrid Kadomtsev/Wesson picture, in which plasma flow patterns have some quasi-interchange structure but reconnection also takes place, becomes possible (See Aydemir et al. [18] for simulations of such Sawteeth). The rapidity of the reconnection event can rather readily be explained by one of the accelerating mechanisms discussed earlier. We conclude that it is possible that somewhat different dynamics controls Sawteeth in different Tokamaks and that $q(0) \approx 1$, with Kadomtsev Sawteeth, in some, while $q(0) \lesssim 0.75$ in others. In the latter cases it appears that the hot plasma ($T \sim 1$–10 keV) in the core of these Tokamaks is not well described by MHD single fluid equations.

When, eventually, a better understanding of the dynamics of Sawteeth is achieved it may throw light on many other phenomena in Tokamak plasmas (such as the disruptive instabilities which sometimes terminate a discharge and the instabilities of the plasma boundary known as Edge Localised Modes, ELMs), but also perhaps on fast reconnection events in the Magnetosphere, the Solar Corona and elsewhere.

Finally, the interested reader should consult the four excellent review articles [14, 65, 73, 74] which cover different aspects of Sawtooth phenomena.

Acknowledgements

I am indebted to many colleagues for sharing ideas and data over many years. In particular, I would like to thank Drs J.A. Wesson, D.J. Campbell, F. Porcelli, F. Pegoraro, R. Gill and B. Alper for many discussions of JET data, Drs C.G. Gimblett, J.W. Connor and Professor J.B. Taylor at Culham Laboratory, and Professor M.G. Haines of Imperial College. I would also like to express my gratitude to the conference organisers for their hospitality in Venezuela and for support. This work was partly supported also by Imperial College, London, the UK Department of Trade and Industry and by Euratom.

References

1. von Goeler, S., Stodiek, W. and Sauthoff, N.: 1974, *Phys. Rev. Lett.* **33**, 1201.
2. Shafranov, V.D.: 1970, *Sov. Phys. Tech. Phys.* **15**, 175.
3. Rosenbluth, M.N., Dagazian, R.Y. and Rutherford, P.H.: 1973, *Phys. Fluids* **16**, 1894.
4. Kadomtsev, B.B.: 1976, *Sov. J. Plasma Phys.* **1**, 389.

5. Kadomtsev, B.B.: 1977, in *Plasma Physics and Controlled Fusion Research* 1976, (Proc. 6th Int. Conf., Berchtesgaden (1976)) IAEA, Vienna, Vol. 1, pp. 555.
6. Sykes, A. and Wesson, J.A.: 1976, *Phys. Rev. Lett.* **37**, 140.
7. Danilov, A.F., Dnestrovski, Yu.N., Kostomarov, D.P. and Popov, A.M.: 1976, *Sov. J. Plasma Phys.* **2**, 93.
8. Dubois, M.A. and Samain, A.: 1980, *Nucl. Fus.* **20**, 1101.
9. Dubois, M.A., Marty, D. and Pochelon, A.: 1980, *Nucl. Fus.* **20**, 1355.
10. Dubois, M.A., Pecquet, A.L. and Reverdin, C.: 1983, *Nucl. Fus.* **23**, 147.
11. Pfeiffer, W.: 1985, *Nucl. Fus.* **25**, 673.
12. Parail, V.V. and Pereversev, G.V.: 1980, *Sov. J. Plasma Phys.* **6**, 14.
13. Campbell, D.J., et al.: 1989, in *Plasma Physics and Controlled Nuclear Fusion Research* 1988 (Proc. 12th Int Conf, Nice (1988)), IAEA, Vienna, Vol. 1, pp. 377.
14. Wesson, J.A.: 1987, in A. Bondeson, E. Sindoni and F. Troyon (eds), *Theory of Fusion Plasmas* 1987, Proc. Varenna Workshop, Editrice Compositori Bologna, 1987, pp. 253.
15. Wesson, J.A.: 1986, in *Plas. Phys. Contr. Fus.* (Proc. 12th European Conf, Budapest, 1985) **28**, 1A, 243.
16. Edwards, A.W., et al.: 1986, *Phys. Rev. Lett.* **57**, 210.
17. Denton, R.E., Drake, J.F. and Kleva, R.G.: *Phys. Fluids* **39**, 1448.
18. Aydemir, A.Y., Wiley, J.C. and Ross, D.W.: *Phys. Fluids B* **1**, 774.
19. Park, W. and Monticello, D.A.: 1990, *Nucl. Fus.* **30**, 2413.
20. Porcelli, F.: 1991, *Phys. Rev. Lett.* **66**, 425.
21. Wesson, J.A.: 1990, *Nucl. Fus.* **30**, 2545.
22. Zakharov, L.E. and Rogers, B.N.: 1992, *Phys. Fluids B* **4**, 3285.
23. Wang, X. and Bhattacharjee, A.: 1995, *Phys. Plasmas* **2**, 171.
24. Aydemir, A.Y.: 1990, *Phys. Fluids B* **2**, 2135.
25. Qingquan Yu: 1995, *Nucl. Fus.* **35**, 1012.
26. Forrest, M.J., Carolan, P.G. and Peacock, N.J.: 1978, *Nature* **271**, 718.
27. Soltwisch, H.: 1986, *Rev. Sci. Instrum* **57**, 1939.
28. Blum, J., et al.: 1990, *Nucl. Fus.* **30**, 1475.
29. Wroblewski, D., Huang, L.K. and Moos, H.W.: 1988, *Phys. Rev. Lett.* **61**, 1724.
30. Levinton, F.M., et al.: 1989, *Phys. Rev. Lett.* **63**, 2060.
31. TEXTOR Team: 1989, in *Plasma Physics and Controlled Nuclear Fusion Research* 1988, (Proc. 12th Int Conf, Nice (1988)), IAEA Vienna, Vol. 1, pp. 331.
32. Yamada, M., et al.: *Phys. Plasmas* **1**, 3269.
33. Bussac, M.N., et al.: 1975, *Phys. Rev. Lett.* **35**, 1638.
34. Porcelli, F.: private communication; Sharapov, S., private communication.
35. Coppi, B., et al.: 1976, *Sov. J. Plasma Phys.* **2**, 533.
36. Bussac, M.N., et al.: 1977, in *Plasma Physics and Controlled Nuclear Fusion Research* 1976 (Proc. 6th Int Conf, Berchtesgaden, 1976) IAEA Vienna, Vol. 1, 607.
37. Soltwisch, H., Stodiek, W., Manickam, J. and Schlüter, J.: 1987, in *Plasma Physics and Controlled Nuclear Fusion Research* 1986 (Proc. 11th Int Conf, Kyoto 1986) IAEA, Vienna, Vol. 1, 263.
38. Rogister, A., Singh, R. and Kaleck, A.: 1989, *Nucl. Fus.* **29**, 1175.
39. Gimblett, C.G., et al.: 1991, in *Controlled Fusion and Plasma Physics*, Proc. 18th European Conf, Berlin 1991, Vol 15C, Part II, European Physical Soc, pp. 21.
40. Migliuolo, S., Pegoraro, F. and Porcelli, F.: 1991, *Phys. Fluids B* **3**, 1338.
41. Zakharov, L.E., Rogers, B. and Migliuolo, S.: 1993, *Phys. Fluids B* **5**, 2498.
42. Levinton, F.M., Zakharov, L., Batha, S.H., Manickam, J. and Zarnstorff, M.C.: 1994, *Phys. Rev. Lett.* **72**, 2895.
43. Campbell, D.J., et al.: 1988, in *Controlled Fusion and Plasma Heating*, Proc. 15th Euro Conf, Dubrovnik (1988), Vol. 12B, part 1, 377, European Physical Soc.

44. Fogaccia, G. and Romanelli, F.: 1995, *Phys Plasmas* **2**, 227.
45. Coppi, B., et al.: 1990, *Phys. Fluids B* **2**, 927.
46. Porcelli, F., Boucher, D. and Rosenbluth, M.N.: 1996, *Plas. phys. Contr. Fus.* **38**, 2163.
47. Mikhailovskii, A.B. and Lakhin, V.P.: 1995, *Plasma Phys. Reps.* **21**, 291.
48. Lakhin, V.P. and Mikhailovskii, A.B.: 1995, *Plasma Phys. Reps.* **21**, 745.
49. Lakhin, V.P., Mikhailovskii, A.B. and Churikov, A.P.: *Plasma Phys. Reps.* **21**, 1112.
50. Weller, A., et al.: 1987, *Phys. Rev. Letts.* **59**, 2303.
51. Gill, R.D., Edwards, A.W. and Weller, A.: 1989, *Nucl. Fus.* **29**, 821.
52. Bobrovskii, G., Esipchuk, V.Yu. and Savrukin, P.: 1987, *Sov. J. Plasma Phys.* **13**, 665.
53. Campbell, D.J., et al.: 1988, *Phys. Rev. Lett.* **60**, 2148.
54. Hender, T.C., et al.: 1992, *Nucl. Fus.* **32**, 2091.
55. Mercier, C.: 1983, *Sov. J. Plasma Phys.* **9**, 82.
56. Lichtenberg, A.J.: 1984, *Nucl. Fus.* **24**, 1277.
57. Lichtenberg, A.J., Itoh, K., Itoh, S.-I. and Fukuyama, A.: *Nucl. Fus.* **32**, 495.
58. Wesson, J.A., et al.: To be published in *Phys. Rev. Lett.*
59. Duperrex, P.A., Pochelon, A., Edwards, A.W. and Snipes, J.A.: 1992, *Nucl. Fus.* **32**, 1161.
60. McGuire, K., et al.: 1983, *Phys. Rev. Lett.* **50**, 891.
61. Chen, L., White, R.B. and Rosenbluth, M.N.: 1984, *Phys. Rev. Lett.* **52**, 1122.
62. Coppi, B. and Porcelli, F.: 1986, *Phys. Rev. Lett.* **57**, 2272.
63. White, R.B., et al.: 1985, *Phys. Fluids* **28**, 278.
64. Pegoraro, F., et al.: 1989, in *Plasma Physics and Controlled Nuclear Fusion Research* 1988 (Proc. 12th Int Conf Nice 1988) IAEA, Vienna. Vol. II, pp. 243.
65. Porcelli, F.: 1991, *Plas. Phys. Contr. Fus.* **33**, 1601.
66. Bussac, M.N. and Pellat, R.: 1987, *Phys. Rev. Lett.* **59**, 2650.
67. Gimblett, C.G. and Hastie R.J.: 1994, Culham Laboratory Report AEA FUS 260.
68. Hennequin, P., et al.: 1995, in *Controlled Fusion and Plasma Physics*, Proc. 22nd Euro Conf, Bournemouth 1995, European Physical Society, Vol. 19C, part III, pp. 61.
69. Haas, F.A. and Thyagaraja, A.: 1995, *Plas. Phys. Contr. Fus.* **37**, 707.
70. Itoh, K., Itoh, S.-I. and Fukuyama, A.: 1995, *Plas. Phys. Contr. Fus.* **37**, 1287.
71. McCormick, K., Eberhagen, A., Murmann, H., et al.: 1988, in *Controlled Fusion and Plasma Heating*, Proc. 15th European Conf, Dubrovnik 1988, European Physical Society, Vol. 12B, part I, pp. 35.
72. Wroblewski, D. and Snider, R.T.: 1993, *Phys. Rev. Letts.* **71**, 859.
73. Migliuolo, S.: 1993, *Nucl. Fus.* **33**, 1721.
74. Kuvshinov, B.N. and Savrukhin, P.V.: 1990, *Sov. J. Plasma Phys.* **16**, 353.

IMPORTANT ISSUES OF PHYSICS OF IMPROVED CONFINEMENT IN TOKAMAKS

M. TENDLER
Alfvén Laboratory, Royal Institute of Technology, Stockholm, Sweden.

Abstract. The review addresses transitions into regimes with improved confinement. The $E \times B$ paradigm is tested on the example of L–H transitions. It is demonstrated that L–H transitions may emerge either due to an amplification of the diamagnetic drift term caused by enhanced pressure gradient or due to an increase poloidal rotation velocity at the separatrix. In general, it is asserted that the emergence and dynamics of transitions is very sensitive to fine details of prelude plasma profiles. The L–H transition occurs provided the ambipolar electric field changes dramatically within a few centimeters. The dynamics of interest consists of threshold conditions for barrier formation, barrier propagation speed, and profile steepening rates, barrier limits and mechanims for their relaxation and termination.

1. Introduction

There is accumulating evidence from tokamak experiments that regimes with improved confinement can achieve higher values of confinement, beta and bootstrap current than had been thought plausible until recently.

Transport barrier dynamics is the key scientific concern at present. The interest is enhanced by the fact that a continuously operated tokamak will not be operated at a steady-state, since for control purposes, it will necessarily require barriers to be created or lowered from time to time in different portions of a plasma to facilitate a reactor operation. Transport barriers will have to be controlled for operation purposes. The location of transport barriers is an important issue because of the magnetohydrodynamic stability constraints. Steep gradients associated with transport barriers can be tolerated only within the robust MHD stable regions of the plasma body.

At present, a large body of evidence obtained on many tokamaks shows that the formation of a transport barrier is closely related with the hypothesis that the increased shear in the $E \times B$ drift leads to a suppression of turbulence, thereby improving confinement.

This paradigm is of fundamental importance for fusion research. Indeed, it has amplified the world-wide efforts put into studies of the origin of the self-consistent electric fields and also encouraged methods to impose electric fields on plasma by various techniques.

Achieving understanding and control of turbulent transport is a necessary prerequisite for the design of an economical fusion reactor. Significant progress in

Astrophysics and Space Science **256:** 205–218, 1998.

this endeavour has been made by exploiting the spontaneous transition to a high confinement regime, H-mode. This occurs provided a certain threshold in power injected into edge plasmas is exceeded. Transitions are envisaged to occur as a multistep process. First, at certain level of injected power, a significant modification of the electric field profile unfolds. Then, the resulting shear of the electric field is responsible for the reduction of fluctuation levels and transport. The bifurcation occurs during the transition as a discontinuous change in the equilibrium quantities such as electric field, temperature and density profiles. The cornestone of the concept is the link between the electric field and improved confinement. At first, we address the issue of the effect of the shear of the electric field on residual turbulence and postpone the more difficult discussion on the origin of the electric field to later sections.

2. Basic Features of Shear Stabilization Models

Important elements of the concept are the option of linear stabilization of a possible instability in a tokamak or an alternative option of turbulence suppression due to enhanced eddy decorrelation induced by the shear of the electric field.

Rotation of a Navier–Stokes fluid resulting in turbulence saturation at a reduced level is a well-known effect in fluid mechanics (Landau and Lifshits, 1988). However, since a sheared flow also provides free energy which can drive the Kelvin–Helmholtz instability the phenomena is not frequently observed in fluids. In contrast, in a magnetized plasma the impact of the sheared flow may obtain two different forms.

The $E \times B$ shear can stabilize an instability within a given range of plasma volume depending on plasma profiles. The complete stabilization of various modes occurs provided the $E \times B$ shearing rate exceeds the linear growth rate of a given instability

$$\gamma_S > \gamma_{\max}, \tag{1}$$

where

$$\gamma_S = \frac{RB_\theta}{B}\frac{\partial(E_r/RB_\theta)}{\partial r} \tag{2}$$

and $\gamma_{\max}$ is the linear growth rate. First, Equation (1) shows that the stabilization is a mode specific feature. The effect stems from the amplified damping of a given unstable mode due to an interaction between this mode with a nearby stable mode. The stabilization results from the enhanced Landau damping of an unstable mode due to this interaction. Furthermore, the form of the shearing rate given by Equation (2) clearly depends not only on the electric field profile, but also on the magnetic shear (Hahm and Burrel, 1995). The dramatic impact of the magnetic shear is a geometric effect occurring because of differences in radial separation of magnetic surfaces. Hence, if the shearing rate significantly exceeds the growth rate of a given instability one might expect stabilization followed by concominant improvements in confinement.

Yet, the constraint given by Equation (1) is rather stringent and can hardly be satisfied for all modes, unstable within the entire plasma body. Therefore, in order to reduce significantly the turbulence level the shearing rate has to be of the order of nonlinear saturatio nof a turbulence level resulting from unstable modes. Quantitatively, in absence of the $E \times B$ shear, the level of turbulence is determined by the balance between the growth rate of an instability and damping provided by an anomalous diffusion. The time scale for nonlinear damping is of the order

$$\tau^{-1} = \omega \sim \gamma = 4D/\delta_0^2, \tag{3}$$

where γ is the growth rate, D is the coefficient of the anomalous diffusion and δ_0 is the radial correlation length of the turbulence. This value has to be compared with a scale δ determined by decorrelation of fluctuating plasma parameters (density, temperature, potential, etc.) arising due to a differential poloidal motion on a scale of the poloidal variation k_θ^{-1}.

Introducing the shearing rate as $\gamma_S \sim k_\theta |dV_0/dr|$, where V_0 is the $E \times B$ drift velocity, one easily obtains the frequency of the combined shearing and turbulent decorrelation rates as (Biglari et al., 1990)

$$\gamma_c = (\gamma_s^2 \Delta\omega_D)^{1/3}, \tag{4}$$

where $\Delta\omega_D$ is the nonlinear decorrelation rate in absence of the $E \times B$ shear.

Both effects are beneficial for the quality of confinement. The former is clearly more beneficial and dramatic, yet the subject to more stringent constraints on the shearing rate. Recently, the full nonlinear numerical analysis of the effect of shearing rate on the ion temperature gradient mode was carried out (Waltz, 1994). Unexpectedly, it has transpired that the ITG modes are completely suppressed already when $\gamma_S \sim \gamma_{\max}$.

This remarkable result indicates that in order to assess quantitatively the impact of the $E \times B$ shear only fully nonlinear models, addressing the effect in detail, suffice.

Models should include synergistic effects of prelude profiles, in their capacity to drive varying modes unstable, the $E \times B$ and magnetic shear profiles. For example, negative or low shear current profiles allow stabilization of high n MHD modes (ballooning), notorious to be harmful to the confinement. Another example, profiles with the safety factor $q > 1$ everywhere within the plasma body stabilize sawteeth oscillations. Hence, lack of these instabilities combined with the localized deposition of particle, momentum and heat sources or sinks allows pressure and/or rotating gradients to build up, thus amplifying locally the radial field gradient. Thus, E_r can be changed locally by particle, momentum and heat sources or sinks, thereby providing for multiple feedback loops leading to turbulence decorrelation and, hence, reduced transport. Therefore, local transport bifurcations might occur resulting in the build-up of transport barriers due to the enhanced $E \times B$ shear decorrelation. Studies of different scenarios of transitions are presented below.

3. Origin of Electric Fields

The origin of electric fields is an important part of the concept. It is complementary to the physics described in the previous section, yet more difficult to assess. In order to address the origin of the electric field the radial component of the momentum balance is invoked

$$E_r = B_\theta V_\phi - B_\phi V_\theta + \frac{1}{eZ_i n_i} \frac{\partial p_i}{\partial r} . \tag{5}$$

Equation (5) is fundamental because it offers insights on how an electric field can be induced within the plasma. Basically, there are three terms ("knobs"), which can be employed. These are diamagnetic drifts, toroidal and poloidal rotations. Obviously, theses three "knobs" can either reinforce or counteract each other. In the ideal situation, when all conditions are favourable all of the three terms can contribute in order to amplify changes in the electric field. To start with, we consider the situation at the edge where the electric field is primarily due to the diamagnetic drift and the poloidal rotation. Toroidal rotation is assumed to be effectively damped there by the anomalous gyroviscosity.

Then, employing the neoclassical value for the poloidal rotation velocity to be the well-known function of the temperature gradient one easily obtains

$$E_r = \frac{T_i}{e} \left[\frac{dlnn}{dr} + (1-k) \frac{dlnT_i}{dr} \right], \tag{6}$$

where $(1-k)$ is the reducing coefficient, well known from the neoclassical theory (Hazeltine and Hinton, 1976). Hence, the electric field is a linear combination of logarithmic derivatives of density and temperature profiles. This approach provides the well-founded estimate for the evolution of the profile of the electric field from first principles. Experimentally, it enables to employ only standard local measurements of density and temperature profiles (without invoking advanced diagnostics such as Heavy Ion Beam Probe, Charge Exchange Recombination, etc.) in order to determine electric field profiles.

Equation (6) demonstrates that the electric field is locally amplified provided density and/or temperature profiles become steeper. The analysis of Equation (6) together with particle, heat, momentum and turbulence balances offers powerful means to follow in detail transitions into regimes with improved confinement (Stabler et al., 1997; Diamond et al., 1997).

However, although Equation (6) yields the convenient approximation to determine the profile of the electric field, it remains rather crude because it does not account for the transport of parallel flows in a tokamak. In general, it is well

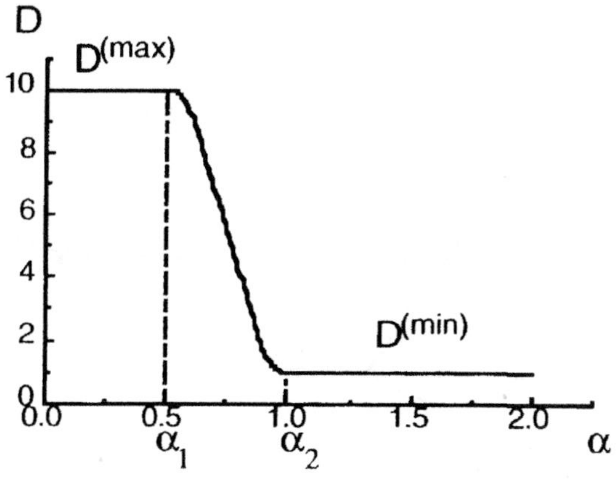

Figure 1. Diffusivity against shear of the electric field α.

known that parallel dynamics is a very important factor for many issues in tokamak physics. Therefore, it is likely that the effect is also important for the issue of electric field profiles. The parallel component of the flux surface average momentum balance yields the equation describing the evolution of the $E \times B$ velocity V_0 (see equation (2.25) in Rozhansky and Tendler, 1996). Obviously, the approach employing Equation (6) cannot provide insights into a cause of the transition because Equation (6) implies the rigid relation between the electric field and density and/or temperature gradients due to its form. Therefore, it is mandatory to invoke the parallel component of the momentum balance equation in order to address the causal relation between the electric field and the effect of steepening of density and/or temperature profiles occurring during the transition.

4. Evolution of Transport Barriers

The formation of the transport barrier is to be modelled by the system of general diffusion equations taking the form

$$\frac{\partial h}{\partial t} - \frac{\partial[D(\alpha)\frac{\partial h}{\partial x} - V(\alpha)h]}{\partial x} = S, \tag{7}$$

where h is the density, energy, toroidal momentum and density of fluctuations, D is the corresponding diffusivity and V is the velocity of the inward pinch. Importantly, all diffusivities and inward pinch velocities are decreasing functions of the shear of the electric field $\alpha \sim |dE_r/dr|$, shown in Figure 1.

Note that only qualitative features of Figure 1 are crucial for analysis of the transition, not fine details of the curve.

The form of the curve shown in Figure 1 brings up the importance of a prelude profile, existing prior to the transition. Indeed, if initially a profile of the electric

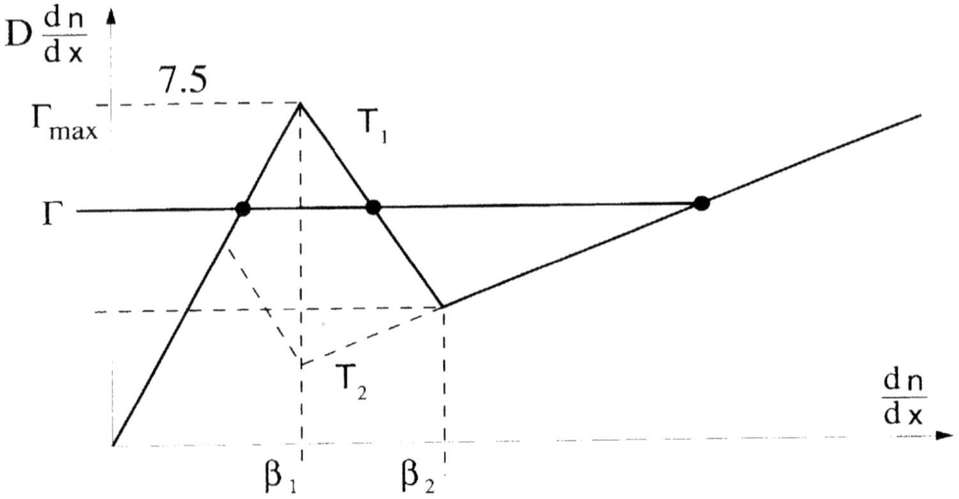

Figure 2. Dependence of a flux *versus* a gradient.

field anywhere in the discharge happens to be right at the point α_1 (this assumption has obviously to be qualified by the form of profiles in existence in the L-mode), then an infinitely small deviation from the equilibrium profile will imply the start of a transition into the regime with improved confinement.

Hence, employing the paradigm of shear stabilization one obtains the dependence of a flux *versus* a gradient shown in Figure 2. The form of this dependence is reminiscent of the S-curves employed in the theory of superconductivity in describing a transition in a superconductive state (Landau and Lifshits, 1990). The physical meaning of Figure 2 is simple. If a profile is on the branch I it corresponds to the L-regime of confinement. In contrast, the branch II describes the H-mode. Therefore, the direct transition occurs provided $\Gamma \geq \Gamma_{max}$.

Among frequently asked questions pertaining to the issue of improved confinement, one may wonder why H-modes were discovered first, long before the other types of transport barriers such as VH-modes and all kinds of internal transport barrier were encountered experimentally. The answer probably lies in the sensitivity of the electric field profile to subtle details of density and temperature profiles. Indeed, employing the formula given by Equation (6) and ignoring, for simplicity, the temperature gradient one obtains

$$\alpha \sim \frac{dE}{dr} \sim \frac{d^2 ln\, n}{dr^2} \sim [-(n'/n)^2 + n''/n]. \tag{8}$$

Focusing on the first term, this simple calculation indicates that for a given gradient of the density profile the constraint $\alpha \geq \alpha_1$ is much less restrictive at the plasma edge, where the absolute values of plasma density are much smaller than in the core. This is why the transport barriers are more resilient at the edge (the case for

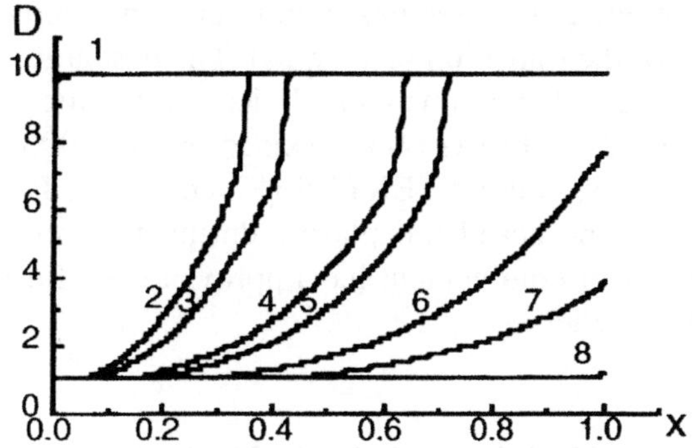

Figure 3a. Evolution of the diffusion coefficient L–H transition $\Gamma = 7.5 > \Gamma_{\mathrm{max}}$.

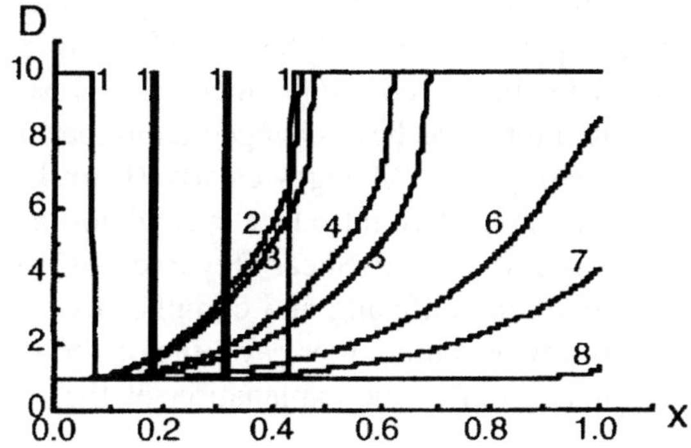

Figure 3b. Evolution of the diffusion coefficient during pellet induced L–H transition $\Gamma = 6.75 < \Gamma_{\mathrm{max}}$ (1, 2, 3 are time slices normalized to confinement time in the L-mode within the barrier).

H modes) than in the centre of the discharge. Another important point following from this argument is the utter sensitivity of the transition to a prelude profile due to the dependence on second derivative, yielded by the second term.

This observation offers the option to trigger the transition below the threshold value ($\Gamma < \Gamma_{\mathrm{max}}$). In other words, to employ the tunnelling in order to materialize the transition. Indeed, the transition unfolds straightforwardly when the influx exceeds a given value $\Gamma \geq \Gamma_{max}$ (see Figure 3a). However, a more fine tune "knob" capable to trigger the transition emerges, provided that density and/or temperature profiles can be dramatically altered locally, without, at the same time changing absolute values of density and temperature. This argument favours strongly localized sources and/or sinks of particles, energy and axial momentum as fine triggers of transitions into regimes with improved confinement. The example of the transition occurring below threshold is shown in Figure 3b where only a very small fraction of particles (around 2%) is added very locally (within poloidal gyroradius), at the edge of the plasma. The transition unfolds on approximately the same timescale. Yet, fine details of the transition differ from the transition triggered by the incoming flux. Indeed, filamentation of regions with improved confinement is observed shortly after the start of the transition ($10^{-3}\tau_{conf}$). The best argument corraborating this point is the fact obtained experimentally. Indeed, experiments along these lines were previously reported on TUMAN-3 (Askinazi et al., 1993). The most recent experiment has been carried out on TEXTOR (Finken et al., 1997), where a small pellet was injected into the peripheral plasma during the improved confinement "RI" mode. The quality of confinement has improved even further in response to the injection of a small pellet.

5. Highlights of Key H-Mode Work

Since there is no future without the past, it is important to review the short history of work carried out, both experimentally and theoretically, aiming at understanding and control of regimes with improved confinement. The history is also important because dominant interpretations have changed a number of times during this period, without actually stating these changes clearly (Burrel et al., 1992, 1994, 1997). Furthermore, theoreticians often introduce special notions such as "Stringer spin up", "Reynolds Stress Dynamo or Shear Reynolds Stress" and so on. The wide use of these terms is often confusing and complicates communications and the mutual understanding of two communities, working on theory and experiments. Therefore, it appears timely to offer the explanation of the real meaning of the terminology.

The original discovery of the H-mode was achieved by F. Wagner et al. (1982) on the ASDEX tokamak in 1982. Since then, the H-mode has proven to be one of the most robust and ubiquitous modes of improved confinement in toroidal magnetic fusion devices. Therefore, the physics of the L-mode to H-mode (L–H) transition has attracted a great deal of interest from both experimental and theoretical communities. The next dramatic step forward for the development of experimental studies was the observation on the DIII–D tokamak that the edge *impurity* ion poloidal velocity changes dramatically and abruptly during the transition reported by Groebner et al. (1989). The demonstration on the CCT tokamak that the H-mode can be triggered by applying internal biasing by means of internal probe has been the next major step forward (Taylor et al., 1989; Weynants and Taylor, 1990). Recently, regimes with internal transport barriers were obtained on all major large tokamaks JET (Hugon et al., 1992), TFTR (Levinton et al., 1995), DIII–D (Strait, 1995) and JT-60 (Kamada, 1994). All these successes in experimental work contributed greatly to the progress of the theoretical studies.

In theory, the idea that the electric field is the controlling parameter for the H-mode was first put forward in (Chiueh et al., 1986). The sign reversal of the poloidal rotation in the H-mode compared to neoclassical predictions and the L-mode observed on DIII–D have led several theorists to put forward and to advocate ions loss mechanism as a steady-state scenario for the transition (Itoh and Itoh,

1988; Shaing and Crume, 1989). These models commanded the full attention of the community until very recently. Yet, six years later, in contrast to previous experiments, measurements of the poloidal rotation velocity of main ions were carried out, employing lines from main species (Kim et al., 1994) and it was demonstrated on DIII–D that the sign of the poloidal rotation of the *main* ions and the impurity ions are *indeed opposite* and, the sign of the poloidal rotation for main ions is in agreement with the neoclassical predictions. In other words, these theories were led to explain the wrong phenomenon.

Yet, to their honour theoreticians were never satisfied with the artificial structure of the ion orbit loss theory for many reasons. First, the electric field caused by the ion orbit loss has not been obtained self-consistently (Tendler et al., 1992). Second, this theory relied upon too many physical assumptions to occur simultaneously. It was obvious that such a coincidence is highly improbable. Third, it did not offer as simple and harmonic picture expected from a plausible theory in physics, in general. Hence, different school of thought (Hinton, 1991; Stabler, 1997) emerged putting forward the bifurcation of equilibrium plasma parameters to result from an S-like dependence of heat fluxes on the neoclassical poloidal rotation velocity (therefore, also on the ion temeprature gradient in accordance with the standard neoclassical theory) in analogy with the Landau–Ginzburg theory of transitions into a superconductive state. However, the same experiment (Kim et al., 1994) has demonstrated that the measured value of the poloidal rotation in the H-mode differs dramatically from neoclassical predictions already after a few milliseconds following the transition. Therefore, the need to address the issue of a torque capable to provide for a fast poloidal rotation velocity remained unanswered. Different models have been put forward adopting the idea that the turbulent transport of parallel flows might provide the poloidal momentum torque by means of off-diagonal terms in the momentum balance (Tendler and Rozhansky, 1989, 1990; Hassam et al., 1991; Carreras et al., 1991; Kim and Diamond, 1991). The issue of parallel dynamics was addressed either from the neoclassical point of view (Tendler and Rozhansky, 1989, 1990; Hassam et al., 1991) or turbulence theory (Hassam et al., 1991; Carreras et al., 1991). As mentioned above, parallel flows are very important. Within the framework of neoclassical theory, they give rise, due to the toroidal geometry to the well-known Pfirsch–Schlueter flows and currents. In general, it is asserted that transport parallel to the magnetic field is in good agreement with neoclassical theory. By contrast, cross-field transport of energy, momentum and particles are grossly anomalous. Furthermore, some models put the special emphasis on the poloidal dependence of anomalous cross-field transport, resulting in additional flux surface average products of harmonics of the anomalous transport of angular momentum with parallel flow harmonics (Hassam et al., 1991). Another important point ignored by most of models then was the impact of the ion diamagnetic drift on the electric field profile following the dependence of neoclassical parallel flows on diamagnetic drifts.

Indeed, since the electric field at the edge within the transport barrier is well known to be negative, the diamagnetic drift is bound to reinforce this effect and therefore has the potential to become a major player, even in absence of both poloidal and toroidal rotations. It should be borne in mind that diamagnetic drift was retained by the model relying upon neoclassical parallel dynamics because this drift is the only cause for the well-known Pfirsch–Schlueter current (Tendler and Rozhansky, 1989, 1990).

Although a complete, detailed theory of confinement improvements does not exist yet, initial works started on a strongly varying range of ideas and assumptions have merged recently in a more coherent path of development.

6. Different Scenarios of L–H Transitions

For some time, there was a lively debate between JET and DIII–D teams about the role of the poloidal rotation velocity in triggering the L–H transition. In more detail, DIII–D team reported (Groebner et al., 1989; Kim et al., 1994) that the poloidal rotation changes dramatically during the L–H transition while the measurements carried out on JET (Hawkes, 1996) did not observe any significant changes in poloidal rotation. The way to reconcile these two seemingly contradicting statements is to consider the formation of transport barriers, in detail. In general, the method to study the evolution of transport barriers employ the system of balance equations (including particle, heat, toroidal and parallel momentum, turbulence) with diffusivities consisting of a small neoclassical part independent of the $E \times B$ shear and the much larger part, which is strongly reduced as this shear is increased. Experimentally, there is a wide variety of range of opinions on the cause of the transition (Hugill, 1994) because on some machines such as JET the poloidal velocity at the edge does not show any change within the accuracy of the measurements. Therefore, it is concluded that any changes of E_r preceding the transport change would have to occur on a timescale shorter than the measurement interval. In contrast, the measurements carried out on DIII–D have demonstrated the significant change in the perpendicular velocity already 5 ms after the transition. Here, it is asserted that it is possible to reconcile these apparently contradicting data by either changing boundary and initial conditions for electric field on the separatrix or the prelude density profile by variation of a particle source invoking a surplus or a depletion locally, in the vicinity of the separatrix. Indeed, for a given source function in particle, energy or toroidal momentum balances the electric field and the resulting from it shearing rate increases reaching the threshold for nonlinear decorrelatioon of turbulent eddies, thereby decreasing diffusivities locally. In this scenario, it is assumed that prelude profiles in the L-mode are close to the threshold for shear stabilization. Thereafter, the avalanche of propagation of the transport barrier unfolds on a very fast time scale of the order of a tenth of a millisecond (see Figure 4). The propagation within the barrier is very fast (much faster than

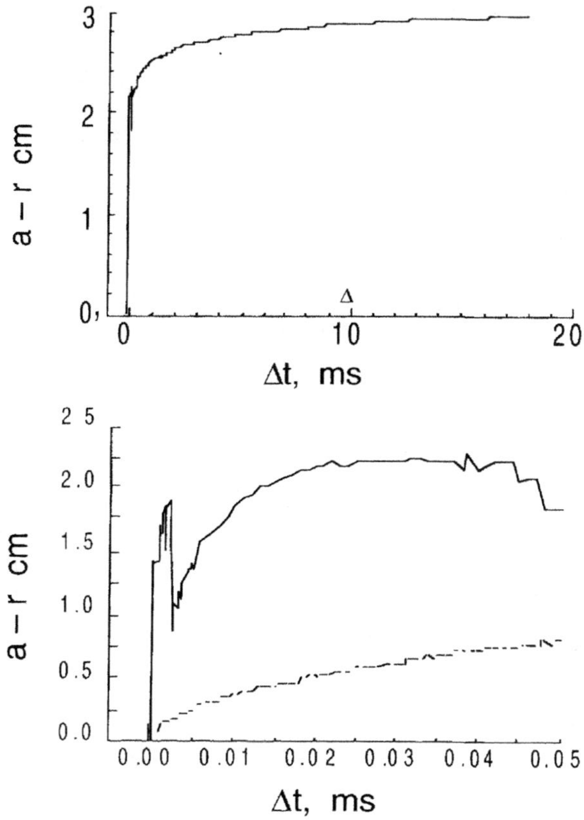

Figure 4. Propagation of the front of the improved confinement is extremely fast due to ballistic nature of governing equations.

the diffusive length of the front) because of the ballistic nature of the solution. Correlations are carried up and down the barrier. The front of reduced diffusivity propagates inwards. The diffusion coefficient drops dramatically, thereby causing either density and/or temperature profiles to steepen (see Figure 5). The formation of the barrier is terminated because the shearing rate decreases as the square of the local density. Once the barrier is formed, it affects the density profile within the entire plasma body. Density profiles are becoming more peaked because of the pedestal whereas transport coefficients are suppressed.

The different scenario may be envisaged if the prelude density profile is not close to the critical shear level. Then, of course, a small perturbation in density cannot trigger the transition. However, the transition can materialise provided the electric field is amplified over the neoclassical level at the separatrix. For example, assuming that the $E \times B$ velocity at the separatrix $V_0 \approx \Theta C_S$ the transition unfolds. The profiles of the poloidal rotation velocity and the density are evolved and broaden quite significantly, thereby reducing transport coefficients within the bar-

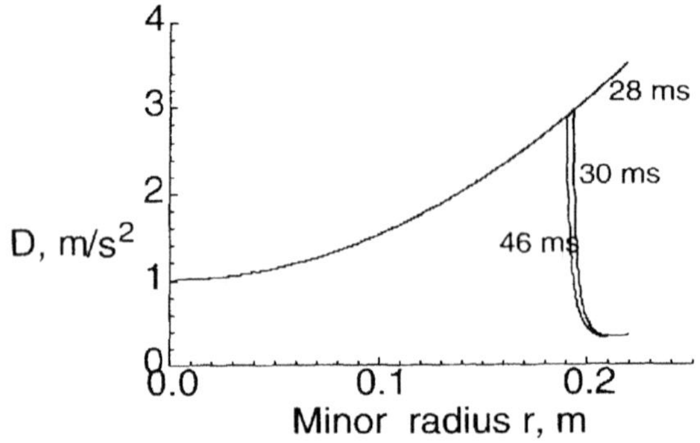

Figure 5. Evolution of the diffusion coefficient due to the density gradient. The poloidal rotation is kept zero $V_\theta = 0$.

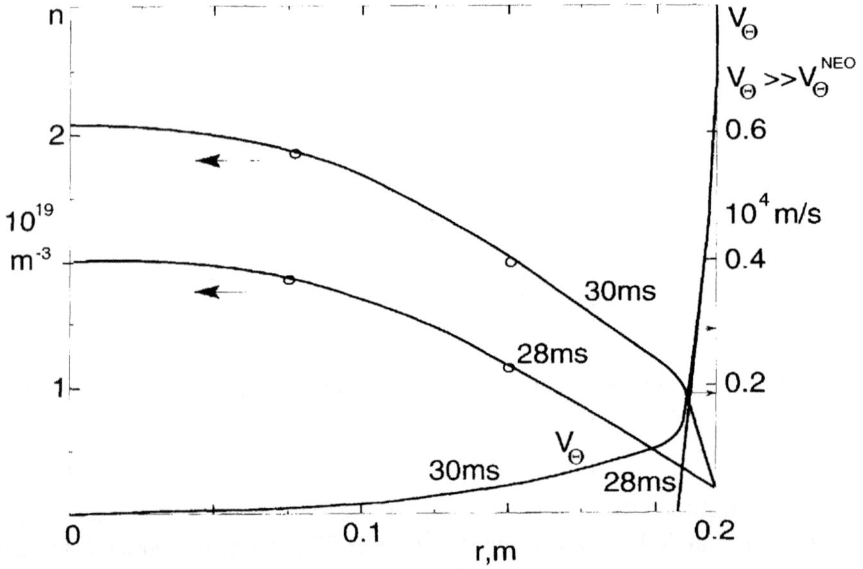

Figure 6. Evolution of the density and the poloidal rotation V_θ during the transition triggered by the enhanced poloidal rotation $V_\theta \gg V_\theta$ NEO at the separatrix.

rier (see Figure 6). The remarkable result is that at the end of the day the evolution of two grossly different initial conditions merges into the same form for profiles of diffusion coefficient, convective velocity, density and the electric field.

7. Conclusions

Discharges with transport barriers are a fascinating laboratory for turbulence studies since the dominant modes are suppressed and subdominant transport mechanisms can be investigated all the way down to neoclassical levels. This calls for further improvements of the standard neoclassical theory with the emphasis on so far academic effects of small aspect ratio portions of the discharge, squeezing of banana orbits by the varying electric fields (Hazeltine, 1989), etc. To adopt power flux to the separatrix as the trigger of the L–H transition appears from the theoretical point of view a "brute force" solution to the issue of the confinement control. The level of understanding of the physical fundamentals has made sufficient progress to guide the present day experiments and extrapolate trends to future devices. Although a complete, detailed theory of confinement improvements does exist yet, initial works started on a strongly varying range of ideas and assumptions have merged recently in a more coherent and consistent path of development, mutually enriching and complementing each other. A few outstanding problems remain to be addressed on the basis of first principles. Specifically, a rigourous equation adequate to describe the propagation of turbulence including the effect of the $E \times B$ shear has to be found. The progress justifies the future work in this area.

References

Askinazi, L., Golant, V.E. et al.: 1993, *Phys. Fluids* **B5**, 2420.
Biglari, H., Diamond, P.H. and Terry, P.: 1990, *Phys. Fluids* **B2**, 1.
Burrel, K.H. et al.: 1992, *Plasma Phys. Control. Fusion* **34**, 1859.
Burrel, K.H. et al.: 1994, *Plasma Phys. Control Fusion* **36** 291.
Burrel, K.H. et al.: 1997, *Phys. Pl.* **4**, 1499.
Carreras, B.A. et al.: 1991, *Phys. Fluids* **B3**, 1438.
Chiueh, T., Terry, P.W. et al.: 1986, *Phys. Fluids* **29**, 231.
Diamond, P.H., Lebedev, V.B. et al.: 1997, *Phys. Rev. Lett.* **78**, 1472.
Finken, K.H. et al.: 1998, Proceedings of this Conference.
Groebner, R. et al.: 1989, in: *Europhysics Conference Abstracts* Vol. 13B, p. 245.
Hahm, T.S. and Burrel, K.H.: 1995, *Phys. Pl.* **2**, 1648.
Hassam, A.B. et al.: 1991, *Phys. Rev. Lett.* **66**, 309.
Hawkes, N.C.: 1996, *Plasma Physics and Controlled Fusion* **38**, 1261.
Hazeltine, R.: 1989, *Phys. Fluids* **B1**, 2031.
Hazeltine, R.D. and Hinton, F.L.: 1976, *Rev. Mod. Phys.* **48**, 239.
Hinton, F.L.: 1991, *Phys. Fluids* **B3**, 696.
Hugill, J.: 1994, *Plasma Physics and Controlled Fusion* **36B**, 173.
Hugon, M. et al.: 1992, *Nuc. Fus.* **32**, 33.
Itoh, S.I. and Itoh, K.: 1988, *Phys. Rev. Lett.* **60**, 2276.
Kamada, Y.: 1994, *Nuc. Fus.* **34**, 1605.
Kim, Y.B. and Diamond, P.H.: 1991, *Phys. Fluids* **B3**, 1626.
Kim, J. et al.: 1994, *Phys. Rev. Lett.* **72**, 2199.
Landau, L. and Lifshits, E.: 1988, *Fluid Mechanics*, Mir, Moscow.
Landau, L.D. and Lifshits, E.M.: 1990, *Statistical Physics II*, Mir, Moscow.
Levinton, M. et al.: 1995, *Phys. Rev. Lett.* **75**, 4417.
Rozhansky, V. and Tendler, M.: 1996, in: B.B. Kadomitsev (ed.), *Reviews of Plasma Physics*, Consultants Bureau, New York/London, Vol. 19.
Shaing, K.C. and Crume, E.C.: 1989, *Phys. Rev. Lett.* **63**, 2365.
Stabler, G.M., Waltz, R.E. and Wiley, J.C.: 1997, *Nuc. Fus.* **37**, 287.

Strait, E.J.: 1995, *Phys. Rev. Lett.* **75**, 4417.
Taylor, R.J. et al.: 1989, *Phys. Rev. Lett.* **63**, 2365.
Tendler, M. and Rozhansky, V.: 1989, *Bull. Am. Phys. Soc.* **34**, 2162
Tendler, M. and Rozhansky, V.: 1990, *Comments Plasma Physics and Controlled Fusion* **13**, 191.
Tendler, M. et al.: 1992, *Plasma Physics and Controll. Nuclear Fusion, IAEA* **2**, 44.
Wagner, F. et al.: 1982, *Phys. Rev. Lett.* **49**, 1408.
Waltz, R.E.: 1994, *Phys. Pl.* **1**, 2229.
Weynants, R. and Taylor, R.J.: 1990, *Nuc. Fus.* **30**, 945.

ROTATION OF TOKAMAK PLASMAS BY ALFVÉN WAVES

V.S. TSYPIN, A.G. ELFIMOV, C.A. DE AZEVEDO and A.S. DE ASSIS
Institute of Physics, State University of Rio de Janeiro, Rua São F. Xavier, 524, Maracanã, Rio de Janeiro, 20550-013, Brazil

Abstract. Externally excited Alfvén waves can generate the poloidal and toroidal rotation in the collisional (edge) and core (weakly collisional) tokamak plasma. The necessary value of these velocities to achieve the L-H transition, can be obtained by manipulating the value of an absorbed power. This rotation linearly depends on dissipated power, decreases in banana regimes and did not depend on toroidal magnetic field. In the plasma layer where the RF power is absorbed, the radial electric field also appears. There is the very satisfactory agreement between our theory and experimental results obtained on Phaedrus-T tokamak. The methods developed in this paper can be applied to other tokamaks with another kind of waves, which can be used to operate the plasma dynamics taking into account the appropriate dispersion relation for that type of waves.

1. Introduction

A plasma rotation has a very strong effect on the dynamics of plasma (equilibrium, stability, transport), which can depend on quasistationary and rf field's characteristics in tokamak plasmas (Callen, 1992; Wong and Hinton, 1984; Tsypin, 1987; Graddock and Diamond, 1991). This problem has the special significance for low-high confinement transition (L-H transition), which can be stimulated by the external forces (Biglari et al., 1990) in the collisional (edge) plasma. The Alfvén waves are among them (Elfimov et al., 1992; Klima, 1968; Fish and Karney, 1984). This wave technique is attractive because of the availability of RF power generators and the simplicity of wave launching systems (Hershkovitz et al., 1994).

In tokamaks, the problem of the plasma rotation produced by ion Bernshtein waves (Biglari et al., 1992) and by fast waves (Wang et al., 1995), by plasma turbulence (Diamond and Kim, 1991) and by a biased electrode (Gao et al., 1994) has already been discussed. In axially-symmetric tokamak plasmas, the rotation has been under study in (Biglari et al., 1992; Wang et al., 1995; Diamond and Kim, 1991). The Reynolds stress term in the RF force has only been taken into account in (Biglari et al., 1992; Diamond and Kim, 1991). It was shown that the poloidal sheared plasma flow can effectively be induced by these effects. Here, we estimate quasistationary poloidal and toroidal velocities and radial electric fields, produced by Alfvén waves, in collisional (edge) plasmas of slightly rippled tokamaks.

Nowadays, the rotation of the bulk (weakly collisional) plasma is supposed to be the wasteful in using of RF energy as a parallel effect of AW plasma heating or AW current drive. That is a reason why the problem of plasma rotation induced by

Astrophysics and Space Science **256:** 219–233, 1998.

RF waves is very important in tokamaks. In this paper as a difference from papers mentioned above, we consider rotation induced by kinetic AW in weakly collisional plasma of slightly rippled tokamaks in plateau and banana regimes. Rippled magnetic field of tokamaks retards the plasma rotation in toroidal direction. Here, we are going to find the quasistationary poloidal and toroidal plasma velocities and the radial electric field taking into account the complete form of ponderomotive forces.

Nowadays, the rotation of the bulk plasma is supposed to be the wasteful in using of RF energy as a parallel effect of AW plasma heating or AW current drive. That is a reason why the problem of plasma rotation induced by RF waves is very important in tokamaks. In this paper as a difference from papers mentioned above, we consider rotation induced by kinetic AW in weakly collisional plasma of slightly rippled tokamaks in plateau and banana regimes. Rippled magnetic field of tokamaks retards the plasma rotation in toroidal direction. Here, we are going to find the quasistationary poloidal and toroidal plasma velocities and the radial electric field taking into account the complete form (Elfimov et al., 1992; Klima, 1968; Fish and Karney, 1984) of ponderomotive forces.

The comprehensive models of the L-H transition have six key elements (Callen, 1992):

- A net radially outward ion current is induced in the edge layer plasma (due to ion orbit losses, radially asymmetric turbulence, externally inserted biased probe, RF forces).
- This radial current leads to a $\mathbf{J}_r \times \mathbf{B}$ force on the plasma that is dampened by the neoclassical viscous force on poloidal flows.
- If the induced poloidal force is large enough it can overcome the viscous damping and induce a large (but usually slightly subsonic) poloidal flow, and shear in the poloidal flow in the edge layer.
- For sufficiently strong poloidal flow shear, turbulent eddies get torn apart.
- The shearing effect reduces the fluctuation level.
- The reduced fluctuation level reduces the anomalous transport in this edge layer and thus produces a "transport" barrier.

Alfvén waves (AW) can be used for the current-drive, RF heating of plasmas, and operating L-H transition in thermonuclear facilities (in tokamaks especially).

Alfvén waves have been suggested for the inducing of the strong poloidal rotation with the strong shear (Graddock and Diamond, 1991; Tsypin et al., 1996).

2. Basic Equations

To find quasistationary quantities of the plasma rotation velocities and radial electric field, we proceed from one fluid motion equations of the plasma (Braginskii, 1965)

$$M_i n \frac{d_i \mathbf{V}_i}{dt} = -\nabla p - \nabla \cdot \hat{\boldsymbol{\pi}}_i + \frac{1}{c} [\mathbf{J} \times \mathbf{B}] + \mathbf{F}^h, \tag{1}$$

where $\mathbf{F}^h$ is the RF force, and the ion viscosity has the form

$$\nabla \cdot \hat{\boldsymbol{\pi}}_i = \nabla \cdot \hat{\boldsymbol{\pi}}_{i\parallel} + \nabla \cdot \hat{\boldsymbol{\pi}}_{ig} + \nabla \cdot \hat{\boldsymbol{\pi}}_{i\perp}. \tag{2}$$

Here, $\hat{\boldsymbol{\pi}}_{i\parallel}, \hat{\boldsymbol{\pi}}_{ig}, \hat{\boldsymbol{\pi}}_{i\perp}$ are parallel, gyro- (magnetic, oblique), and perpendicular viscosities respectively, and last designations of the common use.

We can approximately affirm that the ratio of the viscosities is next

$$\nabla \cdot \hat{\boldsymbol{\pi}}_{i\parallel} : \nabla \cdot \hat{\boldsymbol{\pi}}_{ig} : \nabla \cdot \hat{\boldsymbol{\pi}}_{i\perp} = \frac{\varepsilon^2}{q^2} : \frac{\nu_i}{\omega_{ci}} : \frac{\nu_i^2}{\omega_{ci}^2}, \tag{3}$$

where ε is the inverse aspect ratio, q is the safety factor, ν_i is the ion-ion collision frequency, and ω_{ci} is the ion cyclotron frequency. We suppose the inequalities are fulfilled

$$\frac{\varepsilon^2}{q^2} \gg \frac{\nu_i}{\omega_{ci}}.$$

Thus, we only use the parallel viscosity (Chew et al., 1956)

$$\hat{\boldsymbol{\pi}}_{\parallel} = \frac{3}{2}(\mathbf{hh} - \frac{1}{3}\hat{\mathbf{g}})\pi_{\parallel}, \tag{4}$$

where $\hat{\mathbf{g}}$ is the metric tensor. From Equation (4), we get the expression for $\nabla \cdot \hat{\boldsymbol{\pi}}$

$$\nabla \cdot \hat{\boldsymbol{\pi}} = \frac{3}{2} \left\{ [\mathbf{h}\nabla \cdot \mathbf{h} + (\mathbf{h} \cdot \nabla)\mathbf{h}]\, \pi_{\parallel} + \mathbf{h}(\mathbf{h} \cdot \nabla)\pi_{\parallel} - \frac{1}{3} \nabla \pi_{\parallel} \right\}. \tag{5}$$

The covariant components of this vector are given by

$$(\nabla \cdot \hat{\boldsymbol{\pi}})_\theta = \frac{3}{2} \pi_{\parallel} \frac{\partial}{\partial \theta} \ln B - \frac{1}{2} \frac{\partial \pi_{\parallel}}{\partial \theta}, \tag{6}$$

$$(\nabla \cdot \hat{\boldsymbol{\pi}})_\zeta = -\frac{3}{2} \pi_{\parallel} \frac{\partial}{\partial \zeta} \ln B + \frac{\partial \pi_{\parallel}}{\partial \zeta}. \tag{7}$$

The ambipolarity condition has the well-known form, that follows from the current continuity equation $\nabla \cdot \mathbf{j} = 0$

$$\langle j^r \rangle = 0, \tag{8}$$

where

$$\langle\cdots\rangle \equiv \int_0^{2\pi} (\cdots)\sqrt{g}\, d\theta d\zeta \Big/ \int_0^{2\pi} \sqrt{g}\, d\theta d\zeta.$$

Then, we find from Equations (1), (5–7) in the quasistationary state

$$F_\theta^\pi + F_\theta^h = 0, \quad F_\zeta^\pi + F_\zeta^h = 0. \tag{9}$$

These equations mean that RF driving forces, F_θ^h and F_ζ^h (along the poloidal θ and toroidal ζ angles, respectively), are balanced by the poloidal F_θ^π and toroidal F_ζ^π viscous forces. These forces have the form (Hirshman and Sigmar, 1981; Mikhailovskii and Tsypin, 1984)

$$F_\theta^\pi = \frac{3}{2r}\left(\pi_\parallel \frac{\partial \ln B}{\partial \theta}\right)^0, \quad F_\zeta^\pi = \frac{3}{2R}\left(\pi_\parallel \frac{\partial \ln B}{\partial \zeta}\right)^0, \tag{10}$$

where

$$(\cdots)^0 \equiv \frac{1}{2\pi}\oint(\cdots) d\theta d\zeta.$$

The toroidal magnetic field has the form $B = B_0(1 + \varepsilon\cos\theta + \delta\cos N\zeta)$.

3. RF Forces

We suppose that the quasicylinder approximation can be used to find RF forces, which are similar for collisional and weakly collisional plasmas. The main difference is that calculating oscillating currents in plasma and collisionless wave damping we assume dissipation by way of untrapped and trapped electrons.

$$F_s^h = \frac{1}{4\omega}\left\{\frac{1}{x_s}\frac{\partial}{\partial r}\sum_\alpha\left[x_s j_{\alpha r}^*\left(iE_s - \frac{4\pi\omega}{\omega_{p\alpha}^2} j_{\alpha s}\right)\right] + \frac{i}{L_s} E_k \frac{\partial}{\partial s} j_k^* + \text{c.c.}\right\}, \tag{11}$$

where $s = \theta, \zeta$ and $L_\theta = r$, $L_\zeta = R$, $x_\theta = r^2$, $x_\zeta = r$.

Here, we consider excitation of slow AW (SAW) in the collisional plasma and kinetic AW (KAW) for weakly collisional plasma, which are localized near the conversion point because of strong electron dissipation. That is why the plasma dynamics can be well operated in appropriate place. We are going to use the well-known dispersion relation of these waves in the simplified cylinder geometric optics approximation:

$$N_\perp^2 \approx \frac{\varepsilon_\parallel}{\varepsilon_\perp}(\varepsilon_\perp - N_\parallel^2), \tag{12}$$

where the indices "$\perp$" and "$\|$" mean the orientation to the magnetic field,

$$\varepsilon_\perp = 1 + \sum_\alpha \frac{\omega_{p\alpha}^2}{\omega_{c\alpha}^2 - \omega^2}, \quad N_\perp = \frac{ck_\perp}{\omega}, \quad N_\| = \frac{ck_\|}{\omega}.$$

For slow and kinetic AW, the Landau damping is contained in the longitudinal component of the permeability tensor $\varepsilon_\|$, including the damping on trapped particles (KAW). So, to calculate the RF forces, we can use the expressions for the currents

$$\mathbf{j}_{\alpha\perp} = \frac{\omega_{p\alpha}^2}{4\pi(\omega_{c\alpha}^2 - \omega^2)}(\omega_{c\alpha}[\mathbf{h} \times \mathbf{A}_\alpha] + i\omega\mathbf{A}_\alpha), \tag{13}$$

$$j_\| \approx \frac{\omega_{pe}^2}{4\pi\omega_{ce}k_\|}[\mathbf{E} \times \mathbf{k}]_\| + \frac{i\omega}{4\pi}(1 - \varepsilon_\|)E_\|, \quad j_{\|i} \approx 0.$$

where

$$\mathbf{A}_\alpha = \frac{1}{e_\alpha n_\alpha}\nabla_\perp p_{\perp\alpha} - \mathbf{E}_\perp, \quad p_{\perp e} = -\frac{ie_\alpha n_\alpha}{k_\|}E_\|, \quad p_{\perp i} \approx 0.$$

We used Equation (1), and definitions

$$p_\perp = p - \frac{1}{2}\pi_\|, \quad p_\alpha = \frac{1}{3}M_\alpha \int w^2 f_\alpha d\mathbf{v},$$

$$j_{\alpha\|} = e_\alpha \int d\mathbf{v} v_\| f_\alpha; \quad \pi_{\alpha\|} = M_\alpha \int \left(w_\|^2 - \frac{1}{3}w^2\right) f_\alpha d\mathbf{v}.$$

Here, f_α is the distribution function of ions or electrons, that is determined from the kinetic equation, and, hence, the Landau damping is enclosed in these terms for a weakly collisional plasma case.

To find values $p_{\perp\alpha}$, $\pi_{\|\alpha}$ and $j_{\|\alpha}$ for a weakly collisional plasma in an arbitrary magnetic configuration (and in an arbitrary coordinate system), we should proceed from the drift kinetic equation (Mikhailovskii and Tsypin, 1982)

$$\frac{\partial f}{\partial t} + \frac{d\mathbf{r}}{dt} \cdot \nabla f + \frac{dv_\perp^2}{dt}\frac{\partial f}{\partial v_\perp^2} + \frac{dv_\|}{dt}\frac{\partial f}{\partial v_\|} = \hat{S}t\{f\};$$

where

$$\frac{d\mathbf{r}}{dt} = \mathbf{u} + \frac{1}{\omega_c}\left[\mathbf{h} \times \left(\frac{\partial \mathbf{u}}{\partial t} + (\mathbf{u} \cdot \nabla)\mathbf{u} + \frac{v_\perp^2}{2}\nabla \ln B\right)\right], \quad \mathbf{u} = v_\|\mathbf{h} + \mathbf{V}_E,$$

$$\frac{dv_\perp^2}{dt} = v_\perp^2[v_\|\mathbf{h} \cdot \nabla \ln B - \nabla \cdot \mathbf{V}_E - \mathbf{V}_E \cdot (\mathbf{h} \cdot \nabla)\mathbf{h}],$$

$$\frac{dv_\parallel}{dt} = \frac{e}{M}E_\parallel - \frac{v_\perp^2}{2}\,\mathbf{h}\cdot\nabla\ln B + v_\parallel \mathbf{V}_E\cdot(\mathbf{h}\cdot\nabla)\mathbf{h},$$

where $v_\parallel = w_\parallel + v_0$ and $w_\parallel$ is the chaotic velocity.

We can find the relation between the components of the electric fields, using Maxwell equations

$$E_\perp \approx \frac{N_\perp N_\parallel}{N_\parallel^2 - \varepsilon_\perp} E_\parallel, \quad E_B \approx -\frac{i\varepsilon_\perp \omega_{ci}}{N_\parallel N_\perp \omega} E_\parallel, \tag{14}$$

where $E_\perp$ is the component of the electric field along the propagation vector $\mathbf{k}_\perp$ and the component E_B is along the direction of $[\mathbf{B}\times\mathbf{k}_\perp]$.

After using Equations (12–14), absorbed power in collisional and weakly collisional plasmas can be represented in the form:

$$P_w = \frac{1}{4}(\mathbf{E}\cdot\mathbf{j}^* + \text{c.c.}) \approx \frac{\omega}{8\pi}\,\text{Im}\,\varepsilon_\parallel |E_\parallel|^2. \tag{15}$$

For example, the longitudinal permeability tensor component $\varepsilon_\parallel$ has the form

$$\varepsilon_\parallel = -\frac{\omega_{pe}^2}{\omega(\omega + i\nu_e)}, \tag{16}$$

where ν_e is the electron-ion collision frequency. From Equation (15) we can see that we do not need to find the specific expression for the imaginary part of $\varepsilon_\parallel$ if we express the RF forces through absorbed power P_w. Finally, it helps us to compare the theoretical results with experimental one taking account only the experimental value of the absorbed power. Nevertheless, in the general case the parallel component of the permeability tensor can be found from the expression

$$j_{\parallel e} = \frac{i\omega_{0e}^2}{4\pi\omega_{ce}}(-i\hat{M}_{jb}E_r + \hat{M}_{jr}E_b + \hat{M}_{j\parallel}E_\parallel),$$

where

$$\hat{M}_{jr} = \frac{1}{k_\parallel}\left\{\Lambda - \frac{k_b v_{Te}^2}{2\omega\omega_{ce}}\left[\chi_N\Lambda + \frac{\chi_T}{2}(1 + \Lambda(1+2z^2))\right]\right\}\left(\frac{\partial}{\partial r} + \frac{1}{r}\right),$$

$$\hat{M}_{jb} = \frac{k_b}{k_\parallel}\left\{\Lambda - \frac{k_b v_{Te}^2}{2\omega\omega_{ce}}\left[\chi_N\Lambda + \frac{\chi_T}{2}(1 + \Lambda(1+2z^2))\right]\right\},$$

$$\hat{M}_{j\parallel} = \frac{2\omega\omega_{ce}}{k_\parallel^2 v_{Te}^2}\left\{-\Lambda + \frac{k_b v_{Te}^2}{2\omega\omega_{ce}}\left[\chi_N\Lambda + \frac{\chi_T}{2}(1 + \Lambda(2z^2 - 1))\right]\right\},$$

$$\Lambda = 1 + i\sqrt{\pi}\,zW(z), \quad z = \frac{\omega}{|k_\parallel|v_{Te}},$$

$$W(z) = \exp(-z^2)\left(1 + \frac{2i}{\sqrt{\pi}}\int_0^z \exp(t^2)dt\right), \quad \chi_A = \frac{\partial \ln A}{\partial r}.$$

Now from Equations (11–15), we can obtain the expressions for the RF force θ and ζ components

$$F_\theta^h = \frac{(-k_\parallel h_\theta + k_\theta)}{\omega} P_w, \quad F_\zeta^h = \frac{(-k_\parallel h_\zeta + k_\zeta)}{\omega} P_w. \tag{17}$$

The first term in the parentheses in the numerator equation (17) has its origin from the gradient term (Klima, 1968; Ohkawa, 1989) and the second one is the term with the absorbed power (Fish and Karney, 1984). As the next step we can find the final expressions for components of the RF ponderomotive forces from general expressions for RF forces Equations (17):

$$F_\theta^h \approx \frac{h_\zeta k_b}{\omega} P_w, \quad F_\zeta^h \approx -\frac{h_\theta k_b}{\omega} P_w, \tag{18}$$

where the relation for imaginary part of $k_\perp$ is taken from Equation (12) and $k_\theta = m/r$, $k_\zeta = -n/R$, $k_b = k_\theta h_\zeta - k_\zeta h_\theta$, m and n are the poloidal and toroidal wave numbers. Finally, we can conclude that the longitudinal component of RF force is small.

4. Collisional Plasma

For collisional (edge) plasma, we take the parallel viscosity tensor $\pi_\parallel$ of the Burnett (1935) kind, in the form of Mikhailovskii and Tsypin (1971, 1982, 1984), as for the problem under consideration, it is important to mention, that the viscosity tensor depends on the thermal fluxes $\mathbf{q}_i$. Using the usual viscosity expression (Braginskii, 1965), it would lead to the conclusion that the equilibrium poloidal velocity of plasma rotation is zero. This problem was discussed in several occasions (Hazeltine, 1974; Mikhailovskii and Tsypin, 1982).

$$\pi_{\parallel i} = M_i \int \left(w_\parallel^2 - \frac{1}{3} w^2\right) f_i d\mathbf{w} = -\frac{2}{3}\frac{p_i}{\nu_i}(0.96\beta - 0.59\gamma). \tag{19}$$

Here

$$\beta = 3\left\{\mathbf{h}(\mathbf{h}\cdot\nabla)\mathbf{V}_i + \frac{2}{5p_i}\mathbf{h}(\mathbf{h}\cdot\nabla)\mathbf{q}_i - \frac{1}{3}\left(\nabla\cdot\mathbf{V}_i + \frac{2}{5p_i}\nabla\cdot\mathbf{q}_i\right)\right\}, \tag{20}$$

$$\gamma = -\frac{6}{5}\left\{\mathbf{h}(\mathbf{h}\cdot\nabla)(\mathbf{q}_i - \mathbf{q}_i^*) + \frac{1}{3}(\nabla \ln p_i \cdot \mathbf{q}_i - \nabla\cdot(\mathbf{q}_i + 0.27\mathbf{q}_{\parallel i}))\right\}. \tag{21}$$

To find the ion perturbed velocities in Equation (20) we proceed from frozen-in condition

$$[\nabla \times [\mathbf{V}_i \times \mathbf{B}]] = 0 \tag{22}$$

and the continuity equation. From the ζ-covariant component of Equation (22), we derive

$$\frac{\partial}{\partial\theta}\left[\sqrt{g}(V^\theta B^\zeta - V^\zeta B^\theta)\right] = 0 \tag{23}$$

and, after using Equations (22) and (23)

$$\tilde{V}_i^\zeta = -q\tilde{V}_i^\theta. \tag{24}$$

From the continuity equation, which we rewrite in the form

$$\frac{\partial}{\partial\theta}\left(\sqrt{g}V_i^\theta\right) + \frac{\partial}{\partial\zeta}\left(\sqrt{g}V_i^\zeta\right) = 0, \tag{25}$$

we have

$$\left\langle \frac{\partial \tilde{V}_i^\theta}{\partial\theta} \right\rangle_\zeta = -\frac{U_{\theta i}}{r}\left\langle \frac{\partial}{\partial\theta}\ln(n\sqrt{g})\right\rangle_\zeta, \qquad \langle \tilde{V}_\parallel \rangle_\theta \approx U_{i\zeta}\frac{(\langle B\rangle_\theta - B_s)}{B_s}. \tag{26}$$

Here, $\langle\cdots\rangle_\theta = \oint \ldots d\theta/2\pi$, $\langle\cdots\rangle_\zeta = \oint \ldots d\zeta/2\pi$, and $\tilde{V}$ designates the values, oscillating with the angles θ, ζ. Now, we use the identities

$$\mathbf{h}(\mathbf{h}\cdot\nabla)\mathbf{V}_i = (\mathbf{h}\cdot\nabla)V_{i\parallel} - \mathbf{V}_i(\mathbf{h}\cdot\nabla)\mathbf{h} = (\mathbf{h}\cdot\nabla)V_{i\parallel} - \mathbf{V}_i\cdot\nabla_\perp \ln B, \tag{27}$$

$$\mathbf{h}(\mathbf{h}\cdot\nabla)\mathbf{q}_i = (\mathbf{h}\cdot\nabla)q_{i\parallel} - \mathbf{q}_i\cdot\nabla_\perp \ln B, \qquad \tilde{V}_{i\parallel} \approx g_{22}h^\theta\tilde{V}^\theta + g_{33}h^\zeta\tilde{V}^\zeta.$$

Further, using Equations (20), (21), (24), (26), and (27), we find

$$\beta = \frac{3}{r}\left\{\varepsilon U_{\zeta i}\frac{\partial}{\partial\zeta}\ln B - U_{\theta i}\frac{\partial}{\partial\theta}\ln\left(\sqrt{g}B\right) + U_{Ti}\frac{\partial}{\partial\theta}\ln B\right\}, \tag{28}$$

$$\gamma = -\frac{61}{15r}U_{Ti}\frac{\partial}{\partial\theta}\ln B. \tag{29}$$

As a result, we get the final expressions for the poloidal $U_{\theta i}$ and toroidal $U_{\zeta i}$ velocities of the ion rotation in collisional (edge) plasmas of slightly rippled tokamaks, induced by the ponderomotive RF forces of Alfvén waves

$$U_{i\theta} = -1.83U_{Ti} + 1.39\frac{R^2\nu_i}{m_i n_0 v_{Ti}^2}\frac{h_\zeta k_b}{\omega}P_w, \tag{30}$$

$$U_{i\zeta} = -1.39\frac{R^2\nu_i}{m_i n_0 v_{Ti}^2 (N\delta)^2}\frac{h_\theta k_b}{\omega}P_w, \tag{31}$$

where $k_b = h_\zeta k_\theta - h_\theta k_\zeta$, $k_\theta = m/r$, $k_\zeta = -n/R$.

Except of velocities, the very important value in a tokamak plasma is a quasi-stationary radial electric field E_r. It can be found from the expression for an ion velocity to be obtained from the ion motion equation

$$U_{i\theta} \approx -c\frac{E_r}{B} + U_{i\zeta}h_\theta + U_{pi}, \quad U_{pi} = \frac{c}{e_i n_0 B}\frac{\partial p_i}{\partial r}, \tag{32}$$

if one puts Equations (30) and (31) into it

$$E_r \approx \frac{1}{e_i}\left[2.83\frac{\partial T_i}{\partial r} + \frac{T_i}{n_0}\frac{\partial n_0}{\partial r} - 1.39\frac{R^2 \nu_i \omega_{ci}}{n_0 v_{Ti}^2}\frac{m P_w}{r\omega}\right]. \tag{33}$$

Let us estimate the value of the induced velocities of plasma for the next plasma macroscopic parameters (Phaedrus-T tokamak, see Hershkovitz et al., 1994): $T_i \approx 20$ eV, $n_0 \approx 10^{12}$ cm^{-3}, $B_0 = 0.68$ T, $\tilde{B}_\perp/B_0 = 3 \cdot 10^{-3}$, $r = 26$ cm, $R = 92$ cm, $\omega = 10^7$ rad^{-1}, $m = 1 \div 3$, $n = 2 \div 6$, $N = 20$, $\delta \approx 0.05$, $P_w \approx 0.15$ W/cm^3. If we put these parameters into Equations (32) and (33), we find that

$$U_{i\theta} \approx 6 \cdot 10^4 \text{ cm/s}, \quad U_{i\zeta} \approx 3 \cdot 10^4 \text{ cm/s}, \quad E_r \approx 4 \text{ V/cm}. \tag{34}$$

From Equation (34), we see that in our case the rotation velocities have satisfied the inequalities

$$U_{Ti} < \{U_{i\theta}; U_{i\zeta}\} < c_s, \tag{35}$$

where c_s is the sound velocity. Under the condition $(N\delta)^2 < 1$ the plasma toroidal velocity is greater than the poloidal velocity.

Let us find analogous values for TCA/Br tokamak (Nascimento et al., 1994): $T_i \approx 20$ eV, $n_0 \approx 10^{12}$ cm^{-3}, $B_0 = 1.52$ T, $r = 18$ cm, $R = 61.5$ cm, $\omega = 10^7$ rad^{-1}, $m = 1$, $n = 2 \div 6$, $N = 72$, $\delta \approx 0.015$, $P_w \approx 0.2$ W/cm^3. If we put these parameters into Equations (32) and (33), we find that

$$U_{i\theta} \approx 1.5 \cdot 10^4 \text{ cm/s}, \quad U_{i\zeta} \approx -1.4 \cdot 10^3 \text{ cm/s},$$
$$E_r \approx -2.3 \text{ V/cm}, \quad U_{Ti} \approx 7 \cdot 10^3. \tag{36}$$

We should expect that these parameters will be substantially greater for large tokamaks. The induced velocities Equations (31) and (32) and radial electric field equation (33) exist in the plasma layer

$$\Delta r \approx \frac{c}{\omega N_I} = \frac{c}{\omega_{pe}}\sqrt{\frac{2(\omega^2 + \nu_e^2)}{\omega(\sqrt{\omega^2 + \nu_e^2} - \omega)}}, \tag{37}$$

with the minimum value $(\Delta r)_{\min} = 2.83c/\omega_{pe}$ at $\omega/\nu_e = 0.58$.

Of course, external forces (Equation (18)) stimulate an additional radial transport

$$\Delta V_{er} \approx \frac{mc}{\omega e_e n_0 B_0} P_w, \tag{38}$$

but by an appropriate choosing of the poloidal wave number m sign it is possible to make the plasma additional radial transport to be negative.

5. Weakly Collisional Plasma: Plateau Regime

Here, we consider plateau and banana regimes of weakly collisional plasmas. For the plateau regime the conditions are

$$\frac{\nu_\alpha}{\varepsilon^{3/2}} \geq \frac{v_{T\alpha}}{qR} \geq \nu_\alpha, \qquad \frac{\nu_\alpha}{\delta^{3/2}N} \geq \frac{v_{T\alpha}}{R}, \tag{39}$$

where $v_{T\alpha} = \sqrt{2T_\alpha/M_\alpha}$ and ν_α are the thermal speed and collision frequency of species α respectively, and q is the safety factor. We can write these forces in the form (Mikkailovskii and Tsypin, 1984)

$$\begin{aligned} F_\theta^\pi &= -\frac{\sqrt{\pi}}{2} M_i n_i \frac{q v_{Ti}}{R}\left(U_{\theta i} + \frac{1}{2} U_{Ti}\right), \\ F_\zeta^\pi &= -\frac{\sqrt{\pi}}{2} M_i n_i \frac{N\delta^2 v_{Ti}}{R} U_{\zeta i}, \end{aligned} \tag{40}$$

where $U_{\theta i}$ is the ion poloidal velocity, $U_{\zeta i}$ is the toroidal velocity and $U_{Ti} = (c/e_i B_0)\partial T_i/\partial r$ is the ion drift velocity connected with temperature gradient, M_i and n_i are the ion mass and density, respectively.

Now we can find the quasistationary poloidal and toroidal components of the plasma rotation velocities and the radial component of the electric field. So we have from Equations (9), (10), (18), and (40) for plateau regime

$$U_{\theta i} = -\frac{1}{2} U_{Ti} + \frac{2h_\zeta k_b R P_w}{\sqrt{\pi} M_i n_i q v_{Ti}\omega}, \qquad U_{\zeta i} = -\frac{2h_\theta k_b R P_w}{\sqrt{\pi} M_i n_i N\delta^2 v_{Ti}\omega}. \tag{41}$$

Then, from Equation (32), using Equation (41), we find

$$\bar{E}_r = \frac{B}{c}\left\{U_{ni} + \frac{3}{2} U_{Ti} - \frac{2k_b R P_w}{\sqrt{\pi}\, M_i n_i q v_{Ti}\omega}\left(\frac{\varepsilon h_\theta}{N\delta^2} + h_\zeta\right)\right\}, \tag{42}$$

where $U_{ni} = (cT_i/e_i n_i B)\partial n_i/\partial r$.

Now, we estimate the rotation velocities and the radial electric field for Phaedrus-T, TCA/BR and JET (Joint European Torus) tokamaks. The AW plasma heating and current drive was already demonstrated in Phaedrus-T tokamak, and plasma rotation was experimentally found (Wukitch et al., 1996). This plasma rotation was observed by way of Mirnov coil oscillations induced by magnetic island, which is situated in the surface where $q \approx 2$. Excitations of AW are intensively studied in JET via nonlinear effects. That was the reasons to choose these tokamaks for estimations of the plasma rotation effects.

The Phaedrus-T had the next core plasma parameters: the central electron temperature $T_{e0} = 500$ eV, $T_i \approx 400$ eV, $n_0 \approx 7 \cdot 10^{12} \div 10^{13}$ cm^{-3}, $z_{\text{eff}} \approx 1.7$–2, $B_0 = 0.6$–0.9 T, $\tilde{B}_\perp/B_0 = 3 \cdot 10^{-3}$, $r = 25.5$ cm, $R = 92$ cm, $\omega = 4 \cdot 10^7$ rad^{-1}, $m = 1 \div 3$, $n = 4 \div 9$, $N = 20$, $\delta \approx 0.05$, $P_w \approx 0.65$ W/cm^3. With these data we can see that the core plasma was in the plateau regime for this tokamak. Using Equations (41) and (42), we find

$$U_{Ti} \approx -2 \cdot 10^5 \text{ cm/s}, \quad U_{\theta i} \approx 8 \cdot 10^4 \text{ cm/s}, \quad U_{\zeta i} \approx -3 \cdot 10^5 \text{ cm/s},$$
$$\bar{E}_r \approx -18 \text{ V/cm}, \quad \Delta U_{\theta i} \approx 3 \cdot 10^4 \text{ cm/s},$$
$$\Delta U_{\zeta i} \approx -3 \cdot 10^5 \text{ cm/s}, \quad \Delta \bar{E}_r \approx -4 \text{ V/cm}. \tag{43}$$

Δ means the RF parts of these quantities, e.g., $U_{\theta i}$ is the full poloidal velocity and $\Delta U_{\theta i}$ is the RF part of it. Thus, RF value of the poloidal velocity is of the order or less than this one without the RF field, and RF value of the toroidal velocity is of the order of the ion temperature drift velocity. Comparing these calculations with the experimental results (Wukitch et al., 1996), we find main coincidences. The upshift in frequency (connected with induced plasma flows) of the tearing mode for current drive antenna phasing was linear with RF power, and the experimental toroidal velocity $V_{\zeta i} \approx (3 \pm 1.2) \times 10^5$ cm/s is approximately the same as calculated in this paper. Note that the variation of the poloidal and toroidal rotation velocities induced by RF fields linearly depend on dissipated power and did not depend on toroidal magnetic field as follows from Equations (41) and from experiments. This scaling can be found from figure 1 of the paper of Wukitch et al. (1996). The frequency upshift Δf was flat for $0.65 \leq B_0 \leq 0.8$ T, and the average $\Delta f = 0.92 \pm 19$ kHz and $\Delta f = -0.79 \pm 0.16$ kHz for the phase $\phi = \pm\pi/2$, respectively, between the currents in the two antenna straps.

The TCA/BR will have the next core plasma parameters: the central electron temperature $T_{e0} = 500$ eV, $T_i \approx 400$ eV, $n_0 \approx 1 \cdot 10^{13} \div 2 \cdot 10^{13}$ cm^{-3}, $z_{\text{eff}} \approx 2$, $B_0 = 1.52$ T, $r = 18.0$ cm, $R = 61.5$ cm, $\omega = 4 \cdot 10^7$ rad^{-1}, $m = 1$, $n = 2 \div 6$, $N = 72$, $\delta \approx 0.015$, $P_w \approx 0.2$ W/cm^3. With these data we can see that the core plasma was in the plateau regime for this tokamak. Using Equations (41) and (42), we find

$$U_{Ti} \approx -1.3 \cdot 10^5 \text{ cm/s}, \quad U_{\theta i} \approx 6 \cdot 10^4 \text{ cm/s},$$
$$U_{\zeta i} \approx 10^4 \text{ cm/s}, \quad \bar{E}_r \approx -28 \text{ V/cm},$$
$$\Delta U_{\theta i} \approx -10^3 \text{ cm/s}, \quad \Delta U_{\zeta i} \approx 10^4 \text{ cm/s}, \quad \Delta \bar{E}_r \approx -9 \text{ V/cm}. \tag{44}$$

6. Weakly Collisional Plasma: Banana Regime

Plasma rotation and viscous forces of a tokamak plasma with rippled magnetic field in the conditions of the banana regime,

$$\frac{v_{T\alpha}}{qR} \geq \frac{\nu_\alpha}{\varepsilon^{3/2}}, \quad \frac{v_{T\alpha}}{R} \geq \frac{\nu_\alpha}{\delta^{3/2}N}, \tag{45}$$

had been studied by Stringer (1973), Connor and Hastie (1973), Mikhailovskii (1995). Using results of these authors we can write

$$F_\theta^\pi = -M_i n_i \nu_{i*}\frac{q^2}{\varepsilon^2}\left\{(c_u + c_t)U_{\theta i} + (2.37c_t - 1.17c_u)U_{Ti} - \frac{\varepsilon}{q}c_t U_{\zeta i}\right\}, \tag{46}$$

$$F_\zeta^\pi = -M_i n_t \nu^{lc} U_{\zeta i}, \tag{47}$$

where coefficients in the above equations can be represented as:

$$c_u = 1.46\sqrt{\varepsilon}, \quad c_t = 6.19G(\gamma)\frac{\varepsilon^2 v_{Ti}^2}{q^2R^2\nu_{ii}\nu_{i*}},$$

$$\gamma = \frac{\varepsilon}{qN\delta}, \quad n_t = \frac{n_i}{\pi}(2\delta)^{1/2}G_l(\gamma). \tag{48}$$

The quantity $G_l(\gamma) = 1$ for $\gamma < 1$, $G_l(\gamma) = 1/\gamma$ for $\gamma \geq 1$, and the collisional frequency parameters are equal to

$$\nu_{i*} = 0.75\nu_i, \quad \nu_{ii} = 1.88\nu_i, \quad \nu^{lc} = 1.49\nu_i, \tag{49}$$

where ν_i is the ion-ion collision frequency, $G(\gamma)$ is the function whose explicit form was found (Connor and Hastie, 1973). For the case $\gamma < 1$ (usual condition in tokamaks) $G(\gamma) \approx 1$.

The term with the coefficient c_u in Equation (46) is connected with the friction of the untrapped-untrapped ions, and the term with the coefficient c_t should be taken into account if the friction of the untrapped particles and the particles trapped in the rippled magnetic field is important. The ratio c_t/c_u can be arbitrary in hot tokamak plasmas. In Equation (47), the friction between the untrapped particles and the locally trapped particle is only allowed. The taking account of locally untrapped particles would give in Equation (47) terms of the order $0.1\delta/\varepsilon^{1/2}$.

Analogous quantities for the banana regime can be found from Equations (9), (10), (18), (46), and (47):

$$U_{\theta i} = \frac{1}{c_u + c_t}\left\{U_{Ti}A_{ut} + \frac{\varepsilon k_b h_\theta P_w}{M_i n_i q\omega\nu_i}\left(1.33 - 1.49\frac{c_t}{\delta^{1/2}}\right)\right\}, \tag{50}$$

$$U_{\zeta i} = -\frac{1.49h_\theta k_b P_w}{M_i n_i \delta^{1/2}\omega\nu_i}, \quad A_{ut} = (1.17c_u - 2.37c_t), \tag{51}$$

$$\bar{E}_r = \frac{B}{c}\left\{U_{pi} - \frac{1}{c_u + c_t}\left[U_{Ti}A_{ut} + \frac{\varepsilon k_b h_\theta P_w}{M_i n_i q \omega \nu_i}\left(\frac{1.49c_u}{\delta^{1/2}} + 1.33\right)\right]\right\}. \tag{52}$$

AW were not specially used for the current drive and plasma heating in JET. These waves were exited in JET as result of nonlinear effect. We can suppose that kinetic AW can also be excited in JET. Here, for the estimations we use the next JET plasma parameters (JET Team, 1994) which were during the plasma heating and current drive by low hybrid waves: $T_i \approx 2.5 \cdot 10^3$ eV, $n_0 \approx 2.8 \cdot 10^{13}$ cm^{-3}, $z_{\text{eff}} \approx 2$, $B_0 = 3.2$ T, $r = 95$ cm, $R = 285$ cm, $\omega = 4 \cdot 10^7$ rad^{-1}, $m = 1 \div 3$, $n = 4 \div 9$, $N = 20$, $\delta \approx 0.05$, $P_w \approx 0.7$ W/cm^3. From Equation (45) we see that the JET core plasma is in a banana regime. Thus from Equations (50–52) we have

$$U_{Ti} \approx -7 \cdot 10^4 \text{ cm/s}, \quad U_{\theta i} \approx 1.5 \cdot 10^5 \text{ cm/s}, \quad U_{\zeta i} \approx -1.3 \cdot 10^5 \text{ cm/s},$$
$$\bar{E}_r \approx -50 \text{ V/cm}, \quad \Delta U_{\theta i} \approx -1.3 \cdot 10^4 \text{ cm/s},$$
$$\Delta U_{\zeta i} \approx -1.3 \cdot 10^5 \text{ cm/s}, \quad \Delta \bar{E}_r \approx 0.0 \text{ V/cm}. \tag{53}$$

We see that induced parts of plasma macroscopic velocities are less or on the level of these velocities. These effects take place as the viscosity coefficient η_i for arbitrary collisional regime can be represented in the form

$$\eta_i \sim f(\nu^*, \varepsilon), \tag{54}$$

where $\nu^* = qR\nu_i/v_{Ti}$. The dependence of the functions $f(\nu^*, \varepsilon)$ on the frequency ν^* and on the parameter ε is

$$f(\nu^*, \varepsilon) = \nu^* \varepsilon^{-1.5}/(1 + \nu^* \varepsilon^{-1.5})(1 + \nu^*), \tag{55}$$

(Hirshman and Sigmar, 1981). We find from Equations (53) and (54) that parallel viscosity coefficient grow from plasma edge having maximum in a plasma column corresponding to plateau regime and beginning of banana regime. Then, the parallel viscosity quickly drops to the plasma center.

7. Conclusion

In conclusion, externally excited Alfvén waves can generate the poloidal and toroidal rotation in the collisional (edge) and core (weakly collisional) tokamak plasma. The necessary value of these velocities to achieve the L-H transition, can be obtained by manipulating of the value of an absorbed power. This rotation linearly depends on dissipated power, decreases in banana regimes and did not depend on toroidal magnetic field. In the plasma layer where the RF power is absorbed, the radial electric field also appears.

The parameters of plasma rotation have been estimated for the Phaedrus-T and TCA/Br tokamaks, where the main goal was to study AW plasma heating and current drive, and for JET, where AW are under intensive study. The core plasma in Phaedrus-T tokamak was in plateau regime. For experimental data of this tokamak it was shown that the quasistationary ion rotation velocities and radial electric field are of the order or less than those without the RF field. The core plasma in JET was supposed to be in banana regime according to experimental data of this tokamak. The estimations for experimental data of this tokamak show that the quasistationary ion rotation velocities and radial electric field are weakly changed by AW. This effect can be considered as positive.

There is a very satisfactory agreement between our theory and experimental results obtained on Phaedrus-T tokamak (Wukitch et al., 1996). We find main coincidences: (1) induced plasma flows were linear with RF power; (2) the experimental toroidal velocity $V_{\zeta i} \approx (3 \pm 1.2) \times 10^5$ cm/s is approximately the same as calculated in this paper; and (3) induced plasma flows do not depend on toroidal magnetic field. The methods developed in this paper can be applied to other tokamaks with another kind of waves, which can be used to operate the plasma dynamics taking into account the appropriate dispersion relation for that type of waves.

References

Biglari, H., Diamond, P.H. and Terry, P.W.: 1990, *Phys. Fluids B* **2**, 1.

Biglari, H., Ono, M., Diamond, P.H., and Graddock, G.G.: 1992, in: *Proceedings of 9th Topical Conference on Radio-Frequency Power in Plasmas, Charleston, 1991*, AIP Conference Proceedings 244, New York, p. 376.

Braginskii, S.I.: 1965, in: M.A. Leontovich (ed.), *Review of Plasma Physics*, Consultants Bureau, New York, Vol. 1, p. 205.

Burnett, D.: 1935, *Proc. Lond. Math. Soc.* **39**, 385; **40**, 382.

Callen, J.D.: 1992, *Phys. Fluids B* **4**, 2142.

Chew, G.L., Goldberger, M.L. and Low, F.E.: 1956, *Proc. Royal Soc. A* **236**, 112.

Connor, J.W. and Hastie, R.J.: 1973, *Nuclear Fusion* **13**, 221.

Diamond, P.H. and Kim, Y.-B.: 1991, *Phys. Fluids B* **3**, 1626.

Elfimov, A.G., Kirov, A.G. and Sidorov, V.P.: 1992, in: A.G. Litvak (ed.), *High-Frequency Plasma Heating*, American Institute of Physics Translation series, New York, p. 239.

Fish, N.J. and Karney, F.F.: 1984, *Phys. Fluids* **24**, 27.

Gao, Q., Peng, X., Zhang, J., Wang, S. and Liu, C.: 1994, in: *Proceedings of the 15th IAEA Conference on Plasma Physics and Controlled Fusion, Seville*, IAEA-CN-60/D-P-1-10-2(C).

Graddock, G.G. and Diamond, P.H.: 1991, *Phys. Rev. Lett.* **67**, 1535.

Hazeltine, R.D.: 1974, *Phys. Fluids* **17**, 961.

Hershkovitz, N., Moroz, P., Probert, P., Wang, E.Y., Intrator, T., Diebold, D., Breun, R., Brouchous, D., Doczy, M., Edgel, D., Elfimov, A., Fonk, R., Kishinevsky, M., Litwin, C., Nonn, P., Regan, S., Sorensen, J., Tataronis, J., Vucovich, M., Wang, X. and Wukitch, S.: 1994, in: *Proceedings of the 15th IAEA Conference on Plasma Physics and Controlled Fusion, Seville*, IAEA-CN-60/A3/5-P-14.

Hirshman, S.P.: 1978, *Nuclear Fusion* **18**, 917.

Hirshman, S.P. and Sigmar, D.J.: 1981, *Nuclear Fusion* **21**, 1079.

Klima, R.: 1968, *Czech. J. Phys.* **B 18**, 1280.

Mikhailovskii, A.B.: 1995, *Plasma Phys. Rep.* **21**, 529.

Mikhailovskii, A.B. and Tsypin, V.S.: 1971, *Plasma Physics* **13**, 785.
Mikhailovskii, A.B. and Tsypin, V.S.: 1984a, *Beiträge Plasma Physik* **24**, 317.
Mikhailovskii, A.B. and Tsypin, V.S.: 1984b, *Sov. J. Plasma Phys.* **10**, 142.
Nascimento, I.C., Galvaõ, R.M.O., Tuszel, A.G., Degasperi, F.T., Ruchko, L.F. and TCA/Br team: 1994, in: *Proc. of Xth Int. Conf. on Plasma Physics. Contr. Pap. Brasil, Foz do Igua cu*, Vol. 1, p. 69.
Rosenbluth, M.N., Hazeltine, R.D. and Hinton, F.L.: 1972, *Phys. Fluids* **15**, 116.
Stringer, T.E.: 1972, *Nuclear Fusion* **12**, 689.
Tsypin, V.S.: 1987, *Sov. J. Plasma Phys.* **14**, 466.
Tsypin, V.S., Elfimov, A.G., de Azevedo, C.A. and de Assis, A.S.: 1996, *Phys. Plasmas* **3**, 2818.
Wang, C.Y., Jaeger, E.F., Batchelor, D.B. and Sidikman, K.L.: 1995, *Physics of Plasmas* **1**, 3890.
Wong, S.K. and Hinton, F.L.: 1984, *Phys. Rev. Lett.* **52**, 827.
Wukitch, S., Litwin, C., Harper, M., Parker, R. and Hershkovitz, N.: 1996, *Phys. Rev. Lett.* **77**, 294.

ANISOTROPIC MAGNETIC CONFINEMENT. APPLICATIONS TO FIELD-REVERSED CONFIGURATIONS

ROBERTO ANTONIO CLEMENTE
Instituto de Física Gleb Wataghin, Universidade Estadual de Campinas, 13083-970 Campinas, SP, Brazil

Abstract. Introducing an auxiliary function of the usual poloidal magnetic stream function, it is possible to obtain axisymmetric solutions of the ideal anisotropic magnetohydrodynamic equations for steady rotating plasmas, in terms of solutions of the Maschke and Perrin equation for isotropic plasmas, with temperature as a surface function. For vanishing rotation, the problem is reduced to the classical Grad–Schlüter–Shafranov equation for static equilibria. Some applications of the equilibrium models to the study of tilting stability and classical particle transport in field-reversed configurations are presented.

1. Introduction

For many purposes (e.g. stability and transport calculations, project design) the starting point of magnetic confinement theory is a good equilibrium model. In the case of axisymmetric toroidal configurations, in general, the equilibria are described in terms of ideal MHD equations in which the plasma kinetic pressure is assumed isotropic, no plasma flow is present and the problem is reduced to solving the well known Grad–Schlüter–Shafranov (GSS) equation for the poloidal magnetic stream function. Such equation relies on two hypotheses on plasma pressure and poloidal current profiles.

When auxiliary heating methods are used in toroidal machines, isotropic plasma pressure may not be a good approximation of the real situation. Moreover, some kind of plasma flow may also be present. In general, both toroidal and poloidal flows are present, however, in the following the analysis will be restricted to toroidal flow only and the eventual anisotropy of the kinetic plasma pressure will be taken into account allowing for different pressures along and orthogonally to the magnetic field. The results may be considered appropriated when the kinetic energy associated to the poloidal motion is much smaller than the kinetic energy associated to the toroidal one. Even in the isotropic case, the inclusion of poloidal motion strongly complicates the equilibrium problem, which is reduced to two coupled equations, one partial nonlinear differential equation for the poloidal magnetic stream function and another algebraic Bernoulli equation for the plasma density, together with five hypotheses on different equilibrium quantities (Maschke and Perrin, 1984; Zelazny et al., 1993). In this last case, no exact analytical solutions are available and also numerical ones present several troubles (bifurcations).

Astrophysics and Space Science **256:** 235–246, 1998.

Assuming toroidal rotation and a pressure tensor like $\mathbf{P} = p_\perp \mathbf{I} + \sigma_- \mathbf{BB}$, where $\sigma_- = (p_\parallel - p_\perp)/B^2$, $B = |\mathbf{B}|$ with $\mathbf{B}$ the magnetic field, $\mathbf{I}$ the unit dyadic, $p_\parallel$ and $p_\perp$ the parallel and perpendicular pressures respectively, it is possible to show that the equilibrium problem can be reduced to the scalar pressure case when a special class of σ_- is considered (Clemente, 1993, 1994). The class of equilibria presented here is not the most general one, however, it may be a good starting point. After showing the method to obtain anisotropic rotating equilibria and presenting some examples of analytical solutions, some applications related to the physics of field-reversed configurations (FRCs) will be presented. The first application concerns the study of the tilting mode in FRCs (a still unresolved paradox between theory and experiments). It will be shown that the kind of anisotropy here considered does not introduce any significant stabilizing effect. The second application concerns classical particle diffusion in FRCs, showing that anisotropy should strongly influence their transport properties.

There is no experimental support for the existence of some anisotropy in FRCs, however, it must also be said that no investigations have been carried out in this respect. Depending on ion collisionality and heating processes, pressure anisotropy should be present in FRCs in some special situations, for example:

- at the early stages after formation, specially when low density, high temperature plasmas are obtained and complete thermalization of the ions is delayed;
- in translational experiments when reflection happens and large translational energies are suddenly converted into thermal ones and vice versa.

Theoretical estimates of ion temperature isotropization times for typical FRCs parameters ($T_i \sim$ 100–300 eV; $n_i \sim 10^{15}$–10^{16} cm^{-3}) are in the range of few to some tenths of microseconds, indicating that anisotropy should not be a relevant effect. However, the possibility of the existence of some anisotropy creating mechanism can not be excluded a priori and it may be of interest, in its own right, in some way to treat and quantify the eventual influence of plasma pressure anisotropy on macroscopic stability and transport properties of FRCs.

2. Equilibrium Model

The steady momentum balance equation for an anisotropic ideal plasma reduces to (standard notation and cylindrical coordinates r, ϕ, z will be used throughout):

$$\rho \mathbf{v} \cdot \nabla \mathbf{v} + \nabla \cdot \mathbf{P} = (\nabla \times \mathbf{B}) \times \mathbf{B}; \tag{1}$$

this equation has to be solved together with the assumption that $\mathbf{v} = \Omega r \hat{e}_\phi$ and the ideal Ohm's law:

$$\mathbf{E} + \mathbf{v} \times \mathbf{B} = 0. \tag{2}$$

If axisymmetry is assumed, the magnetic field can be expressed in terms of a poloidal magnetic stream function $\Psi(r, z)$ and a poloidal current function $I(r, z)$ as $\mathbf{B} = \nabla\psi \times \hat{e}_\phi / r + I\hat{e}_\phi / r$ and when this is taken into account in the ideal Ohm's law (2), the result that $\Omega = \Omega(\Psi)$ is obtained. If $\mathbf{P} = p_\perp \mathbf{I} + \sigma_- \mathbf{BB}$ it is possible to write the azimuthal component of Equation (1) as:

$$\mathbf{B} \cdot \nabla[(1 - \sigma_-)I] = 0; \tag{3}$$

which implies that $(1 - \sigma_-)I$ must be a function of Ψ. Unfortunately, this general result does not allow to reduce the problem to a suitable partial differential equation. At this point it is convenient to require that both I and σ_- must be surface functions. The validity of this assumption cannot be supported by any physical model, however, it must be satisfied at the plasma boundary when it coincides with a perfectly conducting wall and it also implies that no torques are acting on the plasma.

From a kinetic point of view, there may be some doubts about the possibility of the existence of such kind of equilibria on a timescale longer than the particle transit time along a magnetic line. In such case a particle distribution function should be constant on a magnetic surface and may be expressed as $F = F(\mu, \varepsilon, \Psi)$, where μ and ε are the magnetic moment and kinetic energy of the particle (per unit mass), respectively, and Ψ labels the magnetic surface (Hastie et al., 1967). Assuming F to be a separable function as $F = H(\mu, \varepsilon)Q(\Psi)$ it follows that:

$$p_\perp = R_\perp(B)Q(\Psi); \quad p_\parallel = R_\parallel(B)Q(\Psi),$$

where

$$R_\perp = m\int d\mu \int d\varepsilon \mu B^2 H(\mu, \varepsilon)/q; \quad R_\parallel = m\int d\mu \int d\varepsilon q B H(\mu, \varepsilon),$$

with $q = [2(\varepsilon - \mu B)]^{1/2}$ and m is the particle mass. Since $H(\mu, \varepsilon)$ is arbitrary, it should always be possible to make $R_\parallel = k_1 + k_2 B^2$ and $R_\perp = k_1 - k_2 B^2$, with k_1 and k_2 constants, in such a way that the equilibrium constraint $\mathbf{B} \cdot \nabla\sigma_- = 0$ is automatically satisfied. Obviously, this makes restrictions on the available distribution functions, but shows that this kind of equilibria is indeed possible.

Introducing an average plasma pressure $\bar{p} = (p_\parallel + p_\perp)/2$ and taking into account $\sigma_- = \sigma_-(\Psi)$ Equation (1) can be written:

$$\nabla\bar{p} + B^2\nabla\sigma_-/2 - \rho\Omega^2 r = (1 - \sigma_-)(\nabla \times \mathbf{B}) \times \mathbf{B}. \tag{4}$$

To further reduce the problem, some additional assumption is needed for $\bar{p}$. In the isotropic case, Maschke and Perrin (MP) treated two possible cases: entropy as a surface function and temperature as a surface function (Maschke and Perrin, 1980). Here, only the second possibility will be considered since in the double adiabatic model for anisotropic MHD, no simple relation between $\bar{p}$ and ρ, like $p = \rho^\gamma S(\Psi)$ for the isotropic case, can be justified. Therefore, assuming that

$\bar{p}$ has a simple relation with Ψ expressed by $\bar{p} = \rho k T(\Psi)/(m_e + m_i)$, where k is the Boltzmann constant and T can be interpreted as the sum of parallel and perpendicular plasma temperatures and taking the cross product of Equation (4) with $\nabla\Psi$, it may be shown that $\bar{p} = f(\Psi)\exp[(m_e + m_i)\Omega^2 r^2/2kT]$ with f an arbitrary function of Ψ. This last relation is exactly the same as that obtained by Maschke and Perrin (1980) for the case of isotropic plasma pressure and T as a surface function, and it allows us to finally reduce Equation (4) to a single scalar nonlinear partial differential equation for Ψ:

$$\Delta^*\Psi - \frac{\sigma_-'}{2(1-\sigma_-)}|\nabla\Psi|^2 = -\frac{1}{1-\sigma_-}\left\{r^2 \exp\left[\frac{(m_e+m_i)\Omega^2 r^2}{2kT}\right] \times\left[f' + f r^2 (m_e - m_i)\left(\frac{\Omega^2}{2kT}\right)'\right] + \frac{[(1-\sigma_-)I^2]'}{2}\right\}, \tag{5}$$

where

$$\Delta^* \equiv r\frac{\partial}{\partial r}\frac{1}{r}\frac{\partial}{\partial r} + \frac{\partial^2}{\partial z^2}$$

is the usual Grad–Shafranov operator and $' \equiv d/d\Psi$. When $\Omega = 0$ Equation (5) reduces to the equation for static anisotropic plasmas first presented by Mercier and Cotsaftis (1961); when $\Omega = \sigma_- = 0$ it reduces to the classical GSS equation. The inherent nonlinearity present in the left-hand side of Equation (5) can be eliminated by a trick first presented for the static case (Clemente, 1993), i.e. introducing an auxiliary function of Ψ, defined by

$$U(\Psi) = \int_0^{\Psi} dx\sqrt{1-\sigma_-(x)}\,. \tag{6}$$

Correspondingly,

$$\Psi(U) = \int_0^{U} dy\Big/\sqrt{1-\sigma_-(y)}$$

and in order to have a proper mathematical definition of U, the condition $\sigma_- < 1$ has to be fulfilled together with the requirement that Ψ should have only one extremum inside the plasma. This in turns automatically satisfies one of the local criteria for stability of static equilibria given by Mercier and Cotsaftis (1961); moreover, if mirror and fire-hose instabilities are considered another necessary condition is $\sigma_- > -1$ (Grad, 1967). This suggests that $|\sigma_-| < 1$ should always hold.

Writing Equation (5) in terms of U gives:

$$\Delta^* U = -r^2 \exp\left[\frac{(m_e + m_i)\Omega^2 r^2}{2kT}\right]\left[\frac{df}{dU} + f r^2 (m_e + m_i)\frac{d}{dU}\left(\frac{\Omega^2}{2kT}\right)\right] - \frac{d}{dU}\left[\frac{(1-\sigma_-)I^2}{2}\right]. \tag{7}$$

Equation (7) has exactly the same form as the MP equation with T as a surface function, its solution depends on five hypotheses on the surface quantities f, Ω, σ_-, T and I (Maschke and Perrin, 1980). It has the advantage that it allows the possibility of obtaining a linear partial differential equation when Ω^2/T does not depend on Ψ (or alternatively U). Once a solution for U is known, Ψ is obtained by inversion, to this end it is necessary to assign $\sigma_-(\Psi)$ explicitly. When tokamak-like equilibria are considered, simple energetic considerations suggest that $|\sigma_-| \ll 1$, in such a way that a linear approximation in Ψ should be sufficient. If $\sigma_- = \sigma_0 + \sigma_1 \Psi$, it follows $\Psi = \{1 - \sigma_0 - [(1-\sigma_0)^{3/2} - 3\sigma_1 U/2]^{2/3}\}/\sigma_1$.

Few analytical solutions of Equation (7) exist in the literature (Maschke and Perrin, 1980; Clemente and Milovich, 1981; Viana et al., 1997) and in general one should solve it numerically. Since the algorithms may be the same as for the GSS equation, this should be a relatively easy task.

In order to illustrate the method, some examples of static anisotropic equilibria will be presented. When $\Omega = 0$ Equation (7) reduces to a GSS equation for U and only three hypotheses on the surface quantities $\bar{p}$, σ, and I are needed. If we consider a tokamak-like equilibrium with square cross section of height L, internal radius L and external radius $2L$, coinciding with the plasma boundary (where Ψ is assumed to vanish), a very simple solution for U can be worked out by assuming (Clemente, 1993):

$$\bar{p} = 2\pi^2 U^2/9L^4; \quad I^2 = (\pi^2 U^2/L^2 + I_0^2)/(1-\sigma_-),$$

where I_0 is a constant depending on the external toroidal magnetic field. In this case, an exact analytical solution for U is given by:

$$U = U_0 \sin[\pi(r^2/L^2 - 1)/3]\cos(\pi z/L), \tag{8}$$

with U_0 a constant related to the maximum value of Ψ at the magnetic axis. In order to obtain Ψ, σ_- has to be assigned; one possible choice is $\sigma_- = \sin^2(\varepsilon\Psi/\Psi_0)$ with constant ε. As a result $U = (\Psi_0/\varepsilon)\sin(\varepsilon\Psi/\Psi_0)$ and $\Psi = (\Psi_0/\varepsilon)\sin^{-1}(\varepsilon U/\Psi_0)$. The corresponding $p_\parallel$ and $p_\perp$ at the equatorial plane $z = 0$, have been plotted in Figure 1 as functions of radius, for the particular case $2\bar{p}/B^2 = 0.05$ and $\sigma_- = 0.01$ at the magnetic axis, in units of $\bar{p}$ $(U = U_0)$.

Another example of exact solution that will be used in subsequent sections, is the Hill's vortex model for FRCs. In FRCs $I = 0$ and if $d\bar{p}/dU$ = constant, an

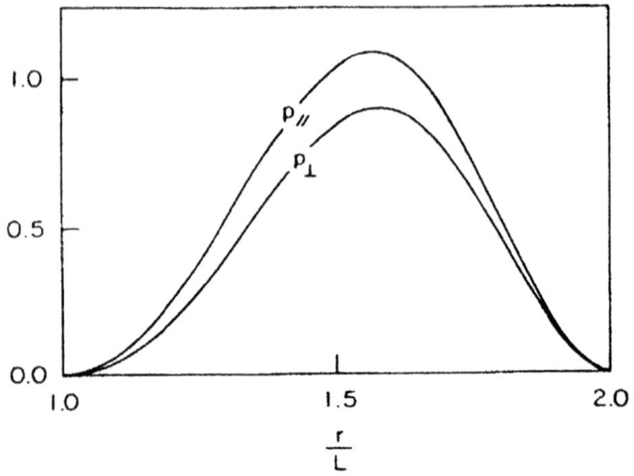

Figure 1. Plots of $p_{\parallel}$ and $p_{\perp}$ at the equatorial plane, corresponding to solution (8), for the particular case $2\bar{p}/B^2 = 0.05$ and $\sigma_- = 0.01$ at the magnetic axis, in units of $\bar{p}$ ($U = U_0$).

exact solution of Equation (7) is given by $U_{\rm HV} = \text{const.} r^2(1 - r^2/a^2 - z^2/b^2)$, a and b being the semi-axes of the ellipsoidal separatrix ($U_{\rm HV} = 0$ surface). To obtain Ψ a reasonable hypothesis for σ_- is needed, the simplest one is $\sigma_- = \sigma_0 \Psi$ with σ_0 constant. The resulting Ψ is:

$$\Psi = [1 - (1 - 3\sigma_0 U_{\rm HV}/2)^{2/3}]/\sigma_0; \tag{9}$$

the constant appearing in $U_{\rm HV}$ and σ_0 are determined once the maximum value of $|\sigma_0 \Psi| < 1$ and $|\Psi|$ are prescribed. Note that at the O-point (circle where **B** vanishes) and at the separatrix, the parallel and perpendicular pressures are equal, while they differ in the bulk of the plasma.

As a general rule, once a solution for U is obtained, an infinity of solutions for Ψ can be obtained simply by changing the hypothesis about $\sigma_-(\Psi)$.

3. Tilting Stability Analysis

According to many theoretical studies (Clemente and Milovich, 1981; Schwarzmeier et al., 1983; Barnes et al., 1986; Ishida, 1988; Steinhauer and Ishida, 1990), FRCs should exhibit a global mode, called tilting instability (toroidal mode number $n = 1$) with growth times of the order of an axial Alfvén transit time. In spite of theoretical predictions, FRC experiments have shown a quite robust stability to such mode (for a general reference to FRC experiments, see Tuszewski, 1988). In order to explain this discrepancy between theory and experiments, several possible stabilizing mechanisms have been considered: plasma rotation (Clemente and Milovich, 1983; Horiuchi and Sato, 1989); Hall's terms in Ohm's law and gyro-viscous effects (Ishida, 1988; Steinhauer and Ishida, 1990); kinetic effects (Barnes et al., 1986); and equilibrium profile effects (Schwarzmeier et al., 1983; Cobb et al., 1993). However, no definite conclusion has been reached and the problem is still open.

As an application, here we will present an analysis of the tilting mode in anisotropic FRCs subjected to the restriction that $\sigma_- = \sigma_-(\Psi)$ at static equilibrium.

Within the framework of the equations of Chew et al. (1956), the stability of an anisotropic static equilibrium can be analyzed through the corresponding extension of the Energy Principle of Bernstein et al. (1958). Limiting the analysis to plasma displacements $\boldsymbol{\xi}$ which do not deform the plasma boundary (i.e. the FRC separatrix), the potential energy associated with the perturbation reduces to (Bernstein et al., 1958; Clemente et al., 1995):

$$\begin{aligned}\delta W = \frac{1}{2}\int_{V_s} d^3x \Big\{ & \mathbf{Q}\cdot\mathbf{Q} - \nabla\times\mathbf{B}\cdot\mathbf{Q}\boldsymbol{\xi} + \nabla\cdot\boldsymbol{\xi}\boldsymbol{\xi}\cdot\nabla p_\perp \\ & - \sigma_-[\mathbf{Q} + \mathbf{B}\nabla\cdot\boldsymbol{\xi} + \mathbf{B}(\boldsymbol{\xi}\cdot\nabla B^2)/2B^2] \\ \cdot\ & [\mathbf{Q} + \mathbf{B}\nabla\cdot\boldsymbol{\xi} + \nabla(\boldsymbol{\xi}\cdot\mathbf{B}) - \boldsymbol{\xi}\times(\nabla\times\mathbf{B})] \\ & + [\nabla\cdot(\sigma_- B^2\boldsymbol{\xi})]\mathbf{B}\cdot(\mathbf{B}\cdot\nabla\boldsymbol{\xi})/B^2 + 5p_\perp(\nabla\cdot\boldsymbol{\xi})^2/3 \\ & + 4\sigma_-[\mathbf{B}\cdot(\mathbf{B}\cdot\nabla\boldsymbol{\xi})/B]^2 + p_\perp[\nabla\cdot\boldsymbol{\xi} - 3\mathbf{B}\cdot(\mathbf{B}\cdot\nabla\boldsymbol{\xi})/B^2]^2/3\Big\}, \end{aligned} \tag{10}$$

where V_s is the volume embedded by the separatrix, $\mathbf{Q} = \nabla\times(\boldsymbol{\xi}\times\mathbf{B})$ and σ_-, $p_\perp$ and $\mathbf{B}$ are equilibrium quantities. If $\delta W < 0$ for some $\boldsymbol{\xi}$ satisfying physical and boundary conditions, the equilibrium is unstable. Looking at expression (10) it is possible to get some insight on the most dangerous displacements:

- if $\nabla\cdot\boldsymbol{\xi} \neq 0$ in a finite volume, it should always be possible to make $\delta W > 0$ by properly adding a finite value to $p_\parallel$ and $p_\perp$ at the separatrix, whilst maintaining all the other equilibrium quantities unchanged;
- even if $\nabla\cdot\boldsymbol{\xi} = 0$, the last term in δW can also be made arbitrarily large and positive when $\mathbf{B}\cdot(\mathbf{B}\cdot\nabla\boldsymbol{\xi}) \neq 0$ in a finite volume by the same previous assumption.

Therefore, in the following analysis it will be assumed that $\nabla\cdot\boldsymbol{\xi} = 0 = \mathbf{B}\cdot(\mathbf{B}\cdot\nabla\boldsymbol{\xi})$, these conditions have already been considered appropriate in some tilting studies (Schwarzmeier et al., 1984; Barnes et al., 1986) and in the case of axisymmetric FRCs they can be satisfied by the following class of trial displacements (Schwarzmeier et al., 1984):

$$\boldsymbol{\xi} = -\nabla[rF(\Psi)\sin(n\phi)]\times\hat{e}_r/n, \tag{11}$$

where n is an integer and $F(\Psi)$ is subject to the conditions of vanishing at the FRC separatrix and being finite and regular throughout the plasma. When $n = 1$, expression (11) represents a global tilting displacement. Assuming that $\sigma_- = \sigma_-(\Psi) = \sigma_-(U)$ and writing the volume element in flux coordinates,

$d^3x = 2\pi dU d\ell/|\nabla U|$, it is possible to reduce δW to the following expression (Clemente et al., 1995):

$$\delta W = \frac{\pi}{2}\int_0^{U_0} dU \oint \frac{d\ell}{|\nabla U|} r$$
$$\times \left\{ \frac{F^2}{r^2}\left[\left(\frac{\partial^2 U}{\partial r\partial z}\right)^2 - \frac{\partial^2 U}{\partial z^2} r \frac{\partial}{\partial r}\frac{1}{r}\frac{\partial U}{\partial r}\right] + \frac{1}{n^2}\left(\frac{dF}{dU}\right)^2\left[\frac{\partial U}{\partial r}\frac{\partial^2 U}{\partial z^2} - \frac{\partial^2 U}{\partial r\partial z}\frac{\partial U}{\partial z}\right]^2\right\}. \quad (12)$$

This expression is completely equivalent to an expression resulting from an isotropic FRC equilibrium with Ψ in place of U, all contributions arising from the special kind of anisotropy here considered cancel out. This disappointing result does not mean that anisotropy has no effect on tilting stability of FRCs, since we have excluded from the analysis case for which σ_- is not a surface function, however, a null result has been found in this case. Similar results have also been obtained in the case of tokamak stability (Salberta et al., 1987). If for given F and n, δW is negative, it will be more negative for $n+1$, while the integral:

$$K = \frac{1}{2}\int_{V_s} d^3x \rho \boldsymbol{\xi}\cdot\boldsymbol{\xi}$$

is a positive decreasing function of n. This together with the variational principle (Bernstein et al., 1958):

$$\omega^2 = \delta W/K; \quad \Delta\omega^2 = 0;$$

indicates that the most dangerous instability should correspond to n tending to infinity. However, even $n = 1$ is not easy to stabilize, this can be seen with a simple example corresponding to the Hill's vortex model for U. Without loss of generality U can be chosen to be normalized to unity in such a way that the Hill's vortex solution can be written in this case as $U = 4(r^2/a^2)(1 - r^2/a^2 - z^2/b^2)$ and the line integrals in Equation (12) can all be performed analytically giving the following expression for δW (Clemente et al., 1995):

$$\delta W = \frac{8\pi}{a^2 b}\int_0^1 dm[A(m)F^2(m) + B(m)(dF/dm)^2],$$

where

$$m = 2\frac{\sqrt{1-U}}{1+\sqrt{1-U}}, \quad A(m) = m[(2-m)K(m) - 3E(m)]/(2-m)^{7/2}$$

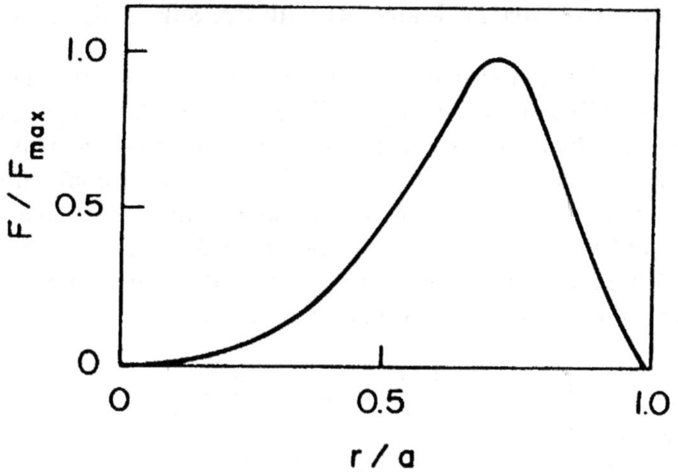

Figure 2. Plot of the critical axial displacement at the FRC midplane, for the Hill's vortex model ($b/a = 2.2$ and $\lambda = 0$).

and

$$B(m) = \frac{16a^2}{n^2b^2}\frac{\sqrt{2-m}}{m}\left\{\left(\frac{(1-m)}{(2-m)}\right)E(m) - 4\frac{(1-m)}{2-m}G_4 + \left[4 + 6\frac{(1-m)}{(2-m)^2}\right]G_6 - \frac{12}{2-m}G_8 + \frac{9}{(2-m)^2}G_{10}\right\}$$

with $G_4 = [2(2-m)E(m) - (1-m)K(m)]/3$; $G_6 = [4(2-m)G_4 - 3(1-m)E(m)]/5$; $G_8 = [6(2-m)G_6 - 5(1-m)G_4]/7$; $G_{10} = [8(2-m)G_8 - 7(1-m)G_6]/9$ and $K(m)$, $E(m)$ are the complete elliptic integrals of the first and second kind, respectively. In order to find the extremum values of δW the Euler–Lagrange method can be used and the following differential equation results:

$$\frac{d}{dm}B\frac{dF}{dm} - (A - \lambda)F = 0, \tag{13}$$

where λ is the Lagrange multiplier. If F is a solution of (13) subject to the conditions $F(1) = 0$ and F regular for $m \le 1$ ($m = 1$ corresponds to the separatrix and $m = 0$ to the O-point), it follows that δW is proportional and has the sign of λ. Negative values of λ (implying instability) have been found numerically when $n \cdot b/a > 2.2$. A plot of F, normalized to unity, as a function of the radius at the FRC midplane, for the critical case $\lambda = 0$ and $n \cdot b/a = 2.2$, is presented in Figure 2. As can be seen the mode is quite extended throughout the plasma and it can be considered a global mode. When $n = 1$, the critical elongation for tilting stability is $b/a = 2.2$, however for $n > 1$ the critical elongation can be as small as one wants.

4. FRC Particle Transport with Pressure Anisotropy

In the following the transport of particles in an axisymmetric and equatorially symmetric FRC will be considered. It will be assumed that it is possible to reach a steady state in which the loss of particles due to plasma resistivity is exactly balanced by some source inside the plasma and the particle confinement time will be defined as $\tau_p = M/\Gamma_s$, where M represents the mass of plasma embedded by the separatix and Γ_s the flux of mass across it. In absence of plasma rotation the steady state Ohm's law simplifies to $\eta_\perp \mathbf{j} = \mathbf{v} \times \mathbf{B}$, from which it is possible to obtain the flux of mass across a constant Ψ surface as:

$$\Gamma(\Psi) = \int\limits_{\Psi=\mathrm{const}} d\mathbf{s} \cdot \rho \mathbf{v} = 4\pi \int\limits_{r_1(\Psi)}^{r_2(\Psi)} drr\rho\eta_\perp \left| \frac{\Delta^* \Psi}{\partial \Psi / \partial z} \right| ,$$

where magnetic field line equations have been used, equatorial symmetry has been taken into account and $r_{1,2}$ are defined in such a way that $\Psi(r_1, 0) = \Psi(r_2, 0)$ with $r_2 > r_1$. In general, $\Gamma(\Psi)$ will be a surface function depending on equilibrium profiles. In previous works (Auerbach and Condit, 1981; Clemente and Grillo, 1984) the source of plasma was limited to a circle at the O-point, in such a way that in order to satisfy the continuity equation, $\Gamma(\Psi)$ must be a constant. This resulted in quite special profiles for density and resistivity, with gradient singularities and also singularities in the diffusion velocity. Here, such difficulties are partially circumvented by allowing for spread plasma sources inside the separatrix and by assuming finite constant density ρ_s and resistivity η_s at the separatrix, in such a way that τ_p reduces to:

$$\tau_p = \frac{\bar{\rho}}{4\pi\rho_s} \frac{V_s}{\eta_s} \left[\int\limits_0^a drr \left| \frac{\Delta^* \Psi}{\partial \Psi / \partial z} \right| \right]^{-1} , \tag{14}$$

where V_s represents the volume embedded by the separatrix, $\bar{\rho}$ is the volume averaged density and a is the separatrix radius at midplane.

In order to compute the integration in the right member of Equation (14), an explicit expression for Ψ is needed, if anisotropic FRCs with σ_- as a surface function are considered, the integration in Equation (14) can be expressed in terms of the auxiliary function U as:

$$I_s = \int\limits_0^a drr \left| \frac{\Delta^* \Psi}{\partial \Psi / \partial z} \right| = \int\limits_0^a \frac{drr}{|\partial U / \partial z|} \left| \Delta^* U + \frac{1}{2(1 - \sigma_-)} \frac{d\sigma_-}{dU} |\nabla U|^2 \right| . \tag{15}$$

In order to perform the integration, the Hill's vortex model can be chosen together with the simplest hypothesis for $\sigma_- = \sigma_0 + \sigma_1 U$. In such case all the integration can be performed analytically and I_s results (Clemente and Cesar, 1995):

$$I_s = b(4 + a^2/b^2) \left[1 - \frac{8}{15} \frac{\sigma_1 U_0}{(1 - \sigma_0)} \right] . \tag{16}$$

Since $V_s = 4\pi a^2 b/3$, the particle confinement time can be expressed as:

$$\tau_p = \frac{\bar{\rho}}{\rho_s} \frac{a^2 b}{3\eta_s I_s} \,. \tag{17}$$

Expression (17) shows the dependence on anisotropy through the term $\sigma_1 U_0/(1 - \sigma_0)$ contained in I_s. $\sigma_1 = \sigma_0 = 0$ corresponds to the isotropic Hill's vortex model already studied by other authors (Auerbach and Condit, 1981; Nguyen and Kammash, 1982) and their results can be recovered. If $|\sigma_-| < 1$ and also $|\sigma_0| < 1$ are considered, it results that $-1 - \sigma_0 < \sigma_1 U_0 < 1 - \sigma_0$. Correspondingly, when σ_0 approaches unity, very small values of τ_p can be obtained. When the upper limit of $\sigma_1 U_0$ is considered, the largest τ_p, about twice the value corresponding to the scalar pressure case for a given η_s, can be obtained. According to Equations (16) and (17) the worst results for τ_p correspond to $\sigma_0 > 0$ and $\sigma_1 U_0 < 0$, this indicates that what should be most relevant is the way σ_- changes with Ψ than its absolute value.

5. Conclusions

An analytical method to obtain anisotropic rotating MHD equilibria has been presented. In spite of some restrictions on the kind of anisotropy, the method allows to construct an infinite number of suitable axisymmetric equilibria starting from solutions of the MP or GSS equations and represents a considerable simplification when compared with previous studies (Salberta et al., 1987; Sestero and Taroni, 1976; Fielding and Haas, 1979; Cooper et al., 1980; Rizk, 1987) where the equilibria were either generated numerically or in an approximate form. Two applications to the study of the tilting mode in FRCs and their particle transport properties have also been presented, indicating that anisotropic effects should not be so relevant with respect to stability but they should be relevant with respect to transport.

In conclusion, all these aspects suggest that anisotropic effects in plasma kinetic pressure should deserve more attention both from a theoretical and experimental point of view, not only in FRCs but also in other confinement schemes when equilibrium and transport properties are concerned.

Acknowledgments

This work was partially supported by Conselho Nacional de Desenvolvimento Científico e Tecnólogico (CNPq, Brazil) and Fundação de Amparo à Pesquisa do Estado de São Paulo (FAPESP, Brazil). Thanks are also due to the Organizing Committee of the VII Latin American Workshop in Plasma Physics for their invitation to present this work.

References

Auerbach, S.P. and Condit, W.C.: 1981, *Nucl. Fusion* **21**, 927.
Barnes, D.C. et al.: 1986, *Phys. Fluids* **29**, 2616.
Bernstein, I.B. et al.: 1958, *Proc. R. Soc.* **A244**, 17.
Chew, G.F. et al.: 1956, *Proc. R. Soc.* **A236**, 112.
Clemente, R.A.: 1993, *Nucl. Fusion* **33**, 963.
Clemente, R.A.: 1994, *Plasma Phys. Control. Fusion* **36**, 707.
Clemente, R.A. and Cesar, R.G.F.: 1995, *Plasma Phys. Control. Fusion* **37**, 137.
Clemente, R.A. and Grillo, C.E.: 1984, *Phys. Fluids* **27**, 658.
Clemente, R.A. and Milovich, J.L.: 1981, *Phys. Lett.* **85A**, 148.
Clemente, R.A. and Milovich, J.L.: 1983, *Phys. Fluids* **26**, 1874.
Clemente, R.A. et al.: 1995, *Plasma Phys. Control. Fusion* **37**, 1381.
Cobb, J.W. et al.: 1993, *Phys. Fluids* **B5**, 3227.
Cooper, W.A. et al.: 1980, *Nucl. Fusion* **20**, 985.
Fielding, P.J. and Haas, A.F.: 1979, *Nucl. Fusion* **19**, 855.
Grad, H.: 1967, in: *Magneto-Fluid and Plasma Dynamics Symposia in Applied Mathematics*, Vol. 18, American Mathematical Society, Providence, RI, p. 162.
Hastie, R.J., Taylor, J.B. and Haas, F.A.: 1967, *Ann. Phys.* **41**, 302.
Horiuchi, R. and Sato, T.: 1989, *Phys. Fluids* **B1**, 581.
Ishida, A.: 1988, *Phys. Fluids* **31**, 3024.
Maschke, E.K. and Perrin, H.J.: 1980, *Plasma Phys.* **22**, 579.
Maschke, E.K. and Perrin, H.J.: 1984, *Phys. Lett.* **102A**, 106.
Mercier, C. and Cotsaftis, M.: 1961, *Nucl. Fusion* **1**, 121.
Nguyen, K. and Kammash, T.: 1982, *Plasma Phys.* **24**, 177.
Rizk, H.M.: 1987, *J. Plasma Phys.* **38**, 209.
Salberta, E.R. et al.: 1987, *Phys. Fluids* **30**, 2796.
Schwarzmeier, J.L. et al.: 1983, *Phys. Fluids* **26**, 535.
Sestero, A. and Taroni, A.: 1976, *Nucl. Fusion* **16**, 164.
Steinhauer, L.C. and Ishida, A.: 1990, *Phys. Fluids* **B2**, 2422.
Tuszewski, M.: 1988, *Nucl. Fusion* **28**, 2033.
Viana, R.L. et al.: 1997, *Plasma Phys. Control. Fusion* **39**, 197.
Zelazny, R. et al.: 1993, *Plasma Phys. Control. Fusion* **35**, 1215.

ENERGY EXHAUST FROM TOKAMAKS: PROBLEMS AND SOLUTIONS

T.K. SOBOLEVA
Instituto de Ciencias Nucleares, Universidad Nacional Autónoma de México, 04510 México, D.F.

Abstract. Investigation of the edge plasma appears to be one of the key problems for tokamak design. The problem of the heat and particle removal from the reactor are supposed to be solved by means of magnetic divertors. Their main missions are to accommodate the power outflow, to control the spatial location of the radiative zone, the helium exhaust, the impurity content in the main plasma, and to withstand the major energy loss in the eventuality of a disruption energy quench.

In the paper different approaches to magnetic divertor design are analyzed. However, emphasis is put on high recycling radiative ITER-like divertor. Results of numerical modeling of the edge plasma are presented along with current experimental results achieved both at tokamaks and linear simulators. It is shown that there is a good agreement between the experimental results obtained at linear simulators and the theoretical predictions obtained by means of "gas box" divertor model, which includes ionization, charge-exchange, and 3-body and molecular advanced recombination.

1. Introduction

Technology R&D for International Thermonuclear Experimental Reactor is now focused on seven critical areas, each the subject of a large project aimed at validating key aspects of the ITER Design, including development and verification of industrial level manufacturing techniques. One of them is the Divertor Cassette Project which is aimed to demonstrate that a divertor can be built with tolerances, and to withstand the thermal and mechanical loads imposed on it during normal operation and during transients such as ELMs and disruptions. Developing of a robust solution to the divertor problem is one of the most important tasks to be undertaken in the R&D supporting the ITER design. In solving the problem experimental resources are used such as tokamaks JET, ASDEX-Upgrade, JT-60U, DIII-D and Alcator C-MOD, linear divertor simulators in Nagoya University (NAGDIS II), Japan, at PPPL and UCSD in the U.S.A. and at Kurchatov Institute in Russia. Also, Edge Plasma Physics is currently a subject of intense theoretical and computational efforts. The following codes are mostly used for divertor modeling: 1.5 dimensional tokamak transport codes GTWHIST and PRETOR, and the 2-dimensional divertor modeling codes B2/EIRENE, B2, EDGE2D/NIMBUS and UEDGE. Modelling provide guidance for such issues as impurity concentration needed to achieve the necessary level of radiation losses, the choice of optimal impurity, the operational range in density, power level, optimal divertor configuration.

Astrophysics and Space Science **256:** 247–262, 1998.

1.5 and 2 dimensional codes are rather cumbersome, so to investigate new physical effects which were neglected in these codes analytical consideration or/and simpler 1-dimensional codes are used.

Structure of the report is as follows. In Section 2 short description of ITER divertor scheme is given, in Section 3 alternative approach is discussed, then after short description of the characteristic features of detached and semi-detached regimes of divertor operation (in Section 4), analysis of experimental data from tokamaks and divertor simulators is presented (Section 5). Basing on these results it is shown that the plasma recombination, the process which was neglected in the most computational models, is essential in explanation of experimental observations. In Section 6 we discuss the mechanisms of plasma recombination relevant to tokamak and divertor simulator conditions. In Section 7 we consider the effects of recombination on the plasma parameters in tokamaks and divertor simulators respectively. In Section 8 the main conclusions are summarized.

2. ITER Divertor Concept

ITER is designed to produce 1.5 GW of power. The major part of it carried by fusion neutrons will be utilized in the reactor blanket. 300 MW of ITER power is α heating power. Considerable part of it, about 100 MW, will be spread over the first wall due to bremsstrahlung and synchrotron radiation. Without special measures the rest amount, about 200 MW, will be deposited at target plates, producing the power densities of the order of 100 MW/m^2, the power flux at the edge of the Sun (ITER, 1996; Wagner, 1997). Modern technology can handle at best with the fluxes of about 5 MW/m^2 in the steady state, so that the problem of safe exhaust is a physical problem rather then engineering one.

The main approach to ITER divertor is based on removal of the bulk of the energy in the diverted plasma channel before it contacts the divertor target plates. Hydrogenic gas, possibly seeded with low and medium-Z impurity, is injected in divertor channel and energy is removed through combination of radiation and charge-exchange losses. It is shown in ITER report (Kukushkin, 1996) that bremsstrahlung losses with intrinsic impurity levels of $\sim$ 10% He and $\sim$ 2% Be or C and with seeded Ne levels in the range 0.12 to 0.88% are 80–135 MW; for seeded Ar levels in the range 0.012 to 0.16% they are 75–100 MW, and for seeded Kr levels in the range 2.4×10^{-3} to 1.3×10^{-2}% they are 75 MW. Impurity line radiation losses from the mantel for seeded Ne, Ar and Kr impurity levels in the above ranges are 20–65 MW, 20–110 MW, and 25–145 MW, respectively. Thus, the total possible radiation losses from inside the separatrix with account for limits imposed by core power balance constraints are $\sim$ 100–200 MW with seeded Ne and $\sim$ 100–260 MW with seeded Ar or Kr. Together with charge-exchange losses it can result in considerable plasma cooling, and the power fluxes approach a level when further optimization can be carried out accounting for plasma recombination which start to be important at high density low temperature regimes.

3. Liquid Lithium Divertor Concept

The above approach to energy exhaust from tokamaks is considered now as a basic one, however it is not free from definite shortcomings and there are alternative approaches. As a good example Liquid Lithium Divertor Concept may be mentioned. The concept was proposed at 16th IAEA Conf. on PP and CNF in the framework of Liquid Lithium Tokamak Reactor project (Evtikhin et al., 1996). Filmy liquid metal shield was proposed for tokamak reactor first wall and divertor in (Karasev et al., 1989). The concept of a DEMO-type reactor (Russian Federation) presently under development is based also on the application of high temperature liquid lithium. The divertor concept is based on capillary porous structure filled with liquid lithium which is used for the divertor plate surface structure (see Figure 1). The authors of the project (Evtikhin et al., 1996) attract attention to the fact that evaporation-condensation devices with liquid metal as working fluid are known to be the most effective means of energy transfer in high temperature facilities. Such a way of eliminating thermal loads can provide excellent performance, achieving hundreds of MW/m^2. The characteristic of capillary-pore system (changing porosity, permitting anisotropy, working surface geometry) are maintained in broad ranges. The design provides sufficient working fluid pressure in supply system without applying external pressure, due to capillary forces only. One of the very attractive features of the concept is that the system is self-sustained and self-regulating because the pressure distribution of the liquid metal coolant in capillary-pore structure reacts to local changes of the thermal loads distribution at divertor plates surface. This type of divertor can operate under steady state and plasma disruption conditions.

Experimental models of liquid lithium capillary structure have been tested at long pulse high heat loads up to 50 MW/m^{-2} during 5 sec (Antonov et al., 1996). Long duration tests have been performed at 20 MW/m^{-2} and 30 MW/m^2 for 60 and 30 sec respectively, and 5 MW/m^2 for 10 minutes. Experiments have shown that in front of the target dense plasma layer, 1–2 cm thick, was formed. The main part of plasma energy flow was absorbed and radiated in this layer. Numerical modeling with 2D-hydrodynamic code DDIC95 was performed. The code provides possibility of self-consistent simulation of a background plasma, neutral gas and impurities. The calculations have shown that thermal loads onto divertor plates were essentially reduced without addition of any heavy impurities.

The project looks very attractive,however, still further experimental and theoretical work is needed to prove if it is really the best solution for tokamak reactor, so in what follows we will concentrate on actual problems of highly recycling gas diver-

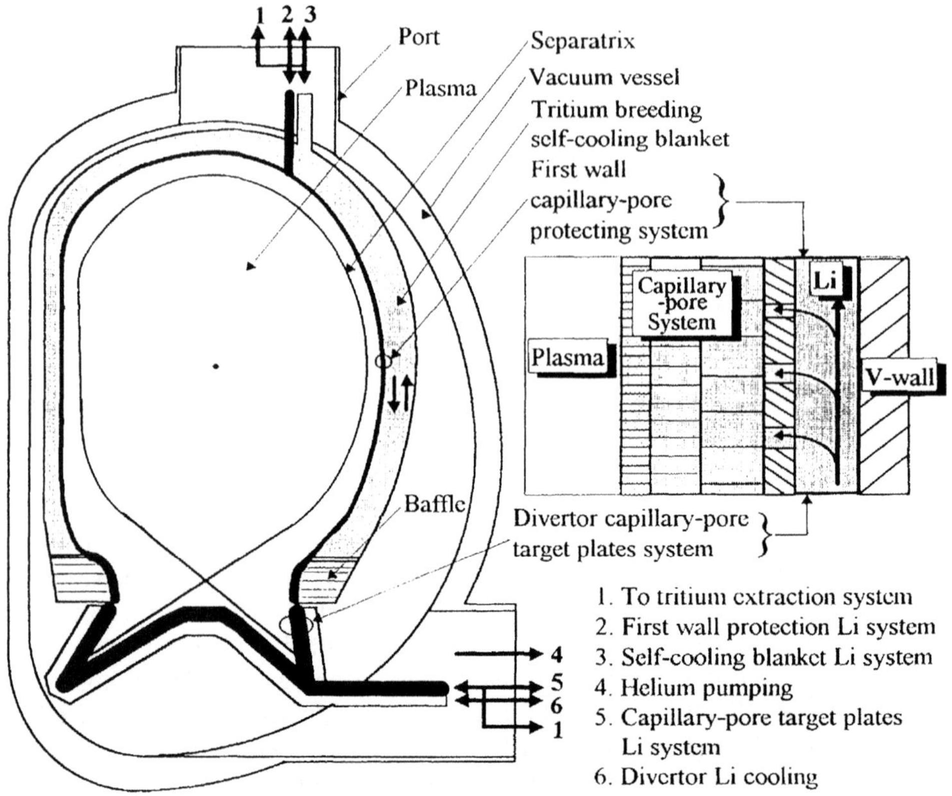

Figure 1. Liquid lithium fusion reactor scheme.

tor which uses solid divertor plates. This divertor project had passed international expertise and is considered as basic for ITER.

4. Detached Regimes in Tokamak Divertors

During the last few years experiments on diverted tokamaks (Hutchinson et al., 1994; Janeschitz et al., 1992; Petrie et al., 1992; Mertens et al., 1993) and divertor simulators (Schmitz et al., 1995; Ohno et al., 1995; Chiu and Cohen, 1996; Nektarov et al., 1996) have focused on the study of the so-called detached divertor regimes. The interest was due, in particular, to reduction of the peak heat flux onto the divertor plates by a factor of five or more even for maximum available heating power. Detached regimes of operation look attractive from the reactor relevant divertor design point of view since it potentially allows the removal of the energy flux and solves the problem of heat overloading of the divertor plates (Krasheninnikov et al., 1997). To make proper scaling of the detached regimes from modern tokamaks to a reactor of ITER scale it is necessary to understand the physical mechanisms giving rise to this mode of operation.

In all experiments the detached regimes in tokamaks has been occurred either in case of strong neutral gas puffing (H, D) or due to combination of the impurity seeding (N, Ne, Ar) and neutral gas puffing. Strong gas puffing increases radiation of intrinsic impurities, whereas impurity seeding often increases both edge and divertor radiation and permit detachment at lower densities. In many early experiments a rapid transition from attached regimes to X-point MARFE was occurred. Recent experiments using feedback control on gas and impurity fuelling demonstrate a more gradual transition and radiation within divertor. These discharges are often detached near to separatrix but remain attached in the outer scrape-off-layer. They are called semi-detached. In modern experiments completely detached regimes often have strongly degraded energy confinement, close to L-mode. Semi-detached regimes demonstrate better confinement but often have high Z_{eff} (> 2.5). A detached discharge that simultaneously has good confinement and low Z_{eff} has not yet been observed experimentally. However modern diverted tokamaks are mostly too open to neutrals, which may cause the negative effects mentioned above. Now experiments are under way to achieve detached, low Z_{eff}, and high confinement regimes.

The detachment regimes are characterized by the following main features: low plasma temperature near the target; strong decrease of the plasma particle flux onto the plates (see Figure 2, where the effect is clearly demonstrated) (Krasheninnikov et al., 1997), and plasma pressure drop along magnetic field lines. Note, that in conduction case (lower density), as well as in high-recycling regimes (medium density) plasma pressure along the magnetic field line was constant. Up until recently research primarily dealt with plasma momentum/pressure loss due to plasma-neutral interactions. However, using only plasma-neutral interactions, it is difficult to explain the loss in plasma particle flux to the plates which occurs as well (Krashenninikov, 1996; Loarte, 1997). More detailed analysis of experimental data suggests that the plasma particle flux reduction in the tokamaks and linear divertor simulators does not occur without a significant role of plasma recombination.

5. Experimental Results on Detached Regimes

5.1. Tokamaks

In tokamak experiments detached divertor regimes were found after reaching high recycling divertor conditions when the plasma flux from the bulk to the Scrape off Layer (SOL) plasma, Γ_s, is much smaller than the plasma source due to neutral ionization in the divertor region, Γ_{ion}, (see Figure 3). Therefore, we can neglect the influence of the particle flux Γ_s by assuming that only a power comes into the SOL from the bulk plasma. This assumes that neutrals are ionized mainly within

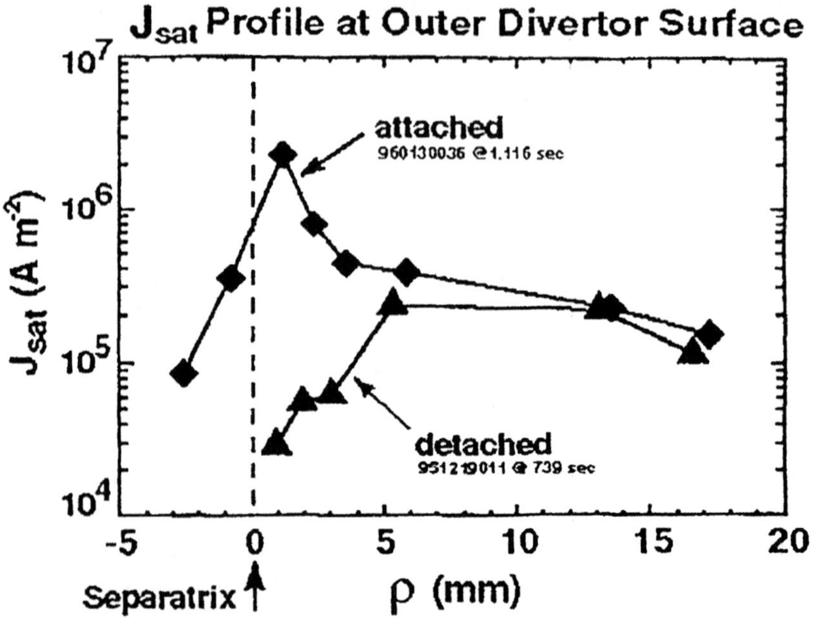

Figure 2. Ion saturation current profile at outer divertor plate in Alcator C-MOD in attached and detached regimes, ρ is the distance of magnetic surface from the separatrix measured at equatorial plane.

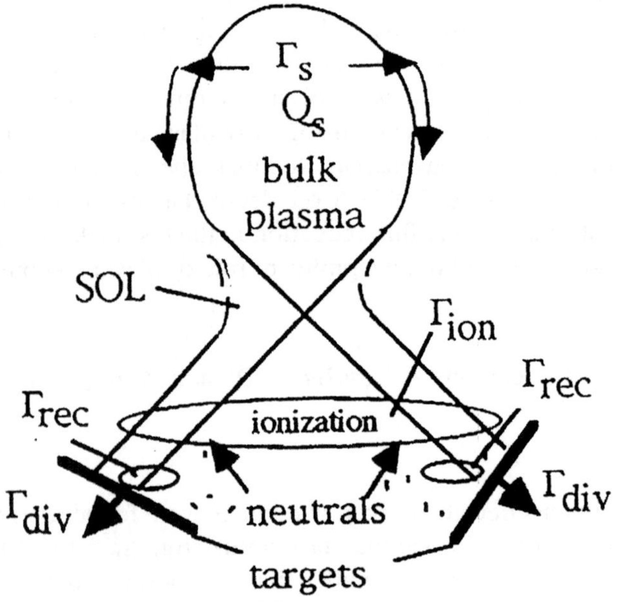

Figure 3. Schematic view of different regions in a tokamak: Γ_S and Q_S are the energy and plasma particle fluxes from the bulk into the SOL, Γ_{ion} and Γ_{rec} are the ionization and recombination plasma particle source and sink, Γ_{div} is the plasma flux onto the targets.

the SOL before they reach the bulk plasma, and therefore the SOL plasma must be dense enough to ensure that this happens. From the plasma particle balance in the SOL, it follows that the plasma flux onto the divertor plates, Γ_d, is defined by the competition between ionization and volumetric recombination. Thus, Γ_d can be reduced either by a decrease in the ionization source Γ_{ion} or by an increase in the plasma recombination sink Γ_{rec}. We will show next that Γ_{ion} does not decrease substantially for detached divertor conditions.

If we assume that plasma volumetric recombination is unimportant, then the only source of the neutrals is the plasma recombination at the surface of the divertor plates. Neutrals originating at the plates penetrate into the hot temperature region of the SOL and are ionized there. Therefore, Γ_{ion} equals to the neutral flux into hot temperature region, Γ_N. When the hot ionization region is screened from the plates by a dense cold plasma buffer (Krashenninikov, 1996) it is possible to alter Γ_N by switching of the neutral flow from free streaming to diffusive. It can only happen when the characteristic width of the buffer, Δ_b, is considerably larger then neutral-ion mean free path, $\lambda_{\text{Ni}} \approx V_{\text{Ti}}/\nu_{\text{Ni}} = V_{\text{Ti}}/K_{\text{Ni}}n_b$ where n_b is the plasma density in the cold plasma buffer, V_{Ti} is the ion thermal velocity, and K_{Ni} is the ion-neutral collision rate constant. In this case, the flux Γ_N is determined by a slow diffusive like neutral flow through the dense cold plasma, and results in the inequality $\Gamma_N = \Gamma_{\text{ion}} \approx \Gamma_N^{\text{fs}}\lambda_{\text{Ni}}/\Delta_b \ll \Gamma_N^{\text{fs}}$, where Γ_N^{fs} is the free streaming flux of the neutrals near the walls, which is proportional to the neutral gas pressure in the divertor region. However, the neutral gas pressure in the private flux zone after detachment occurs remains almost the same as in the attached regime or even increases (Niemczewski, 1997). Consequently, Γ_N^{fs} does not decrease in the detached case.

Then, the reduction of Γ_{ion} in detached regime in neglection of plasma volumetric recombination can only be explained by the influence of the plasma buffer. Existence of cold plasma buffer even in attached regime can be seen from analysis of the energy balance of the scrape-off-layer plasma. It may be shown that for typical experimental conditions in Alcator-C Mod (Hutchinson et al., 1994), energy flux from the bulk plasma $Q_s \approx 0.5$ MW and $\Gamma_N^{\text{fs}} \sim 6 \times 10^{23}$ s^{-1} (for a neutral gas pressure ≈ 30 mTorr) which corresponds to $E_{\text{ion}}\Gamma_N^{\text{fs}} \sim 3$ MW, characteristic width of the buffer considerably exceeds neutral-ion mean free path, $\lambda_{\text{Ni}}/\Delta_b \sim 0.1$.

Cold dense plasma buffer could decrease ionization flux either by increase of plasma density or, else, by increase of characteristic width of the buffer. However, experiments show that in front of the target the plasma density decreases after transition to detached regimes. Also, experimental conditions are such that characteristic buffer width can hardly increase the factor of 100 which is needed to explain experimental observation of strong (~ 100 times) drop in plasma flux onto the target (see Figure 2).

Thus, we see that the cold plasma buffer alone can not explain the strong reduction of the plasma flux onto the plates in detached regimes (although, it may cause some reduction in Γ_{ion} and therefore in Γ_d). So it was concluded that plasma recombination (Krasheninnikov and Pigarov, 1997; Petravic, 1994; Knoll et al.,

1996; Kukushkin et al., 1996; Wising et al., 1996; Post, 1995) must be considered as a crucial element of detachment in current tokamaks.

These important features of neutral transport in tokamaks, which are discussed above, were usually ignored in models of plasma detachment based solely on plasma momentum removal by ion-neutral friction of the cold plasma with stationary neutrals. Moreover, since, as we have seen, $\lambda_{\mathrm{Ni}}/\Delta_b$ is small, the plasma and neutral gas flows are strongly coupled. Therefore, the plasma momentum transport to the walls must be carried by neutral viscosity which is rather weak.

5.2. Divertor Simulators

Let us consider briefly the peculiarities of divertor simulators operation. In the experiments on divertor simulators it is possible to distinguish two different regions: the plasma source, and the working chamber. In some experiments the plasma source region is only linked with the working chamber through a relatively small opening, so that the plasma streams from the source region into the long ($L_{\mathrm{WCh}} \sim 100$ cm) working chamber where the neutral gas can be puffed in. At low neutral gas pressure in the working chamber the plasma stream hits the target. However, with increasing gas pressure, the plasma density in front of the target, n_t, starts to decrease, and when the gas pressure exceeds some critical value n_t exhibits a rather steep drop. This type of plasma behavior is observed for different working gases (Helium, Neon, and Argon).

To discuss this behavior we assume that the plasma source fixes the plasma parameters at the entrance to the working chamber. For a plasma density $n_e \sim 10^{13}$ cm^{-3}, a working chamber radius of about 10 cm, and a neutral gas pressure about 10 mTorr the neutral-ion and neutral-neutral mean free paths exceed the radii of the plasma stream (1 cm) and working chamber, $\lambda_{\mathrm{Ni}} > 1$ cm, $\lambda_{\mathrm{NN}} > 10$ cm (in contrast to a tokamak divertor plasma). Therefore, to lowest order we can assume that the neutrals are in the Knudsen transport regime and their drift velocity is small. For the same parameters, the ion-neutral mean free path $\lambda_{\mathrm{iN}} \approx V_{\mathrm{Ti}}/\nu_{\mathrm{iN}} \approx V_{\mathrm{Ti}}/(K_{\mathrm{iN}}N)$ is smaller than L_{WCh}. Therefore, the plasma flux, j, can be found from the plasma momentum balance equation written in the diffusive approximation as

$$j = -\left[\frac{dP/dx}{M\nu_{\mathrm{iN}}}\right], \tag{1}$$

where M is the ion mass. In the case when there are no plasma sources and sinks (no ionization and recombination) in the working chamber, Equation (1) gives the plasma flux onto the target, j_t, as being determined by the plasma pressure at the entrance to working chamber, P_u, and the neutral gas density:

$$j_t \propto \left[\frac{P_u}{L_{\mathrm{WCh}}N}\right], \tag{2}$$

Expression (2) exhibits a smooth decrease of j_t with increasing neutral gas density and therefore cannot explain the experimental observations. So again, plasma recombination, which can cause a drastic reduction of j_t, seems to be the primary candidate for the explanation of the particle flux drop in divertor simulators. Thus, both tokamak and divertor simulator experiments point to recombination as a crucial element of plasma detachment.

6. Recombination Mechanisms

In low temperature plasmas there are several recombination mechanisms relevant to tokamak and divertor simulator experimental conditions. They are as follows:

(a) radiative recombination: $A^+ + e \Rightarrow A + \hbar\nu$,
(b) three body recombination: $A^+ + e + e \Rightarrow A + e$,
(c) dissociative recombination of the molecular ion $(AB)^+$: $(AB)^+ + e \Rightarrow A + B$
(d) charge exchange recombination of positive and negative ion: $A^+ + B^- \Rightarrow A + B$.

In what follows we will refer to processes (a) and (b) as the Electron-Ion Recombination (EIR) channel.

Although these recombination mechanisms are well known in gas discharge plasmas, until recently, only EIR was employed to interpret data from tokamaks and divertor simulators. The rate constant for EIR, K_{EIR}, is small at high electron temperatures, T_e, but has a very strong temperature dependence which, in a crude approximation, can be described by the scaling $K_{\mathrm{EIR}} \propto n_e/(T_e)^{9/2}$ so that K_{EIR} becomes large at low T_e (usually for $T_e \leq 1$ eV). For $T_e \approx 1$ eV and density $n_e \approx 10^{15}$ cm^{-3} the EIR rate constant is about 5×10^{-12} cm^3/s. Recombination becomes important when the plasma flux onto the target, j_t, is comparable to the plasma particle sink due to recombination, $\nu_{\mathrm{EIR}} n_e \ell_{\mathrm{rec}}$, where $\nu_{\mathrm{EIR}} = K_{\mathrm{EIR}} n_e$, and ℓ_{rec} is the poloidal length of the low temperature region where the recombination occurs. For $j_t \approx 10^{19}$ cm^{-2}/s we find that recombination becomes important when $\ell_{\mathrm{rec}} \geq j_t/(\nu_{\mathrm{EIR}} n_e) \approx 2$ cm, which is quite a realistic value and in reasonable agreement with our estimate of the width of cold plasma buffer.

In (Post, 1995) it was pointed out that the presence of molecular hydrogen in tokamak experiments can switch on the recombination channel (d) via formation of negative hydrogen ions H^- through the dissociative attachment of the electron to ro-vibrationally excited molecular hydrogen and successive charge exchange recombination of negative and positive ions

$$\mathrm{H}_2(j, v) + e \Rightarrow \mathrm{H}^- + \mathrm{H}, \quad \mathrm{H}^- + \mathrm{A}^+ \Rightarrow \mathrm{A} + \mathrm{H}. \tag{3}$$

In (Krasheninnikov et al., 1996) it was shown that the recombination channel (3) can also be activated in the presence of molecular hydrogen via the conversion of an atomic ion into molecular one, followed by dissociative recombination of the molecular ion

$$\mathrm{H}_2(j, v) + \mathrm{A}^+ \Rightarrow (\mathrm{AH})^+ + \mathrm{H}, \quad (\mathrm{AH})^+ + e \Rightarrow \mathrm{A} + \mathrm{H}. \tag{4}$$

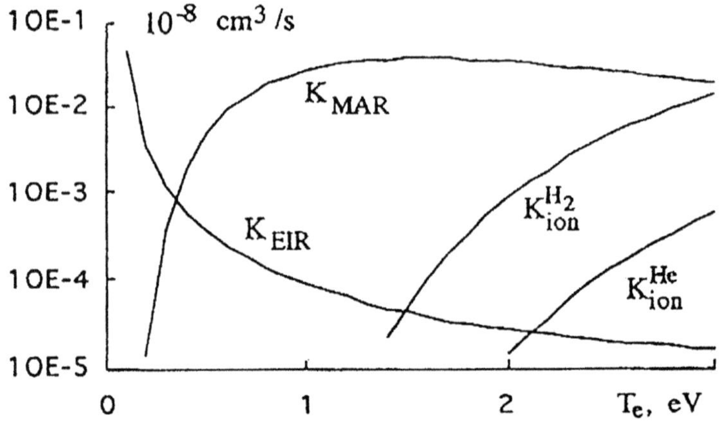

Figure 4. Ionization and recombination rate constants for hydrogen plasma in the presence of molecular hydrogen found in local collisional-radiative model.

The ro-vibrationally excitation of molecular hydrogen is important for both of these Molecular Activated Recombinations (MAR) Equations (3, 4) to make the dissociative attachment of the electron and the ion conversion reactions almost resonant processes. Notice that MAR can be applied for different background ions A^+ (H^+, He^+, Ar^+, etc.) and therefore can also be relevant for the interpretation of experiments with non-hydrogen plasmas and hydrogen puffing. MAR can also involve other molecules available in tokamaks and divertor simulators. In (Pigarov and Krasheninnikov, 1996) MAR for a hydrogen plasma was analyzed with the Collisional-Radiative Atomic-Molecular Data (CRAMD) Code (Pigarov et al., 1995) using the local collisional-radiative (CR) approximation. The results of the calculation for hydrogen plasmas (for $n_e \approx 10^{14}$ cm^{-3}) show that the maximum of $K_{\rm MAR}$ reaches $2 \times 10^{-10} cm^3/s$ in the temperature range of about 1–2 eV (in good agreement with estimates of (Krasheninnikov et al., 1995)), and for $T_e > 2$ eV decreases rather smoothly with increasing T_e, while for T_e below 0.5 eV decreases sharply with decreasing T_e due to weakening of vibrational excitation of molecules (see Figure 4). A similar temperature dependence of $K_{\rm MAR}$ is found with the local CR model of MAR in different plasmas.

Let us now make an estimate of the impact of MAR on plasma recombination in a tokamak divertor. For a plasma temperature $T_e \approx 1$ eV and density $n_e \approx 10^{15}$ cm^{-3} the MAR rate constant is about 10^{-10} cm^3/s (Krasheninnikov et al., 1996) and the ratio of the MAR to EIR plasma sinks is about $20 \times ([H_2]/n_e)$. Assuming that $[H_2]/n_e \sim 0.1$ (which we may expect near the plate), we find that the plasma sink due to MAR can be a few times larger than the EIR plasma sink for tokamak divertor plasma conditions. To find the molecular hydrogen density more accurately we should follow the dynamics of the multispecies gas flow taking into account the formation of molecules at the surface (the volumetric source of H_2 is usually small in a tokamak plasma), and their dissociation and ionization.

7. Effects of Plasma Recombination

7.1. Tokamak Divertors

We saw in Section 6 that plasma recombination can be rather strong at low plasma temperature ($\leq$ 1–2 eV). Intuitively, we expect that the temperature of a divertor plasma decreases with decreasing heating power into the SOL, $Q_{\mathrm{SOL}} = Q_s - Q_{\mathrm{imp}}$, and increasing density of the SOL plasma, n_{SOL}. Therefore, there are two ways to reduce the plasma temperature in the divertor region: (i) impurity puffing which results in an increase of the energy radiation loss Q_{imp} and a concomitant decrease in Q_{SOL}, and (ii) increase the plasma density. These method of detaching are pursued in both experiments and modeling. When the temperature near the divertor plates becomes low enough, recombination takes over and the plasma is almost extinguished in the divertor region in front of the target. In agreement with our qualitative estimates made in Section 6 and with some of the one dimensional models (Krasheninnikov et al., 1995; Krasheninnikov and Soboleva, 1996), the cold plasma buffer alone can only result in a rather modest reduction of $j_d(\Gamma_{\mathrm{ion}})$ in current tokamaks, even though a strong plasma pressure drop near the divertor plate and low plasma temperature ($<$ 1 eV) are found. Results of the modeling of the EIR impact on the ionization/recombination balance in Alcator C-MOD divertor shows that without carbon the plasma temperature is high and the effect of the EIR is negligible, while with 0.5% carbon, when plasma temperature near the plates drops, the ionization source is almost canceled by the EIR (at the magnetic surfaces located below divertor "nose") which results in a strong reduction of the plasma flux onto the target at these flux surfaces. The balancing of ionization by recombination has been confirmed by measurements of the spectrum of hydrogen Balmer lines and the recombination continuum in Alcator C-MOD in the partially detached divertor regime (Lumma et al., 1998). Analysis of the Balmer line spectrum showed that it is that of a plasma dominated by 3-body recombination. The estimates of the plasma density (from Stark broadening of the Balmer lines) and temperature (from bremsstrahlung and radiative recombination continuum) in the recombination region give $n_e \approx 8.8 \times 10^{14}$ cm^{-3} and $T_e \approx 1$ eV. The plasma particle recombination sink, Γ_{rec}, found from spectroscopic analysis is 5.3×10^{22} s^{-1} and almost equals the plasma flux onto the divertor plate $\Gamma_d \approx 7 \times 10^{22}$ s^{-1} found from the probes measurements.

There is no clear experimental evidence of MAR in a tokamak plasma. However, 2D numerical modeling shows that though MAR effect on plasma temperature is small, MAR results in a substantial (about 3 times) reduction in the plasma density (and the plasma flux as well) in front of the plate. Thus, the impact of MAR can be significant and detectable in experiments.

7.2. Divertor Simulators

The plasma density in divertor simulators is lower than the plasma density in tokamak divertors. Therefore, electron-ion recombination becomes important at lower electron temperatures in simulators than in tokamaks. In addition, the plasma and neutral gas flow coupling in a simulators may not be as strong as in tokamaks so that the plasma flux can be described in a diffusive approximation. For a plasma with a temperature 1–2 eV, a density $\sim 3 \times 10^{13}$ cm^{-3} and helium as the working gas with a pressure 10 mTorr, the main ion energy sink is due to ion-neutral interactions, while the electrons are cooling down due to electron-ion collisions. For the conditions we are interested in, the ion-neutral collision frequency is usually higher than the electron-ion equilibration frequency, $\nu = 2(m/M)\nu_{\mathrm{ei}}$, therefore, we can assume that the ion temperature is equal to the neutral gas temperature $T_N = const.$, while T_e can be higher than T_N. Hence we can write the electron energy balance and plasma continuity equations as follows:

$$\partial_x(\kappa_e \partial_x T_e) = (2/3)\nu n_e(T_e - T_N) + (E_{\mathrm{EIR}}\nu_{\mathrm{EIR}} + E_{\mathrm{MAR}}\nu_{\mathrm{MAR}} + E_{\mathrm{ion}}\nu_{\mathrm{ion}})n_e, \quad (5)$$

$$-\partial_x j \equiv \partial_x((M\nu_{\mathrm{iN}})^{-1}\partial_x P) = (\nu_{\mathrm{EIR}} + \nu_{\mathrm{MAR}} - \nu_{\mathrm{ion}})n_e, \quad (6)$$

where $\kappa_e(T_e) = 3.2 n_e T_e / m\nu_{\mathrm{ei}} \propto (T_e)^{5/2}$ is the electron heat conduction coefficient; $\nu_{\mathrm{EIR}} = K_{\mathrm{EIR}} n_e$; $\nu_{\mathrm{MAR}} = K_{\mathrm{MAR}}[H_2]$, and $\nu_{\mathrm{ion}} = K_{\mathrm{ion}} N$ are the electron-ion recombination (EIR), MAR, and neutral ionization frequencies; $[H_2]$ is the molecular hydrogen density; $K_{\mathrm{EIR}}(E_{\mathrm{EIR}})$, $K_{\mathrm{MAR}}\,(E_{\mathrm{MAR}})$, and $K_{\mathrm{ion}}\,(E_{\mathrm{ion}})$ are the rate constants (effective electron energy loss) of EIR, MAR, and neutral ionization. It is to be noted that $E_{\mathrm{ion}} > 0$, while E_{EIR} and E_{MAR} (depending on plasma parameters) can be positive and negative. Analytical consideration of the electron energy balance and plasma continuity equations in assumption of $T_N \ll T_e$, omitting the convective part of the energy flux which is small in the case of subsonic flow, and electron energy loss due to recombination, since typically $\nu_{\mathrm{EIR}} \ll \nu$ shows that the main recombination occurs in the region with $T_e \sim T_* \approx 0.2$ eV (Krasheninnikov et al., 1997).

Before this region the plasma temperature drops sharply, while after it the temperature varies very slowly. The reason for this behavior is the strong dependence of EIR on both plasma temperature and density, and electron cooling due to electron-ion collisions. Indeed, upstream of the recombination region the temperature is higher and recombination is negligible. There the plasma density is relatively large and, therefore, electron energy removal via electron-ion collisions is very efficient causing the electron temperature to drop sharply. As we approach some temperature $T_e \sim T_*$, where the efficiencies of energy and particle transport, and electron cooling and plasma recombination become comparable, plasma recombination sets on. In the process of plasma recombination its density drops and electron cooling via electron-ion collisions becomes inefficient so the electron temperature can only vary slowly. It can be seen that in opposite case, $T_N > T_*$, the electrons cool down to T_N without considerable influence of plasma recombination.

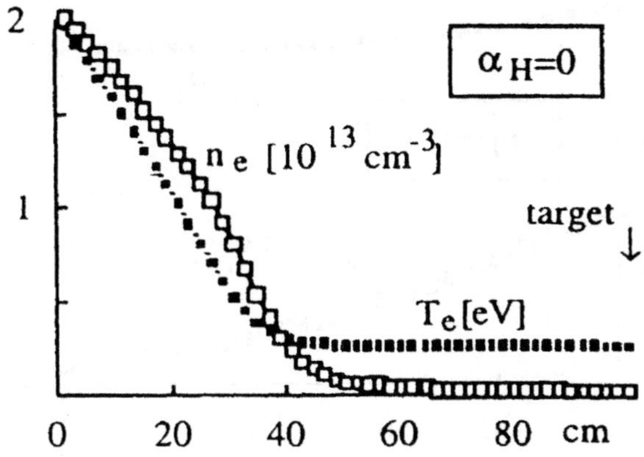

Figure 5. Plasma density and temperature profiles for high (low) $n_u (T_u)$.

Although this simple model does not describe the entire physical picture of the plasma interaction with the neutral gas in divertor simulators, it suggests that the plasma temperature in the recombination region should be about 0.2 eV and have a weak dependence on the experimental conditions. It is interesting that similar temperatures were observed in recombining plasmas in recent experiments on the divertor simulator at Nagoya University (Ezumi et al., 1998).

Let us consider the effects of the MAR. From previous analysis we see that the EIR practically does not affect the plasma parameters at $T_e > T_*$. However, MAR reduces plasma density and slows down electron cooling via electron-ion equilibration. Analysis of Equations (5, 6) for $T_e > T_*$ suggests that the upstream plasma density, n_u, and temperature, T_u, play important roles in the competition of the plasma cooling and recombination due to MAR. It is possible to show that when $\nu_{\rm MAR}$ exceeds the critical value, $(\nu_{\rm MAR})_{\rm crit}$:

$$(\nu_{\rm MAR})_{\rm crit} \approx \left[\frac{\nu_{\rm ei}(n_u, T_u)^2}{\nu_{\rm iN}} (\frac{m}{M})^2 \right],$$

the electron temperature remains higher than T_*, plasma recombines due to MAR, and EIR does not play a significant role. For helium discharge (with neutral gas pressure 10 mTorr, T_u = 1–2 eV, $n_u = 3 \times 10^{13}$ cm^{-3}, and $K_{\rm MAR} = 10^{10}$ cm^3/s) already with 1% of molecular hydrogen in the working chamber (recall $\nu_{\rm MAR} = K_{\rm MAR}[H_2]$) this critical level is exceeded.

To verify the main conclusions of the analytical model we solved Equations (5, 6) numerically. The dependencies of the rate constants and related quantities of EIR, MAR and ionization on T_e and n_e were calculated with the atomic data code CRAMD (Pigarov and Krasheninnikov, 1996). We took $L_{\rm WCh}$ = 100 cm; T_N = 0.1 eV; $N_{\rm He} = 5 \times 10^{13}$ cm^{-3} (5 mTorr pressure); specified the fraction of molecules, $\alpha_H \equiv [H_2]/N_{\rm He}$; electron density. n_u, and temperature, T_u, at the entrance to the working chamber; and applied a sheath boundary conditions at the

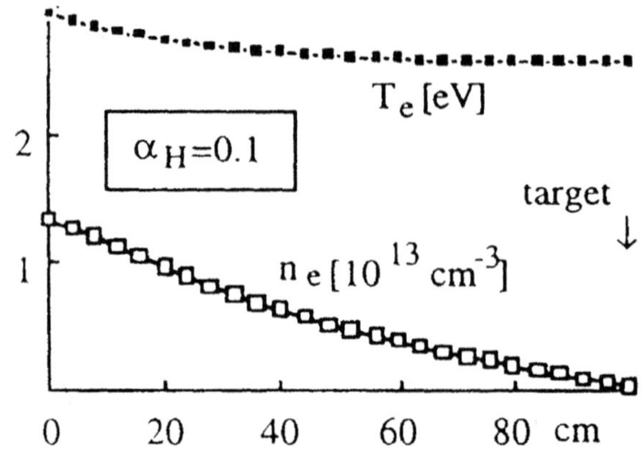

Figure 6. Plasma density and temperature profiles for low (high) $n_u(T_u)$.

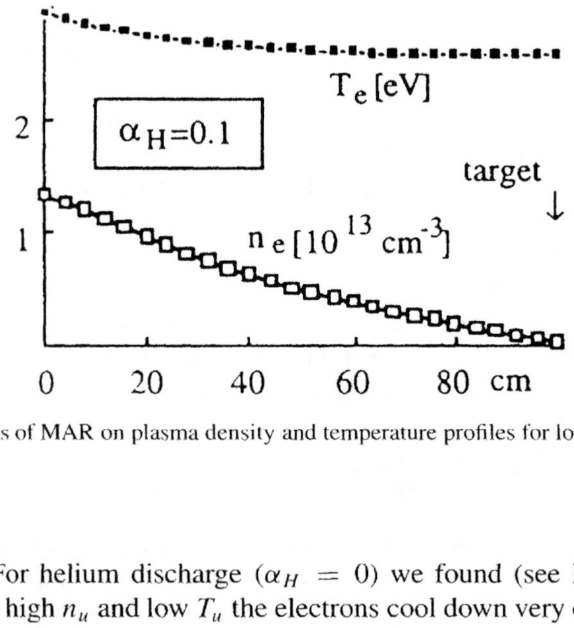

Figure 7. Effects of MAR on plasma density and temperature profiles for low (high) $n_u(T_u)$.

divertor plate. For helium discharge ($\alpha_H = 0$) we found (see Figures 5 and 6) that at relatively high n_u and low T_u the electrons cool down very effectively due to electron-ion coupling,causing fast recombination (Figure 5). In the contrary case of low n_u and higher T_u electron cooling is that efficient and EIR swiches on closer to divertor plates. These results agree qualitatively with analytical estimates, however temperature saturates at a higher level (0.25–0.4 eV) instead of 0.2 eV. This may be due to more acurate account in numerical model of density and temperature dependencies of K_{EIR} as well as boundary conditions. It is worth to mention that at low n_u and high T_u, where EIR is weaker, the influence of MAR can be substantial even for $\alpha_H < 0.1$, since MAR slows down electron cooling (due to reduction of plasma density) and EIR ineffective (compare Figures 6 and 7).

8. Conclusions

We may conclude that during last years considerable progress was achieved in solving the problem of heat and energy removal from large tokamaks. However, the physics of edge plasma is very complicated and for reliable scaling of modern experiment at diverted tokamaks and divertor simulators to ITER-like tokamaks more experimental and theoretical efforts are still needed for better understanding and thus more reliable prediction of the detached regimes.

Analysis of both experimental and modeling results discussed above supports the following conclusions. Plasma recombination in both detached tokamak divertors and divertor simulators is observed and proved to be essential element of plasma detachment. It can cause a strong reduction of plasma flux onto the target and a very sharp variation of plasma parameters at recombination front. In addition to conventional electron-ion recombination, molecular activated recombination can be important for tokamak and divertor simulator experimental conditions. However, so far there is no clear experimental evidence of molecular activated recombination in tokamak and divertor simulator plasmas. Plasma and multispecies neutral gas dynamics are very important in the regimes with strong plasma recombination.

Acknowledgments

This work is supported under CONACyT project 400340-5-241OPE and project ES104393 DGAPA-UNAM at Universidad Nacional Autónoma de México, México.

References

Antonov, N.V., Petrov, V.B. and Pleshakov, A.S. et al.: 1996, *Steady-State Lithium Plasma and Liquid Lithium Divertor Experimental Validation*, Report F1-CN-64/GP-22 at 16th IAEA Conf. on PP and CNF.

Chiu, G.S. and Cohen, S.A.: 1996, *Phys. Rev. Lett.* **76**, 1248.

Evtikhin, V.A., Pistunovich, V.I. et al.: 1996, *Liquid Lithium Tokamak Reactor*, Report F1-CN-64/GP-23 at 16th IAEA Conf. on PP and CNF.

Ezumi, N., Mori, S., Ohno, N., Takagi, M., Takamura, S., Suzuki, H. and Park, J.: 1998, *J. Nucl. Mater.*, to appear.

Hutchinson, I.H. et al.: 1994, *Phys. Plasmas* **1**, 1511.

International Thermonuclear Experimental Reactor: 1996, Preprints, IAEA, 16th IAEA Conf on PP and CNF, Montreal, Canada.

Janeschitz, G. et al.: 1992, in: *Proceedings of 19th European Conference on Controlled Fusion and Plasma Physics, Insbruck, 1992*, European Physical Society, Petit-Lancy, Switzerland, Vol. 16C, Part II, p. 727.

Karasev, B.G., Lavrentjev, I.V., Kolesnichenko, A.F. et al.: 1989, *Fusion Engrg. and Design* **8**, 283–288.

Knoll, D.A., McHuge, P.R., Krasheninnikov, S.I. and Sigmar, D.J.: 1996, *Phys. Plasmas* **3**, 293 and 3358.

Krasheninnikov, S.I.: 1996, *Contribitions to Plasma Phys.* **36**, 293.

Krasheninnikov, S.I. and Pigarov, A.Yu.: 1997, in: *Plasma Physics and Controlled Nuclear Fusion Reasearch 1996*, International Atomic Energy Agency, Vienna, Vol. 1, p. 387.

Krasheninnikov, S.I. and Soboleva, T.K.: 1996, *Phys. Plasmas* **3**, 2280.

Krasheninnikov, S.I., Catto, P.J., Helander, P., Sigmar, D.J. and Soboleva, T.K.: 1995, *Phys. Plasmas* **2**, 2717.

Krasheninnikov, S.I., Pigarov, A.Yu. and Sigmar, D.J.: 1996, *Phys. Lett. A* **214**, 285.

Krasheninnikov, S.I., Pigarov, A.Yu., Knoll, D.A., LaBombard, B., Lipschultz, B., Sigmar, D.J., Soboleva, T.K., Terry, J.L. and Wising, F.: 1997, *Phys. Plasmas* **4**(5), 1638.

Kukushkin, A.: 1996, *Analysis of the Performance of the ITER Divertor and Analysis of the ITER Tokamak Edge Parameter Database*, Report FP-27 at 16th IAEA Conf on PP and CNF, Montreal, Canada.

Kukushkin, A. et al.: 1006, in: *Proceedings of 16th International Atomic Energy Agency Fusion Energy Conference*, Montreal, Canada, 7–11 October 1996, IAEA-CN-64/FP-27.

Loarte, A.: 1997, *J. Nucl. Mater.*, to appear.

Lumma, D., Terry, J.L. and Lipschultz, B.: 1998, *Nucl. Fusion*, submitted.

Mertens, V. et al.: 1993, in: *Proceedings of 20th European Conference on Controlled Fusion and Plasma Physics, Lisboa, 1993*, European Physical Society, Petit-Lancy, Switzerland, Vol. 17C, Part I, p. 267.

Nektarov, E.S., Petrov, V.B., Sychev, V.V., Khripunov, B.I., Shapkin, V.V. and Sholin, G.V.: 1996, *Plasma Phys. Rep.* **22**, 431.

Niemczewski, A., Hutchinson, I.H., LaBombard, B., Lipschultz, B., and McCracken, G.: 1997, *Nucl. Fusion* **37**(2), 151.

Ohno, N. et al.: 1995, *J. Nucl. Mater.* **220–222**, 279.

Petravic, M.: 1994, *Phys. Plasmas* **1**, 2207.

Petrie, T.W. et al.: 1992, *J. Nucl. Mater.* **196–198**, 848.

Pigarov, A.Yu. and Krasheninnikov, S.I.: 1996, *Phys. Lett. A* **222**, 251.

Pigarov, A.Yu., Stotler, D.P. and Karney, C.F.F.: 1995, *Bull. Amer. Phys. Soc.* **40**, 1884.

Post, D.E.: 1995, *J. Nucl. Mater.* **220–222**, 143.

Schmitz, L. et al.: 1995, *J. Nucl. Mater.* **220–222**, 279.

Wagner, F.: 1997, *Plasma Phys. Control Fusion* **39**, A23–A50.

Wising, F., Knoll, D.A., Krasheninnikov, S.I., Rognlien, T.D. and Sigmar, D.J.: 1996, *Contribitions to Plasma Phys.* **36**, 136.

IMPROVED APPROXIMATED SOLUTIONS TO THE NONLINEAR DEBYE POTENTIAL

R.J. PÉREZ
Instituto Universitario Tecnológico Región Capital, Apartado 40347, Caracas 1040 A, Venezuela

P. MARTÍN
Universidad Simón Bolívar, Dpto. de Física, Apdo. 89000, Caracas 1080A, Venezuela

Abstract. The shielding effect of plasma on the electrostatic potential due to a central nucleus is studied. The potential is determined by the Poisson equation with Boltzmann factor for the charge density of ions. The aim of this work is to find approximated analytical solutions, without the usual linearization hypothesis, by means of the technique of quasifractional approximants in two points. In a previous work (Martín and Pérez, 1994) a quasifractional fractional approximation was obtained, but the coefficients were found by means of the numerical solution of the differential equation. Now we find a complete analytical expression. The approximant found is highly precise in the range of validity of the physical model used.

1. Introduction

The shielding effect of plasma on the electrostatic potential due to a central nucleus is studied. We assume that each ion of charge ze and each electron of the cloud surrounding the nucleus of charge $Z_1 e$ interact only via a mean potential $\varphi(r)$. This mean potential is due to the effect of the potential of central charge and that of the equilibrium distribution of the ions surrounding the charge and the electrons that form a uniform neutralizing background. Quantum effects are neglected, therefore the potential $\varphi(r)$ is determined by the Poisson equation with Boltzmann factor for the charge density $\rho(r)$ of ions.

$$\nabla^2\varphi(r) = 4\pi zen_0\left[1 - \exp\left(-ze\varphi(\mathbf{r})/kT\right)\right] - 4\pi Z_1 e\delta(\mathbf{r}). \tag{1}$$

This is a highly nonlinear equation, but the usual solutions presented for this problem start with the linearization of this equation, based in the hypothesis that the electrostatic energy is much smaller than the thermal energy. This assumption does not apply in all the real systems. The aim of this work is to find approximated analytical solutions of Equation (1), without the usual linearization hypothesis, by means of the technique of quasifractional approximants in two points (Martín and Baker, 1991). In a previous work (Martín and Pérez, 1994) a quasifractional approximant was obtained, but the coefficients had to be determined by numerical solution of the differential equation. Now we make an improvement finding a

Astrophysics and Space Science **256:** 263–268, 1998.

complete analytical expression very precise in the range of validity of the physical model used.

2. Statement of the Problem

The potential $\varphi(r)$ can be expressed as the superposition of central nucleus potential and the potential due to the charges around the nucleus. That is $\varphi(r) = (Z_1 e/r) + V(R)$. Non-dimensional quantities Y, x, and F are defined as $Y(r) = \frac{zeV(r)}{kT}$, $x = (r/\lambda_D)$ and $F = (Z_1 ze^2)/(\lambda_D kT)$ where $\lambda_D = \sqrt{(kT)/(4\pi n_0(ze)^2)}$ is the Debye length. Here F is a measure of the electrostatic energy relative to the thermal energy of the plasma, that is, F is the coupling parameter. With these definitions, and taking in account that problem has spherical symmetry, Equation (1) takes the form

$$\frac{1}{x^2}\frac{d}{dx}\left(x^2\frac{dY}{dx}\right) = 1 - \exp\left(-\frac{F}{x} - Y(x)\right), \tag{2}$$

where $Y(x)$ must satisfy the following boundary conditions:

$$Y(0) = c_0 \quad \text{and} \quad \text{when } x \to \infty,\ Y(x) \sim -F/x. \tag{3}$$

3. Method of Solution

The solution of this equation is obtained by means of the transformation $Y(x) = (F/x)\tilde{y}(x)$. This leads to the equation

$$\tilde{y}''(x) = \frac{x}{F}\left[1 - \exp\left(-\frac{F}{x}(1 + \tilde{y}(x))\right)\right]. \tag{4}$$

To construct an approximant valid for all x in the interval $(0, \infty)$ it is necessary to find the solution of the differential equation in the limits of $x \ll F$ and $x \gg F$. Thus, it is convenient to define the variable s and $y(s)$ as $s = x/F$ and $y(s) = \tilde{y}(Fs)$ obtaining

$$y''(s) = F^2 s\left[1 - \exp\left(-\frac{1 + y(s)}{s}\right)\right] \tag{5}$$

with the following boundary conditions:

$$y(s \to \infty) = -1, \quad y(s \to 0) = 0 \quad \text{and} \quad y'(s \to 0) = Y(0) = c_0. \tag{6}$$

For small values of the variable s Equation (5) can be approximated by $y_1''(s) = F^2 s$ whose solution, after impose the boundary conditions, is

$$y_1(s) = \frac{F^2 s^3}{6} + c_0 s. \tag{7}$$

On the other hand, when $s \to \infty$, Equation (5) may be approximated by $y_2''(s) = F^2(1 + y_2(s))$ whose solution is

$$y_2(s) = -1 + c_3 \exp(-Fs). \tag{8}$$

Now we have the necessary information to propose a quasifractional approximant $y_a(s)$ for the exact solution $y(s)$ of Equation (5). The expression for $y_a(s)$ should be chosen in such a way that in the limits $s \ll 1$ and $s \gg 1$ the approximant tends to the functions $y_1(s)$ and $y_2(s)$ respectively. Because of these requirements we propose the form

$$y_a(s) = \frac{1 + p_1 s + p_2 s^2 + p_3 s^3}{(1+s)^3} \exp(-Fs) - 1, \tag{9}$$

where the denominator, in agreement to prior experiences, has been prescribed in such a way that the approximant is free of singularities in the interval $0 < s < \infty$. To find the coefficients p_1, p_2 and p_3 we make third order series expansion around $s = 0$ to the approximant $y_a(s)$. Comparing this series with the solution $y_1(s)$ we obtain:

$$p_1 = 3 + F + c_0, \tag{10}$$

$$p_2 = 3 + 3F + \frac{1}{2} F^2 + (3 + F)c_0, \tag{11}$$

$$p_3 = 1 + 3F + \frac{5}{3} F^2 + \frac{1}{6} F^3 + \left(3 + 3F + \frac{1}{2} F^2\right) c_0. \tag{12}$$

The p's coefficients depend on the coupling parameter F and the constant c_0.

Also it is possible to find a quasifractional approximant for c_0. The procedure consists of finding solutions of Equation (2) in the limit cases $F \to 0$ and $F \to \infty$. When $F \to 0$ Equation (4) can be approximated by $\tilde{y}''(x) = 1 + \tilde{y}(x)$ whose solution is $\tilde{y}(x) = \exp(-x)$ and $Y(x) = (F/x)(\exp(-x) - 1)$ that corresponds to the usual linear solution. Therefore, from this solution and from Equation (8) we deduce that $Y(0) = c_0 \approx -F$ when $F \to 0$.

On the other hand, when x is finite and $F \to \infty$, $x/F = s \to 0$, that corresponds to the solution given in Equation (7). In view of this $Y(x) \approx Y_a(x) = (1/6)x^2 + c_0$. But when $x \to \infty$ the asymptotic boundary condition for this limit may be fulfilled, therefore $Y(x) \sim Y_b(x) = -F/x$. The solutions Y_a and Y_b match in some $x = x^*$. With the equations $Y_a(x^*) = Y_b(x^*)$ and $Y_a'(x^*) = Y_b'(x^*)$ the point x^* and the constant c_0 are determined $c_0 = -(3^{2/3}/2)F^{2/3}$ when $F \to \infty$. This result was previously reported (Salpeter, 1954).

Now with the limit values of c_0 for $F \to 0$ and for $F \to \infty$ we propose the following quasifractional approximant for any value of F:

$$\tilde{c}_0 = -F \Big/ \left(1 + \left(\frac{8}{9} F\right)^{\alpha}\right)^{1/3\alpha},$$

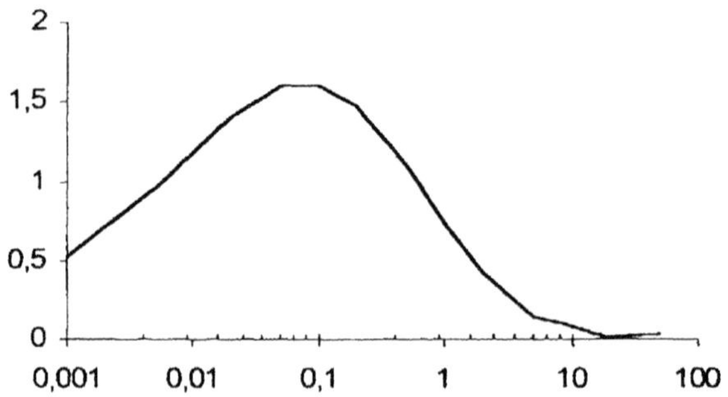

Figure 1. Percentual difference between the approximant $y_a(0)$ and the numerical solution $y(0)$ as a function of the coupling parameter F.

where α is an optimization parameter. Using the minimum squares criteria respect to the numerically determined values of c_0, we have that the optimum value for α is 3/5.

4. Comparison between Analytical and Numerical Results

To evaluate the exactness of the quasifractional approximants proposed in this work it is necessary to compare it with the corresponding numerical solutions. The numerical solution of Equation (5) was found by means of the shooting method. In this way we find the initial condition value of $y'(0) = c_0$. that satisfies the boundary conditions $y(0) = 0$ and $y(s \to \infty) = -1$. For the error analysis shown below, we introduce the approximant $\widetilde{c}_0$ in the approximant $y_a(s)$. In this work we assume that the error of the approximation is the difference between it and the corresponding numerical solution.

Figure 1 shows the percentual difference between the approximant $y_a(0)$ and the numerical solution $y(0)$ as a function of the coupling parameter F. The numerical solution was obtained using a fourth order Runge–Kutta algorithm to Equation (5) and changing the first derivative in the origin until the boundary conditions given by Equation (6) is satisfied. In this way we determine numerically $y(0)$ with six significant digits. As seen in Figure 1, the difference between the approximant $\widetilde{c}_0 = y_a'(0)$ and the $y'(0)$ numerically obtained tends to be very small for large and small values of F, and the greatest difference is 1.6% for $F = 0.08$.

Figure 2 shows the maximum errors of the approximant y_a (continuous line) and the linearized usual solution (dashed line) as a function of the coupling parameter F. The maximum error for the approximant y_a is 4.4% for $F = 11$, in the usual linearized solution the error increase, without bound, as F increase. The physical model used in this work is applicable for F below 5, for values above this one correlation between particles should be included, and the mean field model used here is not valid. According to Vieillefose (1981), for $F = 5$ the difference of the

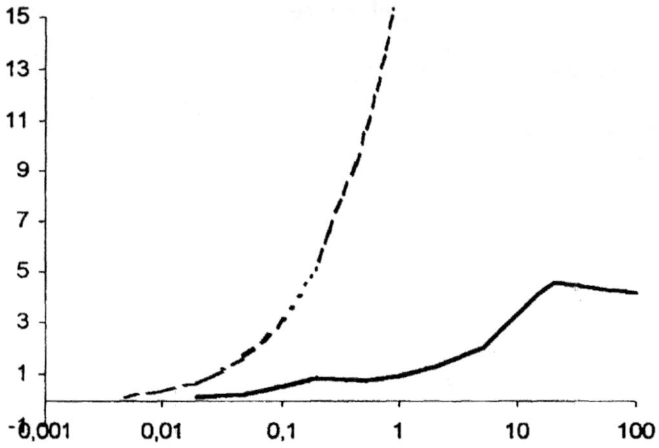

Figure 2. Maximum errors of the approximant y_a (continuous line) and the linearized solution (dashed line) as a function of the coupling parameter F.

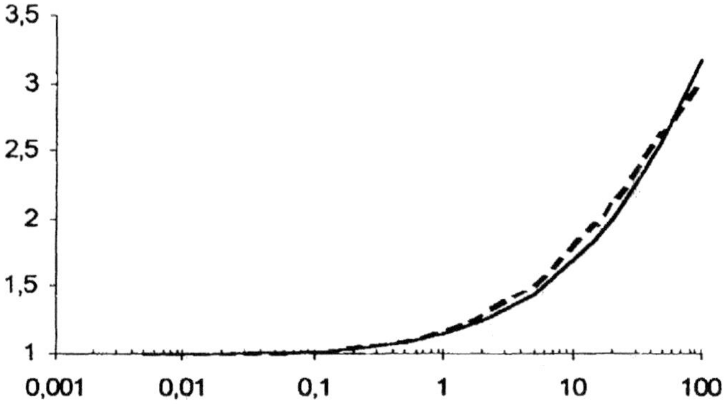

Figure 3. Effective screening length x_{eff} normalized to the Debye length as a function of the coupling parameter.

non-linear numerical solution with the simulations that consider particle correlation is about 5.5%. For $F = 5$ the maximum error or our approximant is 2%. This assures that the approximant here found is enough accurate in the applicability range of the model.

Figure 3 shows the effective screening length x_{eff} normalized to the Debye length as a function of the coupling parameter. x_{eff} is defined as the point where the function $x\varphi(x)$ decreases to $1/e$ the value at $x = 0$. In the usual linearized solution $x_{\text{eff}} = 1$ for any value of F, but Figure 3 shows that x_{eff} depends on the coupling parameter for the nonlinear solution. Dashed line shows the values calculated with our approximant and continuous line shows the values calculated numerically. Note that for $F = 1$ and $F = 5$ the effective screening length is 1.2 and 1.5 times respectively the usual Debye length.

References

Martín, P. and Baker, G.: 1991, *J. Math. Phys.* **32**, 1470.
Martín and Pérez, R.: 1994, in: *Proceedings of the 1994 ICPP*, Vol. 2, p. 103.
Salperter, E.E.: 1954, *Australian J. Phys.* **7**, 353.
Vieillefose, P.: 1981, *J. de Physique* **42**, 723.

ENTROPY AND ENTROPY PRODUCTION IN TRANSITION FROM WEAK TO STRONG TURBULENT PLASMA STATE

S.I. POPEL
Institute for Dynamics of Geospheres, Leninsky pr. 38, Building 6, Moscow 117979, Russia

Abstract. The transition from a weakly turbulent plasma state to a strongly turbulent one is considered using the concept of entropy. It is shown that this transition can be treated as a nonequilibrium phase transition.

The modulational interaction is the key one in transition of a plasma system from the weak turbulent state to the strong turbulent one (Vladimirov et al., 1995). This transition is considered often as a process of self-organization (see, e.g., Tsytovich, 1995; Popel, 1997). For the case of this transition the degree of order of the plasma system has not been considered using the concept of entropy (Klimontovich, 1991). We emphasize the so-called principle of minimum entropy production in processes of self-organization (Klimontovich, 1991). In accordance with this principle the process of self-organization in nonequilibrium media is effected by way of a nonequilibrium phase transition, the system transforming into the state with lower value of the entropy production. We emphasize that the spectrum of the random waves in a plasma is *nonequilibrium*. We note that for the particular case of a decay of the system into a soliton and free (random) waves the transition from the weak turbulent state to the strong turbulent one has been treated as the usual (equilibrium) phase transition (Petviashvili and Yan'kov, 1985). This conclusion has been made under assumption that the wave spectrum is in the state of thermodynamical equilibrium. However, this assumption is not valid. Here we study the transition, considering the plasma system as a self-organizing, nonequilibrium one.

The entropy of a gas described by the distribution function $f_{\mathbf{p}}(\mathbf{r}, t)$ is determined as follows (see, e.g., Klimontovich, 1991):

$$S = -\int f_{\mathbf{p}}(\mathbf{r}, t) \ln f_{\mathbf{p}}(\mathbf{r}, t) \frac{d\mathbf{r}d\mathbf{p}}{(2\pi)^3}, \tag{1}$$

where the distribution function is normalized as $n(\mathbf{r}, t) = \int f_{\mathbf{p}}(\mathbf{r}, t) d\mathbf{p}/(2\pi)^3$ and $n(\mathbf{r}, t)$ is the density of particles. Consider the case of a collisionless plasma in the absence of external fields. In this case the main interaction determining the evolution of the distribution function (and of the entropy) is the resonant interaction of the particles with the waves (Popel, 1998).

Astrophysics and Space Science **256:** 269–272, 1998.

Let us introduce the notion of the "old" state which corresponds to the weak turbulence characterized by the spectrum of random Langmuir waves $W^l_{\mathbf{k}}(\mathbf{r}, t)$ such that their energy density is $W^l(\mathbf{r}, t) = \int W^l_{\mathbf{k}}(\mathbf{r}, t)d\mathbf{k}$. As a result of pumping, the energy density of the random waves increases. It is known (Vladimirov et al., 1995) that there is some threshold value of the energy density of the random waves. We suppose that this threshold value corresponds to the spectrum $W^{(0)}_{\mathbf{k}}(\mathbf{r}, t)$. When the energy density of the random waves exceeds the threshold value, the regular fields are generated in the plasma due to the modulational interaction. The system transforms into the "new" (strong turbulent) state which is characterized by the presence in the plasma of the regular fields in addition to the random ones. We describe the regular fields by the amplitude of their envelope $\mathcal{E}(\mathbf{r}, t)$. Let us introduce also the notion of the extended "old" state which means a hypothetical situation when the system is in the weak turbulent state (the regular fields are absent) but the energy density of the random waves exceeds the threshold value.

To determine the relative degree of order of the weak and strong turbulent states it is necessary to choose the control parameter which characterizes the transition of the system from the "old" state to the "new" one. Furthermore, one has to determine the difference in the entropy of the system in the extended "old" and "new" states under the assumptions of the same magnitudes of the control parameter and of the same values of the mean effective energy of the system in these states. The "new" state is the more ordered one if the entropy difference is positive. This statement is an analogue of the S-theorem (see Klimontovich, 1991) for the situation considered where the system is characterized by the distribution function of the particles and the spectra of the random and regular fields.

It is natural to choose the function which determines the threshold of the modulational interaction as the control parameter. The threshold condition for the random Langmuir waves can be presented in the following form $W^l/n_0(T_e + T_i) > 12\Lambda k^2_{\text{ch}} r^2_{De}$, where n_0 is the unperturbed electron density, $T_{e(i)}$ is the effective temperature of electrons (ions), $\Lambda \sim 1$ is a constant depending on the spectrum, k_{ch} is the characteristic length of the wave vector in the spectrum, r_{De} is the electron Debye length. Thus we choose the function $W^l/\Lambda k^2_{\text{ch}} T_e(T_e + T_i)$ as the control parameter.

Let us assume that the particle distribution in momentum-space can be approximated by the Maxwellian one with the effective temperature T_e. The entropy can be expressed as follows

$$S = \frac{3}{2} N - \int n \ln \left[\frac{(2\pi)^{3/2} n}{(mT_e)^{3/2}} \right] d\mathbf{r}, \tag{2}$$

where N is the total number of electrons in the plasma and m is the electron mass.

Assume also that the ions do not participate in the resonant interaction. In this case the total energy of the electrons and the waves conserves:

$$\int \frac{d\mathbf{r}d\mathbf{p}}{(2\pi)^3} \frac{p^2}{2m} f_{\mathbf{p}} + W^{\text{reg}} + \int d\mathbf{r}d\mathbf{k} W^l_{\mathbf{k}}(\mathbf{r}, t) = C_0, \tag{3}$$

where C_0 is a constant, which characterizes the mean effective energy of the system, and $W^{\text{reg}} \equiv \int d\mathbf{r}|\mathcal{E}(\mathbf{r},t)|^2/2\pi$ is the energy of the regular fields. Under the above assumptions the magnitudes of C_0 should be coincident in both the extended "old" state and the "new" one, i.e.

$$\frac{3}{2}NT_e^{\text{new}} + W^{\text{reg}} + \int d\mathbf{r}W^{l(\text{new})}(\mathbf{r},t) = \frac{3}{2}NT_e^{\text{old}} + \int d\mathbf{r}W^{l(\text{old})}(\mathbf{r},t). \tag{4}$$

Here, the superscripts "new" and "old" are related to the "new" state and the extended "old" one, respectively; the temperature $T_e^{\text{new(old)}}$ is supposed to be independent of $\mathbf{r}$.

We only consider the near-threshold situation when the difference between the energy density of the random waves and its threshold value is significantly less than the threshold one. We have $|\delta T_e| \ll T_e^{\text{new}}$, where $\delta T_e = T_e^{\text{new}} - T_e^{\text{old}}$. The modulational interaction results in a perturbation of the electron density $n = n_0 + \delta n$ ($\int \delta n d\mathbf{r} = 0$). We take into account only the terms linear in δT_e and δn in the expansion of Equation (2) in powers of δT_e and δn. We assume also that the spectra of the random waves in the "new" state and in the extended "old" one differ by the factor α ($W_{\mathbf{k}}^{l(\text{new})} = \alpha W_{\mathbf{k}}^{l(\text{old})}$). The latter assumption allows us to consider the same magnitudes of Λ and k_{ch} in both these states. Using the condition (4) we obtain

$$S_{\text{old}} - S_{\text{new}} \approx -\frac{3}{2}N\frac{\delta T_e}{T_e} \approx \frac{W^{\text{reg}}}{T_e}\,. \tag{5}$$

Here, the index "new" which characterizes T_e is omitted. The relationship between δT_e and W^{reg} used in Equation (5) implies that $\int d\mathbf{r}W^{l(\text{old})} \ll NT_e$.

Let $\delta W^l > 0$ denote the difference between the energy of the random waves in the strong turbulent state and their energy, corresponding to the threshold. Express the right-hand side of Equation (5) in terms of δW^l. We use the equations relating the spectrum of the random waves and the amplitude of the regular field envelope. For the case, when one can neglect the ion thermal motion in description of low-frequency modulational perturbations and the spatial scale of the random fields is significantly less than that of the regular ones, these equations are given in (Tsytovich, 1995, pp. 278–279). We express $\delta W_{\mathbf{k}_1}^l(\mathbf{r},t)$, where $\int d\mathbf{r}d\mathbf{k}\delta W_{\mathbf{k}}^l(\mathbf{r},t) = \delta W^l$, in terms of $\mathcal{E}(\mathbf{r},t)$ with the aid of these equations using the Fourier transform. We obtain

$$\begin{aligned}\delta W^l &= \int d\mathbf{r}d\mathbf{k}d\omega\frac{\left[|\mathcal{E}|^2\right]_{\omega,\mathbf{k}}}{2\pi}e^{-i\omega t+i\mathbf{k}\cdot\mathbf{r}}\\ &\times\left[\frac{k^2v_s^2}{k^2v_s^2-\omega^2}\int d\mathbf{k}_1\frac{\omega_{pe}}{\omega-\mathbf{k}\cdot\mathbf{v}_{\text{gr}}(\mathbf{k}_1)}\left(\mathbf{k}\cdot\frac{\partial}{\partial\mathbf{k}_1}\right)\frac{W_{\mathbf{k}_1}^{(0)}}{4n_0T_e}\right],\end{aligned} \tag{6}$$

where ω_{pe} is the electron plasma frequency, $\mathbf{v}_{\text{gr}}(\mathbf{k})$ is the group velocity of Langmuir waves as a function of the wave vector, $v_s = (T_e/m_i)^{1/2}$ is the ion acoustic speed, m_i is the ion mass, $\left[|\mathcal{E}|^2\right]_{\omega,\mathbf{k}}$ denotes the Fourier component of the function $|\mathcal{E}|^2$. We note that in the case, when $k \ll k_1$ (that corresponds to this consideration) and one can neglect the ion thermal motion, the term in the square brackets of the expression (6) equals unity. This is a consequence of the dispersion relationship for the modulational instability in this case (see Vladimirov et al., 1995, p. 497). Thus we obtain $W^{\text{reg}} = \delta W^l > 0$ and $S_{\text{old}} - S_{\text{new}} \approx \delta W^l / T_e > 0$. We see that the "new" (strong turbulent) plasma state is the more ordered one than the weak turbulent plasma state.

Let us verify the validity of the principle of minium entropy production in processes of self-organization. For this one has to compare expressions for the entropy production $\sigma = dS/dt \geq 0$ in the extended "old" state and in the "new" one under the above assumptions which concern the control parameter and the effective energy of the system. In accordance with the principle the entropy production in the new state, which is formed as a result of the nonequilibrium phase transition, is less than the entropy production in the old (but extended into the unstable region) state (Klimontovich, 1991).

We obtain (Popel, 1998) $\sigma_{\text{new}} \approx \sigma_{\text{old}} - 2\Gamma W^{\text{reg}}/T_e < \sigma_{\text{old}}$, where $\sigma_{\text{new(old)}}$ is the entropy production in the "new" (extended "old") state, and $\Gamma > 0$ is the characteristic growth rate of the modulational instability. The latter inequality means the validity of the principle of minimum entropy production in processes of self-organization for the case of transition from the weak turbulent state to the strong turbulent one. It gives also the possibility of the treatment of this transition as a nonequilibrium phase transition which transforms the system into the more ordered state.

References

Klimontovich, Yu.L.: 1991, *Turbulent Motion and Structure of Chaos*, Kluwer Academic Publishers, Dordrecht.

Petviashvili, V.I. and Yan'kov, V.V.: 1985, in: *Reviews of Plasma Physics*, Consultants Bureau, New York, Vol. 14.

Popel, S.I.: 1997, *J. Plasma Phys.* **57**, 363.

Popel, S.I.: 1998, *Phys. Scripta* **57**, 272.

Tsytovich, V.N.: 1995, *Lectures on Non-Linear Plasma Kinetics*, Springer-Verlag, Berlin.

Vladimirov, S.V., Tsytovich, V.N., Popel, S.I. and Khakimov, F.Kh.: 1995, *Modulational Interactions in Plasmas*, Kluwer Academic Publishers, Dordrecht.

NONLINEAR PROCESSES IN LOWER-HYBRID CURRENT DRIVE AND PLASMA HEATING

S.I. POPEL
Institute for Dynamics of Geospheres, Leninsky pr. 38, Building 6, Moscow 117979, Russia

Abstract. Nonlinear processes in lower-hybrid current drive and plasma heating are considered. An emphasis is made on the role of the modulational interactions.

Here we illustrate how the nonlinear evolution of the lower-hybrid (LH) wave spectra effects LH current drive generation and plasma heating. Current drive appears as a result of momentum transfer to fast electrons when the waves are absorbed according to the Landau mechanism. Investigations of LH current drive are important for the tokamak program because they clear the way to a stationary tokamak reactor.

One often uses the treatment of LH current drive (Fisch, 1987) which considers the quasilinear interactions and effects of binary collisions between plasma particles. However, there is a problem of comparison of the obtained theoretical results and experimental data. Namely, the experimentally observed currents driven by LH waves are two or three orders of magnitude larger than those calculated using the theoretical approach (Fisch, 1987). The above discrepancy is related to the existence of the so-called "spectral gap" (see Fisch, 1987, p. 218), that is the gap between the parallel phase velocities of LH waves necessary for effective Landau damping of LH waves (which are of the order of several electron thermal velocities v_{Te}) and the lowest parallel phase velocity in the spectrum of LH waves injected in a plasma (which significantly exceeds v_{Te}). One of the possible explanations of how the "spectral gap" filling occurs is based on the consideration of the modulational instability (Popel and Tsytovich, 1992; Vladimirov et al., 1995; Popel et al., 1995). We note that in (Popel and Tsytovich, 1992; Vladimirov et al., 1995; Popel et al., 1995) the influence of the modulational instability of monochromatic pumping wave (with fixed frequency and wave vector) has been considered. However, in reality LH wave spectrum is rather wide in **k**-space. The study of packets of LH waves (Popel, 1997) allows us to conclude that the "short-scale" modulational instability, when the length of the waves in the spectrum exceeds significantly the wave length of the modulational perturbations $2\pi/|\mathbf{k}'|$ and the width (in **k**-space) of the wave spectrum δk is less than $|\mathbf{k}'|$, develops analogously to the case of monochromatic pumping. Since it is the case of "short-scale" modulational instability

Astrophysics and Space Science **256:** 273–276, 1998.

that is important for description of the "spectral gap" filling, the above conclusion shows the validity of approximation of monochromatic pumping in application to description of the "spectral gap" filling.

To illustrate how the modulational instability can result in the "spectral gap" filling we should account for the appearance of LH waves with $\omega_{\mathbf{k}}/k_{\parallel} \ll \omega_0/k_{0\parallel}$, where the subscript "0" is related to the monochromatic pumping wave. The significant change in the wave vector may be a result of the modulational instability in the case $|\mathbf{k}'| \gg |\mathbf{k}_0|$. We consider the excitation (due to the modulational instability) of LH waves with parallel phase speeds less or approximately equal to $4.5 v_{Te}$ (i.e., with $|\mathbf{k}| \geq (\omega_0/4.5 v_{Te} \cos\theta_0)$).

For pump levels which are given by

$$\frac{W^{LH}}{4n_0T_e} \gg \left(1 + \frac{\omega_{Be}^2}{\omega_{pe}^2}\right)\sqrt{\frac{m_e}{m_i}}\cos\theta_0, \tag{1}$$

where W^{LH} is the energy density of LH waves, n_0 is the unperturbed plasma density, T_e is the electron temperature, ω_{Be} is the electron gyrofrequency, and ω_{pe} is the electron plasma frequency, LH waves with sufficiently low parallel phase velocities $|\omega/k_{\parallel}|_{\min} \approx |\omega_0/|\Delta\mathbf{k}|_{\max}\cos\theta_0| \leq 3.5 v_{Te}$ can appear. The above result $|\omega/k_{\parallel}|_{\min} \leq 3.5 v_{Te}$ is sufficient to explain LH current drive experimental data. Thus sufficiently intensive LH pumping can provide the "spectral gap" filling by LH waves.

If the LH pump wave is moderate

$$\frac{W^{LH}}{4n_0T_e} \ll \left(1 + \frac{\omega_{Be}^2}{\omega_{pe}^2}\right)\sqrt{\frac{m_e}{m_i}}\frac{\cos\theta_0}{4.5}, \tag{2}$$

then the most important nonlinear effect affecting the LH wave spectrum is the induced scattering (Popel and Tsytovich, 1992).

Above we have considered the influence of the modulational instability of LH waves on current drive generation taking into account only the process with participation of exactly longitudinal virtual fields. In this case the modulational interaction affects the evolution of the electron distribution function (and, correspondingly, the density of LH current drive, efficiency of its generation, etc.) only through the change in the LH wave spectrum. However, the modulational processes with participation of non-longitudinal virtual fields are also possible. Such processes can result in excitation of magnetic field perturbations in plasmas. It is such magneto-modulational processes that significantly influence the efficiency of LH current drive generation for the case of sufficiently strong LH wave pumping (Popel and Elsässer, 1994).

The efficiency of the steady-state LH current drive generation varies (Fisch, 1987) in inverse proportion to the electron collision frequency. Magnetic field perturbations (like binary collisions) tend to randomize the velocity of electrons interacting with LH waves, and as a result these electrons leave the resonant region (the region in the velocity-space where Čerenkov resonance condition is fulfilled). This is caused by scattering of electrons on magnetic field perturbations which results in a change in direction of electron velocity and not in the energy of electron; hence, the component $v_\parallel$ of electron velocity is also changed. Consequently, the presence of chaotic component of magnetic field leads to an increase in effective electron collision frequency. Thus one can expect that the effects of magnetic field perturbations result in a decrease in the efficiency of LH current drive generation.

Under the assumption that $\cos^2\theta_0 \ll B_{0\theta}/B_{0\varphi}$ the effects of magnetic field perturbations are dominant (over binary collision effects) if the following inequality is fulfilled (Popel and Elsässer, 1994):

$$\frac{W^{LH}}{n_0 T_e} \gg \left(\frac{1}{N_D}\right)^{1/2}\left(\frac{c}{v_{\rm ph}}\right)^{3/2}\left(\frac{c}{v_{Te}}\right)^{1/2}\left(\frac{\omega_{Be}}{\omega_{pe}}\right)^{3/2}\frac{B_{0\varphi}}{B_{0\theta}}\cos\theta_0. \tag{3}$$

Here $B_{0\theta}$ is the poloidal component of the magnetic field, $B_{0\varphi}$ is the toroidal one, c is the speed of light, N_D is the number of electrons in the sphere of radius of electron Debye length.

Calculation of the density of steady-state LH current drive (j_z) and the LH wave power absorbed by plasma electrons (P_d) allows us to calculate the efficiency of LH current drive generation ($\eta \equiv j_z/P_d$)

$$\eta \sim \frac{|e|\cos^2\theta_0}{m_e\omega_{Be}}\left(\frac{B_{0\varphi}}{B_{0\theta}}\right)^2\left(\frac{n_0 T_e}{W^{LH}}\right)^2\left(\frac{c}{v_{Te}}\right)^4\left(\frac{\omega_{Be}}{\omega_{pe}}\right)^4\frac{1}{v_{\rm ph}}. \tag{4}$$

Efficiency (4) is considerably less than the one-dimensional efficiency (Fisch, 1987) calculated for the case when the steady-state current drive is maintained as a result of a balance between the quasilinear and collisional effects. We see that excitation of magnetic field perturbations by LH waves results in a decrease in the efficiency of LH current drive generation and in this sense this phenomenon is negative for the process of current drive generation. We also note that even if condition (3) is fulfilled for the parameters averaged over plasma volume, then there is the region at the center of plasma column where condition (3) is not satisfied (because the ratio $B_{0\varphi}/B_{0\theta}$ at the centre of plasma column is always very large). This means that in the central region of plasma column the steady-state current is maintained as a result of a balance of the effects of binary collisions and the quasilinear effects. While outside this region current drive is generated due to competition of the effects of magnetic field perturbations and the quasilinear effects, the efficiency of its generation in the outer region being considerably less than in the central one. Such different character of the steady-state current drive generation in different regions of a plasma can affect the profile of current drive.

In the process of LH current drive generation the modulational interactions can also result in consequences like those taking place in the process of LH plasma heating. For example, density profile modification caused by the ponderomotive force influence (Petržilka, 1991) can lead to the nonlinear generation of higher spatial harmonics in the LH wave spectrum and to the change in the coefficient of the electromagnetic wave reflection.

The LH wave modulational interaction is the main effect resulting in the drift wave excitation in a tokamak plasma under the conditions of experiments on LH plasma heating and current drive (Benkadda et al., 1996). The modulational excitation of long-wavelength drift oscillations (with the wave lengths exceeding the length of the LH pump wave) is described by a dispersion equation similar to the equation for the usual hydrodynamical beam instability. This allows us to treat the excitation of the long-wavelength drift waves as modulational excitation by a beam of the LH waves. The proposed theoretical scheme (Benkadda et al., 1996) has allowed us to explain the nonlinear increase of the density fluctuation level with a rise in the injected LH power, which is in qualitative accordance with the data of experiment on Tore Supra.

Finally, the LH wave modulational interaction can be important in description of the broadening of the injected radiofrequency line in the frequency spectra detected during some LH experiments (see, e.g., Cesario et al., 1992). The main problem to describe theoretically this broadening is to find some nonlinear process which is strong sufficiently to exceed the convective threshold. The LH wave modulational instability can be applied to explain this effect. Indeed, estimates show that for typical data of experiments on LH current drive the instability develops (the covective threshold is overcame) for the LH wave injected power $P_{RF} \gg 10$ W. This condition can be easily fulfilled in experiments.

Acknowledgements

The author is grateful to R. Cesario for the information about the existence of the problem of explanation of the broadening of the injected radiofrequency line. He would also like to thank the Organizing Committee of VII LAWPP 97 for supporting his participation in the Workshop.

References

Benkadda, S., Popel, S.I., Tsytovich, V.N. et al.: 1996, *Phys. Plasmas* **3**, 571.

Cesario, R., Bartiromo, R., Cardinali, A. et al.: 1992, *Nucl. Fusion* **32**, 2127.

Fisch, N.J.: 1987, *Rev. Mod. Phys.* **59**, 175.

Petržilka, V.A.: 1991, *Plasma Phys. and Controlled Fusion* **33**, 365.

Popel, S.I.: 1997, Preprint No. 4, General Physics Institute, Moscow.

Popel, S.I. and Elsässer, K.: 1994, *Comments Plasma Phys. Controlled Fusion* **16**, 79.

Popel, S.I. and Tsytovich, V.N.: 1992, *Contrib. Plasma Phys.* **32**, 77.

Popel, S.I., Vladimirov, S.V. and Tsytovich, V.N.: 1995, *Phys. Rep.* **259**, 327.

Vladimirov, S.V., Tsytovich, V.N., Popel, S.I. and Khakimov, F.Kh.: 1995, *Modulational Interactions in Plasmas*, Kluwer Academic Publishers, Dordrecht.

PLASMA KINETIC THEORY II

M. PSIMOPOULOS
Plasma Group, Physics Department, Imperial College, London SW7 2BZ, U.K.

Abstract. The basic problems occurring in the construction of a convergent kinetic equation describing charged particles are briefly discussed. It is suggested that accelerations of all orders should be included in the equations of evolution of microscopic variables in order to obtain correct macroscopic averages. In this context, the "principle of non-universal time steps" is introduced and the necessity of an exact mechanical description before performing statistical analysis is discussed.

1. Introduction

The development of the collisional plasma kinetic theory can broadly be separated into five stages:

(i) Attempts to use Boltzmann's equation (Boltzmann, 1896) as a basic kinetic equation for particles interacting by Coulomb forces and calculation of transport coefficients of macroscopic equations (Chapman and Cowling, 1960).

(ii) Expansion of the Boltzmann collision integral for small deflexions leading to the Landau equation (Landau, 1936) and the Fokker–Planck approximation (Boyd and Sanderson, 1969) where dynamical friction and velocity diffusion are expressed in terms of Rosenbluth potentials.

(iii) Bogoliubov's adiabatic hypothesis (Bogoliubov, 1946): binary correlations relax faster than distribution functions. Integration of Liouville's equation and closure of the BBGKY hierarchy. Derivation of the BLG equation by Lenard (1960) and Guernsey (1960).

(iv) Liouville's equation (Liouville, 1838) as a new starting point and its formal solution by using diagrammatic techniques (Prigogine, 1962, and the Brussels school). Derivation of the BLG equation by Balescu (1960).

(v) Extension of the above theories to plasmas in the presence of external magnetic fields of various strengths (Rostocker, 1960; Haggerty and de Sobrino, 1964; Psimopoulos and Li, 1992) as well as self-consistent electric fields (Silin, 1960) (non-neutral plasmas). Efforts to eliminate the infinities which exist in the Fokker–Planck and the BLG equation using the screening property of the plasma (Clemmow and Dougherty, 1969) (long distance divergence) as well as quantum mechanical considerations (Kraeft and Stege, 1988; Lifshitz and Pitaevskii, 1981) (short distance divergence).

Astrophysics and Space Science **256:** 277–281, 1998.

2. Criticism of Classical Plasma Kinetic Theory

The basic kinetic equation used today for the theoretical derivation of transport coefficients as well as for computer simulations is the Landau equation in its Fokker–Planck form (Braginskii, 1965). It is well known that both coefficients of dynamical friction and of velocity diffusion of the latter equation are logarithmically divergent for small and large distances. This divergence reflects in all transport coefficients through the factor $\ln \Lambda$ (Ordonez and Molina, 1994) in the collision frequency. Although such divergence is not very strong, the question remains as to whether the present form of the kinetic equation should be retained at all in a correct theory.

An additional objection to the validity of the Fokker–Planck approximation of the Boltzmann integral has been raised by some authors (Shoub, 1987; Li and Petrasso, 1993) who have shown that higher moments of the variation of the velocity cannot be neglected. This is indeed valid criticism that should be further clarified for homogeneous and inhomogeneous plasmas.

It is certainly questionable too whether the Boltzmann equation in itself should form the basis of the kinetic theory of plasmas (Clemmow and Dougherty, 1969). This is mainly because in this equation it is postulated that (i) a universal time step δt exists during which all collisions are complete, (ii) no correlations occur at the beginning of each time step (hypothesis of molecular chaos), and (iii) the system is dilute enough so that only binary collisions take place. It is clear that the first two assumptions are violated in the case of charged particles whereas the third assumption restricts the theory to low density plasmas.

The main thesis of the Brussels school is the formal solution of Liouville's equation by multidimensional Fourier expansion and diagrammatic representation of the terms of the resulting series. In the case of plasmas (Balescu, 1963), various kinetic equations derived earlier by the BBGKY hierarchy (Vlasov, Landau, BLG) were classified according to characteristic time steps.

The first objection to this approach is that the perturbation is "statistical" rather than "mechanical" (Psimopoulos, 1980a), i.e. instead of considering an approximation related to the particle orbits and then perform an exact statistical treatment, the Green's function is expanded in terms of the coupling constant and hence, the change introduced by the approximation into the microscopic picture remains unclear. In this respect, Bogoliubov's system of equations, as derived from the lower part of the BBGKY hierarchy, although less systematic, seems to be more promising than Balescu's theory for further development of plasma kinetic theory.

There is a deeper reason, however, why Liouville's equation should be considered with great care when used as basis for the kinetic theory of charged particles: It is the implicit assumption made in the derivation of this equation that a universal time step δt exists such that, for all systems of the *ensemble* describing the plasma the linear equations

$$r = r' + v'\delta t; \quad v = v' + a_1'\delta t \tag{1}$$

give in the $6N$-dimensional phase space the final positions $r = (\mathbf{r}_1, \mathbf{r}_2, \ldots, \mathbf{r}_N)$ and velocities $v = (\mathbf{v}_1, \mathbf{v}_2, \ldots, \mathbf{v}_N)$ with great accuracy in terms of the initial state (r', v'). Although this assumption is correct for each system of the ensemble (analytic limit of each hypertrajectory), it cannot be generalized to all phase space because of the r^{-2} singularity of Coulomb's law. In particular, the higher order accelerations neglected in Equation (1) are singular functions of (r', v') and become infinite on integration over initial positions. This shows that there are always initial configurations for which higher order accelerations cannot be neglected, and that the projection by integration over (r', v') of Liouville's equation in lower subspaces and the resulting BBGKY hierarchy should not be accepted without careful examination of initial correlations describing the states (r', v').

3. The Principle of Non-Universal Time Steps

On the ground of the above criticism, it is suggested that in order to make progress in constructing the correct kinetic equation describing charged particles, the following general principle should be introduced in the theory.

It is impossible to find a time step during which all possible motions of the atoms constituting a large system can be considered as linear with high degree of accuracy.

The notion of atom here is quite general and means a particle that keeps its parameters fixed during the time of observation. The idea here is that the atoms interact with each other by singular forces which guarantee their existence as independent units. Thus, for any time step δt however small, there are always systems of the ensemble where the movements of atoms *cannot* be described linearly during this time step. In particular, the linear relations $r = r' + v'\delta t$; $v = v' + a_1'\delta t$ do not give in this case the final positions and velocities with great accuracy. These non-linear movements originate when atoms approach each other beyond a certain distance (large scattering angle collisions) and should be taken into account in the construction of macroscopic kinetic equations because they are statistically important. In practice, this is done by considering higher order accelerations for the determination of the final states, viz.

$$
\begin{aligned}
r &= r' + v'\delta t + a_1'\frac{(\delta t)^2}{2!} + a_2'\frac{(\delta t)^3}{3!} + \cdots, \\
v &= v' + a_1'\delta t + a_2'\frac{(\delta t)^2}{2!} + a_3'\frac{(\delta t)^3}{3!} + \cdots,
\end{aligned} \tag{2}
$$

so that mechanical accuracy is improved *before* statistical treatment (i.e. evaluation of moments, etc.) takes place. Note that the above principle is valid for the Boltzmann equation where for each time step δt there are always systems of the ensemble in which collisions take place and break the linearity of particle orbits so that on the average the number of collisions taking place during δt scales as $\delta C = \nu\delta t$. In the case of plasmas governed on the particle level only by Coulombic interactions, earlier studies (Psimopoulos, 1980b, 1986a, 1986b, 1987) have shown that *all* accelerations should be taken into account in order to derive correct macroscopic averages. Also, more recent work (Psimopoulos and Rogister, 1990) suggests that an implicit theory based on the geometry of Keplerian orbits would be preferable than a series expansion in the coupling constant and could throw new light on the derivation of transport coefficients in plasma kinetic theory.

4. Discussion

Extension of the implicit theory of two body collisions in 3-D, giving accurate final positions and velocities of test particles, is expected to provide with the correct time evolution of the various non-linear moments (i.e. moments containing all microscopic accelerations) of the *transition probability* of a system of charged particles. These moments can be used for the construction of the kinetic equation describing weakly coupled plasmas and the relative strength of each moment should be estimated in order to validate a possible "dominant order" Fokker–Planck form.

However, in order to determine correctly the above transition probability, it is also necessary to know the structure of initial correlations near a moving test particle (departure from the hypothesis of molecular chaos). Preliminary work in this direction is the calculation of the number density of untrapped electrons and ions near a test charge of a plasma (i) in thermal equilibrium (Maxwellian velocity distribution at infinity) (Psimopoulos, 1982a, 1982b) and (ii) for an arbitrary velocity distribution at infinity (Psimopoulos and Rogister, 1990). Clearly, self-consistent (screening) effects should be included in order to define the long distance behaviour of correlations. Also, it is expected that in the case of electron-ion interactions, trapped particles will be essential in shaping short distance correlations.

References

Balescu, R.: 1960, *Phys. Fluids* **3**, 52.

Balescu, R.: 1963, *Statistical Mechanics of Charged Particles*, Interscience, New York.

Bogoliubov, N.N.: 1962, *Problems of a Dynamical Theory in Statistical Mechanics* (Engl. trans. in *Studies in Statistical Mechanics*, Vol. 1, J. de Boer and G.E. Uhlenbeck (eds.), North-Holland, 1962).

Boltzmann, L.: 1896, *Lectures on Gas Theory* (Engl. trans. G. Brush, Univ. of California, 1964).

Boyd, T.J.M. and Sanderson, J.H.: 1969, *Plasma Dynamics*, Nelson.

Braginskii, S.I.: 1965, *Rev. of Plasma Phys.* **1**, Consult. Bureau.

Chapman, S. and Cowling, T.G.: 1960, *The Mathematical Theory of Non-Uniform Gases*, Cambridge University Press.

Clemmow, P.C. and Dougherty, J.P.: 1969, *Electrodynamics of Particles and Plasmas*, Addison-Wesley.
Epperlein, E.M. and Haines, M.G.: 1986, *Phys. Fluids* **29**, 1029.
Guernsey, R.L.: 1960, Off. Nav. Res. Contr. No. Nonr., 1124 (15).
Haggerty, M.J. and de Sobrino, L.G.: 1964, *Can. J. Phys.* **42**, 1969.
Khinchin, A.I.: 1949, *Mathematical Foundations of Statistical Mechanics*, Dover, New York.
Kraeft, W.D. and Strege, B.: 1988, *Physica A* **149**, 313.
Landau, L.D.: 1936, *Phys. Z. Sowjetunion* **10**, 154.
Lenard, A.: 1960, *Ann. Phys. (NY)* **10**, 390.
Li, C.K. and Petrasso, R.D.: 1993, *Phys. Rev. Lett.* **70**, 3063.
Lifshitz, E.M. and Pitaevskii, L.P.: 1981, *Course of Theoretical Physics*, Vol. 10, Physical Kinetics, Pergamon, Oxford, p. 184.
Liouville, J.: 1838, *Journal de Math.* **3**, 349.
Ordonez, C.A. and Molina, M.I.: 1994, *Phys. Plasmas* **1**(8), 2515.
Prigogine, I.: 1962, *Non-Equilibrium Statistical Mechanics*, Interscience.
Psimopoulos, M.: 1980a, Ph.D. Thesis, Imperial College.
Psimopoulos, M.: 1980b, *Phys. Lett.* **80A**, 390.
Psimopoulos, M.: 1982a, *Phys. Lett.* **88A**, 172.
Psimopoulos, M.: 1982b, *Phys. Lett.* **88A**, 176.
Psimopoulos, M.: 1986a, *Phys. Lett.* **113A**, 373.
Psimopoulos, M.: 1986b, *Phys. Lett.* **115A**, 274.
Psimopoulos, M.: 1987, *Phys. Lett.* **125A**, 258.
Psimopoulos, M.: 1988, From the complexity of the small to the simplicity of the large, Conference on University and Research, Imola, September 1988.
Psimopoulos, M. and Li, D.: 1992, *Proc. Royal Soc. Lond.* **437A**, 55.
Psimopoulos, M. and Rogister, A.: 1990, *Phys. Lett.* **149A**, 265.
Rostocker, N.: 1960, *Phys. Fluids* **3**, 922.
Shoub, E.C.: 1987, *Phys. Fluids* **30**, 1340.
Silin, V.P.: 1960, *Sov. Phys. JEPT* **11**, 1277.

NONLINEAR ELECTRIC FIELD DIFFUSION IN PLASMAS

Two Point Quasifractional Approximants for Electric Field in Plasmas with Nonlinear Exponential Conductivity Type

E. STELLA
Instituto Universitario Tecnológico Región Capital, Apdo. 40347, Caracas, Venezuela

P. MARTÍN
Dpto. de Física, Universidad Simón Bolivar, Apdo. 89000, Caracas 1080 A, Venezuela

Abstract. Approximated similarity solutions are given for the problem of electric field penetration into a semi-infinite plasma with nonlinear saturating conductivity (exponential type), using the two point quasifractional approximants method.

1. Introduction

The penetration of electric field in a plasma had been analyzed by several authors (Lonngren and Ames, 1974; Ahmadi and Hirose, 1976).

The equation governing the electric field penetration into a semi-infinite plasma is (Anderson and Lisak, 1981)

$$\frac{\partial^2 E}{\partial x^2} = \frac{4\pi}{c^2} \frac{\partial}{\partial t} (\sigma(E)E), \tag{1}$$

where E is the electric field, $\sigma(E)$ the conductivity and x the direction normal the boundary surface. The direction of the electric field is fixed and normal to the x direction.

There are several models for the conductivity, here we will consider the conductivity as given by

$$\sigma(E) = \frac{j_o}{E} (1 - \exp(-\beta E)). \tag{2}$$

This relation was obtained previously (Hirose et al., 1970) considering the experimental results of Hirose et al. The dimensionless electric field is $f = E/E_o$ where E_o is the electric field at the boundary. Equation (1) becomes

$$\frac{\partial^2 f}{\partial x^2} = \frac{4\pi j_o}{E_o c^2} \frac{\partial}{\partial t} (\exp(-\alpha f)), \tag{3}$$

where $\alpha = \beta E_o$.

The boundary and initial conditions are

$$f = 0 \quad \text{for} \quad x \to \infty, \quad t > 0, \tag{4}$$

Astrophysics and Space Science **256:** 283–287, 1998.

$$f = 0 \quad \text{for} \quad t = 0, \quad x \geq 0, \tag{5}$$

$$f = 1 \quad \text{for} \quad x = 0, \quad t > 0. \tag{6}$$

Introducing the similarity variable

$$\eta = x\sqrt{\frac{2\pi jo}{E_o c^2 t}}\,, \tag{7}$$

Equation (3) is

$$\frac{d^2 f(\eta)}{d\eta^2} = -\alpha\eta \exp(-\alpha f)\frac{df(\eta)}{d\eta} \tag{8}$$

with $f(0) = 1$ and $f(\infty) = 0$. In this work, Equation (8) is transformed to

$$\frac{d^2\Phi(s)}{ds^2} = -s\exp(-\Phi(s))\frac{d\Phi(s)}{ds} \tag{9}$$

with $\Phi = \alpha f$, $\Phi(0) = \alpha$, $\eta = s/\sqrt{\alpha}$.

Approximated solutions to (8) have been reported in previous work (Hirose et al., 1970). However, each solution is valid only for some values of α. In this work we present solutions for values of α in the range 0.1–10 to which have the same general form.

2. Method of Solution

To obtain the approximated solution here we will use the two point quasifractional approximation method described recently. The procedure shows how to find solutions to ordinary differential equations using simultaneously the information from the power series and the leading term of the asymptotic expansion coming out from the same equation (Martín and Baker, 1991). The power series for $\Phi(s)$ is

$$\Phi(s) = \alpha - \lambda\, s + \frac{\lambda\, \exp(-\alpha)}{6}\,. \tag{10}$$

The leading term can be obtained only when a first integral of Equation (9) is performed. A first integral is not easy to obtain and several changes of variables must be carried out. Summarizing, the results are

$$s\frac{d\Phi}{ds} - \ln\left(-\left(\frac{d\Phi}{ds}\right)\right)^2 - \Phi - s^2\exp(-\Phi) = C(\alpha, \lambda), \tag{11}$$

where $\mathrm{C}(\alpha, \lambda) = -(2\ln\lambda + \alpha)$.

Using Equation (11), the leading term of the asymptotic expansion is

$$\Phi(s) = \frac{\lambda}{s}\exp\left(\frac{\alpha}{2}\right)\exp\left(-\frac{s^2}{2}\right). \tag{12}$$

The approximant for $\Phi(s)$ is now defined by

$$\Phi(s) = \frac{\alpha}{\exp(\beta s + s^2(\gamma - (1+\delta s)^2)) + H(s)}, \tag{13}$$

where

$$H(s) = \mu s \exp\left(s^2/2\right); \tag{14}$$

the coefficients β, γ, δ and μ are obtained from the power series. From the equations below an analytic form for these coefficients is obtained:

$$\alpha(\beta + \mu) - \lambda = 0, \tag{15}$$

$$\alpha(\gamma - 1 + \beta^2/2) - \lambda(\beta + \mu) = 0, \tag{16}$$

$$\alpha\left(-2\delta + \beta\gamma - \beta + \frac{\beta^3}{6} + \mu/2\right) - \lambda\left(\gamma - 1 + \frac{\beta^2}{2}\right) + \lambda\frac{\exp(-\alpha)}{6} = 0, \tag{17}$$

$$\mu = \frac{\alpha}{\lambda}\exp\left(-\frac{\alpha}{2}\right). \tag{18}$$

The parameter λ is the first derivative at $s = 0$. It is calculated by integrating Equation (9) from zero to infinity. Thus, we obtain the equation for λ as

$$\left(\lambda + \int_0^\infty t\exp(-\Phi(t,\alpha,\lambda))\frac{d\Phi}{dt}dt\right) = 0. \tag{19}$$

Figures 1 and 2 show the approximants f_a and the numerical calculations f by the Runge–Kutta method for $\alpha = 0.1$ and 10. Figures 3 and 4 show the relative error ε of the approximation. In each figure, the function has been plotted from $s = 0$ to the value where the electric field becomes so small that no correlation with experimental results is significant.

As we can see from the figures, both curves, the exact and the approximated, are almost coincident. It is difficult to note any difference, only for $\alpha = 10$, some discrepancy between both curves can be found around the value 0.6 for x. The largest relative errors are in the region of small value of the function.

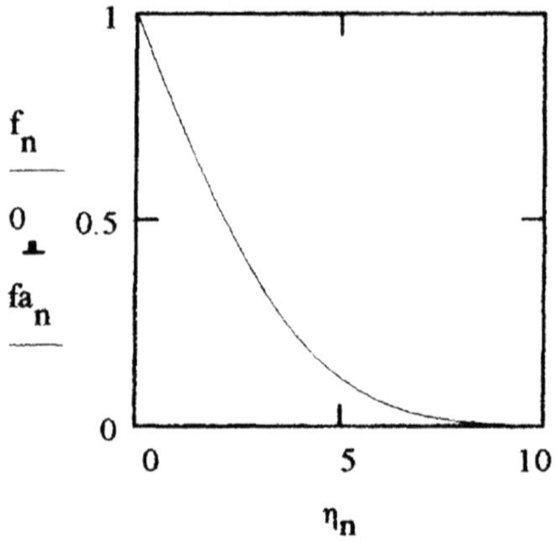

Figure 1. $\alpha = 0.1$.

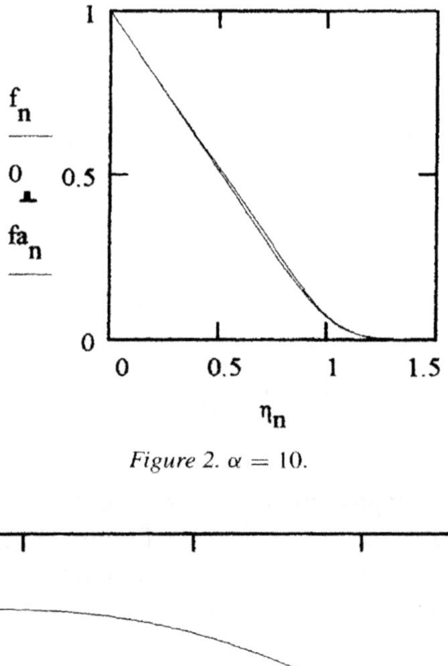

Figure 2. $\alpha = 10$.

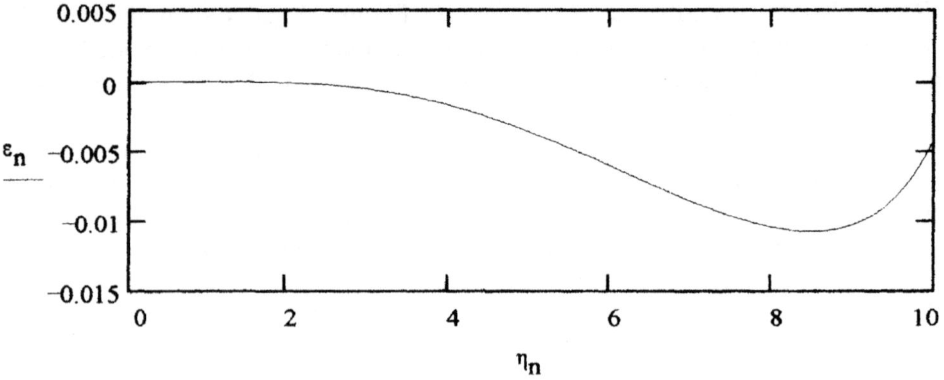

Figure 3. $\alpha = 0.1$.

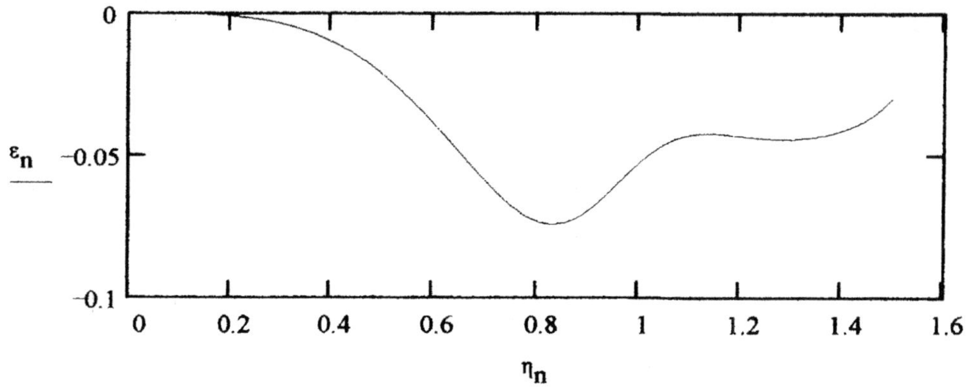

Figure 4. $\alpha = 10$.

3. Conclusion

A unique approximation for Equation (1) has been found for any value of α which are very precise. The accuracy of the approximant here obtained using two point quasifractional approximation method is enough for all the experimental requirements.

References

Ahmadi, G. and Hirose, A.: 1976, *IEEE Trans. Plasma Sci.* **PS-4**, 255.
Anderson, D. and Lisak, M.: 1981, *IEEE Trans. Plasma Sci.* **PS-9**(2).
Hirose, A., Alexeef, I. and Jones, W.D.: 1970, *Phys. Rev. Lett.* **25**, 1563.
Lonngren, K.E. and Ames, W.E.: 1974, *Phys. Fluids* **17**, 1919.
Martín, P. and Baker, G.: 1991, *J. Math. Phys.* **32**(6).

MHD LEAKY WAVES IN A LAYERED PLASMA: I THEORY

A.G. GONZÁLEZ, J. GRATTON and S. FARINA
INFIP-Lab. Física del Plasma, Facultad de Ciencias Exactas y Naturales, Universidad de Buenos Aires, Pab. I, Ciudad Universitaria, (1428) Buenos Aires, Argentina

Abstract. In a three layer compressible plasma we study MHD leaky waves described by Fourier modes of complex frequency. We follow a matrix method that yields four transcendental dispersion relations. The roots (real or complex) make up a spectrum of pseudomodes, consisting of a complicate set of infinite branches. We derive analytical properties that allow to study the topology of the spectrum, the conditions of existence of real and complex roots, the contacts and crossings of branches of different types, etc. Our results allow to analyze any particular case of interest.

1. Introduction

Leakage is a non-dissipative phenomenon that takes place in structured plasmas when a perturbation excited in a layer loses energy transmitting it to another region where it can propagate away. It has been related to the coronal heating of the Sun (Chen and Hasegawa, 1974; Ionson, 1978). Several authors analyzed MHD waves of complex frequency in cylindrical configurations (Roberts and Webb, 1979; Spruit, 1982; Wilson, 1981) (thin tube approximation and arbitrary radius but no external magnetic field). Cally (1986) investigated a cylindrical interface between two plasmas of equal β and found perturbations called *cubic modes* that correspond to leaky waves in H_α fibrils in the Sun (Giovanelli, 1975). Čadež and Okretič (1989) considered three plane layers with different β but their analysis of the leaky waves is faulty. Here we study ideal compressible MHD perturbations of complex frequency in a slab model of three layers. We develop a general method that includes the cubic modes and other kinds of leaky waves, and allows to derive results in closed form for arbitrary equilibrium configurations. The different types of leakage are discussed in a second paper (González et al., 1998a), and in the third paper (González et al., 1998b) of this series two examples show the application of the methods.

We assume a plasma in equilibrium stratified along the y axis, with $\mathbf{B} = B_x(y)\mathbf{e}_x + B_z(y)\mathbf{e}_z$, $\rho = \rho(y)$, $p = p(y)$ ($d(p + B^2/8\pi)/dy = 0$ in the unperturbed state), $C_A^2 = B^2/4\pi\rho$, $C_S^2 = \gamma p/\rho$. Let ζ be the y-component of the displacement of a linear normal adiabatic mode of the type $\delta Q = q(y)e^{i(\mathbf{k}.\mathbf{x}-\omega t)}$, $\mathbf{k} = (k_x, 0, k_z)$, that satisfies (Gratton et al., 1988) $[H(1 - 1/M)\zeta_y]_y - k^2 H\zeta = 0$

Astrophysics and Space Science **256:** 289–296, 1998.

($H = \rho(C_A^2 k^2 \cos^2\psi - \omega^2)$, $M = 1 - (C_A^2 + C_S^2)(k/\omega)^2 + C_A^2 C_S^2 (k/\omega)^4 \cos^2\psi$ and $\psi(y)$ is the angle between **B** and **k**).

In the three layers ($i = 1, 2, 3$) ρ, p, T and **B** are constant, but discontinuous at the interfaces $y = 0, a$. In each layer ρ, p and **B** are arbitrary, subject to the equilibrium conditions $(C_S^2/\gamma + C_A^2/2)\rho = 0$ at the interfaces ($\{Q\}$ is the difference between the value of Q across the interface), that can be written as $1 + \beta_j = f_{ij}(1 + \beta_i)$ ($f_{ij} = p_i/p_j$). The boundary conditions (González and Gratton, 1991) at $y = 0, a$ are $H(1 - 1/M)\zeta_y = 0$ and $\zeta = 0$. In each layer $\zeta_i = \zeta_{-,i}\exp(-k\Gamma_i y) + \zeta_{+,i}\exp(k\Gamma_i y)$, with $\Gamma_i = [(V_i^2 - V_{+,i}^2)(V_i^2 - V_{-,i}^2)/F_i^2(V_{B,i}^2 - V_i^2)]^{1/2}$ ($\mathrm{Re}(\Gamma_i) > 0$ or $\mathrm{Im}(\Gamma_i) > 0$ if $\mathrm{Re}(\Gamma_i) = 0$), $A_i = C_{A,i}/C_{S,i} = \sqrt{2/\beta_i\gamma}$, $F_i^2 = 1 + A_i^2$, $V_{A,i} = A_i\cos\psi_i$, $V_{B,i} = A_i\cos\psi_i/F_i$, $V_{\pm,i}^2 = \frac{1}{2}\left(F_i^2 \pm \sqrt{F_i^4 - 4V_{A,i}^2}\right)$, $V_i = \omega/kC_{S,i}$. Note that $V_i = V_j/m_{ij}$, where $m_{ij} = C_{S,i}/C_{S,j} = \sqrt{T_i/T_j}$. Then in general the problem involves 10 independent dimensionless parameters and variables: two pressure ratios, two temperature ratios, the β of one layer, the three ψ angles, $\alpha \equiv ka$ and V.

Consider ω real. When $0 < V_i < V_{B,i}$ or $V_{-,i} < V_i < V_{+,i}$, G_i is real and the ζ_i depends exponentially on y (evanescent waves, denoted by E). If the region is unbounded the solution may diverge, unless additional constraints are imposed (for example, we must take $\zeta_{+,3} = 0$ if we look for bounded solutions in region 3). When $V_{B,i} < V_i < V_{-,i}$ or $V_i > V_{+,i}$, G_i is imaginary, and the perturbation is a wave that propagates in the y-direction (denoted by P); then $\zeta_{+,i}$ and $\zeta_{-,i}$ are the amplitudes of waves whose y-components of the phase velocity are along $+y$ and $-y$, respectively.

2. Dispersion Relations

The matching of the perturbations of two uniform media i and j at their interface can be described by a matrix equation (González and Gratton, 1991) that relates $\zeta_{\pm,j}$ and $\zeta_{\pm,i}$. The matrix matching the solutions of regions 1 and 3 is the product of the matrix that matches 1 and 2, a matrix that relates the solutions of region 2 at $y = 0$ and $y = a$, and the matrix that matches 2 and 3. Then the matching of the solutions in region 1 with those in 3 is given by:

$$\begin{bmatrix} \zeta_{-,3} \\ \zeta_{+,3} \end{bmatrix} = \frac{1}{Z}\begin{bmatrix} \mathcal{V} & \mathcal{U} \\ \mathcal{S} & \mathcal{T} \end{bmatrix}\begin{bmatrix} \zeta_{-,1} \\ \zeta_{+,1} \end{bmatrix}, \tag{1}$$

where $\mathcal{V} = T_{32}T_{21}\Phi + S_{32}S_{21}$, $\mathcal{U} = T_{32}S_{21}\Phi + S_{32}T_{21}$, $\mathcal{S} = S_{32}T_{21}\Phi + T_{32}S_{21}$, $\mathcal{T} = S_{32}S_{21}\Phi + T_{32}T_{21}$, $Z = 4G_2G_3f_{21}f_{32}\Phi^{1/2}$, $\Phi = e^{-2\alpha\Gamma}$ with $S_{ji} = f_{ji}G_j - G_i$, $T_{ji} = f_{ji}G_j + G_i$, $G_i = (V_{A,i}^2 - V_i^2)/\Gamma_i$.

We shall be looking for complex ω roots that describe the damping of the leaky waves. Then the Γ_i are complex. Four types of solutions may be considered (for $Z \neq 0$ and $\mathrm{Re}(\Gamma_{1,3}) \neq 0$): (a) those bounded for $y \to \pm\infty$ ($\zeta_{+,3} = \zeta_{-,1} = 0$), that satisfy $\mathcal{T} = 0$ ($\mathcal{T}$ branches), (b) those bounded for $y \to +\infty$ but divergent as $y \to -\infty$ ($\zeta_{+,3} = \zeta_{+,1} = 0$) that satisfy $\mathcal{S} = 0$ ($\mathcal{S}$ branches), (c) those that diverge as $y \to \pm\infty$ ($\zeta_{-,3} = \zeta_{+,1} = 0$) so that $\mathcal{V} = 0$ ($\mathcal{V}$ branches), and (d) those divergent as $y \to +\infty$ but bounded for $y \to -\infty$ ($\zeta_{-,3} = \zeta_{-,1} = 0$) so that $\mathcal{U} = 0$ ($\mathcal{U}$ branches). Then we have four different dispersion relations (DR) and four types of branches. All the DR can be written in the general form

$$\mathcal{A}_{32}\mathcal{A}_{21}\Phi + \bar{\mathcal{A}}_{32}\bar{\mathcal{A}}_{21} = 0, \tag{2}$$

where $\mathcal{A}_{ji} = f_{ji}G_j + \sigma_{ji}G_i$, $\bar{\mathcal{A}}_{ji} = f_{ji}G_j - \sigma_{ji}G_i$, $\sigma_{ij} = \sigma_{ji}$ with $\sigma_{32} = -1$, $\sigma_{21} = 1$ for $\mathcal{S} = 0$, $\sigma_{32} = -1$, $\sigma_{21} = -1$ for $\mathcal{T} = 0$, $\sigma_{32} = 1$, $\sigma_{21} = -1$ for $\mathcal{U} = 0$ and $\sigma_{32} = 1$, $\sigma_{21} = 1$ for $\mathcal{V} = 0$. The formal replacement $G_k \leftrightarrow -G_k$ yields $T_{jk} \leftrightarrow S_{jk}$ and $T_{ki} \leftrightarrow -S_{ki}$. Then starting from any DR, the others can be obtained changing the signs of G_1, G_2, G_3. In what follows we present formulae only for the $\mathcal{T}$ branch; those for the other branches can be obtained by the appropriate substitutions.

The normal modes of our problem are of type $\mathcal{T}$. They have been studied for symmetrical configurations (Rae and Roberts, 1983; Roberts, 1981). The solutions of the remaining DR are pseudomodes. Leaky waves are pseudomodes with ω complex, and belong to the $\mathcal{S}$, $\mathcal{V}$ and $\mathcal{U}$ branches. We shall use the variable $\Omega \equiv V_2^2$, since only even powers of ω appear in the DR. The roots are always finite for any $\alpha \neq 0$, but some of them are unbounded for $\alpha \to 0$. As we follow the evolution of a particular branch as α is varied, we find that it may end abruptly. At these points a solution of a certain DR becomes a solution of another DR. We shall see that the transitions between branches of different types occur at predictable values of Ω and α. In addition, real roots may become complex (or vice-versa) at certain points (double real roots). Then, to avoid missing parts of the spectrum, it is better to solve the four DR, including $\mathcal{T} = 0$, and not only for complex frequencies but also for real ones. The spectrum is very complicated, with infinite branches (real or complex). The problem involves many parameters, but our general results allow to unravel the topology of the spectrum in any case of interest.

3. General Properties of the Spectrum

3.1. Real and Complex Ω Roots

When Ω is real, the perturbation is evanescent or propagating in each region. There are 8 combinations, but only EEE and EPE yield real branches. In the case EEE we set $G_1 = g_1$, $G_2 = g_2$, $G_3 = g_3$, $\Gamma_2 = \gamma_2$, with g_1, g_2, g_3, γ_2 real functions of Ω. Then $\mathcal{T} = 0$ leads to:

$$\tanh(\alpha\gamma_2) = -f_{21}g_2(f_{31}g_3 + g_1)/(f_{31}g_1g_3 + f_{21}^2g_2^2), \tag{3}$$

and there can be branches with real $\Omega(\alpha)$. To analyze the case EPE (trapped propagating waves) we set $G_1 = g_1$, $G_2 = ig_2$, $G_3 = g_3$, $\Gamma_2 = i\gamma_2$, then $\mathcal{T} = 0$ yields:

$$\tan(\alpha\gamma_2) = -f_{21}g_2(f_{31}g_3 + g_1)/(f_{31}g_1g_3 - f_{21}^2g_2^2), \tag{4}$$

and again real branches are possible (in this case $\Omega > 0$). The remaining combinations yield only isolated real Ω points unrelated to the leaky waves. In the cases PPE, EPP, they are given by $g_j = 0$ or $g_2 = 0$ ($\mathcal{S} \leftrightarrow \mathcal{T}$, $\mathcal{U} \leftrightarrow \mathcal{V}$ contacts for PPE and $\mathcal{S} \leftrightarrow \mathcal{V}$, $\mathcal{T} \leftrightarrow \mathcal{U}$ for EPP). Likewise isolated solutions exist for the case PEE at the contacts $\mathcal{S} \leftrightarrow \mathcal{T}$ and $\mathcal{U} \leftrightarrow \mathcal{V}$, and for EEP at contacts $\mathcal{S} \leftrightarrow \mathcal{V}$ and $\mathcal{T} \leftrightarrow \mathcal{U}$. Finally, for PEP and PPP isolated real roots exist at contacts $\mathcal{T} \leftrightarrow \mathcal{V}$ and $\mathcal{S} \leftrightarrow \mathcal{U}$.

No complex roots of $\mathcal{T} = 0$ exist: using a well known theorem (Bernstein et al., 1957) it can be shown that Ω is real if the solution is bounded for $y \to \pm\infty$, as occurs for the $\mathcal{T}$ branches. On the other hand, the solutions of $\mathcal{V} = 0$, $\mathcal{U} = 0$ and $\mathcal{S} = 0$ diverge as $y \to +\infty$ and/or $y \to -\infty$, and Ω can be complex. For certain values of α these DR have real double roots, where complex conjugate branches start. Each complex Ω yields two conjugate values of $\mathcal{V}$; leaky waves are those with $\mathrm{Im}(V) < 0$.

3.2. Real and Complex Transitions, Crossings and Periodicity

Different branches may coincide at points (α_0, Ω_0) where a real transition, a crossing or a complex transition occurs. All coincidences occur for real Ω. At a *real transition*, a real branch of a DR becomes a real branch of a different DR, so that there is one branch at each side of (α_0, Ω_0). At a *crossing* two real branches of different type intersect, and continue beyond the coincidence. At a *complex transition* a complex branch and its conjugate make jointly a transition to two complex conjugate branches of a different type (complex transitions occur at isolated real roots Ω_0).

To discuss the coincidences given by $G_j = 0, \infty$ we write Equation (2) in the form

$$f_{j2}G_j(\mathcal{A}_{2i}\Phi + \bar{\mathcal{A}}_{2i}) + \sigma_{j2}G_2(\mathcal{A}_{2i}\Phi - \bar{\mathcal{A}}_{2i}) = 0, \tag{5}$$

with $i = 1, 3$ and $j = 3, 1$ respectively. Consider two branches that coincide at (α_0, Ω_0) where Γ_j has a pole, so that $(\mathcal{A}_{2i}\Phi - \bar{\mathcal{A}}_{2i})_0 = 0$ (the suffix 0 denotes that the quantities are evaluated at (α_0, Ω_0)). Expanding Equation (5) near the coincidence ($G_j = \delta G_j, \alpha = \alpha_0 + \delta\alpha$, $V_j^2 = V_{j0}^2 + \delta V_j^2$) we obtain to the lowest order $\sqrt{-\delta\Omega} = \sigma_{j2}\delta\alpha\ (V_{A.2}^2 - V_{2.0}^2)f_{2j}/C_{j.0}$, where $C_{j.0}(\Omega_0)$ is real. Then as α crosses α_0, the solution of the DR for $\sigma_{j2} = 1$ ceases to exist, while that corresponding to $\sigma_{j2} = -1$ appears, and we have a real transition. Notice that $\delta\Omega \propto \delta\alpha^2$, with $\delta\Omega < 0$ (i.e. $V_j^2 \leq V_{B.j}^2$), i.e. evanescent perturbations in the external region j. The points where $\Gamma_j = 0$ ($V_{j.0}^2 = V_{\pm.j}^2$) and $(\mathcal{A}_{2i}\Phi + \bar{\mathcal{A}}_{2i})_0 = 0$ are also real transitions,

TABLE I

Coincidences of branches and notation for the corresponding Ω.

Conditions	Notation	Branches involved
Real transitions		
$V_1 = V_{+,1}$ ($\Gamma_1 = 0$)	Ω_{+1}	$\mathcal{T} \leftrightarrow \mathcal{S}$, $\mathcal{V} \leftrightarrow \mathcal{U}$
$V_1 = V_{-,1}$ ($\Gamma_1 = 0$)	Ω_{-1}	$\mathcal{T} \leftrightarrow \mathcal{S}$, $\mathcal{V} \leftrightarrow \mathcal{U}$
$V_1 = V_{B,1}$ ($\Gamma_1 = \infty$)	Ω_{b1}	$\mathcal{T} \leftrightarrow \mathcal{S}$, $\mathcal{V} \leftrightarrow \mathcal{U}$
$V_3 = V_{+,3}$ ($\Gamma_3 = 0$)	Ω_{+3}	$\mathcal{T} \leftrightarrow \mathcal{U}$, $\mathcal{V} \leftrightarrow \mathcal{S}$
$V_3 = V_{-,3}(\Gamma_3 = 0)$	Ω_{-3}	$\mathcal{T} \leftrightarrow \mathcal{U}$, $\mathcal{V} \leftrightarrow \mathcal{S}$
$V_3 = V_{B,3}(\Gamma_3 = \infty)$	Ω_{b3}	$\mathcal{T} \leftrightarrow \mathcal{U}$, $\mathcal{V} \leftrightarrow \mathcal{S}$
Crossings		
$V_1 = V_{A,1}$ ($G_1 = 0$)	Ω_{a1}	$\mathcal{T} \leftrightarrow \mathcal{S}$, $\mathcal{V} \leftrightarrow \mathcal{U}$
$V_3 = V_{A,3}$ ($G_3 = 0$)	Ω_{a3}	$\mathcal{T} \leftrightarrow \mathcal{U}$, $\mathcal{V} \leftrightarrow \mathcal{S}$
Crossings/Complex transitions		
$\Phi = 1$, $S_{31} = 0$	Ω_{S+}	$\mathcal{U} \leftrightarrow \mathcal{S}$
$\Phi = -1$, $T_{21}S_{32} = T_{32}S_{21}$	Ω_{S-}	$\mathcal{U} \leftrightarrow \mathcal{S}$
$\Phi = 1$, $T_{31} = 0$	Ω_{T+}	$\mathcal{T} \leftrightarrow \mathcal{V}$
$\Phi = -1$, $T_{21}T_{32} = S_{32}S_{21}$	Ω_{T-}	$\mathcal{T} \leftrightarrow \mathcal{V}$

but now $\delta\Omega \propto \mp\delta\alpha^2$, so that $V_j^2 \leq V_{+,j}^2$ or $V_j^2 \geq V_{-,j}^2$ and again the perturbations are evanescent in the external regions. Summarizing, the real transitions take place at the zeros and poles of Γ_j ($j = 1, 3$). The various possibilities are given in Table I.

Consider the case $V_{j0} = V_{A,j}$. Now $G_j = 0$ and $(\mathcal{A}_{2i}\Phi - \bar{\mathcal{A}}_{2i})_0 = 0$. Expansion of Equation (5) near (α_0, Ω_0), yields $f_{j2}\delta G_j + \sigma_{j2}(V_{2,0}^2 - V_{A,2}^2)\delta\alpha + C_0\delta\Omega = 0$, where $C_0(\Omega_0)$ is real and depends on the DR we are considering. Since $\delta G_j = -\delta\Omega\, m_{2j}^2/\Gamma_{j,0}$, and $\Gamma_{j,0}$ is real, this leads to a crossing since $\delta\Omega = D_0\delta\alpha$ (D_0 real). The remaining crossings and the complex transitions occur for $\Phi = \pm 1$, which requires a propagating perturbation in region 2. They involve $\mathcal{S}$ and $\mathcal{U}$ branches when Ω is a root of $S_{31} = 0$, $\Phi = 1$, or $T_{21}S_{32} - T_{32}S_{21} = 0$, $\Phi = -1$. Alternatively, they involve $\mathcal{T}$ and $\mathcal{V}$ branches when Ω is a root of $T_{31} = 0$, $\Phi = 1$, or $T_{21}T_{32} - S_{21}S_{32} = 0$, $\Phi = -1$. In both instances these conditions can be reduced to polynomials in Ω (in general of degree 5 and 10 respectively), by a rationalization procedure that introduces spurious roots. The non-spurious real roots give the points of coincidence for $\Phi = \pm 1$. The notation for these Ω is given in Table I. These coincidences will be crossings when the perturbation is evanescent in both external regions, or complex transitions if it is propagating everywhere. The $\mathcal{T}$

branches (always real) are not involved in complex transitions. When Ω is real and there are propagating waves in region 2 we have infinite harmonics, since the solutions are periodic in α (period $= \pi/|\Gamma_2|$). Real transitions are periodic if they occur in a domain of propagating waves for region 2 (families of harmonics are related to wave trapping). The complex transitions and the crossings for $\Phi = \pm 1$ are always periodic.

3.3. Limiting Solutions ($\alpha \to \infty$ and $\alpha = 0$)

When $\alpha \to \infty$ the interfaces 1–2 and 2–3 are isolated. If $\mathrm{Re}(\Gamma_2) \neq 0$, $\Phi \to 0$ and $\mathcal{V} \to S_{32}S_{21}$, $\mathcal{U} \to S_{32}T_{21}$, $\mathcal{S} \to T_{32}S_{21}$, $\mathcal{T} \to T_{32}T_{21}$. Then for $\alpha \to \infty$, some $\mathcal{U}$ and $\mathcal{V}$ branches tend to the roots of $S_{32} = 0$, some $\mathcal{S}$ and $\mathcal{T}$ branches tend to those of $T_{32} = 0$, other $\mathcal{S}$ and $\mathcal{V}$ branches approach the roots of $S_{21} = 0$, and other $\mathcal{U}$ and $\mathcal{T}$ branches approach the roots of $T_{21} = 0$. If $\mathrm{Re}(\Gamma_2) = 0$, Φ is indeterminate when $\alpha \to \infty$. The only possibility is EPE, then $\mathcal{T} = 0$ reduces to Equation (4). If $\Omega = V^2_{\pm,2}$, $\gamma_2 \to 0$ but $g_2 \to \infty$ and Equation (4) will be satisfied if $\tan(\alpha\gamma_2) \to 0$. Then there is a family of harmonics such that $\Omega \to V^2_{\pm,2}$ while $\gamma_2 \to 0$ and $\alpha\gamma_2 \to n\pi$ (n integer). Since the other DR yield the same results, we conclude that $\Omega = V^2_{\pm,2}$ are solutions of all four DR for $\alpha \to \infty$.

The case $\alpha = 0$ corresponds to a region 2 of vanishing thickness. In this limit $\mathcal{V}, \mathcal{T} \to G_2T_{31}$, while $\mathcal{U}, \mathcal{S} \to G_2S_{31}$. If $Z \neq 0$, then $G_2 \neq 0$ so that the solutions of $\mathcal{V} = 0$ and $\mathcal{T} = 0$ are roots of $T_{31} = 0$, while the solutions of $\mathcal{U} = 0$ and $\mathcal{S} = 0$ are roots of $S_{31} = 0$. If $Z = 0$, either $G_2 = 0$ or $G_3 = 0$. The case $G_3 = 0$ does not yield solutions. The case $G_2 = 0$ requires $\Omega = V^2_{A,2}$ or $\Omega = V^2_{B,2}$. If $\mathrm{Re}(\Gamma_2) \neq 0$, then Equation (3) holds, $\tanh(\alpha\gamma_2) \to 0$ and there is a solution for which $\Omega \to V^2_{A,2}$. If $\mathrm{Re}(\Gamma_2) = 0$ (case EPE) we find an equivalent result. Similarly, if $\Omega = V^2_{B,2}$ and Γ_2 is real, Equation (3) requires $\tanh(\alpha\gamma_2) \to 0$. Then as $\alpha \to 0$ there is a solution such that $\Omega \to V^2_{B,2}$, $\gamma_2 \to \infty$ but $\alpha\gamma_2 \to 0$. If $\mathrm{Re}(\Gamma_2) = 0$, Equation (4) requires $\tan(\alpha\gamma_2) \to 0$, and a family of harmonics tends to the limiting solution $\Omega \to V^2_{B,2}$, with $\gamma_2 \to \infty$ but $\alpha\gamma_2 \to n\pi$. The other DR lead to the same results. Then $\Omega = V^2_{A,2}$ and $\Omega = V^2_{B,2}$ are solutions of the four DR when $\alpha = 0$.

In addition, when $\alpha \to 0$ there are infinite unbounded complex $\mathcal{V}$ and $\mathcal{S}$ or $\mathcal{U}$ branches, of the form $\Omega = W(1 + iw)$ (W, w real and $W \to \infty$, w finite). Indeed, using this *ansatz* in (2) one finds for $\alpha \to 0$, $W \to \infty$ the equation $\Phi_\infty = \Psi_\mathcal{X}$ ($\mathcal{X} = \mathcal{S}, \mathcal{T}, \mathcal{U}, \mathcal{V}$, $\Phi_\infty = \exp\left(2\alpha(i\sigma_w a - b)\sqrt{W/F_2}\right)$, $\sigma_w = \mathrm{Sign}(w)$, $a = \sqrt{(1/2)\left(\sqrt{1+w^2}+1\right)}$, $b = \sqrt{(1/2)\left(\sqrt{1+w^2}-1\right)}$, $\Psi_\mathcal{X} = \Psi_{-1}\Psi_{-3}/\Psi_{+1}\Psi_{+3}$, $\Psi_{\pm i} = F_2 f_{2i} \pm \sigma_{2i} m_{2i} F_i$). Since $\Psi_\mathcal{X}$ is real this condition implies $\exp(-n\pi b/a) = |\Psi_\mathcal{X}|$ and $2\alpha a\sqrt{W/F_2} = n\pi$, where $n = 1, 3, 5, \ldots$ if $\Psi_\mathcal{X} < 0$ or $n = 2, 4, 6, \ldots$ if $\Psi_\mathcal{X} > 0$. The first condition is a transcendental equation that determines w, and can be satisfied for infinite values of n, provided $|\Psi_\mathcal{X}| \leq 1$. In this case, the second condition determines W. It is easy to verify that $|\Psi_\mathcal{V}| < 1$, $|\Psi_\mathcal{S}| = 1/|\Psi_\mathcal{U}|$

TABLE II

Limiting solutions and notation for the corresponding Ω (for any ij, the total number of roots Ω_{Tij} and Ω_{Sij} is 5).

$\alpha = 0$			$\alpha \to \infty$		
Condition	Notation	Branches involved	Condition	Notation	Branches involved
$T_{31} = 0$	Ω_{T31}	$\mathcal{T} \leftrightarrow \mathcal{V}$	$T_{21} = 0$	Ω_{T21}	$\mathcal{T} \leftrightarrow \mathcal{U}$
$S_{31} = 0$	Ω_{S31}	$\mathcal{U} \leftrightarrow \mathcal{S}$	$S_{21} = 0$	Ω_{S21}	$\mathcal{S} \leftrightarrow \mathcal{V}$
$V_2 = V_{A,2}$	Ω_{a2}	$\mathcal{T} \leftrightarrow \mathcal{V} \leftrightarrow \mathcal{U} \leftrightarrow \mathcal{S}$	$T_{32} = 0$	Ω_{T32}	$\mathcal{T} \leftrightarrow \mathcal{S}$
$V_2 = V_{B,2}$	Ω_{b2}	$\mathcal{T} \leftrightarrow \mathcal{V} \leftrightarrow \mathcal{U} \leftrightarrow \mathcal{S}$	$S_{32} = 0$	Ω_{S32}	$\mathcal{V} \leftrightarrow \mathcal{U}$
$\lvert\Psi_{\mathcal{S}}\rvert < 1$	∞	$\mathcal{V} \leftrightarrow \mathcal{S}$	$V_2 = V_{+,2}$	Ω_{+2}	$\mathcal{T} \leftrightarrow \mathcal{V} \leftrightarrow \mathcal{U} \leftrightarrow \mathcal{S}$
$\lvert\Psi_{\mathcal{U}}\rvert < 1$	∞	$\mathcal{V} \leftrightarrow \mathcal{U}$	$V_2 = V_{-,2}$	Ω_{-2}	$\mathcal{T} \leftrightarrow \mathcal{V} \leftrightarrow \mathcal{U} \leftrightarrow \mathcal{S}$

and $|\Psi_{\mathcal{T}}| > 1$. Then there are always infinite $\mathcal{V}$ branches, as well as infinite $\mathcal{S}$ (if $|\Psi_{\mathcal{S}}| < 1$) or $\mathcal{U}$ branches (if $|\Psi_{\mathcal{U}}| < 1$), of the required form.

In addition to these unbounded complex solutions, bounded complex solutions related to a certain type of leakage (González et al., 1998b) may exist in the limits $\alpha \to 0, \infty$, depending on the parameters of the problem. They may appear among the solutions of $T_{ij} = 0$ and $S_{ij} = 0$. In the general case, they must be found numerically solving an associate polynomial equation of degree 5, and no criteria for the existence of complex roots can be given in closed form.

Table II gives a list of the solutions of the DR for the limiting values of α. To find the solutions of $T_{ij} = 0$ and $S_{ij} = 0$ it is convenient to consider the products $T_{ij}S_{ij} = 0$ ($i, j = 1, 2, 3,\ i \neq j$) that lead to polynomial equations of degree 5 in Ω in the general case. Some of the roots thus obtained will be solutions of $T_{ij} = 0$ (indicated generically as Ω_{Tij}), and the remaining will correspond to $S_{ij} = 0$ (indicated by Ω_{Sij}).

3.4. Multiplicity of the Limiting Solutions and Construction of the Spectrum

To keep track of all the branches we must know how many of them tend to a given limiting solution. The multiplicity of the bounded limiting solutions depends on the kind of perturbation in the intermediate layer as the limit is approached. If it is E, the multiplicity is 2 or 4, according to how many types of branches tend to the limiting solution. If it is P, the multiplicity is $2 \times \infty$ or $4 \times \infty$. The unbounded complex solutions have a $2 \times \infty$ multiplicity.

The properties we derived allow us to find all the roots of the four dispersion relations, among which the leaky waves are found. To this purpose we notice that as α varies, if we follow any one of the infinite roots $\Omega(\alpha)$ we see that its branch type may change at a real or complex transition point, or that it may change from real to complex (or vice-versa) at a double root, but it never disappears (or appears) suddenly nor tend to infinity for finite α. Then starting from the limiting solutions one can follow numerically any root. In addition, the periodic crossings and transitions allow to pick up any desired root among the infinite harmonics. In this way we can find all the roots and no significant solution will be omitted. The branches are analytic functions of the parameters of the problem, except at the transitions and crossings, so that it is not difficult to determine their behavior as the configuration is changed.

Acknowledgement

This work was supported by the CONICET and is part of Project EX 246 of the University of Buenos Aires.

References

Bernstein, I.B., Fireman, E.A., Kruskal, M.D. and Kulsrud, R.M.: 1957, *Proc. Roy. Soc.* **244 A**, 17.
Čadež, V.M. and Okretič, V.K.: 1989, *J. Plasma Phys.* **41**, 23.
Cally, P.S.: 1986, *Solar Phys.* **103**, 277.
Chen, L. and Hasegawa, A.: 1974, *Phys. Fluids* **17**, 1399.
Giovanelli, R.G.: 1975, *Solar Phys.* **44**, 299.
González, A.G. and Gratton, J.: 1991, *Solar Phys.* **134**, 211.
González, A.G., Gratton, J. and Farina, S.: 1998a, *Astrophys. Space Sci.* **256**, 297–301.
González, A.G., Gratton, J. and Farina, S.: 1998b, *Astrophys. Space Sci.* **256**, 303–309.
Gratton, J., Gratton, F. and González, A.G.: 1988, *Plasma Phys. Controlled Fusion* **30**, 435.
Ionson, J.A.: 1978, *Astrophys. J.* **226**, 650.
Rae, I.C. and Roberts, B.: 1983, *Solar Phys.* **84**, 99.
Roberts, B.: 1981, *Solar Phys.* **69**, 39.
Roberts, B. and Webb, A.R.: 1979, *Solar Phys.* **65**, 77.
Spruit, H.C.: 1982, *Solar Phys.* **75**, 3.
Wilson, P.R.: 1981, *Astrophys. J.* **251**, 756.

MHD LEAKY WAVES IN A LAYERED PLASMA: II MECHANISMS OF LEAKAGE

A.G. GONZÁLEZ, J. GRATTON and S. FARINA
INFIP-Lab. Física del Plasma, Facultad de Ciencias Exactas y Naturales, Universidad de Buenos Aires, Pab. I, Ciudad Universitaria, (1428) Buenos Aires, Argentina

Abstract. The application of Fourier analysis to study leaky waves has the advantage of simplicity, but it is not clear why the complex roots of the dispersion relations represent leaky waves, nor how the leakage occurs. We investigate the different kinds of leakage that can occur in a three layer plasma, and to which Fourier pseudomodes they are associated. We find four basic mechanisms, called "surface mode ...", "single interface ...", "trapped wave ..." and "lost insulation leakage". These mechanisms appear in pure form only near certain limiting cases, in which the parameters of the problem take some special values. As soon as the parameters depart significantly from the limiting values, the behavior of the leaky wave complicates and mixtures of the mechanisms occur, in varying amounts. In consequence different points of the same complex branch of the spectrum may correspond to different mechanisms. All the complex branches of the spectrum correspond to leaky waves, but in general it is not possible to classify them according to type of leakage, except close to a limiting case. Since in a three layer configuration there are many of these, the spectrum of leaky waves is very complicated.

1. Introduction

We discuss here the physical mechanisms that lead to the different types of leakage that take place in a three layer configuration. The usual approach to investigate leaky waves is based on Fourier analysis in time. This has the advantage of simplicity, but then the solution of initial value problems is not straightforward. Therefore, it is not obvious within this approach why complex solutions of the DR represent leaky waves, and how the leakage occurs. We address this problem heuristically, to avoid lengthy mathematics. Different types of leakage are possible.

The general case of a three layer plasma has been discussed in the first paper of this series (González et al., 1998a), and is very complicated since it depends on many parameters, and the solutions for arbitrary α must be found numerically. In addition most of the limiting solutions ($\alpha \to 0$ and $\alpha \to \infty$) are found solving high degree polynomial equations so that no closed form expressions are possible and no simple criteria for the existence of complex limiting solutions are available. Complex limiting solutions are related to a certain type of leakage, and may appear among the solutions of $T_{ij} = 0$ and $S_{ij} = 0$, but in the general case, they must be investigated numerically.

A simple case that helps in the present analysis occurs if the magnetic field and the pressure are uniform ($\mathbf{B}_1 = \mathbf{B}_2 = \mathbf{B}_3 \equiv \mathbf{B}$, $p_1 = p_2 = p_3 \equiv p$) so that β

Astrophysics and Space Science **256:** 297–301, 1998.

is uniform. Then $A_1 = A_2 = A_3 \equiv A$, $F_1 = F_2 = F_3 \equiv F$, $V_{\pm,1} = V_{\pm,2} = V_{\pm,3} \equiv V_\pm$, etc. (in this case $T_{ij}S_{ij} = 0$ has a trivial solution $\Omega = 0$). Simple analytical solutions can be found when $\mathbf{k}$ is parallel to $\mathbf{B}$ ($\psi = 0$), since then $V_A = A$, $V_- = \min(1, A)$, $V_+ = \max(1, A)$, and $V_B = A/F \equiv B$ (no confusion with $|\mathbf{B}|$ should arise). This is a degenerate case since V_A coincides with V_- or V_+, and $T_{ij}S_{ij} = 0$ can be reduced to the polynomial equation of degree 3, that after factoring out the trivial root $\Omega = 0$ yields:

$$(V_i^2 - 1)(V_j^2 - 1) + (A^2 - 1)/(A^2 + 1) = 0, \tag{1}$$

whose roots are:

$$V_i^2 = \frac{1}{2}\left[1 + m_{ji}^2 \pm \sqrt{\left(1 + m_{ji}^2\right)^2 - 8B^2 m_{ji}^2}\right]. \tag{2}$$

These roots will be complex if $m_- \leq m_{ji}^2 \leq m_+$, where:

$$m_\pm = F^{-2}\left(3A^2 - 1 \pm \sqrt{8A^2(A^2 - 1)}\right). \tag{3}$$

When $A < 1$ there are no complex limiting solutions. On the other hand, if $A > 1$, $m_\pm$ are real and the limiting solutions are complex if $m_- < m_{ji}^2 < m_+$. Summarizing, for $A > 1$ and appropriate temperature ratios, it is possible to obtain complex roots for $\alpha \to 0, \infty$.

2. Surface Wave Leakage

Consider first a leaky wave that originates from the coupling between a surface wave at the interface (2–3) and a propagating wave in region 1. In the limit $\alpha \to \infty$ these perturbations are decoupled and their frequency is real. When α is finite, the coupling subtracts energy from the surface wave, that is lost because it is carried to $y = -\infty$ by the propagating wave, so that the frequency becomes complex. The corresponding initial value problem consists in exciting the surface wave at the interface (2–3) at a given moment and then letting it decay, while the perturbation in region 1 propagates away from the interface (1–2). The perturbed part of region 1 is bounded by a front, whose distance to the interface increases with time. The front originated at the beginning of the process, when the amplitude of the surface wave was larger. In consequence, at a fixed time, the amplitude of the perturbation in region 1 increases with the distance from the interface and is maximum at the front (Cally, 1986). The solution of this initial value problem is represented by a packet of Fourier modes whose amplitude increases with y in the region 1 (where the propagating wave for $\alpha \to \infty$ is present). The Fourier components diverge in the limit $y = -\infty$ because they exist for any $-\infty < t < \infty$. Then this kind of leakage is related to complex roots of $\mathcal{S} = 0$ that become real as $\alpha \to \infty$. By the same argument one can show that the complex branches of type $\mathcal{U}$ that become real as $\alpha \to \infty$ are related to leaky waves due to the coupling of a surface wave at the interface (1–2) with a propagating wave in the region 3. These types of leakage will be called "surface wave leakage". However, there are other complex branches that become real for finite values of α, or that remain complex in the limit $\alpha \to \infty$. They are due to other types of leakage as we shall presently show.

3. Single Interface Leakage

Let us consider the complex limiting solutions. The existence of these leaky waves depend on the coupling of perturbations on both sides of a single interface i-j (interface 1–3 if $\alpha = 0$, interfaces 1–2 or 2–3 if $\alpha \to \infty$). To understand the physics involved let us consider the case of uniform pressure and magnetic field, with **k** parallel to **B**. The complex roots of the polynomial (1) are due to the matching of a slow propagating wave in one region with a fast one in the other. This can easily be verified by considering the case $A = 1 + \varepsilon$, $m_{ij} = 1 + \delta$ and $V_{i,j}^2 = 1 + \delta_{i,j}$ ($\varepsilon > 0$, ε, $|\delta|$, $|\delta_{i,j}| \ll 1$), when the perturbations in both regions propagate almost parallel to the interface and the leakage is small. To the lowest order one obtains the system $\delta_i \delta_j + \varepsilon = 0$, $\delta_i - \delta_j + 2\delta = 0$, that yields the complex roots

$$\delta_i = -\delta \pm i\lambda, \quad \delta_j = \delta \pm i\lambda \tag{4}$$

if $\lambda = \sqrt{\varepsilon - \delta^2}$ is real. Very close to the double roots ($\lambda \to 0$), $V_i \cong 1 - \delta/2$, $V_j \cong 1 + \delta/2$. If $\delta > 0$ there is a slow propagating wave in region i and a fast propagating wave in j, with $\Gamma_i \cong \lambda - i\delta$, $\Gamma_j \cong \lambda + i\delta$. We recall that the y component of the group velocity and of the energy flow vector of a slow mode has a sign opposite to that of $\alpha Im(\Gamma)$, while for a fast propagating wave the signs are equal. Then, if $\alpha > 0$, the energy flows in the positive y direction (if $j > i$ the flow goes from i to j while if $j < i$ the flow goes from j to i) and the spatial growth of the perturbations has the same direction. On the other hand, if $\alpha < 0$ the energy flow reverses and the perturbations decrease with y. When $\delta < 0$ we have a fast propagating wave in i and a slow one in j, and the directions of the energy flow and of the growth of the perturbations are opposite to those described above.

This type of leakage is related to the *cubic modes* (Cally, 1986) and it does not originate on the existence of a second interface, for this reason we shall call it "single interface leakage". However, the presence of the other interface (when α is finite and nonvanishing) modifies the pseudomodes as shown by numerical examples (González et al., 1998b).

4. Trapped Wave Leakage

A third different type of leakage that we call "trapped wave leakage" occurs when a propagating wave trapped in the intermediate region loses energy by coupling with a perturbation in one of the external regions. It can be studied by considering appropriate limiting values of the parameters and performing perturbation expansions similar to that described above. Thus it can be shown that this kind of leakage leads to complex $\mathcal{S}$, $\mathcal{U}$ and $\mathcal{V}$ branches that appear for α finite and nonvanishing (described in the numerical examples (González et al., 1998b)). We have verified this analytically in the case of uniform pressure and magnetic field, with **k** parallel to **B**. To find all the varieties of this type of leakage several limiting cases must be examined. A particular case that is of interest for a numerical example corresponds to $A < 1, m_{12} > 1$, when there is a fast propagating wave trapped in the intermediate layer, a perturbation that propagates parallel to the interface in region 1 ($V_1 = 1$), and no perturbation penetrates in region 3 ($V_3 = B$), so that $m_{32} = m_{12}/B$. This requires $\alpha = \alpha_0 = [(2n+1)\pi/2]\ [F^2(m_{12}^2 - B^2)/(m_{12}^2 - 1)(m_{12}^2 - A)]^{1/2}$ ($n = 0, 1, 2, \ldots$), so that the perturbation in the intermediate layer is a standing wave with a node at the interface 1–2 and a antinode at the interface 2–3. If we assume a small departure from this limiting case, so that $m_{32} = (m_{12}/B)(1 - \delta m)$, $\Omega = m_{12}^2(1 - \delta\Omega)$, $\alpha = \alpha_0(1 - \delta\alpha)$ ($\delta m, \delta\Omega, \delta\alpha \ll 1$), it can be shown that a damped type $\mathcal{S}$ solution appears, corresponding to leakage of the trapped wave into region 1. The analysis is straightforward but somewhat cumbersome and will not be given here. Since the trapping of propagating waves in the intermediate region always leads to a family of harmonics, their leakage correspondingly yields a family of complex branches. Then sets of leaky waves connected to periodic real branches originate from "trapped wave leakage" in the appropriate limit.

5. Lost Insulation Leakage

A fourth type of leakage can appear if in one of the external regions there is a wave propagating along the interface, but the perturbation cannot penetrate in the intermediate layer ($\Gamma_2 = \infty$). Then the intermediate layer acts as a perfect insulator, and no exchange of energy between the external regions may take place, even if α is very small (but nonzero). If the parameters of the configuration are slightly changed, the perturbation may penetrate in the intermediate region and a leakage to the other external region may occur. An example of this type of leakage happens in a symmetrical configuration, in the limiting case $\alpha \to \varepsilon$ ($0 < \varepsilon \ll 1$), with $V_2^2 \to V_B^2$ and $\Gamma_1 \to 0$, $\Gamma_3 \to 0$. Precisely in the limit the perturbations in each of the external regions are decoupled, but close to it there is a leakage from region 1 to 3 and viceversa, since the perturbation can penetrate in region 2.

6. Mixed Leakage

The mechanisms just discussed illustrate different ways in which leakage may occur. In the simple limiting cases considered above they appear in pure form, but as soon as the parameters depart significantly from the limiting values, mixtures between them may take place in varying amounts. Then a leaky branch can be driven by entirely different mechanisms at its end points. This is the case of the majority of the leaky waves that appear in the spectrum of the numerical examples.

All the complex $\mathcal{S}$, $\mathcal{U}$ and $\mathcal{V}$ roots are related to leaky waves. Since various different types of leakage may occur together in a given configuration, all of them must be considered in applications. Usually it is not possible to classify the leaky waves by a single driving mechanism, except in the neighbourhood of a limiting case. Therefore, when in the examples we ascribe a leakage mechanism to a certain complex branch, it must be understood that we are referring to the neighbourhood of the particular limiting case in consideration. In the three layer configuration there are many limiting cases in which leaky waves originate, and in consequence the spectrum is very complicated. It is not our aim to give a full discussion here, so that when studying the numerical solutions we shall mention the relevant limiting cases only for a few examples. However, all the leaky waves of the spectra that we study in the examples can be obtained by analytical continuation, starting from appropriate limiting cases of the types described above, that for this reason can be considered as the "basic" leakage mechanisms.

Acknowledgment

This work was supported by the CONICET and is part of Project EX 246 of the University of Buenos Aires.

References

Cally, P.S.: 1986, *Solar Phys.* **103**, 277.
González, A.G., Gratton, J. and Farina, S.B.: 1998a, *Astrophys. Space Sci.* **256**, 289.
González, A.G., Gratton, J. and Farina, S.B.: 1998b, *Astrophys. Space Sci.* **256**, 303.

MHD LEAKY WAVES IN A LAYERED PLASMA: III NUMERICAL SOLUTIONS

A.G. GONZÁLEZ, J. GRATTON and S. FARINA
INFIP-Lab. Física del Plasma, Facultad de Ciencias Exactas y Naturales, Universidad de Buenos Aires, Pab. I, Ciudad Universitaria, (1428) Buenos Aires, Argentina

Abstract. We study numerically the Fourier pseudomodes of complex frequency, that represent leaky waves, for two configurations consisting of three layers of uniform plasma in equilibrium, using the methods developed in the previous papers. The spectra consist of an infinite set of branches. We introduce a shorthand notation that allows us to identify the branches and we discuss their behavior. We show examples of leaky waves that in the neighborhood of certain limiting cases correspond to some of the four basic mechanisms discussed in a previous paper.

1. Introduction

The general problem of the MHD leaky waves in a plasma of three layers must be solved numerically, and it is pointless to deal with it, unless with a specific application in mind. However, to give a feeling of the complexity of the spectra and to illustrate the application of our methods we present two cases. We also discuss the connection of our results with those of others (Cally, 1986; Čadež and Okretič, 1989). We use the notation of González et al. (1998a, 1998b) and assume for simplicity $\mathbf{B}_i \equiv \mathbf{B}$, $p_i \equiv p$, then $A_i \equiv A$, $F_i \equiv F$, β is the same in the three layers and $f_{ij} = 1$ $(i, j = 1, 2, 3)$. Only ρ, T are discontinuous at the interfaces. If we fix $m_{ij} = \sqrt{T_i/T_j}$ the ratios ρ_i/ρ_j are given by the equation of state. A configuration is determined by A and, say, m_{21} and m_{32} (then $m_{31} = m_{32}m_{21}$). We further assume $\mathbf{k}$ parallel to $\mathbf{B}$, then $V_{A,i} = V_A = A$, $V_{+,i} = V_+ = \max(1, A)$, $V_{-,i} = V_- = \min(1, A)$, $V_{B,i} = A/F = B$ (no confusion with $|\mathbf{B}|$ shall arise). Since V_A coincides with V_+ (or V_-), a pole of G_i coalesces with the zero at $V_i = A$, and there are no crossings at $V_{1,3} = A$. Furthermore, there are a $\mathcal{U}$ and a $\mathcal{S}$ branch with $\Omega(\alpha) = 0$, and two $\mathcal{V}$ branches tend to $\Omega = 0$ in the limit $\alpha \to \infty$. The propagation bands in region 2 are $\Omega_{b2} < \Omega < \Omega_{-2}$ and $\Omega_{+2} < \Omega$.

We solve the four DR for $\Omega(\alpha) \propto \omega^2$ and plot $\mathrm{Re}(\Omega)$ (full lines) and $\mathrm{Im}(\Omega)$ (dotted lines, only $\mathrm{Im}(\Omega) > 0$ is drawn, $\mathrm{Im}(\Omega) < 0$ represents damping). We label the branches giving the type (underlined if the branch is complex) and the Ω where it starts and ends (in increasing order of α). The symbol "∞" denotes a branch unbounded at $\alpha = 0$ and "0" indicates that $\Omega(\alpha \to \infty) = 0$. Double roots where Ω changes from real to complex are denoted by Ω_{d1}, Ω_{d2}, etc. in increasing order of α. Then $\mathcal{T}[T31, +1]$ is a real $\mathcal{T}$ branch that starts at Ω_{T31} and ends at Ω_{+1},

Astrophysics and Space Science **256:** 303–309, 1998.

and $\underline{\mathcal{S}}[\infty, S-]$ is a complex $\mathcal{S}$ branch with the asymptote $\alpha = 0$ that ends at Ω_{S-}. The harmonics of a family begin and end at the same Ω's (usually the fundamental has different ends) and are distinguished by an index; for example, $\mathcal{U}[b2, -2]_1$ is the first harmonic of a family whose fundamental is the part of $\mathcal{U}[b2, T21]$ within the propagation band $\Omega_{b2} < \Omega < \Omega_{-2}$ (Figure 1). The lines in the graphs are continuous: they begin at $\alpha = 0$, and as α increases they may change from complex to real (and vice versa) or become a branch of another type at a transition point, to arrive at one of the limiting solutions as $\alpha \to \infty$.

2. Case I ($A = 1/5, m_{21} = \sqrt{3/5}, m_{32} = \sqrt{5}$)

With these parameters the limiting solutions are real, $3\Omega_{-1}/5 = \Omega_{-2} = \Omega_{-3}/5 = 1/5$, $3\Omega_{+1}/5 = \Omega_{+2} = \Omega_{+3}/5 = 1$, $\Omega_{a2} = 1/5$, $3\Omega_{b1}/5 = \Omega_{b2} = \Omega_{b3}/5 = 1/6$, crossings occur at $\Omega_{T-} = 1.66$ (period $= 4.274$), complex transitions at $\Omega_{S+} = 6.22$ (period $= 1.51$) and $\Omega_{S-} = 5.44$ (period $= 1.63$), $\Omega_{T31} = 0.44$, $\Omega_{S31} = 6.22$, $\Omega_{T21} = 0.22$, $\Omega_{S21} = 2.44$, $\Omega_{T32} = 0.29$ and $\Omega_{S32} = 5.71$. Real roots (perturbations EEE or EPE) occur for $0 \le \Omega \le \Omega_{b1}$, $\Omega_{-1} \le \Omega \le \Omega_{b3}$ and $\Omega_{-3} = \Omega_{+2} \le \Omega \le \Omega_{+1}$. The spectrum is shown in Figure 1: (a) lower part ($0 \le \mathrm{Re}(\Omega) \le 0.5$), (b) middle part ($0 \le \mathrm{Re}(\Omega) \le 2.5$), omitting some branches shown in (a), (c) upper part ($2.5 \le \mathrm{Re}(\Omega) \le 7.5$). Some branches unrelated to leaky waves are omitted.

Single interface leakage (González et al., 1998b) does not appear ($A < 1$). Pure *surface wave leakage* occurs for $\alpha \to \infty$. For instance, the branch $\underline{\mathcal{S}}[d4, T32]$, that tends to Ω_{T32} as $\alpha \to \infty$, describes near this limit leakage due to the coupling between a surface wave at the interface (2–3) and a propagating wave in 1; the leakage becomes mixed as α decreases. There are many instances of *mixed leakage*, related to limiting cases of pure *trapped wave leakage* (González et al., 1998b). One is $\underline{\mathcal{S}}[d2, d3]$, that is the analytical continuation of a limiting case discussed in (González et al., 1998b) (notice the harmonics in the propagation band $\Omega > \Omega_{+2}$). The fundamentals $\mathcal{S}[d3, +1]$ and $\mathcal{V}[-3, +1]$ change respectively to $\mathcal{T}[+1, +2]$ and $\mathcal{U}[+1, +2]$ at Ω_{+1} (both approach Ω_{+2} as $\alpha \to \infty$. The harmonics of the four families tend to Ω_{+2} as $\alpha \to \infty$; they are connected in pairs $\mathcal{S} \leftrightarrow \mathcal{T}$ and $\mathcal{V} \leftrightarrow \mathcal{U}$ at Ω_{+1} (periodic real transition). To each pair is associated a double root of $\mathcal{S}$ or $\mathcal{V}$ where complex branches (related to trapped wave leakage as $\underline{\mathcal{S}}[d2, d3]$) coalesce. Notice the periodic crossing Ω_{T-} between the $\mathcal{T}$ and $\mathcal{V}$ branches. The $\underline{\mathcal{S}}$ branches of the upper part of the spectrum have a complicate behavior. The first is $\underline{\mathcal{S}}[\infty, d5]$. The second, $\underline{\mathcal{S}}[S-, d7]$, originates from $\underline{\mathcal{U}}[S0, S-]$. It begins at $\alpha = 0$ and suffers complex transitions at Ω_{S-} and Ω_{S+}. The third branch $\underline{\mathcal{S}}[S-, d9]$ and the higher harmonics, connect with branches $\underline{\mathcal{S}}[\infty, S+]$ by means of two complex transitions at Ω_{S+} and Ω_{S-}.

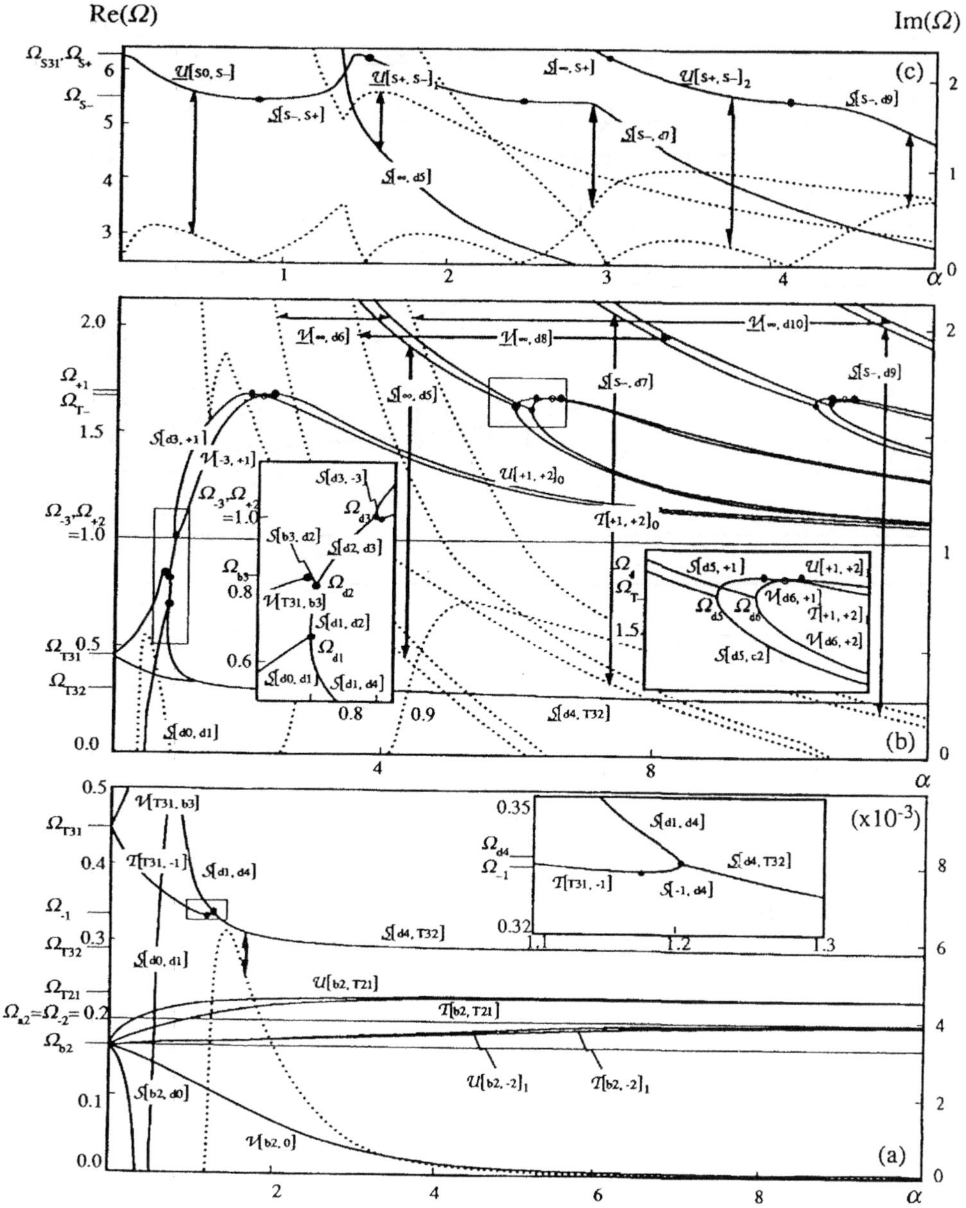

Figure 1. Spectrum of Case I ($A = 1/5$).

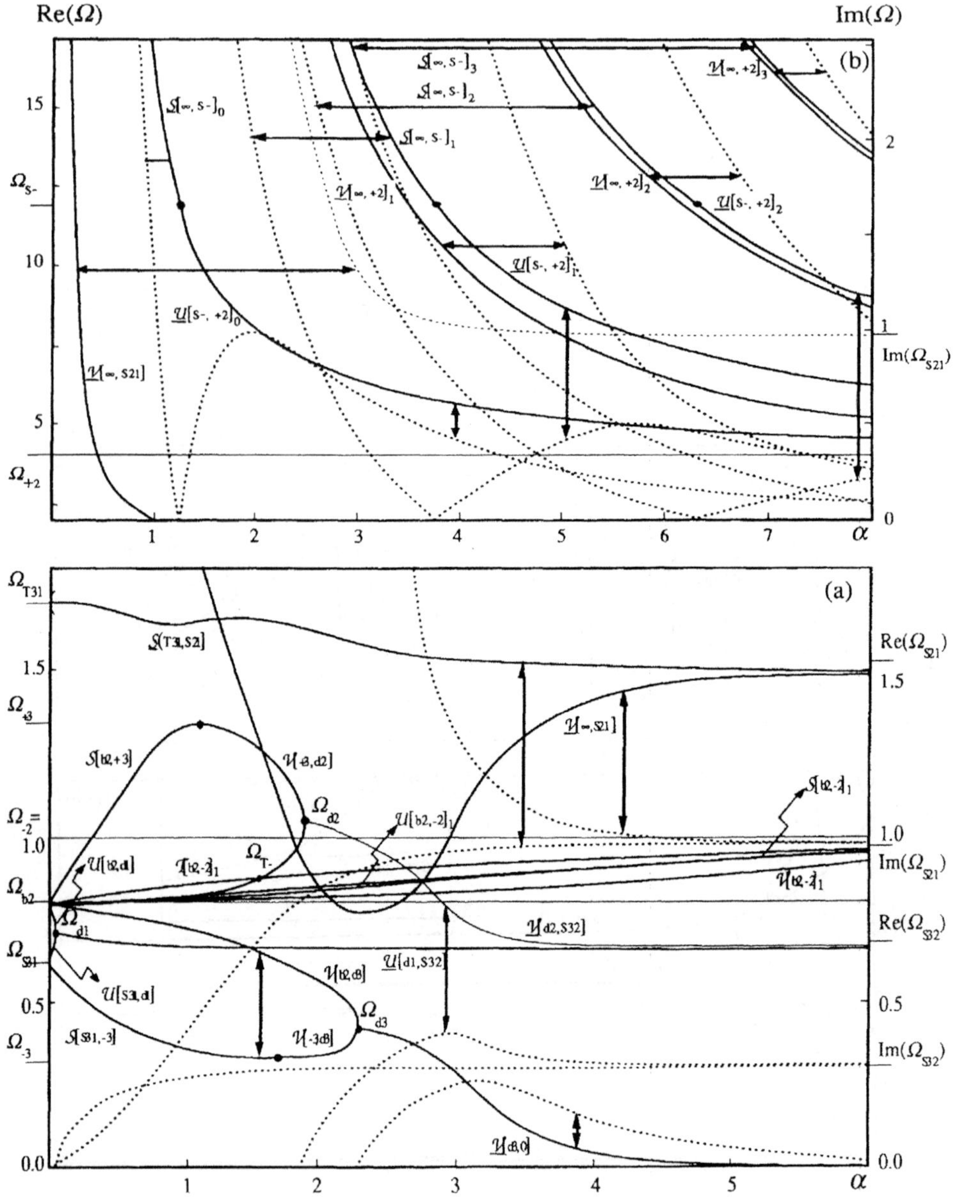

Figure 2. Spectrum of Case II ($A = 2$).

3. Case II ($A = 2, m_{21} = 1/\sqrt{2}, m_{32} = 1/\sqrt{3}$)

In this case $m_- = 0.24$, $m_+ = 4.16$ and m_{21}, m_{32} are such that: (a) there are complex roots when $\alpha \to \infty$ and real roots when $\alpha = 0$, (b) the combination PEE does not occur (according to Čadež and Okretič (1989) we need this combination to have leaky waves). We have $\Omega_{-1}/2 = \Omega_{-2} = 3\Omega_{-3} = 1$, $\Omega_{+1}/2 = \Omega_{+2} = 3\Omega_{+3} = 4$, $\Omega_{a2} = 4$, $\Omega_{b1}/2 = \Omega_{b2} = 3\Omega_{b3} = 4/5$, crossings and complex transitions occur only for $\Phi = -1$, at $\Omega_{T_-} = 0.87$ (period $= 3.08$) and $\Omega_{S_-} = 11.91$ (period $= 2.52$) respectively, there are no roots of $T_{21} = 0$ and $T_{32} = 0$, $\Omega_{T31} = 1.71$, $\Omega_{S31} = 0.62$, $\Omega_{S21} = 1.50 + 0.97i$ and $\Omega_{S32} = 0.66 + 0.30i$. Real roots occur for $0 \leq \Omega \leq \Omega_{b3}$ and $\Omega_{-3} \leq \Omega \leq \Omega_{+3}$. Figure 2 shows the spectrum for: (a) $0 \leq \mathrm{Re}(\Omega) \leq 2$, (b) $2 \leq \mathrm{Re}(\Omega) \leq 20$.

The branches $\underline{\mathcal{U}}[d1, S32]$, $\underline{\mathcal{V}}[d2, S32]$ and $\underline{\mathcal{V}}[\infty, S21]$, $\underline{\mathcal{S}}(T31, S21]$ tend to the complex limiting solutions Ω_{S32} and Ω_{S21} and in this limit represent *single interface leakage*. The branch $\underline{\mathcal{V}}[d3, 0]$ is the analytical continuation of the limiting symmetrical case $\alpha \to \epsilon$, $(0 < \epsilon \ll 1)$, with $\Omega \to B^2$ and $\Gamma_{1,3} \to 0$, in which *lost insulation leakage* takes place (González et al., 1998b). The branch $\underline{\mathcal{S}}(T31, S21]$ tends to Ω_{T31} as $\alpha \to 0$ in a singular way (denoted by using a parenthesis instead of a square bracket). According to González et al. (1998a), at $(0, \Omega_{T31})$ the branches $\mathcal{T}$ and $\mathcal{V}$ coincide, but these branches cannot exist near $\alpha = 0$. Actually Ω_{T31} is real and corresponds to PEP, for which there are no real branches, except in the special case $\Phi = 1$, i.e. $\alpha = 0$. For $\alpha \neq 0$ the solution must be complex, but this is forbidden for $\mathcal{T}$ branches. Then $\underline{\mathcal{S}}(T31, S21]$ exists for a arbitrarily small but non vanishing. In the propagation band $\Omega > \Omega_{+2}$ (upper part of the spectrum) we find complex transitions Ω_{S_-} between pairs of complex branches $\underline{\mathcal{S}}[\infty, S-]$ and $\underline{\mathcal{U}}[S-, +2]$. The $\underline{\mathcal{U}}[S-, +2]$ branches tend to the real limiting solution Ω_{+2} as $\alpha \to \infty$. In this propagation band there are also $\underline{\mathcal{V}}[\infty, +2]$, branches whose imaginary parts vanish as $\alpha \to \infty$, while the real parts tend to Ω_{+2}. We draw the first four harmonics of both families. All these leaky waves tend to a particular degenerate case (since $\Omega_{a2} = \Omega_{+2}$) of *trapped wave leakage* for very large α.

4. Discussion and Conclusions

In a plasma of three plane layers in equilibrium we study complex frequency oscillations that describe the decay of a perturbation due to the transfer of energy between different regions. The dispersion relations corresponding to the different kinds of perturbations are transcendental and very complicated, but relevant information, including the transitions and crossings between branches and the periodicity bands can be obtained in closed form. The complex branches are solutions of $\mathcal{S} = 0$, $\mathcal{U} = 0$ and $\mathcal{V} = 0$ and represent leaky waves. We identify four basic leakage mechanisms, that we call *surface wave ...*, *single interface ...*, *trapped wave* ... and *lost insulation leakage*. Several of these may operate in a given configuration, but they appear in pure form only in some special limiting cases. As the parameters depart significantly from these, mixtures of the mechanisms take place in varying amounts as seen in the examples, and it is impossible to ascribe a single driving mechanism to a leaky branch, except close to a limiting case. Both cases admit leakage for different α but the decay rates vary considerably. We show limiting cases of the basic mechanisms, from which all mixed leaky waves can be obtained by analytical continuation. In a three layer configuration there are many limiting cases and the spectrum is very complex, but our methods allow to investigate any case, since no restriction on the configuration (other than equilibrium) and on **k** is made.

A plane model of three slabs was studied previously (Čadež and Okretič, 1989), with parameters such that perturbations of type PEE exist (it is impossible to verify the calculations, since the full list of parameters is not reported, but it can be checked that the unperturbed state is not in equilibrium). The idea was to excite a surface wave around $y = a$, whose amplitude decays as it transfers its energy to a propagating wave in region 1 (*surface wave leakage*). The authors hoped to find leaky waves among the roots of $\mathcal{T} = 0$. They started with $\alpha \to \infty$ (two uncoupled interfaces), solved the equation $T_{32} = 0$, and used the resulting root as a guess to solve $\mathcal{T} = 0$ for finite a. They actually found complex roots for $3 < \alpha < 7$, that is probably wrong since (if the configuration is in equilibrium!) all the roots of $\mathcal{T} = 0$ are real (González et al., 1998a). The result of Čadež and Okretič (1989) is faulty because: (a) the unperturbed state is not in equilibrium, (b) for an equilibrium, $\mathcal{T} = 0$ does not lead to leaky waves, (c) the dispersion relations that yield leaky waves were not considered, (d) *surface wave leakage* is not the only driving mechanism, and (e) a wrong necessary condition for the existence of leaky waves is given, namely that the configuration must allow PEE perturbations; this is contradicted by case II, where these perturbations are impossible, but leaky waves occur.

The *single interface leakage* is related to the *cubic modes* of Cally (1986) that occur in a cylindrical configuration of two uniform media, when the radius of the interface is much larger than the wavelength of the perturbation. In this limit the interface can be approximated by a plane. In the cylindrical cubic modes the $\mathcal{T}$ and $\mathcal{S}$ branches are weakly coupled, and the leaky waves are roots of $\mathcal{S} = 0$. Then our limiting solutions for $\alpha = 0$ are equivalent to the cubic modes. We checked this by solving $S_{31} = 0$ for the parameters of Cally (1986) and found a perfect agreement.

Acknowledgement

This work was supported by the CONICET and is part of Project EX 246 of the University of Buenos Aires.

References

Cally, P.S.: 1986, *Solar Phys.* **103**, 277.
González, A.G., Gratton, J. and Farina, S.: 1998a, *Astrophys. Space Sci.* **256**, 289.
González, A.G., Gratton, J. and Farina, S.: 1998b, *Astrophys. Space Sci.* **256**, 297.
Čadež, V.M. and Okretič, V.K.: 1989, *J. Plasma Phys.* **41**, 23.

GROWTH RATES OF ENVELOPE MODULATIONS OF ELECTROMAGNETIC WAVES IN RELATIVISTIC TEMPERATURE ELECTRON-POSITRON PLASMAS, STIMULATED BY WEAK OR FINITE PHONON DAMPING

F.T. GRATTON and G. GNAVI
Instituto de Física del Plasma, Consejo Nacional de Investigaciones Científicas y Técnicas, Facultad de Ciencias Exactas y Naturales, Universidad de Buenos Aires, Ciudad Universitaria, Pabellón 1, 1428 Buenos Aires, Argentina

R.M.O. GALVÃO
Instituto de Física, Universidad de São Paulo, CP 66318, 05389-970 São Paulo, SP, Brasil

L. GOMBEROFF
Departamento de Física, Facultad de Ciencias, Universidad de Chile, Casilla 653, Santiago, Chile

Abstract. We study the modulational instability of an electron-positron plasma with relativistic temperatures and phonon damping. It is shown that when the phonon damping is $\mathcal{O}(1)$ or $\mathcal{O}(\varepsilon)$, a modulational instability appears even for classical temperatures. When the phonon damping is $\mathcal{O}(\varepsilon^2)$), we find that in the non-relativistic limit, previous results are recovered, and for ultrarelativistic temperatures, the instability occurs in a wider band around the relativistic plasma frequency.

1. Introduction

The modulational instability of a linearly polarized, electromagnetic wave in a non-magnetized positron-electron plasma, is studied using relativistic two-fluid hydrodynamics to properly account for physical regimes of very high temperatures. These are expected in early Universe epoch extending from about 10^{-4} to 1 s, after the big bang. That is, from the phase of neutrinos decoupling and disappearance of muons, to the time of annihilation of positrons and electrons (Dolgov and Zeldovich, 1981). The electromagnetic wave is nonlinearly coupled to phonons. The acoustic modes are supported by radiation pressure. The theory is also of interest for hot plasma scenarios associated with active galactic nuclei.

The effect of phonon damping is included in the treatment, and it is shown to have a crucial role in the development of the modulational processes. The presence of some amount of sound absorption helps to produce the envelope decay, which may not occur otherwise. The theory shows that for classic temperatures, and in the absence of damping, only a vanishingly small frequency interval (at $\omega \simeq \omega_p$) appears, where the instability is possible in a non-magnetized plasma (Kates and Kaup, 1989a).

Astrophysics and Space Science **256:** 311–319, 1998.

Our theory can be formulated by a pair of extended Zakharov equations for the vector potential of the wave, A, coupled with phonons, i.e. with the sum of perturbed electron and positron density, N (here $A = eA_x/mc^2$, $N = (n_p + n_e - 2n_0)/n_0$, n_0 is the unperturbed density, c: speed of light, $e > 0$, m: positron charge and mass, cgs-ues units are used throughout). These equations are exact to third order in the non-dimensional wave amplitude, which is taken as the expansion parameter ε. They are obtained without the usual time average process, typical of similar equations in ordinary ion-electron plasmas. The envelope modulation is then studied via the corresponding NLS (nonlinear Schrödinger) equation, derived by multi-scale perturbation analysis. According to the intensity of the damping, finite (order 1), or weak (first order in ε), or ultraweak (second order in ε), three different types of NLS are obtained (for details, see Gratton et al., 1997).

Here we examine the interesting case of weak damping, where the NLS is an integro-differential equation. Growth rates and instability regions for the self modulational processes are given, and compared with similar properties of finite damping case. In both cases the instability appears at all frequencies of the electromagnetic wave, in contrast to the case of ultraweak damping, where it is restricted to a narrow frequency band.

The analysis is based on weakly relativistic oscillations, $\gamma \cong 1 + (1/2)(v/c)^2$, of the $e^+ - e^-$ plasma in the field of a nonlinear electromagnetic wave. However, in early time Universe plasma the electron and positron random energy is much larger than the rest mass energy, so that electrons and positrons behave dynamically as photons. At this high temperatures, $T \gg mc^2$, positrons and electrons coexist with a high frequency $h\nu \sim T$ photon gas. This is modelled with a radiation pressure $p_\pm = \text{const}.n_\pm^{4/3}$ to be added to the gas pressure. Positron and electrons follow a similar adiabatic law (Tajima and Taniuti, 1990).

We employed the momentum equation of relativistic hydrodynamics

$$\frac{h}{c^2}\frac{d}{dt}(\gamma \mathbf{v}) = -\frac{1}{\gamma}\nabla p - \frac{\mathbf{v}}{c^2}\gamma\frac{dp}{dt} + nq\left(\mathbf{E} + \frac{1}{c}\mathbf{v}\times\mathbf{B}\right),$$

that shows the enhancement of inertia (h/c^2 instead of nm, h, enthalpy) due to internal energy and pressure, and the extra force term originated from dp/dt on the RHS (there are, of course, two equations for e^+ and e^-).

2. Coupling of ELM Waves and Phonons

For a linearly polarized electromagnetic wave with $E_x = E_x(z,t)$ and $B_y = B_y(z,t)$ propagating along the z axis, we have $A = (A_x(z,t), 0, 0)$ and $v = (v_x(z,t), 0, v_z(z,t))$, where $v_z(z,t)$ is generated by the $v_x B_y$ term of the Lorentz force. Operating on momentum, continuity, and Maxwell equations, with an appropriate ordering in $\varepsilon = |Anmc^2/h|$, the main set of equations of the model can be obtained. Using new variables $\hat{z} \equiv \sqrt{\eta}z$, $\hat{t} \equiv \sqrt{\eta}t$, $\hat{A} \equiv \eta A$, and defining two non-dimensional constants δ and η as $\delta \equiv \Gamma p_0/h_0$, $\eta \equiv n_0 mc^2/h_0$, the coupled equations for the nonlinear wave (correct to order ε^3) can be written as

$$\left(\frac{\partial^2}{\partial \hat{t}^2} - c^2 \frac{\partial^2}{\partial \hat{z}^2}\right) \hat{A} + 2\omega_p^2 \hat{A} = \omega_p^2 (1-\delta)\hat{A}(\hat{A}^2 - \mathcal{N}),$$

$$\left(\frac{1}{c^2}\frac{\partial^2}{\partial \hat{t}^2} - \delta \frac{\partial^2}{\partial \hat{z}^2}\right) \mathcal{N} = (1-\delta)\frac{\partial^2}{\partial \hat{z}^2}\hat{A}^2.$$

Here $\hat{A}$ and $\mathcal{N}$ must be small with respect to the unity; ω_p is the plasma frequency of electrons (or, positrons).

For low thermal energies, case (a), $\eta = 1 - 3/2$, $\delta = v_s^2/(c^2 + (3/2)v_s^2) \ll 1$, where $v_s^2 = (5/3)p_0/(mn_0)$ is the classic speed of sound.

For ultrarelativistic (random) energies, case (b), $\delta = 1/3$, and $\eta \ll 1$. The speed of sound is given now by $v_s^2 = \delta c^2 = (1/3)c^2$.

An important element, a phenomenological phonon damping rate υ, is introduced in the model. To simplify notation we write the last equations as

$$(\partial_t^2 - c^2\partial_z^2 + 2\omega_p^2)A = (1-\delta)\omega_p^2 A(A^2 - \mathcal{N}), \tag{1}$$

$$(\partial^2 - v_s^2\partial_z^2 + \nu\partial_t)\mathcal{N} = (1-\delta)c^2\partial_z^2 A^2. \tag{2}$$

Here, in case (a) (non-relativistic thermal energy) $\delta = v_s^2/c^2$, $v_s^2 = (5/3)p_0/(mn_0)$, t and z are the natural variables and A, $\mathcal{N}$ are defined above. However, for case (b) (ultrarelativistic thermal energy) $\delta = 1/3$ and $v_s^2 = c^2/3$. Moreover, in this case A stands for $\hat{A}$, and t, z represent the rescaled variables $\hat{t}$, $\hat{z}$.

3. Equations for the Amplitude Envelope

Nonlinear Schrödinger equations for (1) and (2) can be derived with time-space multiscale perturbation techniques (Tajima and Taniuti, 1990). We formally write $A = \varepsilon A_1 + \varepsilon^2 A_2 + \varepsilon^3 A_3 + \mathcal{O}(\varepsilon^4), \ldots$, etc., with $\varepsilon \ll 1$. The expansion of ν in powers of ε allows treatment of weaker dampings by taking $\nu_0 = 0$, $\nu_0 = \nu_1 = 0$, etc.

To the lowest significant order $A_1 = ae^{i\theta} + a^* e^{-i\theta}$, $\mathcal{N}_1 = 0$, where $\theta = kz - \omega t$ is the fast variable and k, ω satisfy the dispersion relation $D(\omega, k) = -\omega^2 + k^2c^2 + 2\omega_p^2 = 0$, for the transverse electromagnetic wave. The amplitudes a, a^*, are slow space-time variables and the perturbative expansion assumes that $A_{2,3} = A_{2,3}(a, a^*, \theta)$, $\mathcal{N}_{2,3} = \mathcal{N}_{2,3}(a, a^*, \theta)$, so that the space-time dependence of higher order corrections is through a, a^*, and θ only.

Upon elimination of secularities at each perturbative order (and setting $\partial v_g/\partial k = 2p = 2\omega^2 c^2/\omega^3$) the NLS equations

$$i\frac{\partial a}{\partial \tau} + \frac{1}{2}\frac{\partial v_g}{\partial k}\frac{\partial^2 a}{\partial \xi^2} + \frac{1}{2}(1-\delta)\frac{\omega_p^2}{\omega}a[(3-\alpha)|\alpha|^2 - C] = 0, \tag{3}$$

can be obtained. Here we have introduced new space-time variables $\tau = \varepsilon^2 t$, $\xi = \varepsilon(z - v_g t)$, and α is a complex parameter defined by

$$\alpha = \frac{(1-\delta)c^2k^2}{\omega(\omega + i\nu_0/2) - v_s^2k^2} \,. \tag{4}$$

In Equation (3) $C(a, a^*)$ is a real function of the slow variables that is determined as commented in the following, by the conditions for avoiding secularities.

It can be shown that the value of C depends on the intensity of damping. Basically, there are three different cases: (i) $\nu_0 > 0$ (order-one damping); (ii) $\nu_0 = 0$ and $\nu_1 > 0$ (weak damping); and (iii) $\nu_0 = \nu_1 = 0$ and $\nu_2 > 0$ (ultraweak damping).

In case (i), $\nu_0 > 0$, C cannot depend on a or a^*: it must be an absolute constant. It follows that $C = 0$, which ensures that for $a = 0$ a density perturbation is absent, so that $\mathcal{N}_2 = 0$. We obtain an NLS equation with complex coefficients, since α is complex (Equation (3) with $C = 0$).

In case (ii), when $\nu_0 = 0$, the elimination of secularities follows a different route and $C(a, a^*)$ is no longer zero. Carrying through the analysis, we find that

$$C = C_0|a|^2 - C_0C_1e^{C_1\xi}\int_{\xi}^{\infty} e^{-C_1\zeta}|a|^2 d\zeta, \tag{5}$$

with $C_0 = 2(1-\delta)c^2/(v_g^2 - v_s^2)$, and $C_1 = \nu_1 v_g/(v_g^2 - v_s^2)$. Therefore, the NLS equation takes the form

$$i\frac{\partial a}{\partial t} + \frac{1}{2}\frac{\partial v_g}{\partial k}\frac{\partial^2 a}{\partial \xi^2} + \frac{(1-\delta)\omega_p^2 v_g}{2kc^2} \times \left[\left(3 - \frac{c^2k^2(1-\delta)}{\omega^2 - v_s^2k^2} - C_0\right)|a|^2 + C_0C_1e^{C_1\xi}\int_{\xi}^{\infty} e^{-C_1\zeta}|a|^2 d\zeta\right] a = 0. \tag{6}$$

In case (ii), $C_1 \neq 0$, the NLS becomes an integrodifferential equation with real coefficients.

Finally, for ultraweak damping, case (iii), Equation (5) holds with $C_1 = 0$, so that $C = 2(1-\delta)c^2|a|^2/(v_g^2 - v_s^2)$, and α is real ($\nu_0 = 0$). Ultraweak damping reproduces the NLS equation obtained in (Kates and Kaup, 1989a), which was derived for an $e^+ - e^-$ plasma at non-relativistic temperatures ($v_s^2 \ll c^2$) and without damping effects.

4. Weak Damping

Let us look more closely at the weak damping case ($\nu_0 = 0$, $\nu_1 \neq 0$). The integrodifferential NLS has a solution of constant envelope of the form $\bar{a}(\xi, \tau) =$

$a_0 \exp\{i(k_0\xi - \omega_0\tau)\}$, with a_0 a real constant. This solution has the peculiarity that C given by Equation (5) is zero. Thus the nonlinear wave with constant envelope is described by the same NLS of case (i), but with real α, since here $\nu_0 = 0$.

The small perturbations of this solution are a different matter, since the constant C given by Equation (5) is not zero in this case. Perturbing the solution as

$$\begin{aligned} a(\xi,\tau) &= \bar{a}(\xi,\tau) + \delta a_1 \exp\{i[(K+k_0)\xi - (\Omega+\omega_0)\tau]\} \\ &\quad + \delta a_2 \exp\{-i[(K-k_0)\xi - (\Omega^* - \omega_0)\tau]\}, \end{aligned}$$

we find that in order to satisfy Equation (6), Ω and K must be related by the following dispersion relation

$$\begin{aligned} \Omega &= 2pk_0K \\ &\pm \left(p^2K^4 - 2p\left(q - rC_0\frac{K^2}{C_1^2+K^2}\right)\alpha_0^2K^2 - 2irpC_0a_0^2\frac{C_1K^3}{C_1^2+K^2}\right), \end{aligned} \tag{7}$$

where q is equal to q_r, defined in Section 5, with $v_0 = 0$, and $r = (1-\delta)\omega_p^2 v_g/2kc^2$.

From this equation we conclude that for any given frequency of the wave conditions for the modulational instability can be achieved. The imaginary term in the square root is a source of instability for all values of ω and K. This term depends on the damping ν_1 and disappears when $\nu_1 \to 0$. The destabilizing effect of the imaginary term, for fixed frequency, increases at large values of K, but becomes negligible when $v_g \to 0$ instead, i.e., for large wavelengths, when ω is close to $\sqrt{2}\omega_p$.

On the other hand, an instability can also arise, independently from the damping ν_1, when the sum of the real terms in the square root in negative. This may happen (under restrictive conditions on ω and K) even if C_1 is negligible. We shall comment upon the conditions for this instability in Section 6. Finally, let us note that putting formally $C_0 = 0$ as a control of the dispersion relation, we reobtain the properties of the dispersion relation of finite damping in the limit $\nu_0 \to 0$. In fact setting $C_0 = 0$, Equation (6) coincides with Equation (3) with $\nu_0 = 0$.

5. Finite Damping

Let us now compare these results with those of the envelope instability for damping effects of order $\mathcal{O}(1)$. For consistency we assume $\varepsilon \ll \nu_0/\omega \ll 1$. We have a nonlinear Schrödinger equation where the cubic term is $qa|a|^2$, with complex coefficient $q = q_r + iq_i$,

$$q_r = \frac{(1-\delta)\omega_p^2}{2\omega}\left(3 - \frac{(1-\delta)c^2k^2(\omega^2 - v_s^2k^2)}{(\omega^2 - v_s^2k^2)^2 + (\nu_0\omega/2)^2}\right),$$

and

$$q_i = \frac{\nu_0}{4} \frac{(1-\delta)^2 \omega_p^2 c^2 k^2}{(\omega^2 - v_s^2 k^2)^2 + (\nu_0 \omega/2)^2},$$

where we assume that $|q_i| \ll |q_r|$.

Consider a solution of the form

$$\hat{a}(\xi, \tau) = \frac{a_0}{\sqrt{1 + 2q_i a_0^2 \tau}} \exp\left\{ i(k_0 \xi - p k_0^2 \tau + \int \dot{\phi} d\tau) \right\},$$

where $\dot{\phi}(r) = q_r a_0^2/(1 + 2q_i a_0^2 \tau)$, and a_0 is a real constant. This solution decays slowly (i.e., not exponentially) in time. Introducing small perturbations $\delta a_{1,2}$ to the solution,

$$\begin{aligned} a(\xi, \tau) &= \bar{a}(\xi, \tau) + \delta a_1 \exp\left\{ i[(K + k_0)\xi - p k_0^2 \tau - \int (\Omega - \dot{\phi}) d\tau] \right\} \\ &\quad + \delta a_2 \exp\left\{ -i[(K - k_0)\xi + p k_0^2 \tau - \int (\Omega^* + \dot{\phi}) d\tau] \right\}, \end{aligned}$$

and linearizing the following dispersion equation can be obtained

$$\Omega = 2(p k_0 K - i q_i |\bar{a}|^2) \pm \sqrt{p^2 K^4 - 2 p q_r |\bar{a}|^2 K^2 - q_i^2 |\bar{a}|^4}, \tag{8}$$

where $|\bar{a}|^2 = a_0^2/(1 + 2q_i a_0^2 \tau)$.

Conditions for instability arise for values of K where the expression $S(K) = p^2 K^4 - 2 p q_r |\bar{a}|^2 K^2 + 3 q_i^2 |\bar{a}|^4$ becomes negative. Thus, we find an instability band in the interval (K_1, K_2) with

$$K_{1,2} = \sqrt{|\bar{a}|^2 \frac{q_r}{p} \pm \sqrt{|\bar{a}|^4 \frac{q_r^2}{p^2} - 3|\bar{a}|^4 \frac{q_i^2}{p^2}}}. \tag{9}$$

Note that the growth rate is maximum when $S(K)$ is minimum, that is, for $K_M = (|\bar{a}|^2 q_r/p)^{1/2}$, and its value is $\Im(\Omega) = |\bar{a}|^2(|q| - 2q_i)$.

A stability diagram is shown in Figure 1, where in a semilog graph the ordinates Y_1, Y_2 represent K_1, K_2, measured in units of $\varepsilon|\bar{a}|/d$, *versus* the abscissa $x = \omega/\sqrt{2\eta}\omega_p$ (here $d = c/(\sqrt{2\eta}\omega_p)$). The instability occurs for $Y_1 < Y < Y_2$. In the figure the lines are represented for two damping values, $\nu/\sqrt{2\eta}\omega_p = 0.05$, and 0.5. In addition, the figure shows lines for two temperatures values. The lines are dash-dotted for $v_s/c = 0.01$ and full for $1/\sqrt{3}$.

We can see from Figure 1 that the temperature range, classic or ultrarelativistic, has some influence on the boundaries of the unstable region. The range $Y_1 - Y_2$, at a given frequency, is more narrow at high energies. The lower boundary is also sensitive to changes of the parameter, $\nu/\sqrt{2\eta}\omega_p$, while the effect on the upper

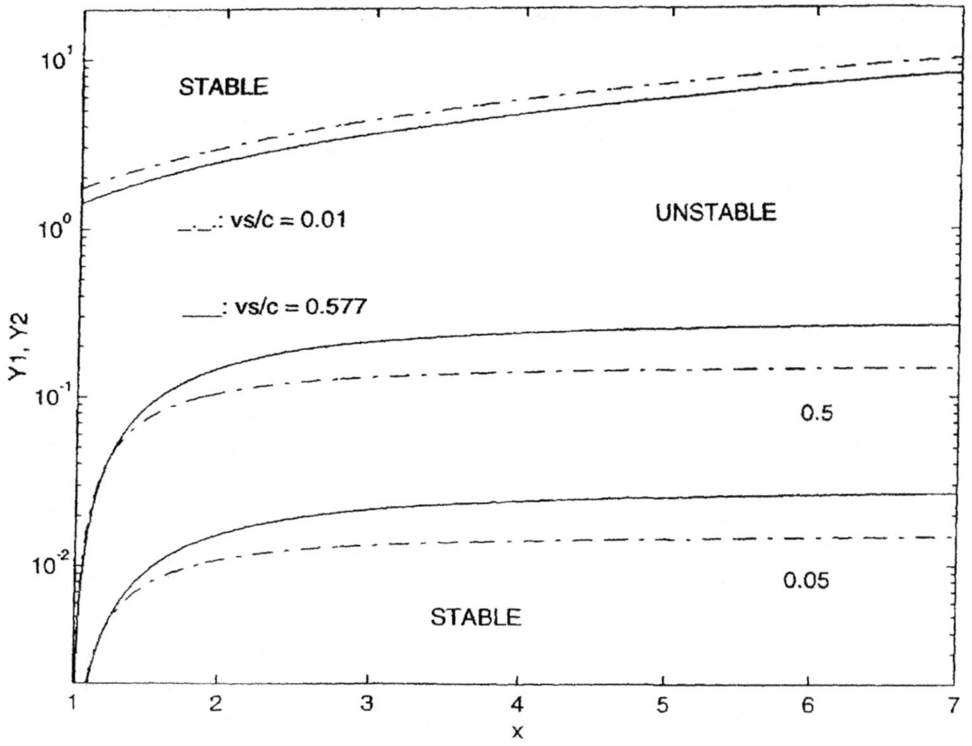

Figure 1. Stability diagram for finite phonon damping, Y_1 and Y_2 *versus* $x = \omega/\sqrt{2\eta}\omega_p$.

boundary is negligible. The boundary Y_1 increases by an order of magnitude when the damping is increased ten times, and so the unstable K interval decreases. The region of instability exists for all frequencies of the wave, but the growth rate tends to zero as $\omega \rightarrow \infty$. Note that, while the characteristics of the instability region depend on the presence of finite damping, the direct effect of ν_0 on the growth rate (in view of the assumption $|q_r| \ll |q_i|$) is not crucial. The instability persists even in the limit $\nu_0/\omega \rightarrow 0$.

6. Ultraweak Damping

Let us consider finally the ultraweak damping case. We find, after rescaling variables, an unstable frequency interval, $\sqrt{2\eta}\omega_p \leq \omega < 1.2247\sqrt{2\eta}\omega_p$. When the temperature is non-relativistic, the unstable frequency interval is, $\sqrt{2}\omega_p \leq \omega\sqrt{2}\omega_p(1 - v_s^2/2c^2)$, instead. Since $v_s^2/c^2 \ll 1$, the unstable frequencies are restricted to a very small interval close to $\sqrt{2}\omega_p$, whose width tends to zero for a cold plasma. Conversely, as the temperature grows the unstable frequency interval increases, and becomes a finite frequency band at ultrarelativistic temperatures.

Taking the limit $\nu_1 \rightarrow 0$ in the dispersion relation of the weak case, it is easy to find that perturbations of the form $\exp\{i(K\xi - \Omega\tau)\}$ are unstable when $0 < K < |a_0|\sqrt{2q/p} = K_c$ for $q > 0$, with a growth rate given by $\Im(\Omega) = \sqrt{2pq|a_0|K^2 - p^2K^4}$. This modulational instability is characteristic of a physical regime with zero phonon damping, since ν_2 does not enter in the equations at this order of the perturbation theory.

7. Conclusions

Comparing the results of Section 6 with those obtained for weak and finite damping cases, we see that a small amount of damping (i.e., $\nu \sim \mathcal{O}(\varepsilon)$) destabilizes the wave in $K - \omega$ regions that are stable in an ideal dissipationless system. Further increase of damping (i.e., $\nu \sim \mathcal{O}(1)$) as in Section 5, restricts again the instability to a particular $K - \omega$ region, which is nevertheless wider than that of the ideal case with zero damping just considered, as it is shown in Figure 1.

Our study shows that if the phonon damping is $\mathcal{O}(\varepsilon^0)$ or $\mathcal{O}(\varepsilon^1)$, a modulational instability appears also in the case of electrons and positrons at classical temperatures. Thus, the presence of some amount of sound absorption is instrumental to the generation of the envelope decay. When phonon damping is very small $\mathcal{O}(\varepsilon^2)$ the NLS of (Kates and Kaup, 1989a) is recovered, and we found that at ultrarelativistic temperatures the instability occurs in a wider band, above and near the relativistic plasma frequency. Our results can be applied to the relativistically hot $e^+ - e^-$ plasmas of early Universe. Growth rates can be used for estimates of characteristic times of formation of envelope solitons, which may then provide structure and non-homogeneous seeds to the primordial plasma.

The self-modulational instability of electromagnetic waves in $e^+ - e^-$ plasmas, has also been proposed to explain very short intensity variations (micropulses) of pulsar radio emmission (see, e.g., Chian, 1990, and references therein). For a pulsar scenario the ambient magnetic field is of paramount importance, but due to radiation cooling the magnetosphere of pulsars is expected to have non-relativistic temperatures. A theory for this physical regime has been given in (Kates and Kaup, 1989b). Our results, however, suggest that to derive the conditions of the instability properly, phonon damping effects must also be taken into account.

Acknowledgments

F.T.G. and G.G. are members of CONICET, Argentina. This work has been partially supported by PID-BID 594/92-CONICET, by INFIP-CONICET, and UBA grant EX-92. We also acknowledge the support of the Brazilian Agencies, CNPq, and FAPESP, and Fundación Andes, Chile.

References

Chian, A.C.L.: 1990, in: Hankins, Ranki and Gil (eds.), *The Magnetosphere Structure and Emission Mechanisms of Radio Pulsars*, Proc. IAU 128, p. 356.
Dolgov, A.D. and Zeldovich, Ya.B.: 1981, *Rev. Mod. Phys.* **53**, 1.
Gratton, F.T., Gnavi, G., Galvão, R.M.O. and Gomberoff, L.: 1997, *Phys. Rev. E* **55**, 3381.
Kates, R.E. and Kaup, D.J.: 1989a, *J. Plasma Phys.* **41**, 507.
Kates, R.E. and Kaup, D.J.: 1989b, *J. Plasma Phys.* **42**, 521.
Tajima, T. and Taniuti, T.: 1990, *Phys. Rev. A.* **42**, 3587.

ELECTROMAGNETIC OSCILLATIONS IN CYLINDRICAL PLASMAS WITH ELECTRON BEAMS INTERACTIONS

A. DEVIA, P.J. ARANGO and H. BARCO
Departamento de Física, Laboratorio de Física del Plasma, Universidad Nacional de Colombia, Sede Manizales AA 127, Colombia

Abstract. The dispersion relation for a radial inhomogeneous cylindrical plasma in a strong magnetic field is studied. In order to determine the fundamental transversal magnetic mode (TM_{01}), we used a numerical algorithm based in expanded Fourier–Bessel functions. Both the effect of the electron beam-plasma interaction and the resonances for a finite transversal cylinder conductor are studied. We used different density profiles and temperatures for the plasma and the beam.

1. Introduction

Using the Vlasov–Maxwell system equations, a theoretical model is developed for the study of the dispersion relations for an inhomogeneous plasma interacting with the electron beam. Some authors have studied the effects of inhomogeneity of plasma and its boundary, using the Vlasov–Poisson equation system. Diaz (1980) obtained a formal solution in this case and calculated a model for a plasma confined in a cylindrical conductor. The space-charge interactions of an electron beam with a cloud collisionless plasma in the presence of an infinite magnetic field in the direction of the beam velocity has been studied by Vlaardingerbroek et al. (1963); they considered the case of both the beam and the plasma filling the wave guide. A review of the excitation of plasma oscillation by a radially inhomogeneous electron beam for a cold axisymmetric beam has been given by Mikhailovskii (1974). Harris (1964) studied the problem of two streams in a cold inhomogeneous plasma in a strong magnetic field with a radial density.

2. Calculation of Relation Dispersion

In order to study the dispersion of the waves in an inhomogeneous and cylindrical plasma we use Vlasov–Maxwell equation system:

$$\frac{\partial f_\mu(\mathbf{r},\mathbf{v},t)}{\partial t} + \mathbf{v}\cdot\frac{\partial f_\mu(\mathbf{r},\mathbf{v},t)}{\partial \mathbf{r}} + \frac{q_\mu}{m_\mu}\left[\mathbf{E} + \frac{\mathbf{v}}{c}\times\mathbf{B}\right]\cdot\frac{\partial f_\mu(\mathbf{r},\mathbf{v},t)}{\partial \mathbf{v}} = 0, \qquad (2.1)$$

$$\nabla\times\mathbf{B} = \frac{4\pi}{c}\sum_\mu q_\mu \bar{n}_\mu \int \mathbf{v}(\mathbf{r},\mathbf{v},t)d^3\mathbf{v} + \frac{1}{c}\frac{\partial \mathbf{E}}{\partial t},$$

Astrophysics and Space Science **256:** 321–326, 1998.

$$\nabla \cdot \mathbf{E} = 4\pi \sum_{\mu} q_\mu \bar{n}_\mu \int f_\mu(\mathbf{r}, \mathbf{v}, t) d^3\mathbf{v}, \tag{2.2}$$

$$\nabla \times \mathbf{E} = -\frac{1}{c}\frac{\partial \mathbf{B}}{\partial t}, \qquad \nabla \cdot \mathbf{B} = 0, \tag{2.3}$$

where $f_\mu(\mathbf{r}, \mathbf{v}, t)$ is the distribution function of the identical particles characterized by the m_μ mass and the q_μ charge of the μ type. $\mathbf{E}$ and $\mathbf{B}$ are the electric and magnetic fields respectively and $\bar{n}_\mu$ is the density particles per unit of volume. We impose the following conditions on the system. (i) Binary collisions are negligible. (ii) The dynamic of ions is neglected. (iii) The beam-plasma system is homogeneous along the guide of the wave; it is axially symmetric, but inhomogeneous in the radial direction. (iv) The method of small perturbations is used. (v) Due to the fact that the system is placed in a strong magnetic field, we only take into account the longitudinal component z of the velocity, i.e. the distribution function takes the following form:

$$f_\mu(r, \theta, z, v, t) = \int dv_\perp f_\mu(\mathbf{r}, \mathbf{v}, t)$$

and the distribution function in the equilibrium is given by

$$f_\mu^0(r, v) = g_\mu(r) F_\mu^0(v).$$

Here $g_\mu(r)$ is a radial function that represents the inhomogeneity of the medium for which $g_\mu(r = 0) = 1$ and $g_\mu(r = a) = 0$, the function $F_\mu^0(v)$ is a velocity distribution function. Equations (1.1) to (1.3) are transformed into the following form:

$$\frac{\partial f_\mu^1(\mathbf{r}, \mathbf{v}, t)}{\partial t} + \mathbf{v} \cdot \frac{\partial f_\mu^1(\mathbf{r}, \mathbf{v}, t)}{\partial z} + g_\mu \frac{q_\mu}{m_\mu} E_z \cdot \frac{\partial F_\mu^0}{\partial v} = 0, \tag{2.4}$$

$$\nabla \times \mathbf{B}^1 = \frac{4\pi}{c} \sum_{\mu} q_\mu \bar{n}_\mu(0) \int v_z f_\mu^1 dv + \frac{1}{c}\frac{\partial E_z}{\partial t},$$

$$\nabla \cdot \mathbf{E}^1 = \sum_{\mu} 4\pi q_\mu \bar{n}_\mu(0) \int f_\mu(\mathbf{r}, \mathbf{v}, t) d^3\mathbf{v}, \tag{2.5}$$

$$\nabla \times \mathbf{E}^1 = -\frac{\partial \mathbf{B}^1}{\partial t}, \qquad \nabla \cdot \mathbf{B}^1 = 0, \tag{2.6}$$

where $\bar{n}_\mu(0)$ is the particle density definite on the z axis. Since there are no perpendicular currents $J_\perp = 0$, therefore

$$J_z(r, \theta, z, t) = \sum_{\mu} q_\mu n_\mu(0) \int v f_\mu(r, v, t) d^3\mathbf{v}.$$

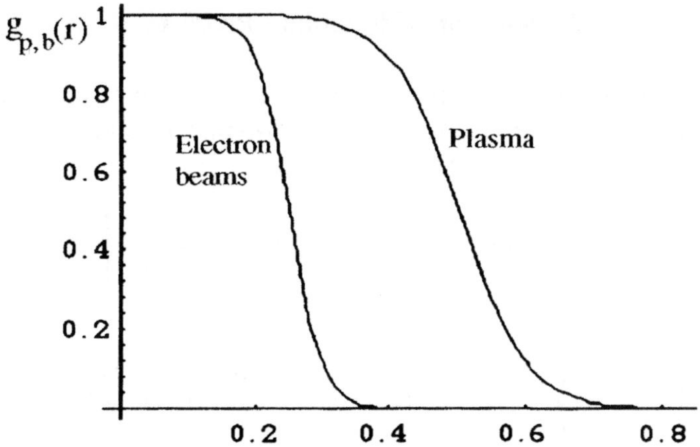

Figure 1. Radial density profile $g(r)$ for the plasma and electron beams.

In order to obtain the dispersion relation it is necessary to first calculate the electric permittivity since this must be in accordance with the previous system equations. This electric permittivity is obtained by expanding Equations (2.4) to (2.6) in terms of the Fourier–Bessel series given by

$$\psi_{m\ell}^{(k)}(r,\theta,z) = \frac{J_m(P_{m\ell}r)}{2aJ_{m+1}} e^{im\theta} e^{ikz},$$

$$m = 0, \pm 1, \pm 2, \ldots; \ell = 1, 2, \ldots; -\infty \leq k \leq \infty$$

for the spatial coordinates. For the temporary variable, we use the Laplace transform, which states that the electric permittivity is given by

$$\tilde{\varepsilon} = \begin{bmatrix} 1 & 0 & 0 \\ 0 & 1 & 0 \\ 0 & 0 & \varepsilon_z \end{bmatrix} \tag{2.7}$$

with

$$\varepsilon_z(r,k,\omega) = 1 - \sum_\mu g_\mu \frac{\omega_{p\mu}^2(0)}{k^2} \int dv \frac{dF_\mu^0/dv}{v - \omega/k},$$

where the plasma frequency $\omega_{p\mu}^2(0)$ is defined as

$$\omega_{p\mu}^2 = (0) = \frac{4\pi q_\mu^2 \bar{n}_\mu(0)}{m_\mu}.$$

3. Solution of Boundary Problem

The wave equation obtained from the Maxwell equations for electric field with dependence azimuth of the form $E_z(r,\theta) = E_{zm}(r)\exp(im\theta)$ and with the boundary condition $E_{zm}(r = a) = 0$, takes the form

$$\frac{d^2}{dr^2}E_{zm}(r) + \frac{1}{r}\left[\frac{d}{dr}E_{zm}(r)\right] - \left\{[1 - g_\mu I_\mu(k,\omega)]\kappa + \frac{m^2}{r^2}\right\}E_{zm}(r) = 0, \quad (3.1)$$

where

$$\kappa^2 = -\left(\frac{\omega^2 - k^2c^2}{c^2}\right) \quad \text{and} \quad I_\mu(k,\omega) = \frac{\omega^2_{p\mu}(0)}{k^2}\int_L dv\frac{dF^0_\mu/dv}{v - \omega/k}.$$

The homogeneous case is obtained by making $g_m(r) = 1$; in the inhomogeneous case $g(r)$ is a function that can be adjusted according to the conditions that are being studied. We considered the case:

$$g_\mu = \frac{1}{1 + \exp\left(\frac{1}{\beta}\left(\frac{r}{s} - 1\right)\right)}, \qquad 0 \le \beta \le 1$$

is a parameter that determines the steepness if the density falls at the plasma edge. We take $\beta = 0.1$ for the plasma and electron beams. $\beta^* s$ is approximately the width of the region where the radial density is different to zero. Figure 1 illustrated the radial density profiles. Besides, we used the Maxwellian velocity distribution functions for the plasma and the beam:

$$F^0_p = \left(\frac{2\pi K_B T_p}{m_e}\right)^{-1/2} e^{-(m_e v^2)/(K_B T_p)},$$

$$F^0_b = \left(\frac{2\pi K_B T_b}{m_e}\right)^{-1/2} e^{-(m_e (v-v_0)^2)/(K_B T_b)}, \quad (3.2)$$

where v_0 is the beam velocity. Equation (3.1) is solved by expanding $E_m(r)$ in the Fourier–Bessel series. Finally, we get the following dispersion relation in dimensionless values: $D_{mn}\underline{\mathbf{x}} = 0$, where $\underline{\mathbf{x}}$ is a vector associated with the eigen functions for the electric field, and the matrix D_{mn} takes the following form:

$$D_{mn} = \frac{\bar{c}^2\bar{k}^2 - \bar{\omega}^2}{\bar{c}^2(\bar{k}^2 + X^2_{m\ell}) - \bar{\omega}^2}\left\{\frac{C^{\ell\ell'}_{mp}Z'(\beta)}{2\bar{\lambda}^2_{dp}\bar{k}^2} + \frac{C^{\ell\ell'}_{mb}Z'(\alpha)}{2\bar{\lambda}^2_{db}\bar{k}^2} - \Pi\right\}, \quad (3.3)$$

$$C^{\ell\ell'}_{mp,b} = \frac{2}{a^2 J_{m+1}(X_{m\ell})J_{m+1}(X_{M\ell'})}\int_0^r dr\, r\, J_m(rp_{m\ell}J_m)rp_{m\ell'})g_{p,b}(r). \quad (3.4)$$

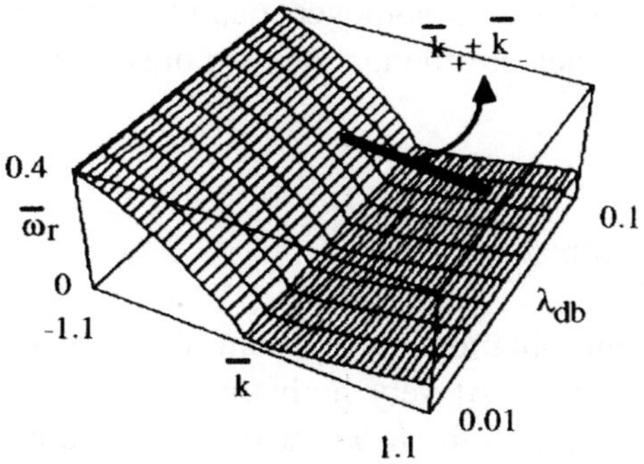

Figure 2a. Variation of the real frequency for the lowest radial model TM_{01}. In this case $\lambda_{dp} = 0.06$.

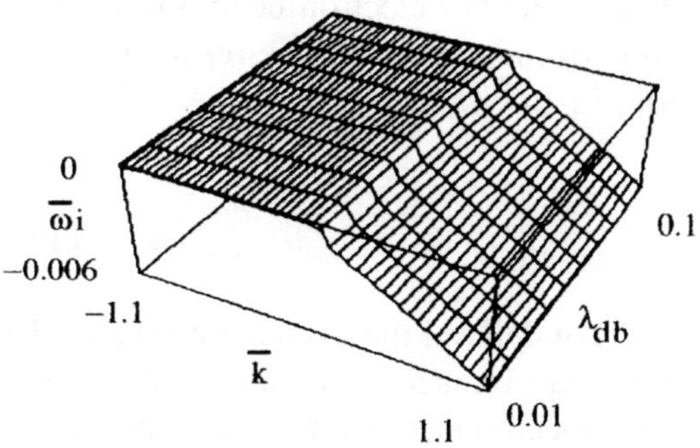

Figure 2b. Variation of the imaginary frequency for the lowest radial model TM_{01}. In this case $\lambda_{dp} = 0.06$.

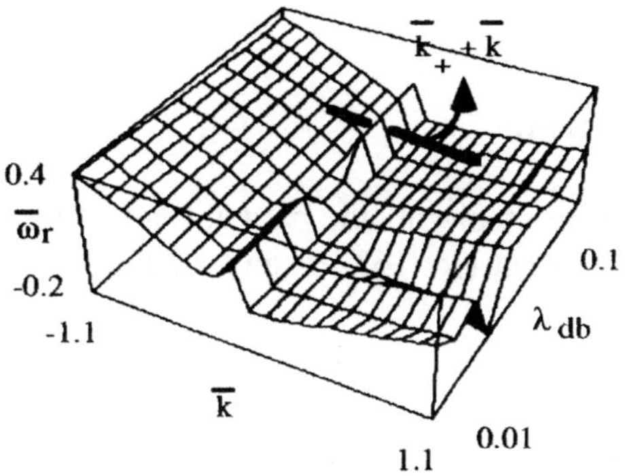

Figure 3a. Variation of the real frequency for the lowest radial model TM_{01}. In this case $\lambda_{dp} = 0.03$.

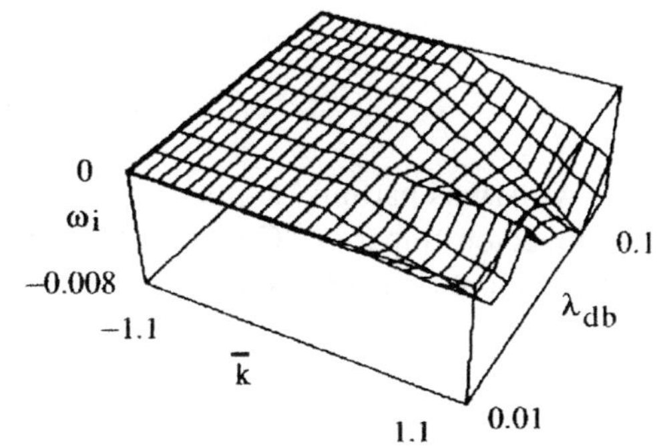

Figure 3b. Variation of the imaginary frequency for the lowest radial model TM_{01}. In this case $\lambda_{dp} = 0.03$.

The values $p_{m\ell} = X_{m\ell}/a$ are the zeros of the Bessel functions J_m. The p and b subscripts are related with the plasma and beam, respectively, a is the wave guide radius, k is the wave number, Π is the unity matrix, E is the beam energy, and Z' is the derivative of the dispersion function.

Due to the transversal boundaries $E_z(z = 0, z = L) = 0$ for the electric field, we obtain the following condition for the wave number, $\bar{k}_+ + \bar{k}_- = (2\tau a\pi)/L_0$, with $\tau = 0, 1, 2, \ldots$ Here $\bar{k}_{+(-)}$ is the wave number when the electromagnetic wave travels to the right (left). Equation (3.3) is solved by using the Zerizawa method (Zerizawa, 1983). The convolution coefficients $c_0^{\ell\ell'}$ were calculated taking into account functions $g_\mu(r)$. Figures 2a and 2b illustrate the electronic plasma waves for an inhomogeneous plasma as a function of temperature. Figures 3a and 3b illustrate the electron beam plasma system for an inhomogeneous plasma and inhomogeneous beam. The resonance frequencies are found when we draw a line from $\bar{k}_+$ to $\bar{k}_-$.

4. Conclusions

We have derived the dispersion relation for an inhomogeneous beam plasma system, and obtained numerical results for this case. At very high frequencies the wave speed tends towards light speed; this is due to the fact that in this range of frequencies the waves ignore the plasma presence, but we show only the branch of the low frequency in this paper. Figures 2 and 3 show the branch of low frequencies, on its lower part, and give the shock-absorbing predicted by Landau (1946). These values are negative and are presented separately make the graph plotting easy. The presence of the electron beam in the system loudly modifies the dispersion relation; for instance, in some lambda beam values the waves are not damping (see Figure 3b).

Acknowledgements

This work was partially supported by the Instituto Colombiano para el Fomento de la Educacion Superior "Colciencias" under contract No. 163-91, and the Universidad Nacional de Colombia.

References

Diaz, C.J.: 1980, *Plasma Physics* **23**, 5.
Harris, E.G.: 1964, *Phys. Fluids* **7**, 10.
Mikhailovskii, A.B.: 1974, *Theory of Plasma Instabilities*, Vol. I, Consultants Bureau, New York.
Vlaardingerbroek, M.T. et al.: 1963, *Philips Res. Rep.* **18**, 95.
Zerizawa, Y. et al.: 1983, *J. Phys. Soc. Japan* **52**, 28.

UV LINE PROFILES OF BIV FROM A 10-PS KRF-LASER-PRODUCED PLASMA

E. IGLESIAS*, H. GRIEM, B. WELCH and J. WEAVER
Institute for Plasma Research, University of Maryland, College Park, MD 20742, U.S.A.

Abstract. We use a KrF laser that generates pulses up to 8×10^{14} watt/cm^2 when focused onto a boron nitride target. We measured the line profile of the near UV, $\Delta n = 0$, $n = 2$ transitions of helium-like boron and we studied the triplet $1s2p^3P$-$1s2s^3S$ at 2825.85 ($J = 1$), 2821.68 ($J = 2$) and 2824.57 ($J = 0$) Å, at various positions from the target surface, in order to correlate details of this profile with the presence of self-generated magnetic fields and/or dynamic turbulence among other processes. We fitted the measured profiles to a 0-D model, that includes, Stark, Doppler, Zeeman and instrumental broadening. The effect of the magnetic field was included using an intermediate field calculations. Preliminary results of these measurements are presented and discussed.

1. Introduction

We study the emitted light of a plasma produced when a KrF laser (Leng et al., 1995), that generates 10 ps pulses, is focused onto a boron nitride target to power densities up 10^{15} watt/cm^2. These short lived plasmas are known to produce self-induced magnetic fields that have been measured by Faraday rotation, in experiments with laser powers $\approx 10^{16}$ watt/cm^2 (Raven et al., 1978; Stamper and Ripin, 1975). Understanding and measuring these fields is of fundamental importance since they sensibly affect the transport properties of fusion-like laser produced plasmas.

The use of Zeeman splitting for measuring such magnetic fields, has the advantage of being a passive measurement that relates simply to the magnitude of the field. In this work we measured the line profile of the near UV, $\Delta n = 0$, $n = 2$ transitions of the helium-like boron triplet at 2825.85 Å ($J = 1$), 2821.68 Å ($J = 2$), and 2824.57 Å ($J = 0$), in order to correlate details of these profiles with the presence of self-generated magnetic fields and/or the ion dynamic turbulence (Wang et al., 1989), among other processes sustained by the plasma.

The $\Delta n = 0$ lines, in the case of helium-like carbon, have already been used (McLean et al., 1984) to determine magnetic field values between 100 and 300 kG (allowing for opacity) with laser powers from 10^{12} to 10^{14} watt/cm^2. These transitions are particularly interesting for their relatively low sensitivity to Stark broadening (fwhm ≈ 0.43 Å at 10^{19} cm^{-3}) and their wavelength (λ), considering that the Zeeman shifts scale with λ^2 ($\Delta\lambda_Z \approx 0.5$ Å at 100 kG). Furthermore, the presence

* On leave from Dep. de Física, Universidad Simón Bolivar, Caracas, Venezuela.

Astrophysics and Space Science **256:** 327–331, 1998.

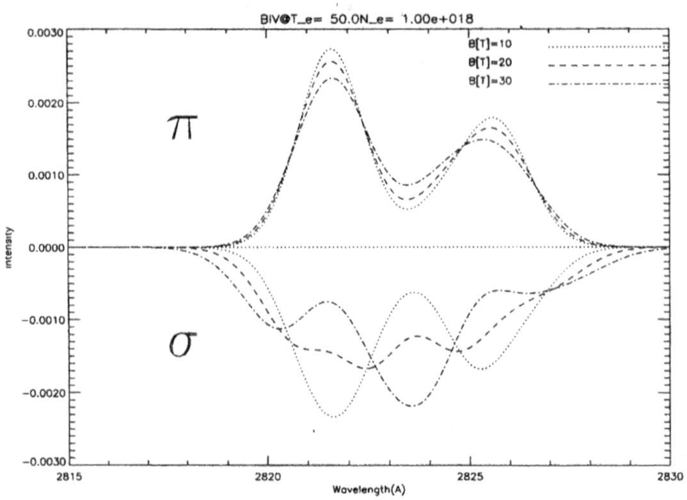

Figure 1. Model calculation for $T_e = 50$ eV, $N_e = 1 \times 10^{18}$ cm^{-3}, $\Delta\lambda_{\mathrm{Inst}} = 1$ Å. For magnetic fields 1, 2, 3 [$\times 10^2$ kG].

of magnetic fields drastically change the appearance of the σ component of the spectrum, starting at magnetic field values of 100 kG, for helium-like boron. This value represents the resolution of the technique, if no other enhancing or deteriorating effects are present, see Figure 1. These lines will also be affected by Doppler related effects, that scale with λ, which originate not just from the thermal effects but also from the macroscopic movement of the ions due to plasma expansion or convective turbulence of the order of the forward flow velocity ($\approx 10^7$ cm/s).

2. The Experiment

We observe the plasma perpendicular to the laser beam axis (side on) using a 2-D detector and a visible spectrometer, see Figure 2. We image the light from the plasma onto the entrance slit and place a Wollaston prism within the light path inside the spectrometer, in order to separate the π and σ components of the radiation, perpendicular to the dispersion direction.

We measured the line profile of the triplet at different positions (Z) from the target surface, from $Z = 0.2$ to 1 mm. The experimental line profiles were analyzed using a 0-D model, that includes Stark, Doppler effects and instrumental broadening. The effect of the magnetic field was included using and intermediate field calculation, assuming observation perpendicular to the magnetic field lines.

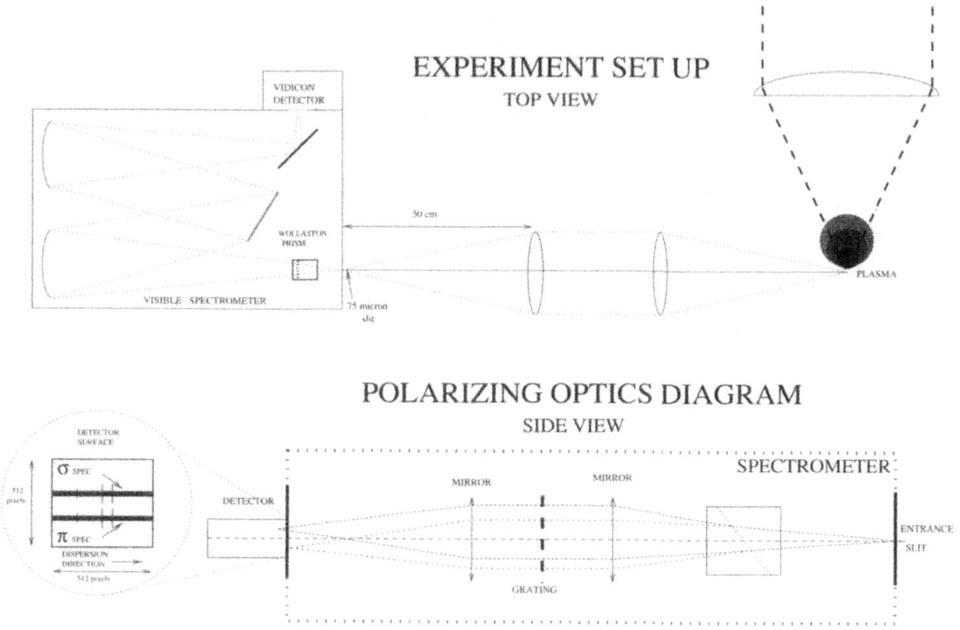

Figure 2. The experimental set up.

TABLE I
Results

	RUN 1:10 ps: Int = 8×10^{14} W/cm^2			RUN 2:20 ps: Int = 4×10^{14} W/cm^2		
	Z [mm]	N_e [10^{18} cm^{-3}]	B_{field} [kG]	Z [mm]	N_e [10^{18} cm^{-3}]	B_{field} [kG]
1	0.2	20	800	0.45	10	100
2	0.33	16	800	0.58	10	100
3	0.45	16	100	0.7	10	100
4	0.58	10	100	0.83	2	100
5	0.83	10	100	0.95	2	100

3. Results

Preliminary results of these measurements are presented in Table I.

The spectra at every position Z ($\Delta Z = \pm 0.13$ mm) is the result of the average (after background substraction) of $\approx$ 70 shots and electronically adding 35 rows of pixels, see Figures 3a and 3b.

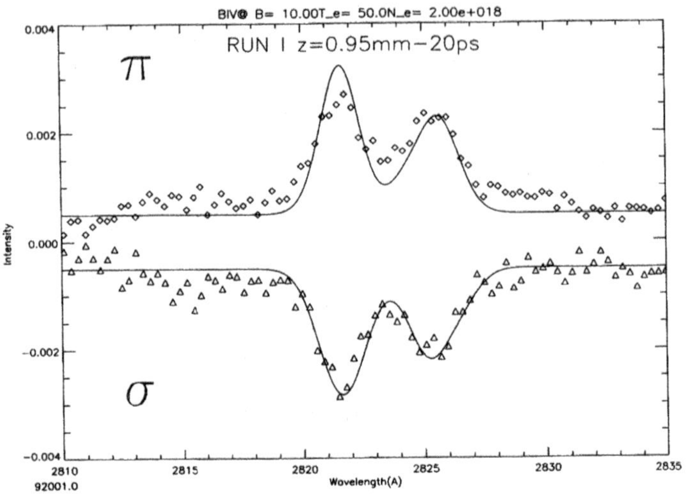

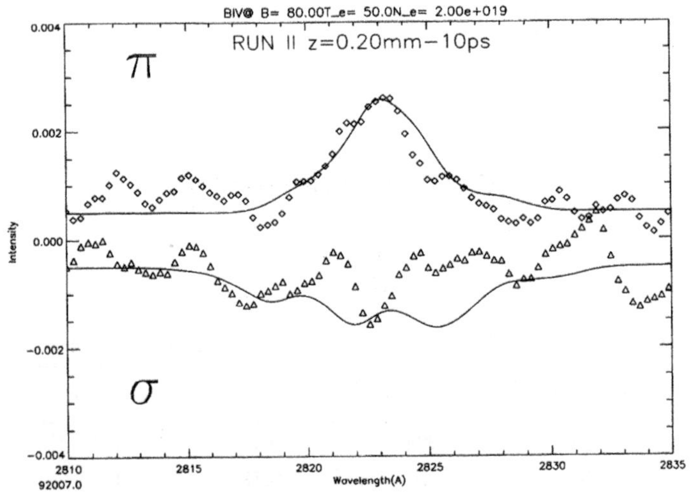

Figure 3. Dots = experimental, solid line = model. (a) $Z = 0.95$ mm, fit $B = 10^2$ kG, $T_e = 50$ eV, $N_e = 2 \times 10^{18}$ cm^{-3}, (b) $Z = 0.2$ mm, fit $B = 8 \times 10^2$ kG, $T_e = 50$ eV, $N_e = 2 \times 10^{19}$ cm^{-3}.

The fits to the data were adjusted using a nominal electron temperature of 50 eV. The effective Doppler broadening amounted to ≈ 1.1 Å. The instrument effect 1 Å fwhm was included as an extra Gaussian width.

4. Conclusions

In these preliminary observations there is some evidence of high fields, up to 800 kG at distances $0.2 \leq Z \leq 0.43$ mm. The value was inferred from the particular line shape of the π component of the spectrum (see Figure 3b), and it could be affected by the neighboring background noise. The σ component is merged with the background noise, therefore it did not contributed to our conclusion. For $Z \geq 0.4$ mm, the field was within the resolution of our technique 100 kG (for a typical spectrum, see Figure 3a).

References

Leng, Y., Yun, V.E., Goldhar, J. and Griem, H.R.: 1995, *Rev. Sci. Instrum.* **66**, 4045.

McLean, E.A., Stamper, J.A., Manka, C.K., Griem, H.R., Droemer, D.W. and Ripin, B.H.: 1984, *Phys. Fluids* **27**, 1327.

Raven, A., Willi, O. and Rumsby, P.T.: 1978, *Phys. Rev. Lett.* **41**, 554.

Stamper, J.A. and Ripin, B.H.: 1975, *Phys. Rev. Lett.* **34**, 138.

Wang, J.S., Griem, H.R. and Iglesias, E.J.: 1989, *Phys. Rev. A* **40**, 4115.

EXPERIMENTAL ANALYSIS OF HYDRODYNAMIC SOLITONS

GERMÁN DA COSTA
Laboratorio de Optica y Fluidos, Dpto. de Física, Universidad Simón Bolivar, Apdo. 89000, Caracas 1080 A, Venezuela

Abstract. Nonlinear interactions between solitary waves currently observed in Plasma Physics are simulated in one and two dimensional water channels.

1. Optical Visualization of Solitons Propagating in a 1D Water Channel

The experimental setup in Figure 1 shows a collimated laser beam (i) incident on a point (P) of a water channel. The trace of the reflected light beam (r) on the observation plane (OP) is recorded by means of a TV-CCD camera linked to a computer and image digitizer. Two counter-propagating surface waves are created by pouring a certain amount of water at a given point of the channel. Both waves travel back and forth undergoing multiple reflections at the channel ends. In an infinite length channel KdV equation previews formation of a finite number of solitons (Sagdeev et al., 1988). In the present case channel ends introduce a non-linear reflection which essentially changes the boundary condition to be imposed to KdV equation. Existence of solitonic solutions in that case is, to our knowledge, an open subject for theoretical and experimental research. The setup of Figure 1 is used in two different modes:

(a) When the incident light beam (i) is optically enlarged up to a diameter of about 2 cm and the angle between beam (i) and the water surface is small enough a region of the water surface several decimeters long is illuminated. In that case the light caustics of the curvilinear mirror formed by the water surface are observed in the plane (OP). A typical record is shown in the upper photograph of Figure 1. Bright fringes are the traces of the light caustics. In the first few seconds after introducing the initial perturbation the structure of caustics is random and rapidly time-varying. Later the caustics traces are formed by a couple of bright straight lines contained in plane (OP) and normal to the plane of the drawing. Each bright line proceeds from a point on the water surfaces where the "focal potential" (Da Costa, 1997) of the curvilinear mirror formed by the water free surface is stationary. The ensemble travels in the vertical direction at a constant speed. The vertical distance between bright lines remains constant. This shows that these are the caustics of a solitary wave with invariant shape and travelling at a constant speed. The speed of the travelling wave along the water channel equals the speed of the caustic system divided by the trigonometric tangent $\tan(\alpha)$ of the angle

Astrophysics and Space Science **256:** 333–336, 1998.

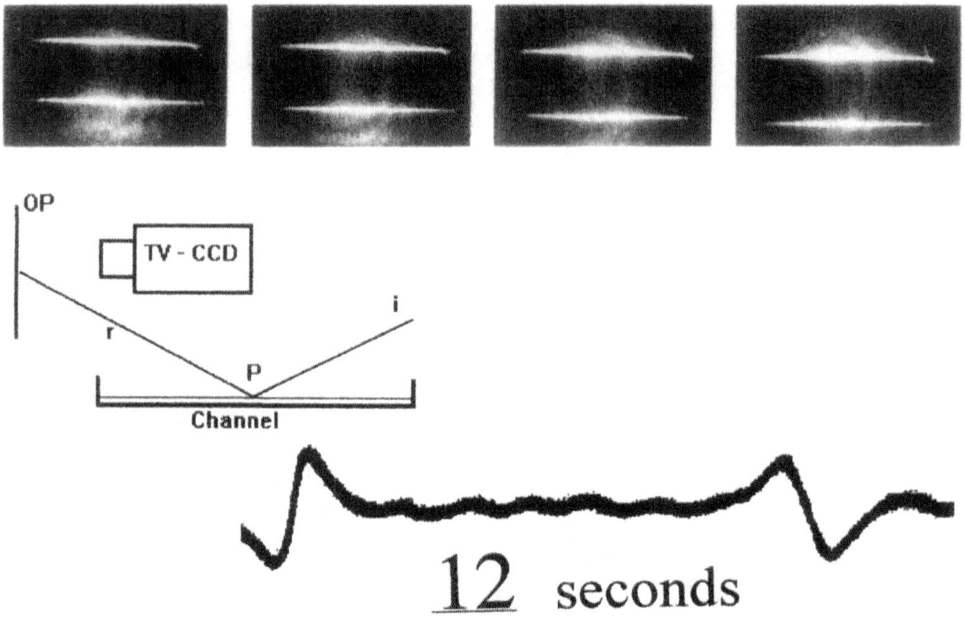

Figure 1. Above photographic sequence shows the light caustics (projected on a vertical screen OP) of a hydrodynamic soliton travelling along a horizontal water channel. The lower graph plots the slope of the free water surface at a fixed point (P) of the channel as a function of time.

between the laser beam (i) and the undeformed water surface. The approximate length of the solitary wave (about 30 cm) equals the vertical distance between two consecutive bright lines divided by $\tan(\alpha)$. This structure is observed after a propagation time of about 2 minutes. As the approximate propagation speed is 60 cm/s and the channel length is 3 m, we conclude that the solitary wave is well defined after travelling about 70 m and after having suffered 24 collisions with the channel ends. Note that this technique allows: (1) confirmation of the asymptotic formation of a shape-invariant solitary wave; and (2) approximate determination of its length and propagation speed. However, it does not provide any information about the actual shape of the solitary wave. This is achieved by using the system in mode (b).

(b) If the light beam is not expanded the light spot projected over the water surface has a small diameter. In such small region the slope of the water surface is almost constant with respect to the horizontal coordinate (x) measured along the channel at each instant of time. Therefore, the light spot observed in plane (OP) has itself a small diameter (about 2 mm) and travels vertically following the time variation of the local slope of the water surface at point (P) in Figure 1. Let $y = f(x)$ be the equation of the water surface profile. The slope $s(x)$ of the reflected beam (r) is (in the usual case ($f'(x^2) \ll 1, mf'(x) \ll 1$)) equal to:

$$s(x) \cong -m + 2m^2 f'(x), \tag{1}$$

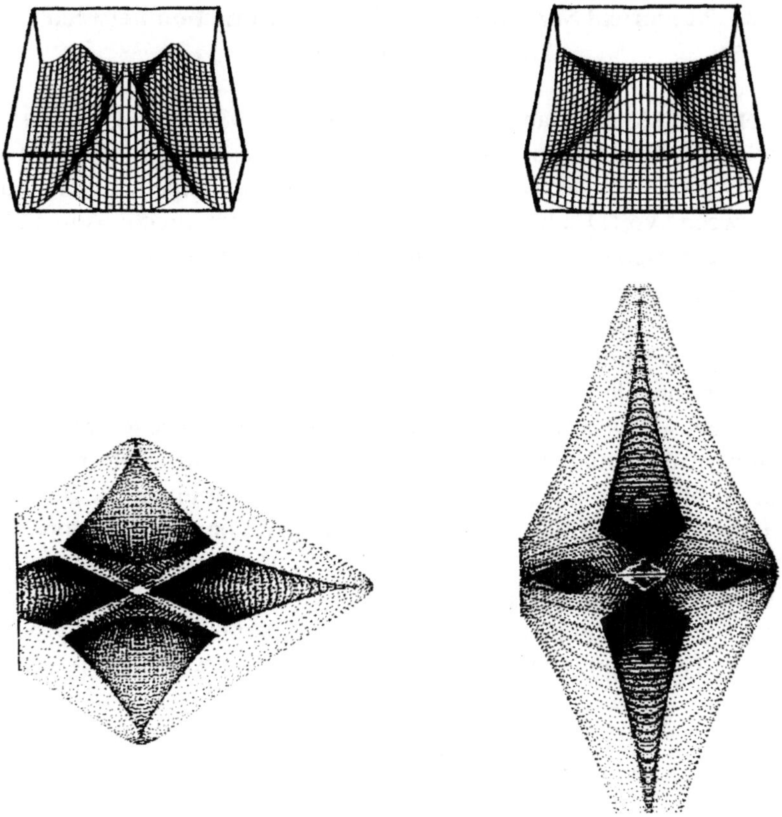

Figure 2. Upper pictures show collisions between 1D hydrodynamic solitons in a 2D pool at two different angles between the incident wave vectors. Lower pictures are negative portrays of caustics of light rays reflected from the water surface.

where (m) is the slope of the incident laser beam. This shows that the light spot in the observation plane performs vertical displacements proportional to the local slope of the water surface at point (P). The time-evolution of the light spot is recorded as a function of time by means of the CCD camera working in TDI (time-delayed image) mode. This mode is currently used to record objects in movement at a constant speed, allowing obtention of unblurred images. The technique consists in electronically displacing the charge collection in the photodiode matrix at a constant speed, which would be equivalent to displace by hand a common photographic camera at the same angular speed as the moving object to be recorded. In the actual case the displacement of the charge collection introduces a linear time-scanning of the position of the light spot, which results in obtention of a Cartesian graph of its vertical position vs. time. The lower graph in Figure 1 is an experimental record of the slope of the surface profile at point (P) as a function of time. Note the couple of solitary waves coming from opposite directions.

2. Numerical Simulation of Oblique Interaction between 1D Solitons in a 2D Pool

The equation of Kadomtsev and Petviashvili (Segur, 1982, eq. 14) model the evolution of relatively long waves of moderate amplitude as they propagate in shallow water in a 2D pool. We used the two-soliton formula to numerically visualize the collision between two 1D solitons propagating in different directions in a 2D water pool. Results corresponding to different angles between the wave vectors of the incident 1D solitons are shown in the left and right sides of Figure 2. The upper graph in each side represent the liquid surface profile. Note the formation of the "third wave" propagating in the direction of the bisector of the incident waves vectors in the right side graph. The latter corresponds to the case when the angle between incident waves is smaller than a certain threshold. The lower dotted graph shows a numerical simulation of the light caustics of a laser beam reflected from each surface profile. The incident laser beam is contained in a plane normal to the drawing. Local density of black points in the observation plane is proportional to local light intensity. Formation of the "third wave" is announced by an elongation of the reflected light pattern in the vertical sense. These theoretical light intensity distributions are to be compared with results of experiments currently performed at our Laboratory of Optics and Fluids.

References

Da Costa, G.: 1997, Light caustics of hydrodynamic solitons illuminated by a laser beam, *Applied Optics* **36**, 4069.

Sagdeev, R.Z., Usikov, D.A. and Zaslavsky, G.M.: 1988, *Nonlinear Physics: From the Pendulum to Turbulence and Chaos*, Harwood Academic Publishers.

Segur, H.: 1982, Some physical applications of solitons, in: K.B. Wolf (ed.), *Nonlinear Phenomena*, Proceedings of the CIFMO School and Workshop held at Oaxtepec, Mexico, November 29–December 17, 1982, Lecture Notes in Physics, Springer-Verlag, 1983, pp. 210–232.

PRE-BREAKDOWN PROCESSES IN THE HOLLOW CATHODE REGION OF A TRANSIENT HOLLOW CATHODE DISCHARGE

M. FAVRE, J. MORENO, H. CHUAQUI and E. WYNDHAM
Facultad de Física, Pontificia Universidad Católica de Chile, Casilla 306, Santiago 22, Chile
M. ZAMBRA
Comisión Chilena de Energía Nuclear, Casilla 188-D, Santiago, Chile
P. CHOI and C. DUMITRESCU-ZOITA
LPMI, Ecole Polytechnique, Palaiseau 91128, France

Abstract. Discharge formation in the A–K space of transient hollow cathode discharge (**THCD**) is causally linked with the emission of high energy electron beams originating in the hollow cathode region (**HCR**). Ionization in the A–K gap proceeds through the formation of a moving virtual anode, whose time evolution is strongly correlated with different periods in the electron beam activity. Here, we report on time and space resolved observations of different ionization events inside the HCR, which are time correlated with ionization processes inside the A–K gap. The experiments have been performed in Hydrogen, at pressures between 50 and 400 mTorr. A statistical study of the characteristic times associated with the different ionization events, based on von Laue plots, shows that the time distribution of events is well described by a single Gaussian distribution.

1. Introduction

It has been found, both theoretically (Boeuf and Pitchford, 1991) and experimentally (Choi et al., 1993), that electric breakdown in a transient hollow cathode discharge (THCD) proceeds through the formation of a moving virtual anode, which propagates from anode to cathode, under the assistance of electron beams emitted from the hollow cathode region (HCR). In a previous publication (Choi et al., 1995) we have presented a statistical study of breakdown formation in transient hollow cathode discharges, where three distinct regimes of ionization growth were identified. (1) Initial ionization growth with plasma formation close to the anode, (2) extension of this plasma region towards the cathode, and (3) ionization growth inside the HCR under enhanced field due to the close proximity of the anode potential. Regime (1) was found to be essentially random, whereas regime (2) exhibited a deterministic nature. The statistics of regime (3) indicated more than one competing ionization processes, whose nature was not possible to infer from the available data. In this paper we present further studies of ionization processes associated with the final breakdown of a THCD, with particular emphasis in the situations associated with regime (3). The arrival of the virtual anode at the proximity of the cathode, the formation of plasma inside the HCR, the penetration of the

Astrophysics and Space Science **256:** 337–342, 1998.

anode potential into the HCR through the cathode aperture and the peak in the high energy component of the electron beams, are time correlated with respect to electric breakdown in the A–K gap. A well defined sequence of events is established and some properties of the characteristic statistics are identified. From the statistical behavior, the dominant role played by the hollow cathode in the final breakdown events is clearly identified over a certain pressure range.

2. Experimental Details

The experiments were performed in Hydrogen, at pressures between 50 and 400 mTorr and 30 kV applied voltage. The A-K separation was kept at 10 cm, and a 5 mm cathode aperture was used. Further details of the experimental apparatus have been published elsewhere (Favre et al., 1995). The previous study was based on measurements with a capacitive probe array, whose signal is proportional to the time derivative of the product between the plasma potential and the coupling capacitance of the probe relative to the plasma (Choi and Favre, 1994). In this case we have used two capacitive Probes, A_C located at 15 mm from the cathode face, inside the A–K gap, and A_{HC}, located inside the HCR, at 4 mm from the back face of the cathode. Plasma formation is monitored by collecting light from a small region inside the HCR, on the axis and at the same axial position of A_{HC}. The light is fed through an H_α filter into a fiber optic-photomultiplier combination. The high energy component of the electron beams is measured with an optical fiber scintillator-photomultiplier combination. A 30 cm focal length spectrometer, coupled to a 1–D, gated intensifier was used to record 20 ns exposure time spectra of the HCR emission, next to the cathode aperture. Measurements of the electron density in the HCR are obtained from Stark broadening of the H_β line in the recorded spectra (Wise, 1965).

3. Experimental Results

Figure 1 shows characteristic signals at 100 and 400 mTorr. From top to bottom, the signals are the capacitive probe inside the A–K gap (A_C), the capacitive probe inside the HCR (A_{HC}), the electron beam (e-b), and the H_α emission. To characterize the different ionization events associated with the signals shown in Figure 1, the following convention has been adopted: the arrival of the virtual anode at the vicinity of the cathode is marked by the peak value in A_C, the initial growth of a significant plasma density inside the HCR is marked by the onset of signal in H_α emission, the penetration of the anode potential inside the HCR is marked by the peak value in A_{HC}, and the enhanced period of electron beam activity is marked by the peak value in e-b. Figure 2 shows characteristic average times and standard deviations associated with these four ionization events, in the 100 to 400 mTorr

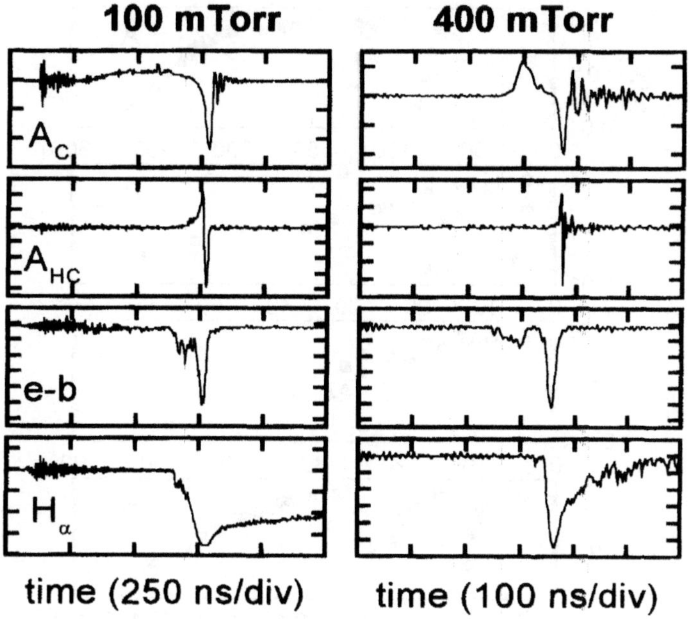

Figure 1. Characteristic signature of different ionization events.

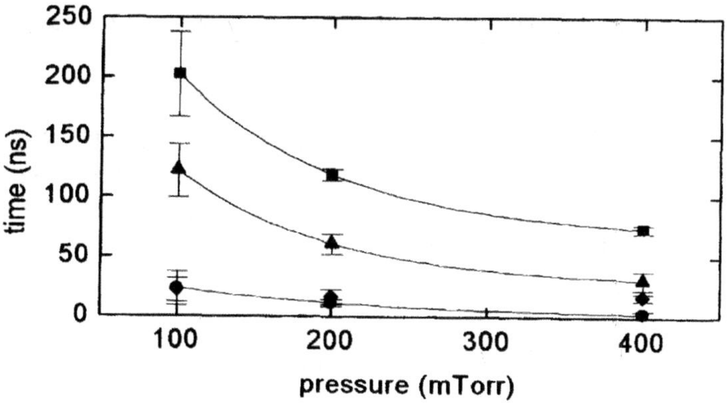

Figure 2. Characteristic time intervals for the different ionization events: ■ A_C, ▲ H_α, ● A_{HC} and ◆ e-b.

pressure range, measured over 150 shots. The size of the distribution is of the same order of the 128 points used in previous studies, which was found to be statistically significative (Choi et al., 1995). To visualize the statistical nature of the different characteristic times, the cumulative probability of the time intervals were plotted using the von Laue formalism (1925). The result for the different ionization events is shown in Figure 3. At low operating pressure it becomes difficult to identify the defining features for each particular event. This is shown in Figure 4, for two different situations at 50 mTorr. Preliminary measurements based on Stark broad-

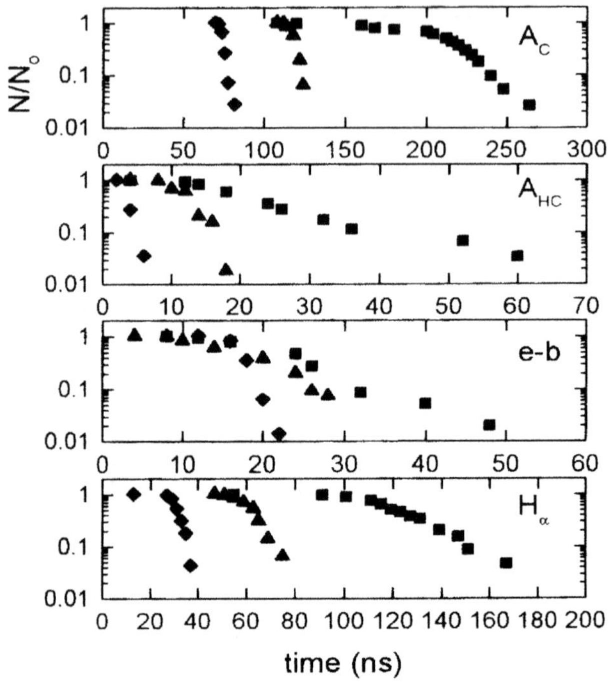

Figure 3. Von Laue plots of the different characteristic times: ■ 100 mTorr, ▲ 200 mTorr, and ♦ 400 mTorr.

ening of the H_β line indicate that the electron density inside the HCR grows from a non detectable value up to a few times 10^{15} cm^{-3}, in a very short time, just before electric breakdown. The available data are still not significant enough to describe the time evolution of the electron density inside the HCR.

4. Discussion

For an exponential distribution, the von Laue plot fits a straight line and the intercept on the unit probability is related with the characteristic time interval. In the case of a Gaussian distribution, the von Laue plot fits a parabolic envelope, and the characteristic time for the particular process is given by the 50% cumulative data point in the plot. If the 50% cumulative data point intersect of a simple spline fit to the von Laue plots in Figure 3 is compared with the average values shown in Figure 2, it is found that the corresponding values are almost identical. This is a good indication that the statistical distribution of times for the different ionization events follows a Gaussian distribution. From the average data presented in Figure 2 a clear sequence of events, which gives rise to regime (3) presented above, can be established. First: the virtual anode arrives to the vicinity of the cathode under the

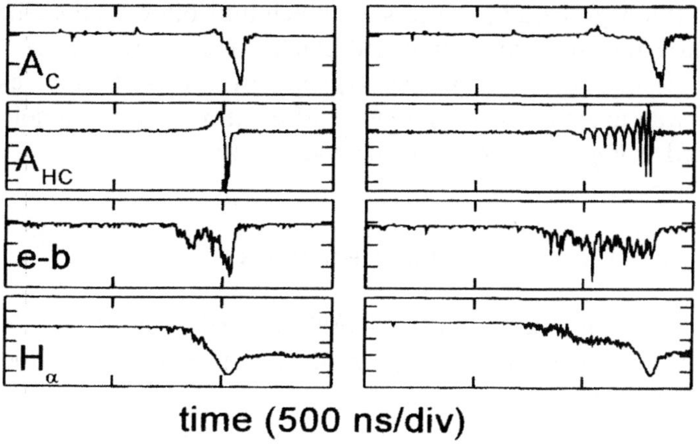

Figure 4. Different ionization events observed at 50 mTorr.

assistance of electron beams originated in the HCR; second: significant ionization develops inside the HCR, close to the cathode aperture, under an enhanced electric field due to the proximity of the anode potential; third: the anode potential diffuses inside the HCR through the conducting HCR plasma; and fourth: enhanced electron beam emission takes place from the HCR, due to electron acceleration in the enhanced field and across the sheaths in the back of the cathode. Some of the features of this sequence have also been observed in computer based simulations of the THCD (Boeuf and Pitchford, 1991). When the pressure becomes sufficiently low or a small cathode aperture is used, the hollow cathode does not play the dominant role in the ionization events, as it is shown in Figure 4. For the same operational pressure two distinctive situations are observed. On the left-hand plot the situation is similar to that shown at 200 mTorr in Figure 1, with a well defined sequence of ionization events. However, in the right-hand plot a totally different situation is observed. A_{HC} exhibits a pulsing feature, with a well defined periodicity. This periodic fluctuation takes place simultaneously with an enhanced period of electron beam activity, together with a nearly constant excursion in the H_α emission. This situation cannot be described by simple Gaussian statistics and a plausible explanation is still not available, although it can possible explain the mixed statistics observed at low operating pressure in previous investigations (Choi et al., 1995).

5. Conclusion

We have presented a statistical study of ionization events in the final electric breakdown phase of a transient hollow cathode discharge. We have found that, in a given pressure range, possible dependent on the applied voltage and discharge geometry, the statistics describing the different related processes can be fitted by a Gaussian distribution. At lower operational pressure, when the hollow cathode becomes less effective in assisting final ionization and electric breakdown in the A–K gap, a simple statistics can not describe the sequence of events. Further experimental work is underway to extend these studies to different operational conditions, which include applied voltage and size of the cathode aperture.

Acknowledgments

This investigation has been funded by FONDECYT grant No. 1950798 and a CNRS-CONICYT collaboration program. J. Moreno holds a postgraduate scholarship from CONICYT.

References

Boeuf, J.P. and Pitchford, L.: 1991, *IEEE Trans. Plasma Sci.* **19**, 286.
Choi, P. and Favre, M.: 1994, *Rev. Sci. Instrum.* **65**, 2281.
Choi, P. et al.: 1993, *Appl. Phys. Lett.* **63**, 2750.
Choi, P. et al.: 1995, *IEEE Trans. Plasma Sci.* **23**, 221.
Favre, M. et al.: 1995, *IEEE Trans. Plasma Sci.* **23**, 212.
von Laue, M.: 1925, *Ann. Phys.* **76**, 721.
Wise, W.L.: 1965, in: R. Huddlestone and S. Leonard (eds.), *Plasma Diagnostic Techniques*, Academic Press, New York, p. 309.

RAPID RECONNECTION AND COMPRESSIBLE PLASMA WAVES

M.F. HEYN
Institut für Theoretische Physik, Technische Universität Graz, Petersgasse 16, A-8010 Graz, Austria

Abstract. Magnetic reconnection causes the local non-linear decay of an infinitely thin current sheet, i.e., a tangential discontinuity, into large amplitude magnetohydrodynamic waves. This decay process propagates along the initial current sheet and induces linear wave perturbations in the surroundings. The Green's function for the self-consistent linear compressible plane wave problem is constructed and it is shown how to perform the convolution with some source function in form of a local reconnection pulse in an efficient way.

1. Introduction

Magnetic field line reconnection is an important concept in our understanding of basic plasma processes such as the solar wind interaction with planetary magnetospheres or the energy release in solar flares. First attempts to explain the rapid energy release in solar flares were based on pure resistive magnetic field dissipation (Sweet, 1958; Parker, 1963) but the energy conversion time for typical conditions, i.e., extremely large magnetic Reynolds numbers, is orders of magnitudes to small when compared with observational estimates.

In 1964, Petschek proposed a model for magnetic field reconnection including non-linear magnetohydrodynamic (MHD) waves (Petschek, 1964). In this model the boundary conditions and the global geometry define the energy conversion rate rather than the global conductivity of the plasma. In order to reconnect magnetic field lines, some finite conductivity is needed but only in a local region, say along a reconnection line located in a thin current layer. The electric field along the reconnection lines launches MHD waves, which propagate along the current layer. Energy conversion can take place now across the large amplitude waves (shocks) and is essentially independent of the actual dissipative mechanism within the shocks.

Since then, Petschek's idea of non-linear reconnection has been elaborated and extended in many ways (see Heyn and Semenov, 1996, and references therein). Of particular interest is the inclusion of transient features of magnetic reconnection. Consider a prescribed time varying reconnection electric field tangential to a thin current sheet. This electric field will induce the non-linear decay of the tangential discontinuity into large amplitude MHD waves. These waves will propagate with nearly constant speed, the deHoffman–Teller velocity, along the initial current

Astrophysics and Space Science **256:** 343–348, 1998.

sheet and transport the reconnection electric field into a much larger region. The non-linear waves naturally excite linear perturbations in the upper and lower half spaces. In this paper it is outlined how to construct the Green's function for these linear perturbations. In particular it is shown how to make use of complex contour integration in order to perform the inversion from Fourier–Laplace space to real time-coordinate space.

2. Linear Waves Radiated by Reconnection

In the following the linear MHD wave problem is considered in the two half spaces, say $z > 0$ and $z < 0$. The governing equations are obtained from the ideal MHD equations linearized with respect to a constant background. First order velocity and magnetic field can be derived from the displacement vector $\boldsymbol{\xi}(t, \mathbf{x})$,

$$\mathbf{v}^{(1)}(t, \mathbf{x}) = \left(\frac{\partial}{\partial t} + \mathbf{v}^{(0)} \cdot \nabla\right) \boldsymbol{\xi}(t, \mathbf{x}), \tag{1}$$

$$\mathbf{B}^{(1)}(t, \mathbf{x}) = \mathbf{B}^{(0)} \cdot \nabla \boldsymbol{\xi}(t, \mathbf{x}) - \mathbf{B}^{(0)} \nabla \cdot \boldsymbol{\xi}(t, \mathbf{x}). \tag{2}$$

Rewriting the linearized ideal MHD equations using the displacement vector and Fourier transforming the equations with respect to time and the xy-coordinates, an ordinary differential equation for the z-displacement $\zeta(\omega, \mathbf{k}, z)$ with $\mathbf{k} = (k_x, k_y)$ is obtained,

$$\frac{\mathrm{d}^2\zeta}{\mathrm{d}z^2} - q^2(\bar{\omega}, \mathbf{k})\zeta = 0, \quad q^2 = -\frac{\bar{\omega}^4 - (v_A^2 + c_s^2)k^2\bar{\omega}^2 + (\mathbf{k} \cdot \mathbf{v}_A)^2 c_s^2 k^2}{\bar{\omega}^2 \left(v_A^2 + c_s^2\right) - c_s^2 (\mathbf{k} \cdot \mathbf{v}_A)^2}, \tag{3}$$

where $\mathbf{v}_A$ and c_s are background Alfvén velocity and sound speed, and $\bar{\omega} = \omega - \mathbf{v}^{(0)} \cdot \mathbf{k}$ is the Doppler shifted frequency. The total pressure expressed in terms of $\zeta(\omega, \mathbf{k}, z)$ is

$$p^{(1)} = \frac{\varepsilon}{q^2} \frac{\mathrm{d}\zeta}{\mathrm{d}z}, \qquad \varepsilon = \rho^{(0)} \left[\bar{\omega}^2 - (\mathbf{k} \cdot \mathbf{v}_A)^2\right]. \tag{4}$$

The solutions of Equation (3) which vanish at infinity and satisfy first order total pressure balance at $z = 0$ are

$$\zeta(\omega, \mathbf{k}, z) = \frac{\tilde{L} e^{-qz}}{L + \tilde{L}} \left(\zeta_0 - \tilde{\zeta}_0\right), \quad z > 0, \tag{5}$$

$$\tilde{\zeta}(\omega, \mathbf{k}, z) = -\frac{L e^{\tilde{q}z}}{L + \tilde{L}} \left(\zeta_0 - \tilde{\zeta}_0\right), \quad z < 0, \tag{6}$$

with $L = \varepsilon(\bar{\omega}, \mathbf{k})/q(\bar{\omega}, \mathbf{k})$ and $\zeta_0(\omega, \mathbf{k}) = \zeta(\omega, \mathbf{k}, 0)$. Tildes indicate the respective functions for half space $z < 0$. Notice that here $\zeta_0 \neq \tilde{\zeta}_0$ in contrast to Kelvin–Helmholtz (KH) analysis. This results from the finite boundary layer at $z = 0$

due to reconnection. KH stability is assumed in the present context. In (Heyn and Semenov, 1996) it is shown how to derive the source function $\zeta_0 - \tilde{\zeta}_0$ from the boundary layer analysis appropriate for reconnection.

3. Inversion of Laplace–Fourier Transform

Specializing to plane geometry ($v_y \equiv B_y \equiv 0$) without background flow ($v_x = 0$) and using Laplace rather than Fourier transform for the time component, i.e., $\omega = ip$, the relevant quantities in Equations (5) and (6) are (all velocities now normalized to some v_0)

$$q = \frac{p}{v_0}\sqrt{\frac{(s^2v_A^2+1)(s^2c_s^2+1)}{s^2v_A^2c_s^2+v_A^2+c_s^2}}, \tag{7}$$

$$L = -\rho^{(0)}v_0 p\sqrt{\frac{(s^2v_A^2+1)(s^2v_A^2c_s^2+v_A^2+c_s^2)}{s^2c_s^2+1}}, \tag{8}$$

$$\zeta_0 - \tilde{\zeta}_0 = \frac{v_0}{p}Q(s^2)\Phi_0(p), \quad Q(s^2) = \sum_{i=A,S,\tilde{S}} \alpha_i \frac{w_i}{s^2w_i^2+1}, \tag{9}$$

where $s = kv_0/p$ and $w = v^{(0)} - (m/\rho B_n)B^{(0)}$ are the zero order deHoffman–Teller velocities of the corresponding shock waves (for details, see Heyn and Semenov, 1996).

The inversion for the displacement vector to time-coordinate space can be accomplished in closed form. What follows is an extension of a method discussed in Hudson (1980) and Rikitake et al. (1987). The reality condition implies that $\zeta(p,-k,z) = \zeta^*(p,k,z)$ and the inverse transformation with respect to the wave-number k can be written as

$$\zeta(p,x,z) = \Re\frac{1}{\pi}\int_0^\infty \mathrm{d}k\, \mathrm{e}^{ikx}\zeta(p,k,z) = \Re\frac{1}{\pi}\int_0^\infty \mathrm{d}s\, \mathrm{e}^{-p\tau(s)}F(s^2)\Phi_0(p), \tag{10}$$

$$\tau(s) = \sqrt{\frac{(s^2v_A^2+1)(s^2c_s^2+1)}{s^2v_A^2c_s^2+v_A^2+c_s^2}}\frac{z}{v_0} - is\frac{x}{v_0},$$

$$F(s^2) = \frac{\tilde{L}(s^2)}{L(s^2)+\tilde{L}(s^2)}Q(s^2). \tag{11}$$

If the contour in the complex s-plane can be analytically deformed in such a way that $\tau(s)$ becomes real along the whole path of integration, the shift theorem of Laplace transform, $\mathcal{L}^{-1}\{\mathrm{e}^{-p\tau}\Phi_0(p); p\} = \Phi_0(t-\tau)$, gives immediately the real time-coordinate representation of the z-displacement,

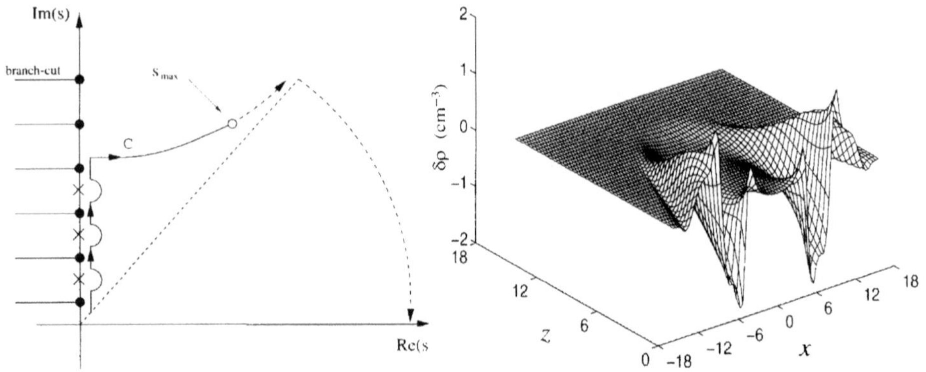

Figure 1. On the left, a sketch of the integration contour $\mathcal{C}$ in the complex s-plane where $\tau(s)$ is real. On the solid part $0 \le \tau \le t$, on the dashed part $\tau > t$. On the right, the first order density distribution in the upper half space $z > 0$ corresponding to the right panel of Figure 2.

$$\zeta(t, x, z) = \Re \frac{1}{\pi} \int_{\mathcal{C}} \mathrm{d}s\, F(s^2) \Phi_0 (t - \tau(s)), \tag{12}$$

where the contour $\mathcal{C}$ is such that $\tau(s)$ is real for $0 \le s \le s_{\max}$ and $\tau(s_{\max}) = t$.

For $x > 0$ and $z > 0$, the first quadrant of the complex s plane has to be taken so that there are no contributions to the integral from the arc at infinity when closing the contour to the real s-axis and applying Cauchy's theorem. Along the positive imaginary s-axis there are six branch points and three poles respectively two poles for the respective integral for $-\tilde{\zeta}(p, k, -z)$. The branch cuts are chosen in the half space $\Re(s) < 0$ whereas the poles have to be circumvented by small semi-circles lying in the first quadrant when necessary (see Figure 1). Besides of that there are no restriction on how to choose the integration contour in the first quadrant from $s = 0$ to $s = s_{\max}$. There are no contribution from $s_{\max}$ to infinity because $\Phi_0(t) \equiv 0$ for $t \le 0$. From the numerical point of view, it is more advantageous to perform the integration along a suitable contour where $\tau(s)$ will be complex but one can stay away from branch points and singularities.

In the limit $z \to 0$ the integration contour $\mathcal{C}$ with $\tau(s)$ real is along the positive imaginary axis. Transforming the integration variable from s to $\tau = -isx/v_0$, the result for $z = 0$ can be written as the sum of principal value integrations and contributions in form of half residua,

$$\begin{aligned} \zeta(t, x, 0) &= -\frac{v_0}{\pi x} \mathcal{P} \int_0^t \mathrm{d}\tau\, \Im \left\{ \left. \frac{\tilde{L}}{L + \tilde{L}} \right|_{s^2 = -v_0^2 \tau^2 / x^2} \right\} Q\left(-\frac{v_0^2 \tau^2}{x^2} \right) \Phi_0(t - \tau) \\ &+ \sum_{i = A, S, \tilde{S}} \Re \left\{ \left. \frac{\tilde{L}}{L + \tilde{L}} \right|_{s^2 = -1/w_i^2} \right\} \frac{\alpha_i}{2} \Phi_0 \left(t - \frac{w_i}{v_0} x \right). \end{aligned} \tag{13}$$

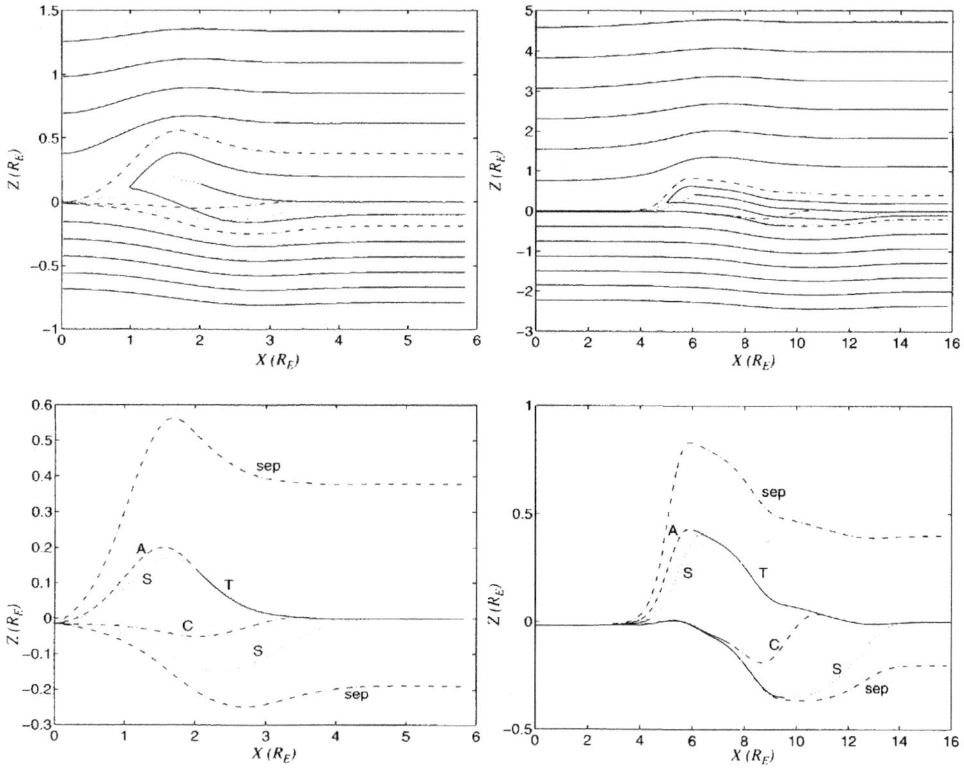

Figure 2. The evolution of the non-linear waves initiated by an electric field pulse at $x = 0$ at the end of active reconnection (left panel) and at some later time (right panel). Shown are the separatrix (sep), the Alfvén wave (A), the slow shocks (S), the contact discontinuity (C), and the tangential discontinuity (T). The lower panel shows the magnetic field perturbations and one reconnected field inside the separatrices. Conditions are very asymmetric with a much higher Alfvén speed for $z < 0$.

This result for the displacement $\zeta(t, x, 0)$ defines the actual location of the large amplitude waves when propagating along the current sheet. An example for asymmetric background conditions in upper and lower half space is shown in Figure 2. The shock waves and discontinuities generated by the reconnection electric field at $x = 0$ move to the right (and, of course, similarly to the left not shown in the figure) along the current sheet at $z = 0$. Because of the asymmetric conditions all waves move with different speed. The linear wave modes (fast wave, slow wave and Alfvén wave) in the upper and lower half spaces are self-consistently coupled by pressure balance across the boundary at $z = 0$. This effect can be clearly observed in the upper right panel of Figure 2 in the behavior of T between $9 < x < 13$. The faster waves in the lower half space $z < 0$ produce a side-wave in the upper half space $z > 0$. In the right panel of Figure 1 the corresponding density distribution is shown. Similar to a moving object, the density increases in the front part $x \approx 12$

and decreases in the wake $x \approx 6$. Also, slow and Alfvén wave produce additional density perturbations near $x \approx 4$.

References

Heyn, M.F. and Semenov, V.S.: 1996, *Phys. Plasmas* **3**, 2725.
Hudson, J.A.: 1980, *The Excitation and Propagation of Elastic Waves*, Cambridge University Press, London.
Parker, E.: 1963, *Astrophys. J.*, Suppl., Ser. 77, **8**, 177.
Petschek, H.E.: 1964, in: W.N. Hess (ed.), *AAS-NASA Symposium of the Physics of Solar Flares, NASA-SP 50*, National Aeronautics and Space Administration, Washington, pp. 425–439.
Rikitake, T., Sato, R. and Hagiwara, Y.: 1987, *Applied Mathematics for Earth Scientists*, Terra Scientific Publishing Company, Tokyo.
Sweet, P.A.: 1958, in: B. Lehnert (ed.), *Electromagnetic Phenomena in Cosmical Physics*, Cambridge University Press, London, p. 123.

NEW CALCULATION OF THE ELECTRICAL MICROFIELD DISTRIBUTION IN PLASMA

J. PUERTA
Departamento de Física, Universidad Simón Bolívar, Apdo. 89000, Caracas, Venezuela

C. CERECEDA
Laboratoire de Physique Gaz et des Plasmas, Unité de Recherche au CNRS, Université Paris-Sud, Bât. 212, 91405 Orsay Cedex, France

Abstract. Recently new calculations for the probability field distribution in plasma generating a new expression for the Debye screening fields was carried out by Martin and Perez (1996). This field is a very accurate solution of the Poisson equation for a point test charge in a plasma described by a uniform charge density at Boltzmann thermal equilibrium. The method that we here present is based mainly on the introduction of this adequate field considering the dependence of the potential with the magnitude of the test charge in the calculus of the distribution. First we find the spectral function $F(\mathbf{k})$ carrying out the Fourier transform of the field which is numerically calculated for each representative value of the variable in $\mathbf{k}$ space and then fitting these data in order to get $F(\mathbf{k})$. Thus, after the usual numerical integration on the wave number space $d^3\mathbf{k}$, the field strength distribution is achieved. This approximation recovers the Holtsmark distribution in the limit of high temperature, and/or, small densities as also these where $\delta \neq 0$. As in previous works (Ecker, 1972; Spatschek, 1990) the most probable value of the field decreases as the temperature is reduced, and/or, the density increased.

1. Introduction

The probability distribution function for the electric field in a plasma is a very important problem in order to find solutions for several situations as, for example, the broadening of spectral lines emitted by atoms. Such a problem was already attacked by many authors (Baranger and Mozer, 1959; Chandrasekhar, 1943; Ecker, 1972; Martin and Perez, 1996; Spatschek, 1990). The conditions under which this problem is solved are well known, but the important physical situation argument, in our case, is that the volume in consideration of the cloud which surround each charged particle and shields it should be large compared to the volume per particle. In order to overcome these limitations we use an analytical approximated – but very accurate – solution for the nonlinear Debye Equation, where the problem is solved in the case that the test charge q_T is high compared with the volume and charge of the clouds obtaining a potential ϕ that match the real potential very well. On the other hand, each linearization of the Debye Equation in $e\phi/kT$ is consistent with the physical situation where correlations of higher order are not considered.

Astrophysics and Space Science **256:** 349–353, 1998.

Using this effective potential we overcome this problem. In that sense our approach is a generalization for the simulation of the electric microfield field distribution.

2. Theoretical Treatment

In order to calculate the electric field $\mathbf{E}$ produced at a given point by a particles located at $\mathbf{r_1}, \mathbf{r_2}, \ldots, \mathbf{r_N}$ we put

$$\mathbf{E} = \mathbf{E_1} + \mathbf{E_2} + \cdots + \mathbf{E_N}. \tag{1}$$

We are interested in the probability distribution $W(\mathbf{E})$ of the electric field. For this calculation we take the Fourier Transform

$$F(\mathbf{k}) = \int e^{(i\mathbf{k}.\mathbf{E})} W(\mathbf{E}) d\mathbf{E}. \tag{2}$$

Now in order to calculate $W(\mathbf{E})$ we find the inverse of the Fourier Transform to get

$$W(E) = \frac{1}{(2\pi)^3} \int e^{i\mathbf{k}.\mathbf{E}} F(\mathbf{k}) d^3\mathbf{k}, \tag{3}$$

where the function $F(\mathbf{k})$ defined in Equation (2) is called the Spectral Function and can be expressed in the form

$$F(\mathbf{k}) = \exp(nh_1(\mathbf{k})), \tag{4}$$

with the function h_1 calculated using the effective microfield in the form

$$E = -d\phi/dr$$

with ϕ defined by Pérez and Martín (1998)

$$\phi = q_T en\lambda_D^2 \frac{Q}{s} \frac{1 + as + b^2 s^2}{(1 + bs)^2} e^{-s\sqrt{2}}, \tag{5}$$

where Q is given by $q_T\sqrt{2}/en\lambda_D^3$, and

$$s = \frac{r\sqrt{2}}{\lambda_D}.$$

With these definitions we finally obtain

$$h_1(k) = 2\pi \int_0^\pi \int_0^\infty (e^{ikE\cos(\theta)} - 1) r^2 dr d(\cos(\theta)),$$

$$h_1(k) = 2\pi \int_0^\infty \left(\frac{\sin(kE)}{kE} - 1 \right) r^2 dr. \tag{6}$$

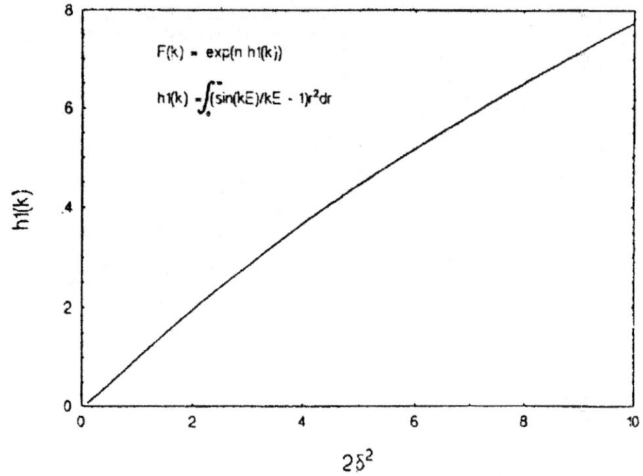

Figure 1. Effective Spectral function $h_1(k)$.

Calculating numerically this integral for a large enough representative range of k values and making the change of variable

$$x = k\frac{k_B T_e \lambda_D}{e r_0^2 \sqrt{2}} \tag{7}$$

with

$$r_0 = 15/4(2\pi)^{3/2} n$$

we obtain the fitted function, $h_1(x)$ in the form

$$nh_1(x) = -x^{3/2} - cx^{(3/2+d)}\delta^{2d} \Rightarrow x \geq 1/2\delta^2$$

$$nh_1(x) = \frac{-15}{\sqrt{\pi}} c' 2^{(d'-3)} x^{d'} \delta^{(2d'-3)} \Rightarrow x < 1/2\delta^2$$

as shown in Figure 1, with $\delta = r_0/\lambda_D$ and the parameters c, d, c' and d' determined from the fits as follows:

$$c = -0.7085,$$

$$d = 0.2304,$$

$$c' = 0.1138,$$

$$d' = 0.8377,$$

Now defining

$$\beta = \frac{E}{E_0} \tag{8}$$

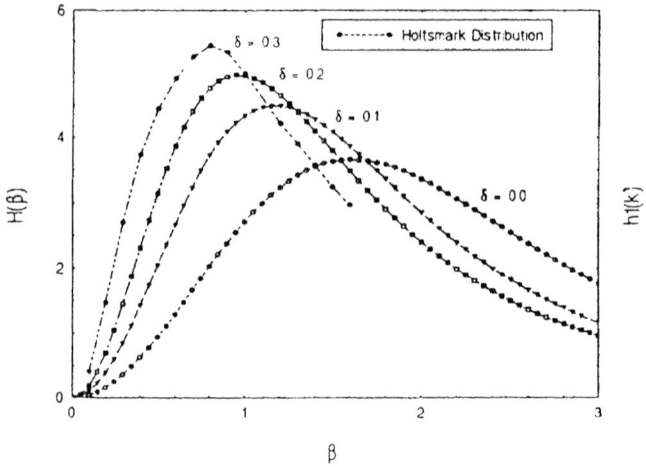

Figure 2. Electrical Microfield Distribution for $Q = 1$.

with

$$E_0 = \frac{er_0^2}{k_B T_e \lambda_D \sqrt{2}},$$

and since $W(E)$ does not depend on angles, we can write the dimensionless distribution function for the field in the form (see Figure 2):

$$H(\beta) = \frac{2\beta}{\pi} \int_0^{\infty} \sin(\beta x) F(x) x dx. \tag{9}$$

3. Results and Discussion

For several values of δ (small δ where our approximations is valid) the integral (6) is calculated numerically for a test charge $Q = 1$. For high temperatures and/or small densities (i.e., $\delta = 0$) the Holtsmark approximation is recovered.

It should be seen that the most probable value of the electric field get smaller as the temperature is decreased and/or the density is increased. In conclusion, we have obtained a field distribution function that matches very good the screening potential considering all the correlations.

References

Baranger, M. and Mozer, B.: 1959, *Phys. Rev.* **115**, 521.

Chandrasekhar, S.: 1943, *Rev. Modern Phys.* **15**, 73.
Ecker, G.: 1972, *Theory of Fully Ionized Plasmas*, Academic Press, New York.
Pérez, R.J. and Martín, P.: 1998, *Astrophys. Space Sci.* **256**, 263–268.
Spatschek, K.H.: 1990, *Theoretische Plasmaphysik*, Teubner Studienbüecher, Stuttgart.

ANALYTICAL APPROXIMANT OF THE UNIVERSAL LINE SHAPE

J. PUERTA and P. MARTÍN
Departamento de Física, Universidad Simón Bolívar, Apdo. 89000, Caracas, Venezuela

Abstract. The pressure broadening of spectral lines based in the Unified Franc–Condon (UCF) treatment, leads to universal line shape function $L(z)$ which is a highly complicated integral of the Airy function and exponentials

$$L(z) = \int_0^\infty \xi^{-2}[Ai(-z\xi)]^2 \exp(-\xi^{-3}) d\xi. \tag{1}$$

An approximated analytical expression to this function has been found in terms of elementary functions, which is more precise than any previous one, giving also enough accuracy for most of the applications in molecular and plasma spectroscopy. This approximant has been obtained using the recently developed technique of two-point quasifractional approximation. A unique approximant has been found with better accuracy for both sides of the variable. The method used here is an improvement of previous procedures leaving enough free parameters in order to determine exactly the value of the maximum. On the other hand, since there are two different asymptotic forms for large negative and positive values of z, a new way to impose this additional condition has been found.

1. Introduction

The analysis of the pressure broadening of spectral lines is of interest from both an experimental and theoretical point of view (Bekefi, 1976). The treatment using the quantum-mechanical treatment adiabatic approach (unified Frank–Condon treatment) leads to good results (Szudy and Baylis, 1975). However, the universal line shape is rather complicated due to a combination of powers of Airy function and elaborated exponentials. Therefore, the calculations of this kind of integrals become regularly cumbersome, and complicated programs using 24-point Gauss–Laguerre integration have to be used. On the other hand, recent techniques of approach by simultaneous use of power and asymptotic have allowed us to obtain approximants to complicated functions, using elementary functions in a better way than any previous method. This technique, so called *two-point quasifractional approximants*, seems very appropriate to treat the present problems (Martín and Baker, 1991). There are, however, some new difficulties. One of these is that a unique approximant for both the positive and the negative part are different (Stokes phenomenon) from that of the negative part. However, here we have managed to find a unique approximant that reproduces both asymptotic expansions together with the power series near zero.

Astrophysics and Space Science **256:** 355–358, 1998.

In spite of the new problems discussed in detail in Section 2, the approximant obtained here is simple to calculate and better than the previous one.

2. Theoretical Treatment

Using the quantum mechanical theory of pressure broadening of spectral lines, and the adiabatic approximation, a Universal (unified) Spectral Profile $I(\omega)$ can be obtained in terms of single-perturbers radial overlap integrals taking in to account several Condon points (Szudy and Baylis, 1975) in the form

$$I(\omega) = \frac{n\xi^2}{\pi|\nu|^2} J(\xi), \tag{2}$$

and

$$J(\xi) \approx \frac{1}{2}\sum_c J_c^{(p)}(\xi),$$

with the sum over any number of Condon points, where n = density of perturbers; $\xi = \omega - \omega_0 \pm nk$ with real quantity; and $\nu = \omega - \omega_0 - \Delta - i(\gamma/2)$, defined in an appropriate way in the paper given before.

On the other hand, $J_c(\xi)$ is also defined by

$$J_c(\xi) = \frac{48\pi^2(\mathrm{Re}\, r_c)^2}{|\Delta V'(r_c)/\hbar|} |\pi z_c|^{1/2} L(z_c) \exp[-V_i(\mathrm{Re}\, r_c)/kT]$$

and also

$$z_c = \frac{1}{2}\left(\frac{\mu}{kT}\right)^{1/3} [\Delta V'(r_c)/\hbar]^2 |\Delta V''/\hbar|^{-4/3},$$

where V_i and V_f are appropriate potentials in the Schrödinger Equation. Here is also $L(z)$ defined in Equation (1). The behavior of this function in zero is very peculiar because the function is finite but the derivative is infinite:

$$L(0) = [Ai(0)]^2\Gamma(1/3) = 0.11255553, \tag{3}$$

$$L'(0) = \infty,$$

$$\lim_{z\to 0} L'(z) \approx 2A_i'(0) \ln|z|, \tag{4}$$

$$a = 0.11255553, \tag{5}$$

$$b = -0.183777629 = 2Ai(0).$$

On the other hand, the asymptotic forms for $z \gg 1$ and $z \ll -1$ are different:

$$L(z) \sim [36\pi z]^{-1/2}, \tag{6}$$

$$L(z) \sim [-108\pi z]^{-1/2} \exp[12^{1/3} z]. \tag{7}$$

In order to keep all of these conditions, and considering the power series with the leading term in order to apply the asymptotic Padè method, explained elsewhere (Pérez and Martin, 1994) we obtain for the approximant the form with two free parameters λ and μ:

$$\hat{f} = \frac{a + \frac{1}{2} bz \ln z^2 + P_0 z + P_1 z^2}{(1+\lambda z^2)^{1/4}\{1 + \mu z^2[1 + \sqrt{3}\, \exp((-12)^{1/3} z)]\}}.$$

Now choosing

$$\lambda = 2.0, \tag{8}$$

$$P_1 = \frac{\lambda\mu}{6\sqrt{\pi}}, \tag{9}$$

it remains to determine only two parameters μ and P_0. We have found that the best procedure to determine these two parameters is to impose the conditions on the maximum of the function. Thus, if we denote by z_m the point where $L(z)$ is maximum, then we get two additional equations

$$\hat{f}(z_m) = L(z_m) = 0.176373857, \tag{10}$$

$$\hat{f}'(z_m) = L'(z_m) = 0, \tag{11}$$

with $z_m = 0.328814$.

Using these equations, μ and P_0 are determined given

$$\mu = 0.731067, \tag{12}$$

$$P_0 = 0.0378948. \tag{13}$$

A plot of the function and the approximant are shown in Figure 1 together with the asymptotic forms. With this scale we can see that only for large values of z the approximant departs a little from the exact function $L(z)$.

3. Conclusions

In this paper we have found an analytic approximant for the Universal Line Shape, which gives good accuracy for all values of the variable z. Furthermore, the form

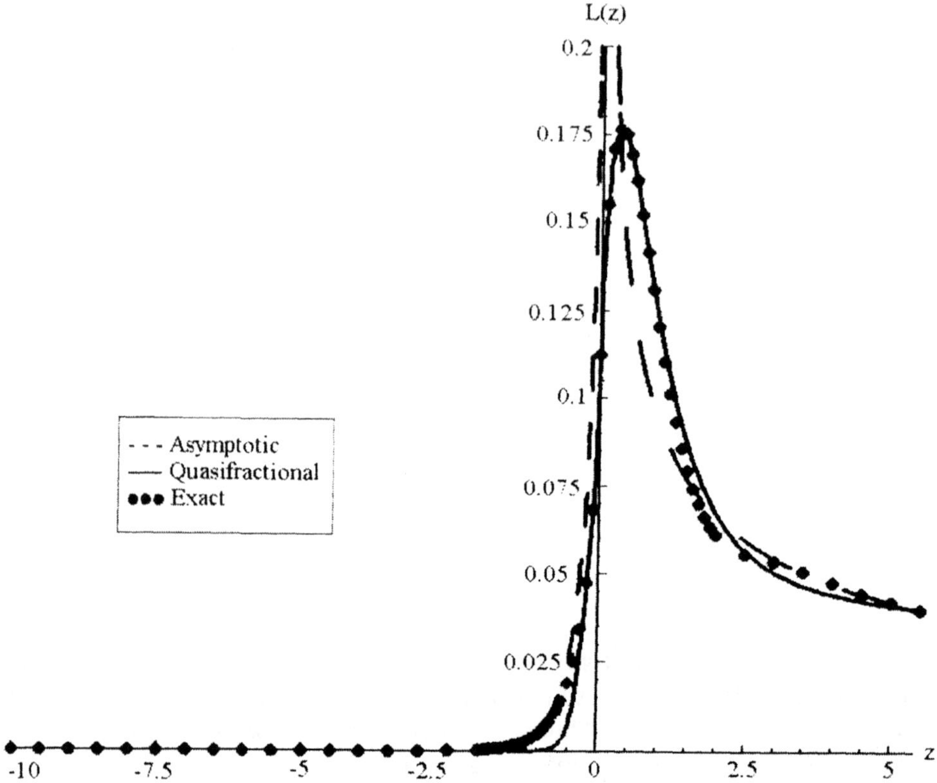

Figure 1. Universal Spectral Line and their approximations.

of the approximant is rather simple considering the complex form of the function $L(z)$.

The main problem is to obtain a good approximant to this function and to define the form of the approximant, which is now a peculiar combination of logarithmic, exponential and power functions.

References

Bekefi, G.: 1976, *Principles of Laser Plasma*, Wiley, New York.
Martín, P. and Baker, G.: 1991, *J. Math. Phys.* **32**, 1740.
Pérez, R. and Martín, P.: 1994, *Phys. Plasmas* **1**, 2105.
Szudy, J. and Baylis, W.E.: 1975, *Spectrosc. Radiat. Transfer* **15**, 641.

SPECTRAL STUDY OF IONIZED XENON USING A CAPILLARY PULSED DISCHARGE

R. SARMIENTO
Centro de Investigaciones Opticas, (CIOp), Casilla de Correo 124, 1900 La Plata, Argentina; Departamento de Física, Universidad del Atlántico, A.A. 1890 Barranquilla, Colombia

M. GALLARDO, A. LAQUIDARA and J. REYNA ALMANDOS
Centro de Investigaciones Opticas, (CIOp), Casilla de Correo 124, 1900 La Plata, Argentina

Abstract. This work presents preliminaries spectroscopic results about ionized xenon and xenon-helium mixture using a capillary pulsed discharge under several experimental results.

1. Introduction

The study of systems with two optical electrons as well as hydrogen-type structures provides relevant data on many characteristics of physical plasmas. Thus, the information derived from medium and highly ionized xenon spectra can be applied in diagnostic techniques for high temperature confined plasmas (Martinson, 1989) as well as in laser physics and other fields.

This work presents results on ionized xenon spectra in the region between 400 and 650 nm, obtained with a capillary pulsed discharge source. In some cases, an axial pulsed magnetic field was applied to the source and the inner pressure was varied. The results of our recently classified (Gallardo et al., in press) laser transitions belonging to XeV are discussed, as well as certain non-classified XeVIII lines that fall in this wavelength region (private communication).

2. Experimental Setup

The experimental setup is sketched in Figure 1. The source is a Pyrex tube ended in Brewster-angle windows. Two side-arm tungsten electrodes close the electric discharge inside the tube. The gas pressure was varied between 10 and 150 mTorr.

The excitation of the gas was achieved through fast risetime discharges of low inductance capacitors (30 nF) charged at 15–20 kV (Gallardo et al., 1988). The magnetic field was obtained discharging a low-inductance capacitor bank (C2) charged at 2 kV through a 100 coil inductor wounded around the source tube. In order to overlap the gas discharge with the maximum intensity interval of the magnetic field pulse, the former is delayed with respect to the latter through a delay

Astrophysics and Space Science **256:** 359–363, 1998.

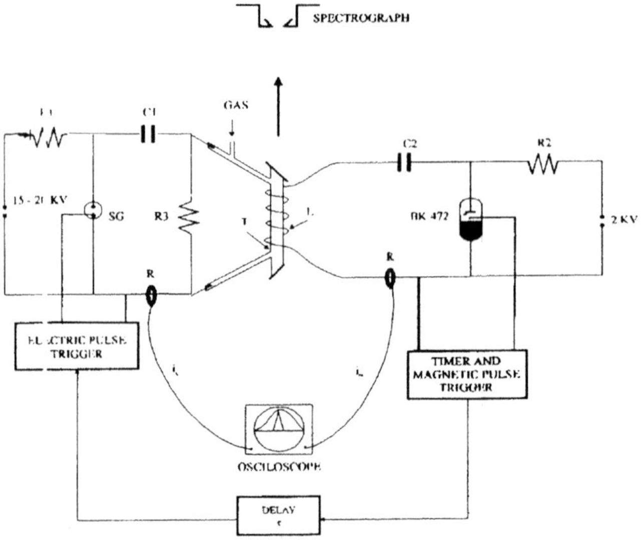

Figure 1. Schematic view of electric and magnetic pulse discharge circuit.

circuit (Gallardo et al., 1988). Both the magnetic and the gas discharge pulse were plotted with the aid of a digital oscilloscope Tektronix 100MHz, and a plotter, as shown in Figure 2. Besides, the light pulse (a) and the electric discharge (b) pulse were also plotted (Figure 3). Since the duration of the magnetic pulse was about 100 μs FWHM, it appears basically as a stationary field during the interval corresponding to the gas discharge pulse (25 μs FWHM). Although some jitters in the latter pulse was observed (mainly due to RF interference of the spark-gap), there was a high percentage of synchronized discharges, providing a splitting of some spectral lines. Both the magnetic and the gas discharge pulses were measured using calibrated Rogowski coil couple to a 100 MHz bandwidth oscilloscope.

3. Results

Table I shows some of the Xe lines emitted by our non-conventional spectroscopic source, in the wavelength region of 430 to 650 nm.

The analysis of the emitted spectra shows the presence of once ionized xenon (XeII) up to seven times ionized xenon (XeVII). The appearance of highly ionized species corresponds to decreasing values of gas pressure, under 80 mTorr. This shows that this type of source is useful for plasma spectroscopic studies. The applied magnetic field on the plasma allowed us to observe Zeeman splitting on some lines belonging to XeII and XeIII, reported firstly by Humphreys et al. (1939) and later by Hansen and Persson (1987). In our work we report Zeeman splitting on the lines 5007.80 Å(XeV) recently classified and on the line 5714.67 (XeVIII), which, to our knowledge, was not reported yet. Besides, difference in behavior

TABLE I

L: laser, Z: splitted, Z*: splitted observed, a: private comm., b: not reported – classified.

λ_{air}(Å)	Ion	Obs	Reference
4305.73	V	L	a
4325.20	V		a
4334.15	VII		a
4422.07	V		a
4438.43	V		a
4558.51	V	L	a
4634.50	V		a
4651.90	II	Z	(Hansen and Persson, 1987)
4653.44	V		a
4683.55	III	Z	(Humphreys et al., 1939)
4723.62	III	Z	(Humphreys et al., 1939)
4746.65	?		
4849.30	V		a
4954.13	V	L	a
5007.80	V	L,Z*	a
5159.08	V	L	a
5191.34	V	L	a
5260.19	V	L	a
5339.31	II	Z	(Hansen and Persson, 1987)
5343.33	VIII?		a
5352.92	V	L	a
5372.39	II	Z	(Hansen and Persson, 1987)
5394.61	V	L	a
5472.58	II	Z	(Hansen and Persson, 1987)
5531.03	II	Z	(Hansen and Persson, 1987)
5602.83	V		a
5614.03	V		a
5632.19	?		
5714.67	VIII?	Z*	b
5778.40	?		
5955.67	V	L	a
5976.44	II	Z	(Hansen and Persson, 1987)
6132.58	?		b
6200.18	VIII?	b	
6507.88	?		

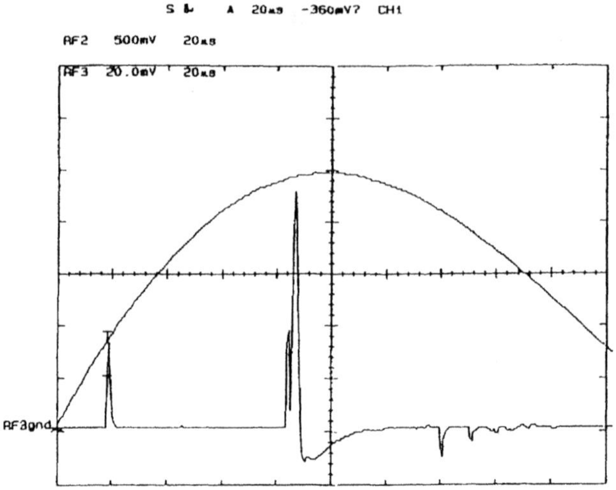

Figure 2.

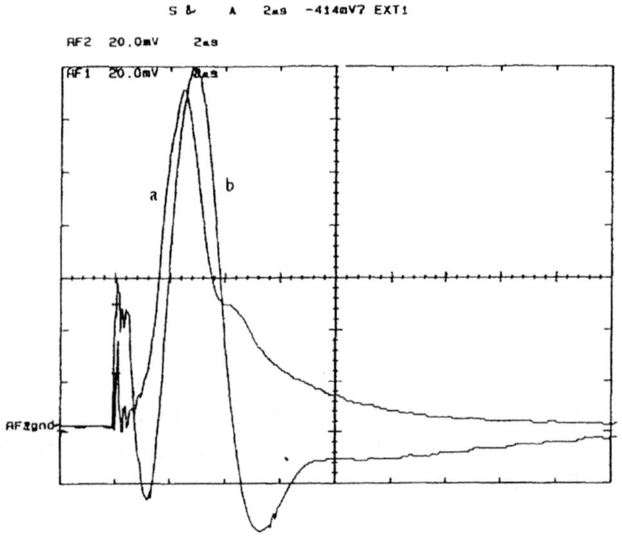

Figure 3.

of lines belonging to different stages of ionization when acted upon by this kind of discharge was observed; XeII and XeIIII lines appear broader at the ends of the lines, but lines belongings to higher degrees of ionization appear broader at the center of the lines. The measurement of the splitting of some of the lines of spectra taken in first diffraction order allowed us to determine the peak magnetic field intensity to be about 10 kGauss. Because of the pulsed nature of the light source, long times of exposure are required when high diffraction orders are used, producing heating of the discharge tube and instabilities of the pulse synchronism, which makes difficult the recording of high resolution spectra.

Discharges were also tried in Xe-He mixtures. Preliminary results show an increase in the intensity of certain Xe lines, which are weak in pure Xe discharges, as well as a better discrimination of the studied ions.

Acknowledgments

The authors wish to express their acknowledgments to: Dirección Nacional de Cooperación Internacional del Ministerio de Cultura y Educación de la República de Argentina, Consejo Nacional de Investigaciones Científicas y Técnicas and the Comisión de Investigaciones Científicas de la Provincia de Buenos Aires (Argentina) and the Universidad del Atlántico of Colombia.

References

Gallardo, M. et al.: 1988, *Applied Opt.* **27**, 2371.
Gallardo, M. et al.: *J. Quant. Spectrosc. Radiat. Transfer* (in press).
Hansen, J.E. and Persson, W.: 1987, *Physica Scripta* **36**, 602.
Humphreys, C.J. et al.: 1939, *Jour. Res. Nat. Bur. Stand.* **23**, 683.
Martinson, I.: 1989, *Rep. Prog. Phys.* **52**, 157.

ON THE THERMOCHEMICAL INSTABILITY OF AN IONIZED PLASMA COOLED BY OXYGEN FORBIDDEN LINES

M.C. BESSEGA and M.H. IBÁÑEZ S.

Centro de Astrofísica Teórica, Facultad de Ciencias, Universidad de los Andes, Apartado de Correos 26, Ipostel, La Hechicera, Mérida, Venezuela

Abstract. The thermal stability of an ionized hydrogen plasma in which the reaction OIII + $(h\nu) \rightleftharpoons$ OIV + e^- proceeds is analyzed. The gas is assumed to be heated by a diluted stellar radiation of effective temperature T_*. The stability analysis is carried out taking into account the chemical relaxation time effects from which a fifth order dispersion relation is obtained instead the fourth order polynomial obtained when such effects are neglected. The effects of the thermal conduction are also considered.

1. Introduction

The thermal (or more generally the thermochemical) stability of optically thin plasmas is a crucial subject for understanding the origin of astrophysical structures at very different scales (Field, 1965). In particular, the onset of thermal instabilities can directly induce the formation of inhomogeneities in different regions of the interstellar medium, specifically, in planetary nebulae and in HII regions where the heating of the plasma is provided by radiation from a hot star and the cooling by emission of forbidden lines of metal atoms highly ionized.

In this paper, the attention will be focussed on the case when cooling is dominated by emission of OIII and the effective ionizing radiation is about 55 eV which corresponds to the ionizing energy for OIII $\rightarrow$ OIV. The relaxation time effects will be considered (Ibáñez, 1998) and the results compared with those when such effects are neglected (Yoneyama, 1973). Additionally, the effects of the thermal conduction are also examined.

2. Basic Equations

When the initial thermochemical equilibrium is considered to be frozen, the dispersion equation for linear disturbances is given by

$$N^4 + \bar{a}_1 N^3 + \bar{a}_2 N^2 + \bar{a}_3 N + \bar{a}_4 = 0\,, \tag{1}$$

where the real coefficients $\bar{a}_j$ are defined by equations (21–23) in Ibáñez and Parravano (1983). Instead, when the effects of the chemical relaxation time are taken into account the dispersion relation becomes

Astrophysics and Space Science **256:** 365–368, 1998.

$$a_0N^5 + a_1N^4 + a_2N^3 + a_3N^2 + a_4N + a_5 = 0 . \tag{2}$$

For the definitions of coefficients a_j and symbols, see Ibáñez (1998).

For a hydrogen plasma optically thin with oxygen abundance 4.4×10^{-4}, heated and ionized by a background radiation field of a star of effective temperature T_*, the net rate function $X(\rho, T, \xi)$ and the net cooling rate per unit mass $\mathcal{L}(\rho, \mathcal{T}, \xi)$ are respectively given by

$$\begin{aligned} X(\rho, T, \xi) &= 2.07 \times 10^{-11} J^2 n\xi T^{-1/2}\phi_0(T) \\ &- 6.26 \times 10^8 J^4 \frac{(1-\xi)W_*}{n_0^5} \int_{y_0}^{\infty} \frac{dy}{y(\exp y - 1)}, \end{aligned} \tag{3}$$

$$\begin{aligned} \mathcal{L}(\rho, T, \xi) &= N_0 n \Bigg\{ 3.8 \times 10^{-9} \frac{\Omega\chi_\epsilon}{g} (1-\xi)T^{-1/2} \exp\left(\frac{-\chi_\epsilon}{k_B T}\right) \\ &- 2.85 \times 10^{-27}(J-1)^2[T_*\psi_0 T^{-1/2}\phi(T) - T^{1/2}\chi_0] \Bigg\}, \end{aligned} \tag{4}$$

where n, T, ξ are the hydrogen number density, the plasma temperature and the concentration of the OIV, respectively. Additionally, $y = h\nu/k_B T$ and $y_0 = h\nu_0/k_B T = 54.934$ eV is the ionization potential of OIII. The remaining symbols are those defined by Yoneyama (1973) (see also Allen, 1973).

On the other hand, the electronic thermal conduction coefficient is given by (Spitzer, 1962)

$$k(\rho, T, \xi) = 1.84 \times 10^{-5} \frac{\xi T^{5/2}}{\ln \Lambda(\rho, T, \xi)} , \tag{5}$$

where

$$\begin{aligned} \ln \Lambda &= 23.24 + \ln\left[\left(\frac{T}{10^4}\right)^3 \left(\frac{1}{N_0\rho\xi}\right)^{1/2}\right], \quad \text{if} \quad T < 4.2 \times 10^5 \text{ K}, \\ \ln \Lambda &= 29.71 + \ln\left[\frac{T}{10^6(N_0\rho\xi)^{1/2}}\right], \quad \text{if} \quad T > 4.2 \times 10^5 \text{ K}. \end{aligned} \tag{6}$$

3. Results

Figures 1a and 1b show the rates $|\mathrm{Re}\{N_j\}|$ (yr^{-1}) resulting from numerically solving the dispersion relations (1) and (2), respectively, as functions of the wave number k (in units of 10^{-18} cm^{-1}) for a plasma with number density $n = 10^3$ cm^{-3}, at the equilibrium temperatures $T = 9.15 \times 10^4$ K ($\xi = 9.901 \times 10^{-1}$) and neglecting thermal conduction effects. Instead, Figures 2a and 2b show the corresponding rates when the thermal conduction is taken into account.

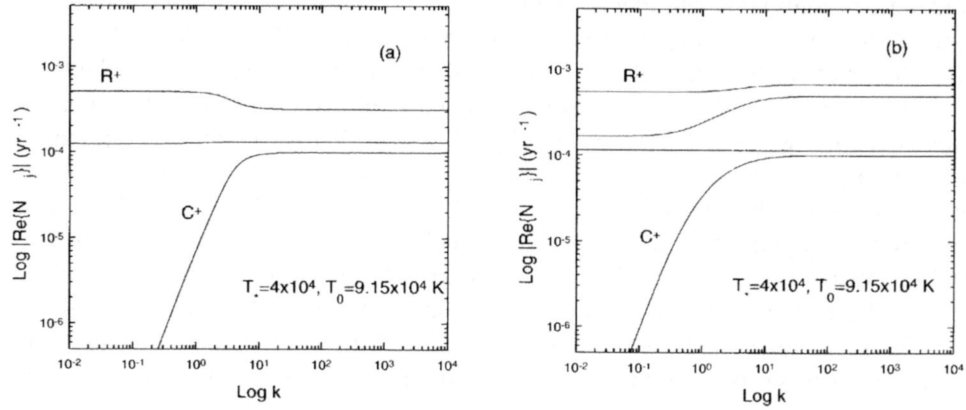

Figure 1. (a) The rates resulting from solving de fourth order polynomial (Equation (1)) neglecting thermal conduction, as functions of the wave number for $T_* = 4 \times 10^4$ K, $T_0 = 9.15 \times 10^4$ K and $n = 10^3$ cm^{-3}. The wave number k is in units of 10^{-18} cm^{-1}. The range where the oscillating and real modes are unstable are indicated by C^+ and R, respectively. (b) As Figure 1a for the rates resulting from solving the fifth order polynomial (Equation (2)).

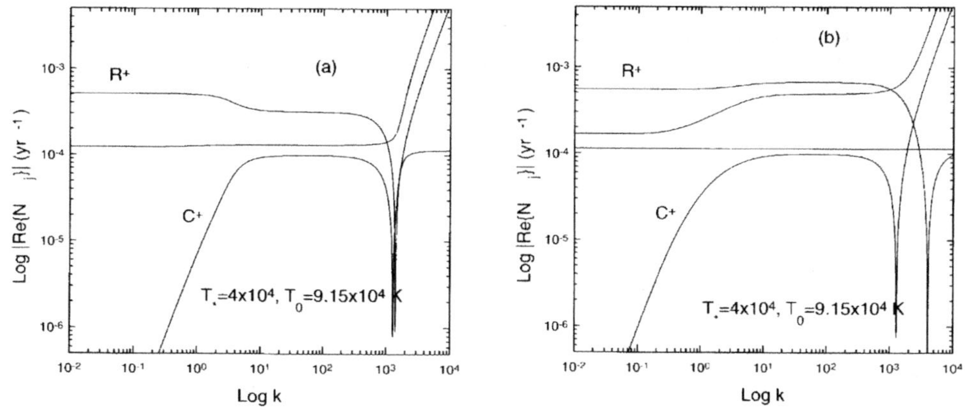

Figure 2. (a) As Figure 1a but including thermal conduction. (b) As Figure 1b but including thermal conduction.

In addition to the small changes in the values of the rates of the real modes, the fifth order polynomial gives an a new damped real mode which shows strong dispersion at values of $k \approx 1$. However, except by a shift towards smaller values of k, the oscillating unstable mode remains unmodified, for the above values of free parameters. See Figures 1a and 1b.

The thermal conduction (Figures 2a and 2b) strongly affect both, real and oscillating modes at large wave numbers ($k > 10^3$). It stabilizes the unstable oscillating and real (thermal) modes resulting from the fourth degree polynomial (Figure 2a) for $k \geq k_c = 1.29 \times 10^3$ and 1.41×10^3, respectively. Therefore, the maximum growing rate of the oscillating mode becomes $\mathrm{Re}\{N_{\max}\} = 9.93 \times 10^{-5}$ yr^{-1} for fluctuations with wave number $k \approx 60$ (i.e., 5.4×10^{-3} pc). Similar conclusions can be drawn for the modes resulting from solving the fifth order polynomial (2). However, in this case the instability of the thermal mode is quenched for $k \geq k_c = 4.07 \times 10^3$. Additionally, the maximum growing rate of the thermal mode becomes $\mathrm{Re}\{N_{\max}\} = 6.73 \times 10^{-4}$ yr^{-1} at $k \approx 70$. Instead, the corresponding maximum growing rate for this unstable mode resulting from solving the fourth order polynomial becomes 5.08×10^{-4} yr^{-1} for $k \ll 1$ (see Figure 2a). Note also the strong damping introduced by the thermal conduction on the oscillating and real modes for $k > 3 \times 10^3$ (Figures 2a and 2b). All the above results hold for the values of the free parameters under consideration, the results for another values, as well as a detailed analysis introducing more realistic cooling and rate equations (considering other ions, for instance the NII and NIII) will be published elsewhere.

References

Allen, C.W.: 1973, *Astrophysical Quantities*, Athlone Press, London.
Field, G.B.: 1965, *ApJ* **142**, 531.
Ibáñez, S.M.H.: 1998, *ApJ* (in press).
Ibáñez, S.M.H. and Parravano, A.: 1983, *ApJ* **275**, 181.
Spitzer, L.: 1962, *Physics of Fully Ionized Gases*, Wiley, New York.
Yoneyama, T.: 1973, *PASJ* **25**, 349.

SPACE CHARGE POTENTIALS IN CYLINDRICAL GEOMETRIES INCLUDING TEMPERATURE EFFECTS

E. VALDEBLANQUEZ
Facultad de Ingeniería, Universidad del Zulia, Apartado 98, Marcacaibo, Venezuela

P. MARTÍN
Departamento de Física, Universidad Simón Bolívar, Apartado 89000, Caracas, Venezuela

Abstract. The analysis of space-charge in cylindrical geometries has been performed using kinetic theory. The current collected by central electrodes inmersed in plasmas with a discriminant grid has been determined through the Poisson equation and truncated Maxwellian distribution functions for the ions. The electrons are assumed to be repelled by the entrance grid. The non-linear adimensional equations have been solved approximately using linear expansion around the point with zero potential.

1. Introduction

The space-charge effects for planar electrodes have been extensively treated using both fluid equations and kinetic theory (Page and Adams, 1958; Braun et al., 1973) for planar velocity analyzers, thermoionic diodes, and other engineering problems. For cold plasmas and simple boundary conditions, the fluid equations can be integrated and the result is known as the Langmuir–Child equation (Langmuir, 1923). This equation does not include temperature effects and can only be considered as a first approximation of the Martin–Donoso's paper (1988). The analysis using kinetic theory in cylindrical geometries leads to a non-linear second-order differential equation which can be solved using a linear expansion around the point of zero potential. In kinetic theory treatment, we have assumed truncated Maxwellians depending on whether the trapped charge particles in the entrance grid are included or not. Recently experiments using velocity analyzers of variable interelectrodic distance seem to confirm that the trapped particles participate in the space-charge phenomena and they should be included (Donoso and Martín, 1987; Donoso et al., 1986), thus the analysis done in detail here seems to be the most appropriate one.

2. Theoretical Analysis

Here the problem of space charge is considered for the particular structure of a velocity analyzer with cylindrical electrodes $c < \rho < a$ inmersed in a plasma with

Astrophysics and Space Science **256:** 369–372, 1998.

only one kind of ions. The distribution function for an unlimited homogeneous ion plasma with potential $V(\rho)$ is

$$F(\rho, \dot{\rho}, \dot{\phi}) = \eta_0(\rho)\left(\frac{m}{2\pi T}\right)^{1/2} \exp\left\{-\left(\frac{1}{2}\, m(\dot{\rho}^2 + \rho^2\dot{\phi}^2) + qV(\rho)\right)/T\right\}, \quad (1)$$

where m and q are the mass and charge of the particles respectively, and T is measured in electron volts. The distribution function inside the electrodes are truncated Maxwellians and the Poisson equation in adimensional variables (see Martín and Donoso, 1988, for details in plane grids) is written as

$$\frac{d^2\phi}{d\rho^2} + \frac{1}{\rho}\frac{d\phi}{d\rho} = \frac{1}{2}\,\eta_0(\rho)e^{(\phi-\phi_p)}(1 \pm erf(\phi^{1/2})), \quad \phi = \phi_p - q\frac{V(\rho_p)}{T}. \quad (2)$$

For $\rho_p < \rho < a$ we now perform an expansion in power series around ϕ_p:

$$\frac{d^2R_+}{d\rho^2} + \frac{1}{2}\frac{dR_+}{d\rho} - \beta_0^2 R_+ = 0, \quad (3)$$

where

$$R_\pm(\rho) = \alpha_{0\pm} + \beta_0^2[\phi(\rho) - \phi_p],$$

$$\alpha_{0\pm} = \frac{1}{2}\,(1 \pm erf(\phi_p^{1/2})), \quad (4)$$

and

$$\beta_0^2 = \frac{1}{2}\left(1 + erf\,\phi_p^{1/2} + \frac{e^{\phi_p}}{2\phi_p^{1/2}}\right). \quad (5)$$

The solution is given by the modified Bessel function

$$\phi(\rho) = CI_0(\beta_0\rho) + DK_0(\beta_0\rho) + C_1^+, \quad (6)$$

where

$$C_1^\pm = \frac{\alpha_{0\pm}}{\beta_0^2} \mp \phi_p. \quad (7)$$

Now for $c < \rho < \rho_p$ an expansion in series around $\phi(\rho_0) = \phi_p$, that is $V(\rho_0) = 0$, is also performed. This point is usually out of the geometry of the system.

$$\frac{d^2R_-}{d\rho^2} + \frac{1}{\rho}\frac{dR_-}{d\rho} + \beta_0^2 R_- = 0. \quad (8)$$

The solution can now be expressed in terms of Bessel function as

$$\phi(\rho) = AJ_0(\beta_0\rho) + BY_0(\beta_0\rho) + C_1^-. \quad (9)$$

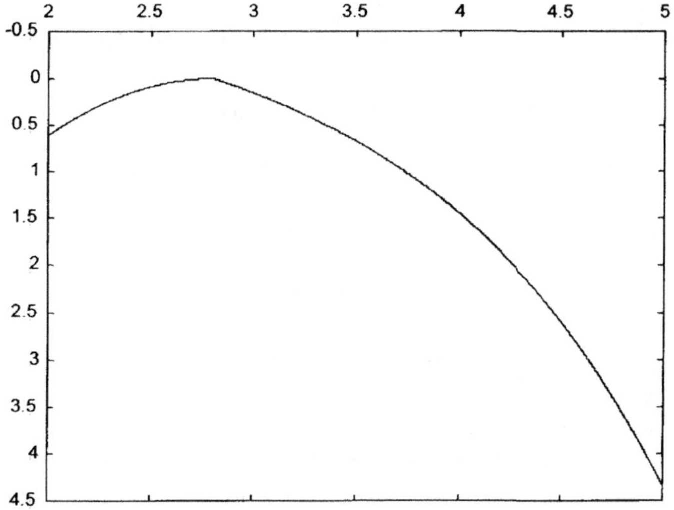

Figure 1. Electric potential as a function of the interelectrodic distance. The numbers on the top line are the radii.

Now the boundary conditions of the solutions are

$$\phi_a = \phi_p - \frac{qV(a)}{T}, \quad \phi_c = \phi_p - \frac{qV(c)}{T} \quad \text{and} \quad \left.\frac{d\phi}{d\rho}\right|_{\rho=\rho_p} = 0. \tag{10}$$

We have the system of equations

$$AJ_0(\beta_0 c) + BY_0(\beta_0 c) + C_1^- = \phi(\rho = c) = \phi_c, \tag{11}$$

$$CI_0(\beta_0 a) + DK_0(\beta_0 a) + C_1^+ = \phi(\rho = a) = \phi_a, \tag{12}$$

$$0 = AJ_0(\chi_p) + BY_0(\chi_p) + C_1^- = CI_0(\chi_p) + DK_0(\chi_p) + C_1^+, \tag{13}$$

$$A\beta_0 J_0'(\chi_p) + B\beta_0 Y_0'(\chi_p) = 0, \tag{14}$$

$$C\beta_0 I_0'(\chi_p) + D\beta_0 K_0'(\chi_p) = 0, \tag{15}$$

where $\chi_p = \beta_0 \rho_p$, and the solution of this system of equation is given by

$$\begin{aligned} A &= C_1^- \frac{\pi}{2} x_p Y_1(x_p), \quad B = -C_1^- \frac{\pi}{2} x_p J_1(x_p), \\ C &= C_1^+ x_p K_1(x_p) \quad \text{and} \quad D = C_1^+ x_p I_1(x_p). \end{aligned} \tag{16}$$

Finally, we obtain the pair of equations for the quantities ϕ_p and ρ_p

$$\phi_a = C_1^+ x_p K_1(x_p) I_0(\beta_0 a) + C_1^+ x_p I_1(x_p) K_0(\beta_0 a) + C_1^+, \tag{17}$$

$$\phi_R = \phi_C = \frac{C_1^-}{2} \pi x_p Y_1(x_p) J_0(\beta_0 c) - \frac{C_1^-}{2} \pi x_p J_1(x_p) Y_0(\beta_0 c) + C_1^- . \qquad (18)$$

Figure 1 shows these potentials for the particular case

$$\phi_a = 4.3, \quad \text{and} \quad \phi_c = 0.6 \quad \text{then} \quad \rho_p = 2.8 \quad \text{and} \quad \phi_p = 2. \qquad (19)$$

3. Conclusions

The space-charge analysis including temperature effects has been done using kinetic theory. The dimensionless equations in this case are strongly non-linear. Here we look for approximated analytic solutions, which could lead to a generalization of the classical Langmuir–Child equation obtained by using fluid treatment with plane grids. In order to do that, expansions around the zero potential ($\phi = \phi_p$) have been performed. In this way the solutions can now be obtained using cylindrical functions. Our results lead to two non-algebraic equations for ρ_p and ϕ_p. Using some values for the entranced and discriminant potentials, the interelectrodic potential curve has been determined.

References

Braun, J.C., Richard, M. and Felden, M.: 1973, *J. Physique (Paris)* **34**, 859.
Donoso, G. and Martín, P.: 1987, *Bull. Am. Phys. Soc.* **32**, 1817.
Donoso, G., Martín, P. and Puerta, J.: 1986, *Sci. Instrument.* **57**, 1507.
Langmuir, I.: 1923, *Phys. Rev.* **21**, 419; **22**, 347.
Martín, P. and Donoso, G.: 1988, *Phys. Fluids B* **1**, 247.
Page, L. and Adams, Jr., N.I.: 1958, *Principles of Electricity*, 3rd ed., Van Nostrand, Englewood Cliffs, NJ, pp. 253ff.

CME PHENOMENOLOGY IN DIFFERENT SOLAR MAGNETIC SCENARIOS

R. RODRÍGUEZ TABOADA, G. GIL MORENO and A. MÉNDEZ BERHONDO
Institute of Geophysics and Astronomy, Calle 212 No. 2906, La Coronela, Havana, Cuba

R. PÉREZ-ENRIQUEZ
Institute of Geophysics, Universidad Nacional Autónoma de México, Ciudad Universitaria, Mexico D.F.

Abstract. The existence of three magnetic scenarios with magnetic field intensities one order apart was considered. CMEs were associated to different solar activity phenomena and solar features showing different characteristics. Those associated to intense magnetic fields were more complex, having either more components or multiple emissions and a greater magnetic energy to particle acceleration conversion rate. On the other hand, simple CMEs were found to associate to weak magnetic fields. No evidence was found that the process that leads to CME starts before their detection at 1.6 solar radii. This indicates that the whole volume moves upward simultaneously, no matter how deep it reaches into the solar atmosphere.

1. Introduction

Statistical studies of CME and solar flares as related events (Munro et al., 1979) underlay two widely diverging viewpoints. One considers solar flares as CME generating agents (Pneuman, 1980), while the other studies CMEs as solar flares generators (Harrison, 1986). Another approach considers CMEs as a result of the eruption of prominences or of magnetic structures capable of supporting a prominence and their overlaying coronal structures (Webb and Hundhausen, 1987). Harrison (1990) links CME activity to active regions.

We consider CMEs as independent phenomena linked to sheared magnetic field structures overlying inversion lines. In this paper we consider the existence of different scenarios in which CME and associated solar activity occur. Active regions, plages and the large-scale magnetic field pattern, which has not been considered earlier, are included. These scenarios characterized by magnetic field intensities one order apart further the clarification of the role of magnetic field intensity on CME characteristics and on associated solar activity phenomena. A comprehensive formulation of CME phenomena will contribute to our understanding of the why and wherefore of solar activity and will improve our prediction capabilities.

2. Data

The data used for this study is compiled in "A Catalogue of Solar Features and Active Phenomena Associated to Coronal Mass Ejections in 1980" (Gil et al., 1995)

Astrophysics and Space Science **256:** 373–376, 1998.

which is available upon request and provides a wide range of observational information on the main characteristics of the more than one hundred CME observed by the SMM coronograph.

The main parameters used while looking for associations between the CMEs and other solar features were:

- the location of the CME estimated from its central position angle and the corresponding limb coordinate; and
- the time difference between the onset of the solar activity feature under consideration and the time when the CME was first detected with the solar coronograph onboard the SMM satellite.

3. Results and Discussion

After exploring the data on CME characteristics we found we could group all SMM reports consisting of more than one emission into two classes:

- simultaneous emissions from different sites reported as one event which we defined as a multiple CME; and
- several emissions at different times at the same site reported as one event, which we considered as components.

Different type CMEs were associated to sunspot groups, plagues and polarity inversion lines. A weak association of multiple and 2 component CME to plagues and sunspot groups is suggested (20% significance level), pointing to multiple and two component CME preference to intense magnetic field and probably more complex topologies.

However, it is very difficult to estimate the significance of CME association to other solar features particularly with magnetic field polarity inversion lines. The reliability of the 94 association percentage obtained for the association between CMEs and magnetic field polarity inversion lines was analyzed indirectly.

The association percentage obtained between this network and the magnetic field polarity inversion lines for solar rotation No. 1702 (from the Solar Synoptic Chart of Active Features published in Solnichnie Dannie No. 12 1980) was 85–90%. Therefore, the hypothesis of CME association to magnetic field polarity inversion lines was not rejected on the basis of the 94 association percentage obtained for one component CMEs, with a 10×10 degrees window.

Considering endurance for simple CMEs was noticed that loops associated to lines present a well defined maximum in the 1–2 hour interval while those associated to sunspot groups have a maximum in the 3–5 hours interval. These distribution are shown to be different with a 10% significance level by a Kolmogorov–Smirnoff test.

Considering CME duration an estimate of its propagation speed, we can interpret the longer duration of loops associated to sunspot groups as corresponding to a lower propagation speed than that of those associated to lines. If the total stored magnetic energy is approximately the same in both scenarios then, we can assume a greater conversion of magnetic energy into particle acceleration in the sunspot scenario which leaves less energy for the coronal mass ejection, rendering a much slower CME propagation speed.

Different shape CMEs showed no significant latitudinal groupings. Plage associated CMEs and simple CMEs grouped around 120 and 300 degrees of longitude, with a significance level of 1 and 10% respectively. These preferential bands occupy the "gaps" between the giant fast rotating active longitude positions for 1980 (Rodriguez and Gil, 1993).

Webb and Hundhausen (1987) found a significant relation between mass ejections and H_α flares and radio bursts. We could only associate 13 CMEs to H_α flares. All flares were of importance 1 or less, with a significant proportion of subflares. This might be due to the sunspot group locations, close to the limb, where flare importance evaluation has greater uncertainty.

Proton flare association was investigated using the listing given by Molchanov (1984) from satellite data. Although, this phenomenon is often referred to as tightly linked to CME (Akinyan et al. 1979), only three cases could be associated to CMEs. This seems to be a clear symptom that particle acceleration is not the main process in CME generation.

The morphological distribution of the 49 radio burst associated CMEs was done. Dividing them into metric, centimetric or both; we find a distribution of 39, 45 and 16% respectively. It is very difficult to observe isolated metric bursts in the foreground of a noise storm, yet 39% of the associated bursts fell into this category. This fact together with the low H_α flare CME association (8%) points to the CME generation process as an essentially coronal rearrangement process, in which the energy is not mainly released by particle acceleration, but by large scale volume movement. That is, a process without characteristic longitude change.

Only 9 CMEs were found to associate to type II and IV metric radio bursts. They represent approximately 5% of the total, which is similar to the association percentage reported by Gergely (1986). Their association to plages and groups was 44%, larger (but not significantly so) than the 32% found for all CMEs. We found that the association to type II metric bursts increases with the number of components of the CME. This might indicate a bias for intense magnetic field scenarios.

If the CME onset occurs at photospheric heights and propagates upward with an average velocity of 300 Km/s, we should expect a 1 hour delay until its first observation by the SMM coronograph at 1.6 radii. If radio bursts are associated to the particle acceleration process at CME onset, a 1 hour delay should appear on the associated radio burst temporal distribution.

The temporal distribution of the associated radio bursts centered on CME starting time does not show any significant lag between CME commencement (first observation by the SMM coronograph) and burst starting time. We conclude that the CME commencement occurs very close to the time of the reported SMM starting time. The whole volume involved moves simultaneously no matter how deep it reaches into the solar atmosphere.

For further investigation on this subject we used the Toyokawa observations in 1000, 2000, 3750 and 9400 MHz. We calculated the temporal distribution of the differences between the CME starting time and the start of the nearest burst. We did the same for a non related random process and obtained the distributions.

We compared for each frequency both distributions grouping into two classes: "before" and "after" the CME or the random process commencement. The distributions associated to the random process were found to satisfy a $p = 0.5$ binomial distribution as expected. The bursts associated to CMEs were found to distribute significantly different from a $p = 0.5$ binomial distribution ($p < 0.08$), there were more coincidences "after" the CME starting time than "before".

4. Conclusion

Particle acceleration processes are not what we should search for when looking for CME signatures in other solar phenomena. The associations found are scarce and appear mainly after the CME onset.

We propose two scenarios for the appearance of CMEs. Neutral lines of the background magnetic field (weak magnetic fields) and active regions: sunspot groups and plagues (intense magnetic fields). Those associated to intense magnetic fields were more complex having more components or multiple emissions. They were also found to have a greater conversion rate of magnetic energy into particle acceleration.

No evidence was found that the process that leads to coronal mass ejections starts before their detection by the SMM coronograph at 1.6 solar radii. That is, the whole volume involved moves simultaneously upward no matter how deep it reaches into the solar atmosphere.

References

Akinyan, S.T., Fomichef, V.V., Chertok, I.M. and Kruger, A.: 1979, *Phys. Solariterr.* **11**, 5.

Gergely, T.E.: 1986, *Solar Phys.* **104**, 175.

Gil, G., Méndez, A. and Rodríguez, R.E.: 1994, Available upon request. Institute of Geophysics and Astronomy, Calle 221 No. 2905, La Coronela, Havana Cuba.

Harrison, R.A.: 1986, *Astron. Astrophys.* **162**, 283.

Harrison, R.A.: 1990, *Solar Phys.* **126**, 185.

Molchanov, A.P. (ed.): *Forecasting of Solar Flares and Their Consequences, Radioemission of the Sun*, No. 5, 1984.

Munro, R.H., Gosling, R.T., Hildner, E., MacQueen, R.M., Poland, A.I. and Ross, C.L.: 1979, *Solar Phys.* **61**, 201.

Pneuman, G.W.: 1980, *Solar Phys.* **65**, 369.

Rodriguez, R.E. and Gil, G.: 1993, *Solar Phys.* **144**, 399.

ON THE THERMOCHEMICAL INSTABILITY OF OPTICALLY THIN PLASMAS

M.H. IBÁÑEZ S.
Centro de Astrofísica Teórica, Facultad de Ciencias, Universidad de los Andes, Apartado de Correos 26, Ipostel, La Hechicera, Mérida, Venezuela

Abstract. The effects of the thermal conduction on the linear stability analysis of an optically thin plasma with metallicity $Z = 1$ and including chemical relaxation time effects, is carried out. Generally, a further stabilizing effect on the unstable modes and an increasing of the damping of stable modes appear for disturbances with values of wave number large enough. However, the oscillating mode appearing at low temperature remain unaffected by the thermal conduction.

1. Introduction

The thermal (or more generally the thermochemical) stability of optically thin plasmas is a crucial subject for understanding the origin of astrophysical structures at very different scales. In fact, the onset of thermal instabilities can directly induce the formation of inhomogeneities or can set in physical conditions appropriated for developing other kind of instabilities, say for instance, the gravitational one. As a direct agent in forming condensations it has been considered in studies of the solar atmosphere, the interstellar medium, gas at high latitude and in outer disks of galaxies, Lyα clouds and quasars gas, and as an indirect promoter of the formation of stars, globular clusters and galaxies.

Previous works (Goldsmith, 1970; Defouw, 1970; Yoneyama, 1973; Flannery and Press, 1979; Ibáñez and Parravano, 1983; Ibáñez and Mendoza, 1990; Corbelli and Ferrara, 1995) on the generalization of Field's (1965) basic study on thermal instability for including effects of chemical reactions have not taken into account the fact that when the temperature, density and the other thermodynamical quantities change, the position of the chemical equilibrium also varies (Landau and Lifshitz, 1987). Recently, Ibáñez (1998) has considered such effects in a general reacting gas and applied his results, in particular, to the photoionized plasma with metallicity Z studied by Corbelli and Ferrara (1995). In the above work the attention was focused on the relaxation time effects only. However, it would be interesting to examine the role played by the thermal conduction. Therefore, the present paper is aimed to show how the Ibáñez (1998) results are modified by the thermal conduction for the particular case of a plasma with metallicity $Z = 1$ and heated and ionized by a ration field with mean photon energy $E = 10^2$ eV.

Astrophysics and Space Science **256:** 377–380, 1998.

2. Basic Equations

According to Ibáñez (1998) the dispersion relation for linear disturbances is given by

$$a_0 N^5 + a_1 N^4 + a_2 N^3 + a_3 N^2 + a_4 N + a_5 = 0. \tag{1}$$

For the definitions of coefficients and symbols, see Ibáñez (1998).

For an optically thin hydrogen plasma with metallicity Z heated and ionized by a background radiation field of mean photon energy E and ionization rate ζ, the net rate function $X(\rho, T, \xi)$ and the net cooling rate per unit mass $\mathcal{L}(\rho, \mathcal{T}, \xi)$ are respectively given by

$$X(\rho, T, \xi) = N_0 \rho [\xi^2 \alpha - (1 - \xi)\xi \gamma_c] - (1 - \xi)(1 + \phi)\zeta, \tag{2}$$

$$\begin{aligned} \mathcal{L}(\rho, T, \xi) &= N_0^2 \rho [(1 - \xi) Z \Lambda_{HZ} + \xi Z \Lambda_{eZ} + (1 - \xi)\xi \Lambda_{eH} + \xi^2 \Lambda_{eH^+}] \\ &\quad - N_0 (1 - \xi)\zeta [E_h + (1 + \phi)\chi], \end{aligned} \tag{3}$$

where ϕ and E_h are the number of secondary electrons and the heat released per photoionization respectively, Λ_{ij} are: the cooling efficiencies by collisions neutral hydrogen-ions and metal atoms Λ_{HZ}, by electrons-ions and metal atoms Λ_{eZ}, by Lyα emission by neutral hydrogen Λ_{eH} and by hydrogen recombination Λ_{eH^+} on the spot approximation (see Ibáñez, 1998).

On the other hand, the thermal conduction coefficient can be taken in the form

$$\kappa(\rho, T, \xi) = 2.5 \times 10^3 (1 - \xi) T^{1/2} + 1.84 \times 10^{-5} \frac{\xi T^{5/2}}{\ln \Lambda(\rho, T, \xi)}, \tag{4}$$

where the first term on the right-hand side is the heat conduction by neutral atoms (Parker, 1953) and the second one is the Spitzer (1962) thermal conduction by electrons. Additionally,

$$\begin{aligned} \ln \Lambda &= 23.24 + \ln \left[\left(\frac{T}{10^4} \right)^3 \left(\frac{1}{N_0 \rho \xi} \right)^{1/2} \right], \quad \text{if} \quad T < 4.2 \times 10^5 \text{ K}, \\ \ln \Lambda &= 29.71 + \ln \left[\frac{T}{10^6 (N_0 \rho \xi)^{1/2}} \right], \quad \text{if} \quad T > 4.2 \times 10^5 \text{ K}. \end{aligned} \tag{5}$$

3. Results

Figures 1a–3a show the rates ($|\mathrm{Re}\{N_j\}|$) resulting from solving numerically the dispersion relation (1) for a plasma with $Z = 1$, and photon energy $E = 10^2$ eV, neglecting thermal conduction effects and for plasma temperatures $T = 2 \times 10^2$,

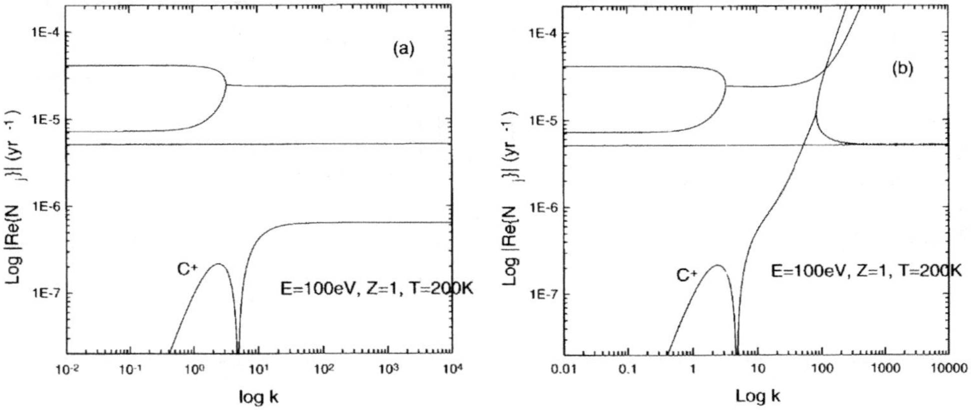

Figure 1. (a) The rates resulting from solving the fifth order polynomial (Equation (1)) neglecting thermal conduction, as functions of the wave number for a photon energy $E = 100$ eV, metallicity $Z = 1$, $N_0\rho = 1\ \text{cm}^{-3}$ and $T = 200$ K. The wave number k is in units of $10^{-18}\ \text{cm}^{-1}$. The range where the oscillating mode is unstable is indicated by C^+. (b) As Figure 1a but including thermal conduction.

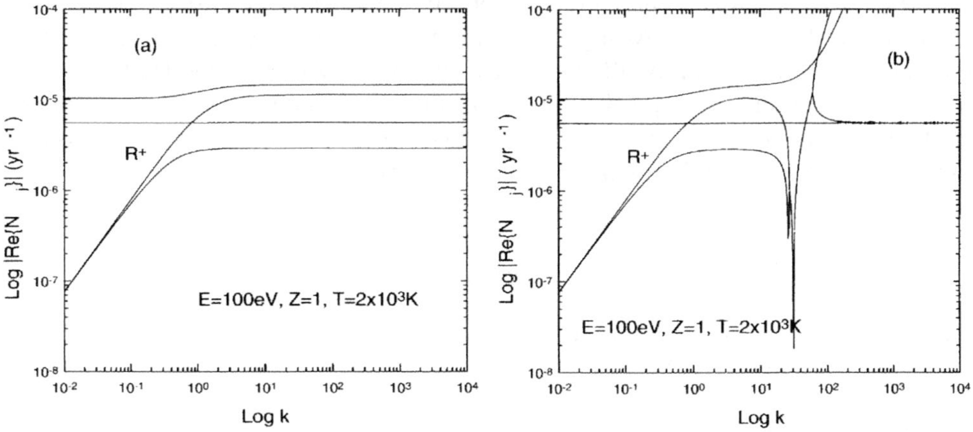

Figure 2. (a) As Figure 1a for $T = 2 \times 10^3$ K. The range where the real (thermal) mode is unstable is indicated by R^+. (b) As Figure 1b for $T = 2 \times 10^3$ K. The range where the real (thermal) mode is unstable is indicated by R^+.

2×10^3 and 2×10^4 K, respectively. Instead, Figures 1b–3b show the corresponding rates when the thermal conduction is taken into account. The thermal conduction does not affect at all the oscillating unstable mode appearing at low temperatures (Figures 1a–1b), but it does strongly affect both damped oscillating modes at large wave numbers ($k > 10$). At middle values of temperature (2×10^3 K, Figures 2a–2b) thermal conduction stabilizes the real (thermal) unstable mode for fluctuations with wave numbers $k > k_c$ ($= 31$). Therefore, the maximum value of the growing rate of the unstable thermal mode becomes 1.05×10^{-5} (yr^{-1}). Note

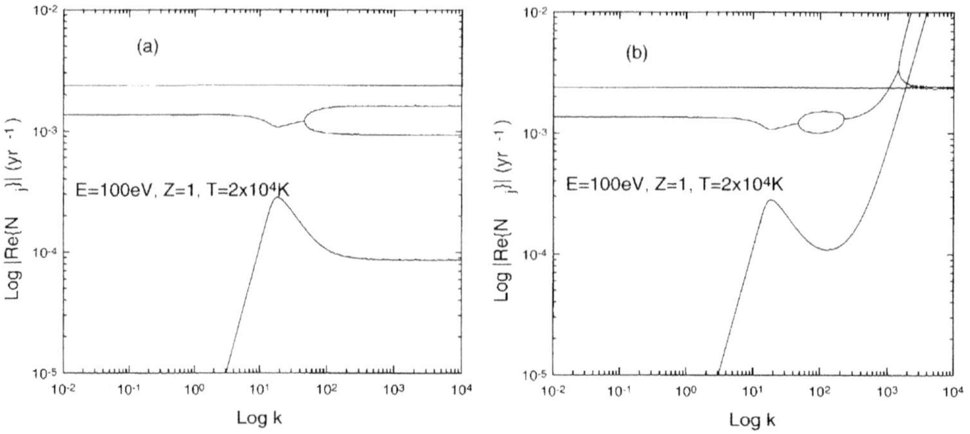

Figure 3. (a) As Figure 1a for $T = 2 \times 10^4$ K. (b) As Figure 1b for $T = 2 \times 10^4$ K.

that in absence of thermal conductivity this unstable mode shows a maximum rate $\mathrm{Re}\{N\} \approx 1.1 \times 10^{-5}$ (yr^{-1}) for all disturbances with wave number $k > 5$ (strictly speaking there is a very weak increasing of the rate when k increases undetectable at the scale of Figure 2a). At high temperatures (2×10^4 K, Figures 3a–3b) the thermal conduction strongly affects the two damped oscillating modes for values of wave number large enough ($k > 80$). Note the conspicuous effects, in particular, the thermal conduction introduces two additional bifurcation points for the fastest mode and increases the damping rates of both stable oscillating modes.

References

Corbelli, E. and Ferrara, A.: 1995, *ApJ* **447**, 720.
Defouw, R.J.: 1970, *ApJ* **161**, 55.
Field, G.B.: 1965, *ApJ* **142**, 531.
Flannery, B.P. and Press, W.H.: 1979, *ApJ* **231**, 688.
Goldsmith, G.B.: 1970, *ApJ* **161**, 41.
Ibáñez, S.M.H. and Parravano, A.: 1983, *ApJ* **275**, 181.
Ibáñez, S.M.H. and Mendoza, B.C.A.: 1990, *Ap&SS* **164**, 193.
Ibáñez, S.M.H.: 1998, *ApJ* (in press).
Landau, L.D. and Lifshitz, E.M.: 1987, *Fluid Mechanics*, Pergamon Press, London.
Parker, E.: 1953, *ApJ* **117**, 431.
Spitzer, L.: 1962, Physics of Fully Ionized Gases, Wiley, New York.
Yoneyama, T.: 1973, *PASJ* **25**, 349.

NONLINEAR EVOLUTION OF THERMAL STRUCTURES. NUMERICAL APPROACH

MARÍA CRISTINA TREVISAN and FELIPE CORDERO
Departamento de Matemáticas, Facultad de Ciencias, Universidad de los Andes, Mérida, Venezuela

MIGUEL H. IBÁÑEZ S.
Centro de Astrofísica Teórica (CAT), Facultad de Ciencias, Universidad de los Andes, Mérida, Venezuela

Abstract. A numerical algorithm using a two stage, two level difference scheme has been developed to solve the heat transfer equation with nonlinear heat diffusion and bulk energy losses. The algorithm is an extension of the scheme developed by Meek and Norbury (1982). The first stage calculates an intermediate value which is used in a second stage to estimate a new value. The scheme is consistent, second-order convergent in space and almost second order in time. It has been applied to the nonlinear stability and time evolution of thermal structures constituted by optically thin plasmas with solar abundances. The configuration has been assumed to be heated at a rate $\sim T^m$, cooled at a rate $\sim T^n$ and a thermal conduction coefficient $\sim T^k$. In particular, the second order analytical approximation considered in previous papers (Ibáñez and Rosenzweig, 1995; Steele and Ibáñez, 1997) has been worked out for arbitrary amplitude of the initial temperature disturbance. Particular cases of interest in Astrophysics are considered.

1. Introduction

A problem of particular importance in Astrophysics as well as in laboratory plasmas is concerned with the stability of the steady solutions of the M.H.D. equations. One may assume that the above problem, taking into account many different physical effects (Parker, 1953; Field, 1965; Heyvaerts, 1974), is well understood as far as the linear approximation is concerned. In contrast, the nonlinear regime is not far well understood (even when one only consider the bary thermal effects) and much of the progress in this area has been done from numerical simulations of particular structures (Oran et al., 1982, 1987; Klimchuck et al., 1987; Mok et al., 1991).

In previous works (Ibáñez and Rosenzweig, 1995; Steele and Ibáñez, 1997) the stability up to the second order of thermal structures initially in thermal equilibrium or in steady states was carried out and explicit conditions for asymptotic and supercritical stability, and for superexponential and subcritical instability were obtained. The departure point was the successive approximations method proposed by Landau and Lifshitz (1987) to study turbulence, and the main interest was focused on the onset of the nonlinear regime rather than to follow the evolution of the structure in a well advanced nonlinear stage. In particular, the attention was addressed to

Astrophysics and Space Science **256:** 381–386, 1998.

study the scope of the linear approximation, and the pattern of structures which would tend to develop just at the onset of the nonlinear regime.

In the present paper a numerical algorithm is used to solve the energy equation taken into account heat/loss processes and nonlinear heat diffusion. Notwithstanding the small amount of physics involved in the above approach, the understanding of such restricted problem allows one to clarify not only the effects from the bare thermal field, but also to gain insight into more complex thermal structures and into the coupling of the thermal field with other physical processes (Zel'dovich and Pikel'ner, 1970; Doroshkevich and Zel'dovich, 1981; Velarde et al., 1994).

2. Basic Equations

In absence of dynamics, the equation of conservation of energy can be written as (Landau and Lifshitz, 1987)

$$\rho c_i \frac{\partial T}{\partial t} = \nabla \cdot [\kappa(\rho, T)\nabla T] + Q(\rho, T), \tag{1}$$

where $\kappa(\rho, T)$ is a thermal conduction coefficient and $Q(\rho, T)$ is the difference between the effects of heat sources and sinks per unit volume and time – generally a function of density ρ and temperature T and c_i is the specific heat per unit mass: c_p or c_v accordingly one assumes pressure or density to be a constant, respectively. In the above two cases, the heat diffusion coefficient, and the heat gain/loss term simplify to the form

$$\kappa(T) = \kappa_0' T^k, \quad Q(T) = \Gamma_0 T^m - \Lambda_0 T^n. \tag{2}$$

For constant pressure for instance, and plane symmetry Equation (1) can be written in dimensionless form as

$$\frac{\partial \theta}{\partial \tilde{t}} = \theta \left\{ \frac{\partial}{\partial z} \left[\tilde{\kappa}(\theta) \frac{\partial \theta}{\partial z} \right] + G(\theta) \right\}, \tag{3}$$

where $z = z/L$, L being the half-thickness of the slab, $\theta = T/T_*$, $\tilde{t} = t/t_\chi$, t_χ being the relaxation time, and χ_* the thermometric conductivity at the normalizing temperature T_*. Additionally,

$$\tilde{\kappa}(\theta) = \kappa(T)/\kappa_* = \theta^k; \quad G(\theta) = \lambda_*(\theta^m - \theta^n), \tag{4}$$

$$\lambda_* = \frac{L^2 \Gamma_0 T_*^{m-k-1}}{\kappa_0'}, \tag{5}$$

and appropriated boundary and initial conditions have to be provided.

The values of the power n of the cooling function in Equation (2) are well known for optically thin plasmas with solar abundances (Vesecky et al., 1979).

For instance, assuming a constant pressure, $k = 5/2$, $k = -5/2$ and $k = 1/2$ for heat diffusion by electrons in the direction parallel and perpendicular to a magnetic field (Spitzer, 1962), and diffusion by neutral particles (Parker, 1953), respectively. On the other hand, $m = 0$ for a constant heating rate per unit volume, heating by coronal current dissipation, or heating by Alfvén mode/mode conversion; and $m = -1$ for a constant heating rate per unit mass or heating by Alfvén mode/anomalous damping (Vesecky et al., 1979).

3. Numerical Algorithm

At the present section a brief description of the numerical approximation will be given. The equation to be solved is the nonlinear parabolic equation

$$b(z,\theta)\frac{\partial\theta}{\partial t} = \frac{\partial}{\partial z}\left(K(z,\theta)\frac{\partial\theta}{\partial z}\right) + c(z,\theta), \tag{6}$$

in the region $\alpha \leq z \leq \beta$, $t \geq 0$ with the initial condition

$$\theta(z,0) = \theta_0(z), \quad \alpha \leq z \leq \beta, \tag{7}$$

and boundary conditions

$$\begin{aligned} \theta(\alpha,t) &= \theta_\alpha(t), \quad t \geq 0, \\ \theta(\beta,t) &= \theta_\beta(t), \quad t \geq 0. \end{aligned} \tag{8}$$

Additionally, it is assumed that $\theta = \theta(z,t)$ is a smooth solution of Equations (6), (7) and (8) and that $K(z,\theta)$ and $b(z,\theta)$ are also smooth and strictly positive, i.e. $b(z,\theta) \geq \nu > 0$, and $K(z,\theta) \geq \eta > 0$. It follows that the boundary value problem is uniformly parabolic and θ is unique.

A two stage, two level finite difference scheme based on the algorithm of Meek and Norbury (1982) will be used. The region is covered by a rectilinear grid with sides parallel to the z- and t-axes with h and k being the grid spacings in the z- and t-directions, respectively. u_i^j denotes the function satisfying the difference equations at the grid points $z_i = \alpha + ih$, $t = jk$. The scheme is obtained by linearizing the standard Crank–Nicolson discretization, i.e.

First Stage

$$\begin{aligned} b(z_i, u_j^i)\left(\frac{u_i^{j+1/2} - u_i^j}{k/2}\right) &= \frac{1}{h^2}\Big[K\left(z_{i+1/2}, \frac{1}{2}\left(u_{i+1}^j + u_i^j\right)\right)\left(u_{i+1}^{j+1/2} - u_i^{j+1/2}\right) \\ &- K\left(z_{i-1/2}, \frac{1}{2}\left(u_i^j + u_{i-1}^j\right)\right)\left(u_i^{j+1/2} - u_{i-1}^{j+1/2}\right)\Big] + c(z_i, u_i^j). \end{aligned} \tag{9}$$

Second Stage

$$b\left(z_i, u_i^{j+1/2}\right)\left(\frac{u_i^{j+1} - u_i^j}{k}\right) = \frac{1}{2h^2}\left[K\left(z_{i+1/2}, \frac{1}{2}\left(u_{i+1}^{j+1/2} + u_i^{j+1/2}\right)\right)\right.$$

$$\times\left(u_{i+1}^{j+1} + u_{i+1}^j - u_i^{j+1} - u_i^j\right)$$

$$\left. - K\left(z_{i-1/2}, \frac{1}{2}\left(u_i^{j+1/2} + u_{i-1}^{j+1/2}\right)\right)\left(u_i^{j+1} + u_i^j - u_{i-1}^{j+1} - u_{i-1}^j\right)\right]$$

$$+ c\left(z_i, u_i^{j+1/2}\right). \tag{10}$$

These discretizations involve two adjacent time levels and lead to systems of linear algebraic equations

$$\mathbf{A}^{(s)}\mathbf{u} = \mathbf{d}^{(s)}, \tag{11}$$

where $\mathbf{A}^{(s)}$ is a tridiagonal symmetric and strictly diagonally dominant matrix and $\mathbf{u}$ is the solution in time level (s). The scheme is consistent, convergent and second order accurate in space and order 3/2 in time. Numerical experiments show that this scheme has good stability properties.

4. Results and Conclusions

The above numerical solution of Equation (6) with initial condition (7) and boundary conditions (8) allows one to study a great deal of different physical problems, in particular heat propagation in structures with a heat/loss function which can be considered either to a constant pressure or to a constant density. The values of the parameters m, n and k depend on the heat/loss function and the kind of heat diffusion going into play, respectively (see Ibáñez and Rosenzweig, 1995). In the remaining part of this section a particular example of the above cases will be analyzed.

Figure 1a shows the evolution of a strong negative temperature disturbance ($A\cos(\pi z/2)$, where $\mathbf{A} = -0.9$) to a trivial first order stable solution ($\lambda_* = \lambda_c/2$, λ_c the critical value) for the values $m = -1$, $n = -3/2$ (bremsstrahlung cooling) and $k = -5/2$ (electronic heat conduction perpendicular to a magnetic field). The trivial solution corresponds to the thermal equilibrium solution of Equation (1). It has been found that just the above value of the amplitude corresponds to the threshold value, i.e. the trivial solution becomes unstable for negative A with $|A| > 0.9$. Instead, it becomes stable for even very strong positive perturbations ($A \leq 34$) as can be seen in Figure 1b.

The corresponding evolution of a small disturbance ($A = -0.01$) in a linearly unstable structure ($\lambda_* = 2\lambda_c$) is shown in Figure 1c. Figure 1d shows the evolution of a linearly unstable structure ($\lambda_* = 2\lambda_c$) initially in thermal equilibrium, under

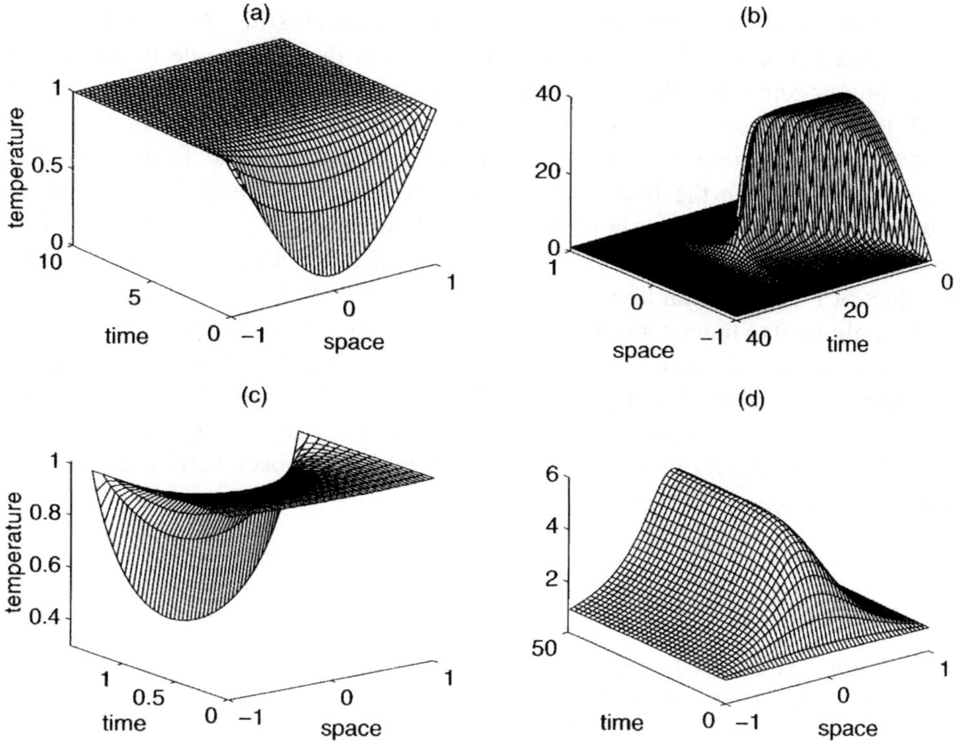

Figure 1. Time evolution of temperature disturbance $A\cos(\pi z/2)$ to a trivial solution for $k = -5/2$, $m = -1$, $n = -3/2$, $\lambda_* = \lambda_c/2 = \pi^2/4$ for (a) and (b), $\lambda_* = 2\lambda_c = \pi^2$ for (c) and (d).

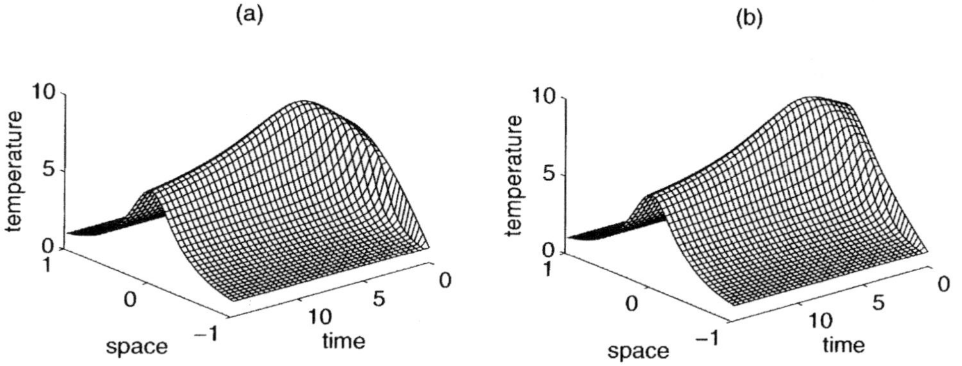

Figure 2. Time evolution of temperature disturbance $A\cos(\pi z/2)$ to (a) a trivial solution and (b) to a nontrivial solution for $k = -5/2$, $m = -1$, $n = -3/2$, $\lambda_* = 2\lambda_c = \pi^2$.

the action of very small positive disturbances ($A = 0.01$), for the same values of m, n and k of Figure 1a. The saturation to "the nearest" stable nonhomogeneous steady state solution is apparent. Such a saturation occurs in a time of the order of 15 relaxation times. It is interesting to note that the fastest rate of change of temperature occurs at early time (of the order of 5 relaxation times). On the other hand, Figure 2a shows the time evolution of the trivial solution disturbed initially so strongly that its temperature is larger than the above steady state solution. The structure initially (in the first 5 relaxation times) overheats and thereafter cools down to reach the above nonhomogeneous state in about 10 relaxation times. It is interesting to remark that this conspicuous way of thermal evolution seems to be an effect of the strong nonlinearity and is not manifested in asymptotic studies of weak nonlinearity (Ibáñez and Rosenzweig, 1995; Steele and Ibáñez, 1997).

Similar effects are found when one analyzes the stability of nontrivial steady solutions of Equation (1). In particular, the stability of the steady solution found above is illustrated in Figure 2b, where the overheating effect is apparent.

A detailed study of other cases of physical interest has been carried out and the results will be reported elsewhere.

References

Doroshkevich, A.G. and Zel'dovich, Ya.B.: 1981, *Sov. Phys. JETP* **53**, 405.
Field, G.B.: 1965, *Astrophys. J.* **142**, 531.
Heyvaerts, J.: 1974, *Astron. Astrophys.* **37**, 65.
Ibáñez, S., M.H. and Rosenzweig, P.: 1995, *Phys. Plasmas* **2**, 4127.
Klimchuck, J.A., Antiochos, S.K. and Mariska, J.T.: 1987, *Astrophys. J.* **320**, 409.
Landau, L., and Lifshitz, E.M.: 1987, *Fluid Mechanics*, Pergamon, London.
Meek, P.C. and Norbury, J.: 1982, *IMA J. Num. Analysis* **2**, 335.
Mok, Y., Schnak, D.D. and Van Hoven, G.: 1991, *Solar Physics* **132**, 95.
Oran, E.S., Mariska, J.T. and Boris, J.P.: 1982, *Astrophys. J.* **254**, 349.
Oran, E.S., Mariska, J.T. and Boris, J.P.: 1987, *Astrophys. J.* **315**, 385.
Parker, E.N.: 1953, *Astrophys. J.* **117**, 431.
Spitzer, L.: 1962, *Physics of Fully Ionized Gases*, Interscience, New York, p. 146.
Steele, C. and Ibáñez, S., M.H.: 1997, *Phys. Plasmas* **4**, 618.
Velarde, M.G., Chu, X. and Ross, J.: 1994, *Phys. Fluids* **6**, 550.
Vesecky, J.F., Antiochos, S.K. and Underwood, J.H.: 1979, *Astrophys. J.* **233**, 987.
Zel'dovich, Ya.B. and Pikel'ner, S.B.: 1970, *Sov. Phys. JETP* **29**, 170.

ON THE NONLINEAR THERMAL EVOLUTION OF SOLAR CORONA STRUCTURES

C.A. MENDOZA-BRICEÑO
Centro de Astrofísica Teórica (CAT), Facultad de Ciencias, Universidad de los Andes, Mérida, Venezuela

Abstract. The thermal evolution of structures is investigated for different values of the size of structure. A simplified cooling function and a constant heating mechanism are assumed. Analytical criteria for thermal instability are obtained. It is found that the response of the thermal structure not only depends on the amplitude of the disturbances, but also on whether the disturbance increases or decreases the initial steady temperature. Additionally, the evolution of the structure is examined numerically by using a time-dependent code under the assumption that the inertia terms are small. In particular, the analytical criteria obtained for thermal instability are verified.

1. Introduction

Observations in X-ray and EUV have shown that the solar corona consists of a large number of loops which outline the coronal magnetic field in form of magnetic flux tubes that confine the plasma. These loops have temperatures of the order of 1.4–4 $\times$ 10^6 K. Much cooler structures are also visible in the corona, these are prominences. They are normally very long-lived surviving for up to 9 months.

Recently, Mendoza and Hood (1996, 1997) have examined the prominence formation taking into account spatial dependence of the heating, studying the thermal equilibrium in order to find the critical parameters which can induce a quasistatic evolution up to the point where condensation forms. The thermal stability of the equilibrium up to a second order approximation was also studied analytically. It was found the response of the structure not only depends on the amplitude of the disturbance but also on whether the disturbance increases or decreases the equilibrium temperature.

In this paper the thermal evolution of loops will be examined numerically for different values of their size L_*. Here a simplified cooling function and a constant heating are considered.

2. Basic Equations

The governing equations in dimensionless quantities are the well known equations of magnetohydrodynamics in which viscosity, resistivity and gravity are neglected. It is assumed that the coronal magnetic field is so strong (i.e. the coronal plasma

Astrophysics and Space Science **256:** 387–392, 1998.

β is small) that the subsequent motions are restricted to those along the field. In addition, the radiative timescale in the corona is much longer than the sound travel time, allowing the system to evolve *isobarically* (Mendoza-Briceño, 1996). Thus, the MHD equations reduce to

$$\frac{\partial v}{\partial s} = \frac{\partial}{\partial s}\left(T^{5/2}\frac{\partial T}{\partial s}\right) - L_*^2[Q(T) - H], \tag{1}$$

$$\frac{\partial T}{\partial t} = -v\frac{\partial T}{\partial s} + T\frac{\partial v}{\partial s}, \tag{2}$$

where the density has been eliminated by using gas law equation, s is the distance along a field line, v is the plasma velocity parallel to the fixed magnetic field, t is the time, T is the temperature, H is the unknown heating function taken as constant and L_*^2 is ratio of conductive time scale to the radiative timescale and is proportional to the loop length squared. $Q(T)$ is the cooling function which is assumed to be a two range piecewise function. It is assumed that the loop is symmetric about the apex and therefore one needs to consider only one-half of the loop, i.e. from a footpoint to the summit. Thus, Equations (1) and (2) are solved numerically with the boundary conditions that temperature gradient at the loop summit is zero and that T_b is the temperature at the loop footpoint. The initial conditions are $v(s) = 0$ and $T(s) = T_e$ where T_e is the steady state temperature profile. In the next section the numerical results will be presented.

3. Equilibrium Solutions

This section will discuss the thermal equilibrium of the system (1)–(2) by considering a simplification to the cooling function. The cooling function will be assumed to be a two range piecewise function:

$$Q(T) = \chi T^{\alpha-2} = \begin{cases} T_a^{-7}\,T^{7/2}, & T < T_a, \\ T^{-7/2}, & T > T_a, \end{cases} \tag{3}$$

where T_a is the temperature at which the powers in the radiative loss function change value; for a maximum in the loss function at about 10^5, $T_a = 10^5/T_0$. This function satisfies the basic requirement of the general cooling function, namely that the radiation has a peak at 10^5 K (Hildner, 1974).

Figure 1 is a plot of summit temperature as a function of the boundary temperature resulting from the numerical integration in the equilibrium of Equations (1) and (2) for a constant heating function. The different curves are labelled with the corresponding value of L_*^2. As is seen in this figure two points in which all the curves intersect are found at $T_b = 0.07$ and $T_b = 0.138$. These points correspond to two thermal equilibrium states given by the solutions of $\mathcal{L} = 0$ (= cooling-heating).

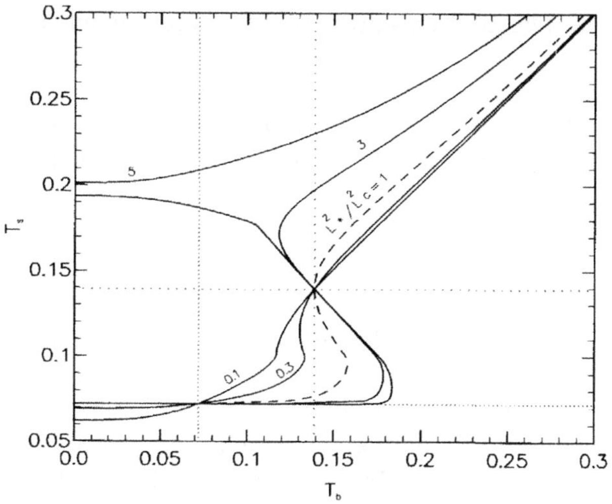

Figure 1. The summit temperature T_s as a function of the boundary temperature T_b for different values of the loop length L_*^2. The close spaced dot and wide spaced dot lines indicate the two thermal equilibria as a result of $\mathcal{L} = 0$.

For L_*^2 small ($< 0.1L_c^2$) the relation $T_s(T_b)$ is a single-valued function while for $L_*^2 > L_c''^2$ ($\approx 0.1L_c^2$) it becomes a three valued function with an S-type characteristic. In particular, at $T_b = T_e$, T_e being the temperature at equilibrium (in this case $T_e = 0.138$) and for $L_*^2 > L_c'^2$ ($\approx 0.39L_c^2$), the multiple solutions are the thermal equilibrium state and two additional states in non-thermal equilibrium. For large values of L_*^2 ($L_*^2 > L_{*0}^2$, L_{*0}^2 being the value for which $T_b = 0$ at the turning point) a gap for the summit temperature T_s appears where no solution exists.

For values of $L_*^2 < L_c''^2$ the summit temperature is a monotonically increasing function of T_b. For any value of L_*^2 in the range $L_c''^2 < L_*^2 < L_{*0}^2$, however, there are two turning points (or catastrophe points) that divide the curve $T_s(T_b, L_*^2)$ in the branches corresponding to three different solution of equilibrium equations (1) and (2). In the upper and lower branches, the summit temperature T_s increases when the boundary temperature T_b increases, and in the middle branch the summit temperature decreases when the boundary temperature increases.

When $L_*^2 > L_{*o}^2$ only the lower turning point remains, and it separates the middle and lower branches and, as was pointed out above, a gap for T_s exists. The width of such a gap increases when L_*^2 increases. When the value of L_*^2 is considered to be equal to L_c^2 (the dashed curve in Figure 1) one can see that the upper turning point coincides with the temperature at thermal equilibrium. This value of L_*^2 is the critical one found by Mendoza and Hood (1997) making use of a simple first order stability analysis.

4. Thermal Evolution of a Coronal Structure

In this section the evolution of a thermal structure will be examined numerically for different values of the parameter L_*. This study allows us to introduce a physical quantity which can be expressed in terms of its unperturbed and perturbed quantity as

$$T = T_e + T_p, \tag{4}$$

where T_e is the temperature equilibrium considered as the trivial solution $T_e = 0.138$ and $T_p = \varepsilon\, T_e \cos(\pi s/2)$ (ε being the perturbation amplitude which can be positive or negative).

As mentioned in Section 3, when $L_* > L_c$, three possible solutions exist. One of the solutions is the trivial solution T_e, which is located in the middle branch. The first approximation of the stability analysis predicts that the trivial solution is unstable, but the second order approximation presents two possibilities for the evolution depending on whether the perturbation is positive or negative (Mendoza and Hood, 1997). The perturbed solution would then evolve towards its saturated states i.e., to the stable solution.

Figure 2a shows the plot of the summit temperature as a function of time for an increase (solid line) or decrease (dotted line) of temperature. When $\varepsilon = +0.1$ the summit temperature T_s starts to rise until it reaches a saturated state which corresponds to the static solution found in Section 3. The value of ε was increased and reduced from this value but the separation from the trivial solution T_e (dotted line) was still observed.

However, when the temperature was reduced by considering $\varepsilon = -0.1$ in Equation (4) the summit temperature began to decrease further from the trivial solution reaching a lower temperature value. This value corresponds to the lowest temperature found in Section 3 which represents a non-isothermal profile. Therefore, the isothermal solution T_e is unstable for $L_* > L_c$.

For $L'_c < L_* < L_c$, the thermal structure again presents three possible solutions; one isothermal solution T_e, and two non-isothermal with summit temperature $T_s < T_e$.

If the temperature is increased when $\varepsilon = +0.1$, the summit temperature decreases with time, approaching asymptotically to the trivial solution T_e as can be seen in Figure 2b for the case plotted with solid line. On the other hand, if $\varepsilon = -0.1$, the dashed line in Figure 2b shows that the temperature increases, approaching the trivial solution T_e (dotted line).

The magnitude value of ε was reduced and a threshold value was found for which the temperature instead of increasing, decreased towards a lower value (dotted-dash line). This lower value of the summit temperature corresponds to the value found in the lower branch in Figure 1 of Section 3. Therefore, the trivial solution as well as the lowest temperature are stable solutions to the energy equation.

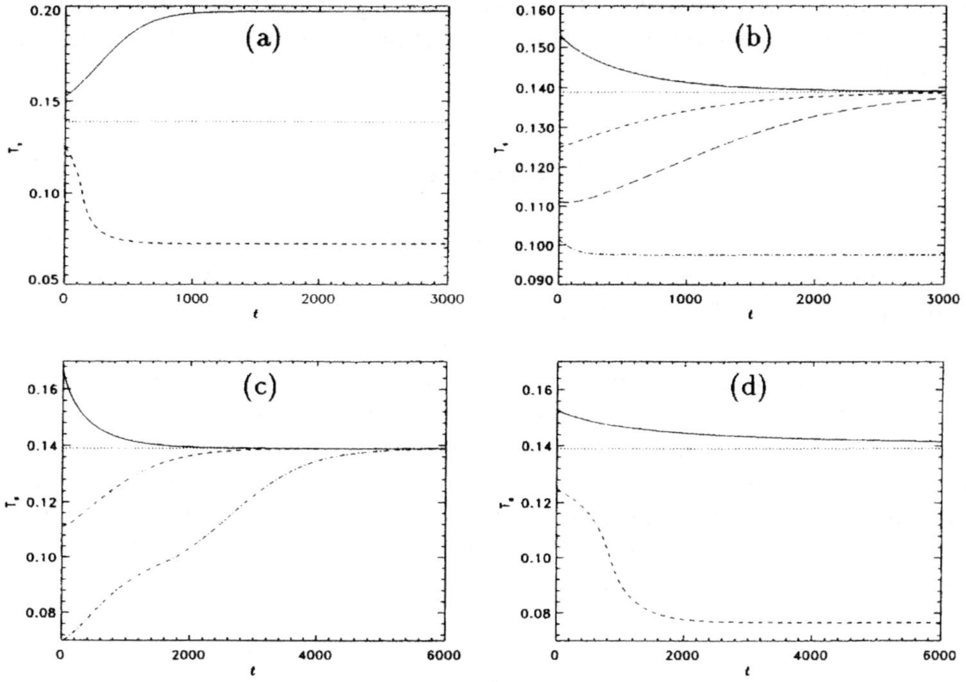

Figure 2. The summit temperature T_s as a function of time for (a) $L_* > L_c$, (b) $L'_c < L_* < L_c$, (c) $L_* < L'_c$ and (d) $L_* = L_c$.

When $L_* < L'_c$, the graph $T_s(T_b)$ plotted in Figure 1 shows a monotonic increasing function of T_b. If one perturbs the temperature positively, it decreases asymptotically to the trivial solution as can be seen in Figure 2c. Furthermore, if larger values of ε are considered, the same result is obtained. On the other hand, when $\varepsilon = -0.1$ is considered, the summit temperature increases towards the trivial solution. This value of ε was increased negatively and the summit temperature increased asymptotically to the trivial solution. Therefore, T_e presents a stable behaviour when positive or negative disturbances of different amplitude are considered.

For $L_* = L_c$, the trivial solution is disturbed by increasing the temperature above T_e with $\varepsilon \geq +0.1$. Figure 2d shows that the summit temperature decreases asymptotically, returning to the equilibrium value T_e. But, if it is decreased below T_e, namely with $\varepsilon \leq -0.1$, the temperature will fall further reaching a low value of T_s.

This case was studied analytically up to the second order approximation by Mendoza-Briceño (1996) and Mendoza and Hood (1997). This trivial solution is marginally stable to the linear approximation, but, if one considers a second order approximation, it is quadratically unstable (Hood and Priest, 1979). The Landau constant a_2 for a positive perturbation is always negative and the temperature will return to the isothermal solution. However, when negative disturbances are considered, the Landau constant a_2 is positive, so that the temperature will drop and the gas cools.

5. Conclusions

In this paper we have investigated the thermal evolution of structures in a symmetric coronal loop when the heating function is constant and the radiation losses is a two range piecewise function. Initially, the thermal equilibrium was investigated. It was found multiple solutions for the same value of the parameter L_*.

The thermal evolution of these solutions up to a second order approximation predicting the evolution of the thermal structure against disturbances was studied by Mendoza-Briceño and Hood (1997). This evolution was studied by using numerical simulations and the results obtained verify their work. It is also important to mention that the evolution of the structure depends of the initial temperature profile considered. The temperature increased or decreased according to whether the initial profile was above or below a threshold one.

References

Hildner, E.: 1974, *Solar Phys.* **35**, 123.
Hood, A.W. and Priest, E.R.: 1979, *Astron. Astrophys.* **77**, 233.
Mendoza-Briceño, C.A.: 1996, Ph.D. Thesis, University of St. Andrews, St. Andrews, Scotland.
Mendoza, C.A. and Hood, A.W.: 1996, *Astro. Lett. and Communications* **34**, 107.
Mendoza-Briceño, C.A. and Hood, A.W.: 1997, *Astron. Astrophys.* **325**, 791.

PHOTO-IONIZING FLASH AS A TRIGGER FOR EFFICIENT H_2 COOLING IN METAL-FREE CLOUDS

A. PARRAVANO

Centro de Astrofísica Teórica, Facultad de Ciencias, Universidad de los Andes, Apartado de Correos 26, Ipostel, La Hechicera, Mérida, Venezuela

Abstract. The effect of a photo-ionizing radiation flash as a trigger of an efficient H_2 cooling process in metal-free clouds is investigated. We show that a cloud near thermo-chemical equilibrium at ~ 10000 K could increase its H_2 content by many orders of magnitude after the passing of the flash. Furthermore, the cloud cools rapidly to temperatures as low as 100 K. The temperature drop occurs in a fraction (~ 0.2) of the free fall time causing a rapid decrease of the Jeans mass. This mechanism is effective even in the case when this process is simultaneous to a photo-dissociating flash with photo-dissociating flux bigger than the critical value for thermo-chemical equilibrium near 10000 K in the post-flash period. The detailed form of the flash is not important if the ionizing flux decrease to background values in a time short compared to the hydrogen recombination time, once the cloud is ionized. These flashes can be produced by massive population III stars, or by an active galactic nucleus. Rapid exhaustion of the ionization source or late phase evolution of Stromgen spheres may induce the required rapid decrease of the ionizing flux. This mechanism might have implications in the context of the induced formation of globular clusters.

1. Introduction

There are many scenarios in which a radiation flash affects the evolution of gas clouds. In particular, the ionizing flash effect, might be a key step in the sequence of events that have conduced to the formation of Globular Clusters. During the proto-galactic collapse, if the gas is assumed to be lumpy, the overdense regions will cool more rapidly than the underdense regions producing a two-phase medium (Fall and Rees, 1985). But the developing rate of this overdense regions depend on the initial conditions (i.e. the denser regions develop faster; Murray and Lin, 1990) and on their interaction with the system. Then, the overdense regions located in the denser regions of the proto-galaxy (i.e. near the center of the original perturbation) will be the first structures to form stars. As a consequence of the birth of this first generation of stars in the galactic nucleus, the evolution of the surrounding cloudy medium must be drastically affected. If a large group of stars is formed in a short period of time, the stellar formation process may be temporally interrupted (Parravano et al., 1990) producing the ionizing flash. In this note, we consider metal-free clouds which are initially near thermo-chemical equilibrium at 10^4 K. This kind of clouds and initial conditions might be relevant at the epoch of galaxy formation. The effect of the ionizing flash might act as a selective effect over the clouds that finally will form stars. In this evolutionary picture we will concentrate

Astrophysics and Space Science **256:** 393–397, 1998.

our attention over the effects of this hypothetical initial flash on the evolution of a single cloud.

2. Basic Equations

In order to follow the thermal and chemical evolution of a metal-free cloud in a presence of a variable radiation flux, we consider an idealized uniform cloud initially near thermo-chemical equilibrium at $\sim 10^4$ K. The evolution of the cloud is calculated by solving the system of kinetic equations and the equation of energy conservation.

The kinetic follows a general equation of the form

$$\frac{dX_i}{dt} = n\left(\sum_{j,k}\alpha_{jk}X_jX_k - X_i\sum_l \alpha_{il}X_l\right) + \sum_m \alpha_m X_m - \alpha_i X_i, \tag{1}$$

where n is the total density number of hydrogen nuclei and X_i is the relative number density of the i-th specie. We consider a gas composed by: H, H^+, e, H^-, H_2^+, and H_2. In Equation (1) α_{jk} and α_{il} are the formation and destruction rates of the i-th specie due to double collisions, whereas, α_m and α_i are the production and destruction rates of the i-th specie due to interaction of particles with radiation, respectively. In the constant pressure approach the energy conservation equation is expressed as:

$$\frac{dT}{dt} = \frac{T}{\mu}\frac{d\mu}{dt} - \frac{2}{5}\frac{\mu}{R}\mathcal{L}, \tag{2}$$

where $\mu = (1 + X_e - X_{H_2} - X^+_{H_2})^{-1}$, R is the gas constant, $\mathcal{L} = (\Lambda - \Gamma)/\rho$ is the net cooling rate by unit mass, ρ is the mass density, and T is the temperature. In addition to the processes of formation and destruction of the six species and the cooling rates adopted by Izotov (1989), we include the photoionization of the hydrogen, the H_2 photo-dissociation, and, its associated heating rates as in Rosenzweig et al. (1994). In the Lyman continuum we assume a quasar-like external flux (Shapiro and Kang, 1987)

$$n_{ph-i}(\nu, t) = 4.6 \times 10^{-11}\,\varepsilon_{ph-i}(t)\left(\frac{\nu}{\nu_H}\right)^{-1.7}\ \mathrm{cm^{-2}\,Hz^{-1}\,s^{-1}}, \tag{3}$$

where ν_H is the H Lyman-edge frequency and ε_{ph-i} is introduced to represent the temporal changes of the flux level outside the cloud due to both, the variations of the ionizing sources and the attenuation and remission by the material between the sources and the cloud. Inside the cloud the ionizing radiation flux suffers a selective extinction. Since we assume an homogeneous cloud, the optical depth at its center is

$$\tau_\nu = n\,\sigma_\nu\,X_H\,R_c, \tag{4}$$

where R_c is the cloud radius and σ_ν, the absorption cross section, is calculated from

$$\sigma_\nu = A_o \left(\frac{\nu_H}{\nu}\right)^4 \exp\left[4 - \frac{4 \arctan \varepsilon}{\varepsilon}\right] / \left[1 - \exp(-2\pi/\varepsilon)\right] \text{ cm}^2, \quad (5)$$

where $A_o = 6.3 \times 10^{-18}$ and $\varepsilon = \left[(\nu/\nu_H) - 1\right]^{1/2}$. Disregarding the diffuse emission by the cloud gas, the radiation flux at its center is given by

$$\overline{n}_{ph-i}(\nu, t) = n_{ph-i}(\nu, t) \exp(-\tau_\nu). \quad (6)$$

The photoionization rate can be calculated as (Shapiro and Kang, 1987)

$$\alpha_{ph-i} = \sum_k \overline{n}_{ph-i,k}\, \sigma_k\, \Delta\nu_k \text{ s}^{-1}, \quad (7)$$

where $\overline{n}_{ph-i,k}$ is the radiation flux at frequency ν_k and $\Delta\,\nu_k = \nu_{k+1} - \nu_k$ is the width of the k-th frequency bin. In this equation is not considered the secondary ionization by energetic photons ($\nu > 2\,\nu_H$).

The corresponding heating rate is

$$\Gamma_{ph-i} = n\, X_H \sum_k \overline{n}_{ph-i,k}\, \sigma_k\, (h\nu - h\nu_H)\, \Delta\nu_k \text{ erg cm}^{-3} \text{ s}^{-1}. \quad (8)$$

For the photo-dissociating photons it is also assumed that the external flux is given by

$$n_{ph-d}(\nu, t) = 4.6 \times 10^{-11}\, \varepsilon_{ph-d}(t) \left(\frac{\nu}{\nu_H}\right)^{-1.7} \text{ cm}^{-2} \text{ Hz}^{-1} \text{ s}^{-1}, \quad (9)$$

where ε_{ph-d} is a function introduced to adjust the temporal changes of the photo-dissociating flux level.

The photo-dissociating rate can be calculated as

$$\alpha_{ph-d} = 1.9 \times 10^{-3} n_{ph-d}\, L(\tau_{H_2}) \text{ s}^{-1}, \quad (10)$$

where $\tau_{H_2} = K_0\, n\, R_c\, X_{H_2}$, $K_0 = 1.0 \times 10^{-15}$ cm^2, and $L(\tau_{H_2})$ is the fraction of n_{ph-d} which penetrates up to optical depth τ_{H_2}.

The corresponding heating rate might be approximate as

$$\Gamma_{ph-d} = 0.4\, n\, X_{H_2}\, \alpha_{ph-d} \text{ ev cm}^{-3} \text{ s}^{-1}. \quad (11)$$

3. Results

For a given initial state of the cloud, there is a critical background flux level, ε_{cri}, above which the cloud cannot produce enough H_2 to self-shield the dissociating radiation. At this stage, the cloud evolves to a thermo chemical equilibrium near

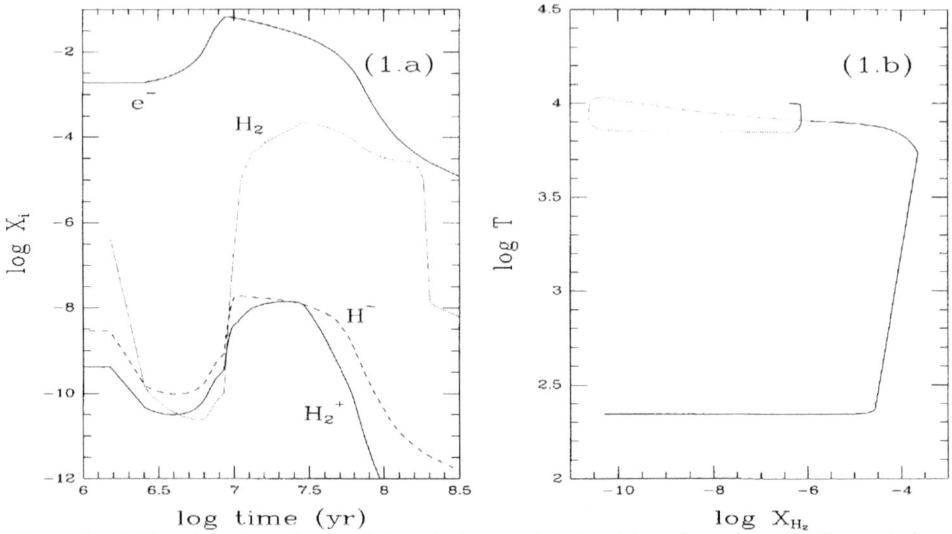

Figure 1. (a) The time dependence of the relative number densities of species. (b) The evolution of the cloud in the plane (T–X_{H_2}).

8000 K. However, as we will see later, the cloud might be induced to cool to a temperature of the order of 100 K by an ionizing flash, even when the pre- and post-flash values of the external radiation flux are above the critical flux level ε_{cri}. To illustrate this effect, in Figure 1a we show the evolution of the relative number densities of e^-, H^-, H_2^+, and H_2 when the cloud is subject to the following flash: the pre-flash value are $\varepsilon_{ph-i} = \varepsilon_{ph-d} = 1.5 \times \varepsilon_{cri}$; at $t = 2 \times 10^8$ yr and by a period of 5×10^6 yr, the flux levels ε_{ph-i} and ε_{ph-d} increase linearly up to $220 \times \varepsilon_{cri}$; both ε_{ph-i} and ε_{ph-d} remains at this value during 2×10^6 yr; and finally, ε_{ph-i} decreases exponentially to its initial value in a characteristic time 2×10^6 yr, but ε_{ph-d} decreases to a level twice its initial value. The results shown in Figure 1a correspond to a 10^7 $M_\odot$ cloud at the initial conditions: $n = 0.05$ cm^{-3}, $X_{H^+} = 3.2 \times 10^{-3}$, $X_{H_2} = 4.0 \times 10^{-7}$, $X_{H_2^+} = 6.5 \times 10^{-10}$, and $X_{H^-} = 4.1 \times 10^{-9}$. For the same initial and external conditions as in Figure 1a, Figure 1b shows the evolution of the cloud in the plane (T–X_{H_2}). The dotted segment of the curve in Figure 1b corresponds to the flash period. Note that, even when the post-flash value of the dissociating flux level ε_{ph-d} is twice the pre-flash value, the abundance of H_2 may reach a large enough value to produce considerable self-shielding allowing an efficient cooling of the cloud. At this point, the H_2 concentration grows in several orders of magnitude.

4. Conclusions

It have been showed that a cloud near thermo-chemical equilibrium at $\sim$ 10000 K could increase its H_2 content by many orders of magnitude after the passing of an ionizing flash. Furthermore, the cloud cools rapidly to temperatures as low as 100 K. This mechanism is effective even in the case when the flash is simultaneous accompanied by a photo-dissociating flash with a post-flash flux bigger than the critical value. The results indicate that the mechanism of induced condensation might play a relevant role in the gas evolution of a diffuse gas in the pregalactic conditions.

Acknowledgements

This work has been supported by CDCHT of the Universidad de Los Andes through project C-717-95. The author is very grateful to Dr. Pedro Colmenares for improving the English manuscript.

References

Fall, S.M. and Rees, M.J.: 1985, *Ap. J.* **298**, 18.
Izotov, Yu.I.: 1989, Kiev Inst. Teoret. Fiz. ITP-89-17E.
Murray, S.D. and Lin, D.N.C.: 1990, *Ap. J.* **363**, 5.
Parravano, A., Rosenzweig, P. and Terán, M.: 1990, *Ap. J.* **356**, 100.
Rosenzweig, P., Parravano, A., Ibáñez, M.H. and Izotov, Yu.I., 1994: *Ap. J.* **432**, 485.
Shapiro, P.R. and Kang, H.: 1987, *Ap. J.* **318**, 32.

THERMAL INSTABILITY FOR CONVECTION IN ASTROPHYSICAL PLASMAS

N. FALCÓN
Dpto. de Física, Facultad de Ciencias y Tecnología, Universidad de Carabobo, Apdo. Postal 129, Valencia 2001 Edo. Carabobo, Venezuela

Abstract. The plasma thermal stability condition under convection is discutided. Using Cattaneo law and mixing length theory before relaxation lets show a new thermal stability condition for plasmas in astrophysics. Application in neutron stars and low mass x-ray binaries stars is mentioned.

1. Introduction

Convection is a small scale dynamical instability, of considerable importance for stars. The interaction of convective motion and magnetic fields poses a number of interesting problems in astrophysical scenarios (Chandrasekhar, 1961). Idealized studies with a large scale magnetic field threads a convectively unstable layer sandwiched between convectively stable region have been made (Moss, 1986).

The energy transport by convection changes the plasma thermal stability condition. The formulation of Walen and Chandrasekhar for plasma thermal stability, based on the Maxwell–Fourier law for thermal energy transport equation (Section 2), is approximated when the relaxation time is not negligible. The Cattaneo law predicts that the convective fluxes and thermal stability condition are very different (Section 3). This result insinuates a new thermal stability condition for plasmas in neutron stars and low mass x-ray binaries stars (Section 4).

2. Thermal and Dynamical Instability

We consider a fluid distribution with vertical temperature gradient (β) and horizontal magnetic field. One can easily imagine a fluctuation in the position of some element (blob) in the fluid, which is displaced vertically (Δr) and its temperature increases for $\beta\Delta r$ respect to the temperature T in the surroundings.

If α is the dilatation volumetric coeficient thereby the change density is $\Delta\rho = \alpha\beta\rho_0\Delta r$, therefore the buoyancy volumetric force is

$$K_r = g\alpha\beta\rho_0\Delta r. \tag{1}$$

The force lines of the magnetic field are constrained in the surrounding, and the initial convection changes the force lines (initially straight) in sinusoidal curves with wavelength (λ), for the displacement (Δr) of every point.

Astrophysics and Space Science **256:** 399–402, 1998.

The mechanical effect generated by the magnetic field (H) are the tension $\mu H^2/4\pi$ directed along the force lines (Cowling, 1968), and the curving cause at volumetric force, opposite to this displacement, is

$$F = \pi \mu \Delta r (H/\lambda)^2. \tag{2}$$

Although the blob have a displacement under the pressure balance with its surroundings $(DP = 0)$, in all instant the hydrostatic field pressure $\mu H^2/8\pi$ is equilibrated for the fluid pressure.

The force (F) created by the field (H) is opposite to the buoyancy force (K_r) and to the displacement of the blob, as soon as instability condition when:

$$g\rho_0\beta\alpha\lambda^2 > \pi \mu H^2 \tag{3}$$

and the stability condition occurs in the other case.

This criterion propose by Walen (1949) insinuates that the thermal instability changes by the magnetic field. This result was extended by Chandrasekhar (1954), both criteria are equivalent if H exchanges by H_0 (the vertical component of the unperturbed field) and $2L = \lambda$ for wavelength perturbation in the field direction.

3. Convection and Relaxation Time

In the deduction of the Chandrasekhhar criterion we have included the dissipation effect by the heat propagation equation in the diffusion approximation (Cowling, 1968):

$$\frac{dT}{dt} = \chi \nabla^2 T, \tag{4}$$

where χ is the calorific diffusibility.

This is a parabolic equation whose implicit acausality comes from the Maxwell–Fourier (MF) law:

$$F(x,t) = -k\nabla T(x,t), \tag{5}$$

where k is the thermal conductivity.

The origin of this non-causal behavior found in MF law, is the assumption that the energy flux appears at the same time the temperature gradient is switched on. But in dense degenerate material where thermal conductivity is dominated by electrons, one expects the relaxation time (t) is not neglecting due to larger mean free paths of electrons.

Cattaneo was the first to propuse the generalization of Equation (5) as (Joseph and Preziosi, 1989):

$$\tau \frac{\partial F(x,t)}{\partial t} + F(x,t) = -k\nabla T(x,t). \tag{6}$$

This heat flux equation leads to a hyperbolic equation, the well known telegraph equation (see Joseph and Preziosi, 1989, and references therein).

Changing the MF law by Cattaneo law in the well known Mixing Length Theory by standard treatment of convective motions we obtain that the buoyancy force is given by (Herrera and Falcón, 1995):

$$K_r = g\delta T^{-1} DT(0) \exp(-t/\tau_d) f(x, \omega), \tag{7}$$

with

$$f(x, w) = \exp[x(\omega^2 - 1)/2]\{\cos(\omega x) - (3 + \omega^2)/2\omega] \sin(\omega x)\}, \tag{8}$$

with

$$x \equiv \frac{t}{2\tau}, \quad \omega \equiv \sqrt{\left(\frac{4\tau}{d\tau} - 1\right)}, \quad d = \rho d^2 c_p (12k)^{-1},$$

where d denotes the diameter of the convective element, c_p the specific heat at constant pressure, $DT(0)$ is the initial temperature difference and δ is a thermodynamic coefficient (see Herrera and Falcón, 1995, for details). Also, for the local flux of convective energy we have (Herrera and Falcón, 1995):

$$F_{\text{con}} = \rho v c_p DT^{(\text{MF})} f(x, \omega). \tag{9}$$

4. Result and Discussion

Equation (9) shows that the convective blob does not always move down the temperature gradient before relaxation, but in the opposite direction during short periods of time.

As a consequence of this instability, conditions are momentarily enhanced by changing the sign periodically in $f(x, \omega)$. Therefore, using Equations (1), (3) and (7) we obtain:

$$g\delta\rho_0\lambda^2 T^{-1} DT^{(\text{MF})} f(x, w) > \pi\mu H^2. \tag{10}$$

This shows that the Chandrasekhar (or Walen) condition is not sufficient for guarantee the thermal stability under convection in time scale less than relaxation time.

This instability transforms energy of the heat flux into energy of the magnetic field and hydrodynamical motion (Cowling, 1968).

One scenario where we could expect this instability to play an important role is in neutron stars (NS). According to the NS model (Sieber and Wielebinski, 1981) an NS has a liquid core, which may be superfluid and superconductive, and a solid crust. In these materials the relaxation time is in the range of 10^{-4} to 100 seconds (Herrera and Falcón, 1995).

Blandford et al. (1983) suggest that the magnetic field in NS arises as a consequence of the thermomagnetic instability in their outer crust.

Another scenario is the quasi-periodic oscillations (QPO) in the emission in low mass x-ray binaries (Lewin et al., 1988). Some models for QPO suggest that magnetic fields and variation are generated locally due to turbulent dynamo-action in the NS envelope, where the convection and thermomagnetic instability are present.

The quasi-periodic oscillations of the charged blobs in the NS magnetosphere could be transversal waves (MHD) in the emission of the low mass binaries stars x-ray.

Notice that quasi-periodic oscillations observed in QPO could be "naturally" modeled by instability criterion (see Equation (10)) whereas the blob (and hot spots) are generated by convection in star atmospheres before thermal relaxation. Also, this quasi-periodic oscillation in the millisecond range is in the order of relaxations time for NS.

References

Blandford, R., Applegate, J.H. and Hernquist, L.: 1983, *Mon. Not. R. Ast. Soc.* **204**, 1025.
Chandrasekhar, S.: 1954, *Phyl. Mag.* **45**, 1117.
Chandrasekhar, S.: 1961, *Hydrodynamic and Hydromagnetic Stability*, OUP, Oxford.
Cowling, T.G.: 1968, *Magnetohidrodinamica*, Alhambra De, Madrid.
Herrera, L. and Falcón, N.: 1995, *Astrophys. Space Sci.* **234**, 139.
Joseph, D. and Preziosi, L.: 1989, *Rev. Mod. Phys.* **61**, 41.
Lewin, W., Paradijs, J. and Der Klis, M.: 1988, *Space Sci. Reviews* **46**, 274.
Moss, D.: 1986, *Phys. Rep.* **140**, 1.
Sieber, W. and Wielebinski, R.: 1981, *Pulsars*, Reidel, Dordrecht.
Walen, C.: 1949, *On the Vibratory Rotation of the Sun*, Lindstahl Bookhandel, Estocolmo.

PARAMETRIC DECAY OF A CIRCULARLY POLARIZED ELECTROMAGNETIC WAVE IN AN ELECTRON-POSITRON MAGNETIZED PLASMA

L. GOMBEROFF and V. MUÑOZ
Department of Physics, Facultad de Ciencias, Universidad de Chile, Casilla 653, Santiago, Chile

Abstract. We study the parametric decays of an electromagnetic wave propagating along an external magnetic field in an electron-positron plasma. We include weakly relativistic effects on the particle motions in the wave field, and the nonlinear ponderomotive force. We find resonant and nonresonant wave couplings. These include, ordinary decay instabilities, in which the pump wave decays into an electro-acoustic mode and a sideband wave. There are also nonresonant couplings involving two sideband waves, and a nonresonant modulational instability in which the pump wave decays into two sideband modes. Depending on the parameters involved, there is a resonant modulational instability involving a forward propagating electro-acoustic mode and a sideband daughter wave.

1. Introduction

The subject of the nonlinear behaviour of an electromagnetic wave propagating along an external magnetic field in an electron-positron plasma, has been studied by several authors (Shukla et al., 1986). However, most of the effort has been devoted to the study of possible self modulation of the waves. The reason lies in the observational fact that radiation emitted from pulsars magnetospheres shows short intensity variations (Curtis, 1991; Hewish, 1970; Smirnova, 1988; Smirnova et al., 1986). To account for these micropulses, Chian and Kennel (1983) proposed a self modulational instability of the electromagnetic waves. To this end, these authors considered a non linear Schrödinger equation, whose coefficients were shown to be incorrect (Kates and Kaup, 1989a) The reason for this is that Chian and Kennel [6] omitted two sources for nonlinearity, namely, harmonic generation and ponderomotive effects. Kates and Kaup (1989a), by using a multiple time-space scale perturbation theory, solved the problem consistently. They showed that in a nonmagnetized electron-positron cold plasma, the system is modulationally stable for both linear and circular polarization. When thermal effects are included, the system becomes modulationally unstable in a very narrow frequency region. It has recently been shown that in the ultrarelativistic case, the wave can be self modulated due to relativistic temperatures and phonon damping (Gratton et al., 1997).

In the case when the plasma is strongly magnetized, Kates and Kaup (1989b), showed that the plasma is unstable for frequencies below $\omega_p/\sqrt{2}$, where $\omega_p =$

Astrophysics and Space Science **256:** 403–410, 1998.

$(4\pi n_0 e^2/m)^{1/2}$ is the electron (positron) plasma frequency. Note that the actual plasma frequency of the system is $\sqrt{2}$ times the electron plasma frequency.

Here, we study the general structure of parametric instabilities of a large amplitude electromagnetic wave propagating along an external magnetic field. We include weakly relativistic effects on the particle motions in the wave field, the ponderomotive force, and nonrelativistic thermal effects. Note that in the case of a circularly polarized wave there is no harmonic generation (Kates and Kaup, 1989a; Gratton et al., 1997).

Thus, in Section 2 we describe the model and derive the nonlinear dispersion relation. In Section 3, we study numerically the dispersion relation. In Section 4, the results are summarized and discussed.

2. The Model

We assume that the electron-positron plasma is described by the following set of equations:

$$\frac{\partial n_j}{\partial t} = -\nabla \cdot (n_j \mathbf{v}_j), \tag{1}$$

$$\left(\frac{\partial}{\partial t} + \mathbf{v}_j \cdot \nabla\right)(\Gamma_j \mathbf{v}_j) = \frac{q_j}{m}\left(\mathbf{E} + \frac{1}{c}\mathbf{v}_j \times \mathbf{B}\right) - \frac{\gamma K T}{m n_0} \nabla n_j, \tag{2}$$

$$\nabla \cdot \mathbf{E} = 4\pi\rho, \tag{3}$$

$$\nabla \times \mathbf{E} = -\frac{1}{c}\frac{\partial \mathbf{B}}{\partial t}, \tag{4}$$

$$\nabla \times \mathbf{B} = \frac{4\pi}{c}\mathbf{J} + \frac{1}{c}\frac{\partial \mathbf{E}}{\partial t}, \tag{5}$$

where

$$\mathbf{J} = \sum_j q_j n_j \mathbf{v}_j, \quad \rho = \sum_j q_j n_j, \quad \Gamma_j = \left(1 - \frac{\mathbf{v}_j{}^2}{c^2}\right)^{-1/2}, \tag{6}$$

where n_j is the density of each fluid, $\mathbf{v}_j$ is the bulk velocity of each fluid ($j = e$ for electrons and $j = p$ for positrons), $\mathbf{E}$ and $\mathbf{B}$ are the electric and magnetic fields, respectively, K is the Boltzmann constant, $\mathbf{J}$ is the total current, γ is the adiabatic coefficient, T is the common temperature, m is the particle mass, and c is the speed of light.

We assume that a circularly polarized electromagnetic pump wave propagates along the z-axis, as well as the existence of a constant magnetic field in the same direction, so that the total zeroth order electric and magnetic fields are given by:

$$\mathbf{B}_0 = B[\hat{x}\cos(k_0 z - \omega_0 t) + \hat{y}\sin(k_0 z - \omega_0 t)] + B_{0z}\hat{z}, \tag{7}$$

$$\mathbf{E}_0 = E[\hat{x}\sin(k_0 z - \omega_0 t) - \hat{y}\cos(k_0 z - \omega_0 t)], \tag{8}$$

$$B = \frac{ck_0}{\omega_0}E. \tag{9}$$

Adopting the notation for transversal magnitudes:

$$A_{\perp 0} = A_{0x} + iA_{0y} = Ae^{i(k_0 z - \omega_0 t)}, \quad A \text{ real}, \tag{10}$$

we find that this wave induces a particle transversal velocity given by:

$$v_j = \frac{q_j B}{mck_0}\eta_j\left(1 - \frac{\alpha^2}{2}\eta_j^3\right), \tag{11}$$

where

$$\alpha = \frac{eB}{mc^2 k_0}, \qquad \eta_j = \frac{\omega_0}{\omega_0 - \omega_{jc}}, \qquad \omega_{jc} = \frac{q_j B}{mc}. \tag{12}$$

Note that the small parameter is α. So, since v_j is proportional to α (see Equation (11)), the theory is weakly relativistic and the factor Γ can be expanded around α^2. The dispersion relation of the electromagnetic wave, including weakly relativistic effects, is given by

$$\omega_0^2 = c^2 k_0^2 + \omega_p^2 \sum_j \eta_j\left(1 - \frac{\alpha^2}{2}\eta_j^3\right), \tag{13}$$

where $\omega_p = (4\pi n_0 e^2/m)^{1/2}$ is the electron plasma frequency.

Assuming that the system consists now of electrons, positrons and a circularly polarized electromagnetic wave satisfying Equation (13) – the pump wave – we perturb the system again. Longitudinal and transversal perturbations are of the form:

$$\delta A_z = \mathrm{Re}\big[\tilde{A}e^{i(kz-\omega t)}\big] \quad \text{and} \quad \delta A_\perp = a_+ e^{i(k_+ z - \omega_+ t)} + a_- e^{i(k_- z - \omega_- t)},$$

respectively, with $k_\pm = k_0 \pm k$, $\omega_\pm = \omega_0 \pm \omega$. Linearizing Equations (1–5), and with the further condition of charge quasineutrality $\delta n_p = \delta n_e$, we find a set of equations for perturbed quantities (δn_j, $\delta v_{j\perp}$, δv_{jz}, $\delta E_\perp$, δE_z and $\delta B_\perp$). Upon elimination of all of them we obtain the following nonlinear dispersion relation:

$$\begin{aligned} 0 = {} & \left[I_+ + I_1\left(\frac{H_{p++}}{D_p} + \frac{H_{e++}}{D_e}\right) + \frac{1}{2}I_2\frac{H_{p0+}}{D_p}\right] \\ & \times \left[I_- + I_1\left(\frac{H_{p--}}{D_p} + \frac{H_{e--}}{D_e}\right) - \frac{1}{2}I_2\frac{H_{p0-}}{D_p}\right] \\ & - \left[-\frac{1}{2}I_1\left(\frac{H_{p-+}}{D_p} + \frac{H_{e-+}}{D_e}\right) + \frac{1}{2}I_2\frac{H_{p0+}}{D_p}\right] \\ & \times \left[-\frac{1}{2}I_1\left(\frac{H_{p+-}}{D_p} + \frac{H_{e+-}}{D_e}\right) - \frac{1}{2}I_2\frac{H_{p0-}}{D_p}\right], \end{aligned} \tag{14}$$

where

$$
\begin{aligned}
H_{j\pm\pm} &= \omega_{\pm}\left\{-S_j[\omega_{\mp}(1+\alpha^2\eta_j^2)-\omega_{jc}] \pm \frac{1}{2}\,\alpha^2 c^2 k_0^2 \omega \frac{\omega_{jc}}{\omega_0}\;\eta_j\left(1-\frac{1}{2}\,\alpha^2\eta_j^3\right)\right\} \\
&\pm \frac{1}{2}\,\alpha^2\omega\frac{\omega_{jc}}{\omega_0}\;k_0 k_{\pm} c^2\eta_j^2\left(1-\frac{1}{2}\,\alpha^2\eta_j^3\right)^2 \\
&\times \left[\omega_{\pm}\frac{1}{2}\,\alpha^2\eta_j^2-\omega_{\mp}(1+\alpha^2\eta_j^2)+\omega_{jc}\right], \qquad (15)
\end{aligned}
$$

$$
\begin{aligned}
H_{j\pm\mp} &= \omega_{\mp}\left[-\omega_{\pm}\alpha^2\eta_j^2 S_j \pm \alpha^2 c^2 k_0^2\omega\frac{\omega_{jc}}{\omega_0}\;\eta_j\left(1-\frac{1}{2}\,\alpha^2\eta_j^3\right)\right] \\
&\pm \alpha^2 c^2 k_0 k_{\mp}\omega\frac{\omega_{jc}}{\omega_0}\;\eta_j^2\left(1-\frac{1}{2}\,\alpha^2\eta_j^3\right)^2 \\
&\times \left[\omega_{\pm}\frac{1}{2}\,\alpha^2\eta_j^2-\omega_{\mp}(1+\alpha^2\eta_j^2)+\omega_{jc}\right], \qquad (16)
\end{aligned}
$$

$$
\begin{aligned}
H_{j0\pm} &= \omega_{\pm}\left[\omega_{\mp}\frac{1}{2}\,\alpha^2\eta_j^2+\omega_{\mp}(1+\alpha^2\eta_j^2)-\omega_{jc}\right]-\frac{k_{\pm}}{k_0}\;\eta_j\left(1-\frac{1}{2}\,\alpha^2\eta_j^3\right) \\
&\times \left\{[\omega_{+}(1+\alpha^2\eta_j^2)-\omega_{jc}][\omega_{-}(1+\alpha^2\eta_j^2)-\omega_{jc}]\right. \\
&\left.-\,\omega_{-}\omega_{+}\left(\frac{1}{2}\,\alpha^2\eta_j^2\right)^2\right\}, \qquad (17)
\end{aligned}
$$

$$
\begin{aligned}
D_j &= [\omega_{+}(1+\alpha^2\eta_j^2)-\omega_{jc}] \\
&\times \left\{-[\omega_{-}(1+\alpha^2\eta_j^2)-\omega_{jc}]S_j+\frac{1}{2}\,\alpha^2 c^2 k_0^2\omega\frac{\omega_{jc}}{\omega_0}\;\eta_j\left(1-\frac{1}{2}\,\alpha^2\eta_j^3\right)\right\} \\
&-\,\omega_{-}\frac{1}{4}\,\alpha^2\eta_j^2\left[-S_j\omega_{+}\alpha^2\eta_j^2+\alpha^2 c^2 k_0^2\omega\frac{\omega_{jc}}{\omega_0}\;\eta_j\left(1-\frac{1}{2}\,\alpha^2\eta_j^3\right)\right] \\
&+\frac{1}{2}\,\alpha^2 c^2 k_0^2\omega\frac{\omega_{jc}}{\omega_0}\;\eta_j\left(1-\frac{1}{2}\,\alpha^2\eta_j^3\right) \\
&\times \left[\omega_{+}\frac{1}{2}\,\alpha^2\eta_j^2-\omega_{-}(1+\alpha^2\eta_j^2)+\omega_{jc}\right], \qquad (18)
\end{aligned}
$$

$$
I_{\pm} = c^2 k_{\pm}^2 - \omega_{\pm}^2, \qquad (19)
$$

$$
I_1 = \omega_p^2 \qquad (20)
$$

$$
I_2 = \omega_p^2 k k_0 c^2\left(\frac{eB}{mc^2 k_0}\right)^2\left[\eta_p\left(1-\frac{1}{2}\,\alpha^2\eta_p^3\right)+\eta_e\left(1-\frac{1}{2}\,\alpha^2\eta_e^3\right)\right] \qquad (21)
$$

$$
S_j = \omega^2\left(1+\frac{1}{2}\alpha^2\eta_j^2\right)-v_s^2 k^2. \qquad (22)
$$

When $\alpha = 0$, Equation (14) reduces to

$$0 = S_p S_e D_+ D_-, \tag{23}$$

where

$$D_\pm = c^2 k_\pm^2 - \omega_\pm^2 + \omega_p^2 \left(\frac{\omega_\pm}{\omega_\pm - \omega_{pc}} + \frac{\omega_\pm}{\omega_\pm - \omega_{ec}} \right). \tag{24}$$

The solutions $D_\pm = 0$ and $S_p = S_e = 0$ correspond to the electromagnetic waves and the electro-acoustic modes. When $\alpha \neq 0$, the modes are coupled. A necessary condition for wave coupling is that $n\omega_0 = \omega_1 + \omega_2$, with $n = 0, 1, 2, 3 \ldots$, where ω_1 and ω_2 are the frequencies of the daughter waves.

3. Numerical Analysis of the Dispersion Relation

We now solve numerically the nonlinear dispersion relation, Equation (14), by using a method first derived by Longtin and Sonnerup (1986) (see also Hollweg et al., 1993; Galvão et al., 1994; Gomberoff et al., 1995; Gnavi et al., 1996). To this end, we choose a value for the pump wave frequency, $\omega_0/\omega_{pc} = 0.5$, and from Equation (13) we determine the corresponding wavenumber, $ck_0/\omega_{pc} = 0.6455$, for $\alpha = 0$. In Figure 1a, we show the solutions of Equation (14) for $\omega_p/\omega_{pc} = 1$, $v_s/c = 0.1$, and $\alpha = 0$. It is sufficient to consider the upper half of the (x, y) plane, since the figure is symmetric to reflections through the origin. There are 12 lines in the figure: the 8 real solutions of $D_\pm = 0$ (labeled in Figure 1a as $D_\pm$], and 4, straight, lines corresponding to the electro-acoustic modes, 2 propagating in the direction of the pump wave (S_+), and 2 in the opposite direction (S_-). When $\alpha = 0$ there are only 2 such lines because we have assumed equal electron and positron temperature. However, in the presence of the pump wave, the acoustic modes associated to the electrons and positrons separate out from each other giving rise to 4 lines.

From Figure 1a, it follows that there are a number of crossings between the solutions of Equation (14). Some of these crossings can give rise to wave couplings when the pump wave is switched on. Only crossings between modes satisfying the conditions mentioned in Section 2 can lead to wave coupling. In Figure 1b, we have switched on the pump wave by setting $\alpha = 0.1$. From the figure it follows that some crossings have developed gaps – labeled in Figure 1 from (1) to (4). If we draw a horizontal line passing through one of the gaps, this line will cut only 10 lines, instead of 12. Therefore, in such cases there are only ten real solutions of Equation (14), which means that two of them have become complex conjugate. Hence, at these gaps we have instabilities, indicating thereby, wave coupling.

From Figure 1b, it follows that there are several possible couplings. Gap (1) is an ordinary decay instability, in which the pump wave decays into a forward propagating electro-acoustic mode and a sideband wave. Note that it is only one

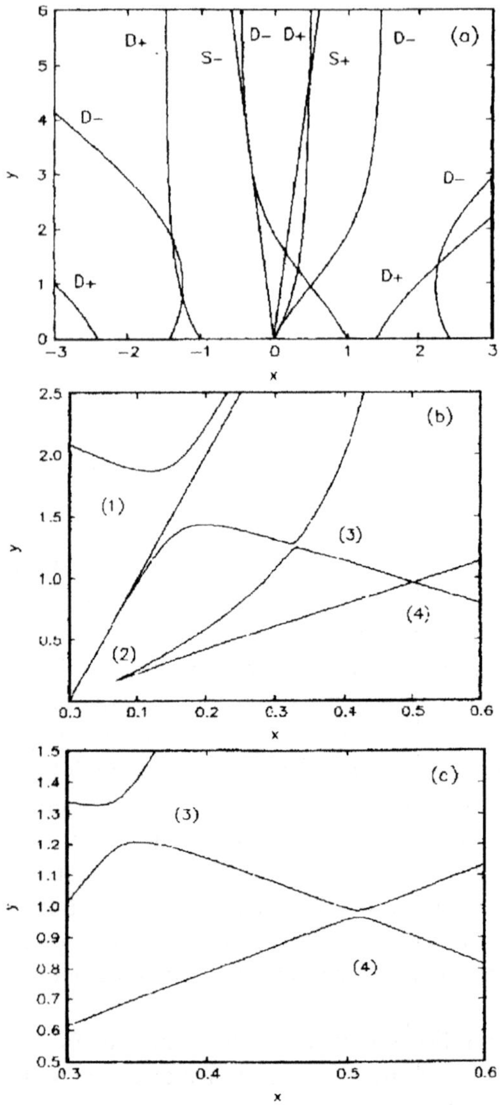

Figure 1. Dispersion relation, Equation (14), normalized wavenumber $y = ck/\omega_{pc}$ vs. normalized frequency $x = \omega/\omega_{pc}$, for $\omega_p/\omega_{pc} = 1$, $v_s/c = 0.1$, $x_0 = 0.5$, $y_0 = 0.6455$, for (a) $\alpha = 0$, (b) $\alpha = 0.1$, (c) $\alpha = 0.2$.

of the electro-acoustic modes which is involved in this coupling. There is also a crossing at the origin between a solution of $D_- = 0$ and one of $D_+ = 0$. This crossing gives rise to gap (2), starting at the origin, and corresponds to a nonresonant modulational instability. Gaps (3) and (4), detailed in Figure 1c, are nonresonant couplings in which the pump decays into two sideband waves, one solution of $D_- = 0$ and one of $D_+ = 0$ [gap (3)], or two solutions of $D_- = 0$ [gap

(4)]. Finally, not shown in Figure 1, the first crossing from left to right in Figure 1a, between a solution of $D_- = 0$ and one of $D_+ = 0$, gives rise to a fifth gap.

4. Summary

We have studied parametric decays of a circularly polarized wave in an electron-positron plasma, propagating in the direction of an external magnetic field. In general, there exist several instabilities, both resonant and nonresonant. The resonant instabilities are decay instabilities in which the pump wave decays into a forward propagating electro-acoustic wave and a sideband wave. The nonresonant instabilities are essentially electromagnetic, in which the pump wave decays into two sideband waves. Depending on the sound velocity, there are two types of modulational instabilities. For $v_s/c \ll 1$, the modulational instability is nonresonant and gives rise to two sideband daughter waves. For larger values of v_s/c, the modulational instability is resonant and involves a forward propagating electro-acoustic wave and a sideband wave.

The particular case of pulsars magnetospheres corresponds to a situation in which $\omega_p/\omega_{pc} \ll 1$, since very strong magnetic fields are present. In general, the type of instabilities involved are the same as those shown here, except that larger pump wave amplitudes are needed to trigger the nonresonant modulational instability.

It is important to point out that, since our model is a fluid model, there are important effects, such as Landau damping and resonance absorption, that have not been considered. They can change the results and provide thresholds which are not present in our model. Therefore, a similar approach to the one used here, but using kinetic theory, is required.

Acknowledgements

This work has been partially supported by FONDECYT, grant No. 1960874, and Proyecto Trinacional, Fundación Andes, No. C-12999/6.

References

Chian, A. and Kennel, C.: 1983, *Astrophys. Space Sci.* **97**, 9.

Curtis, M.F.: 1991, *The Theory of Neutron Stars Magnetospheres*, University of Chicago Press, Chicago.

Galvão, R.M.O., Gnavi, G., Gomberoff, L. and Gratton, F.T.: 1994, *Plasma Phys. Control. Fusion* **36**, 1679.

Gnavi, G., Galvão, R.M.O., Gratton, F.T. and Gomberoff, L.: 1996, *Phys. Rev. E* **54**, 4112.

Gomberoff, L., Gratton, F.T. and Gnavi, G.: 1995, *J. Geophys. Res.* **100**, 17, 221.

Gratton, F.T., Gnavi, G., Galvão, R.M.O. and Gomberoff, L.: 1997, *Phys. Rev. E* **55**, 3381.
Hewish, A.: 1970, *Ann. Rev. Astron. Ap.* **8**, 265.
Hollweg, J.V., Esser, R. and Jayanti, V.: 1993, *J. Geophys. Res.* **98**, 3491.
Kates, R.E. and Kaup, D.J.: 1989a, *J. Plasma Phys.* **41**, 507.
Kates, R.E. and Kaup, D.J.: 1989b, *J. Plasma Phys.* **42**, 521.
Longtin, M. and Sonnerup, B.U.Ö.: 1986, *J. Geophys. Res.* **91**, 6816.
Shukla, P.K., Rao, N.N., Yu, M.Y. and Tsintsadze, N.L.: 1986, *Phys. Rep.* **138**, 1.
Smirnova, T.V.: 1988, *Soviet Astron. Lett.* **14**, 20.
Smirnova, T.V., Soglasnov, V.A., Popov, M.V. and Novikov, A.Yu.: 1986, *Soviet Astron.* **30**, 51.

NEW SYSTEM OF COORDINATES FOR TOKAMAKS

P. MARTÍN

Departamento de Física, Universidad Simón Bolívar, Caracas 89000, Venezuela

Abstract. A new system of coordinates for tokamak is presented here. They are described by the usual toroidal angle ϕ and coordinates in the meridian plane using the projection of the magnetic field lines and families of orthogonal lines. The number of families depends on the number of magnetic axis, and in this way the number of regions in the meridian plane are also defined. The differential operators are obtained using the curvatures κ_s and κ_σ of each family. Grad–Safranov equation, poloidal field equations and Pfirsch–Schlüter transport are described in a general way using this kind of coordinates.

1. Introduction

The problem concerning what kind of coordinates to use in tokamak has been treated extensively (D'Haeseler et al., 1991; Wesson, 1987). Some of them are more useful in equilibrium or in treating some instabilities, or in writing codes. Thus, several kinds of coordinates have been used: toroidal, Boozer, bipolar, Hamada, etc. Some of the coordinates are orthogonal and others are not; however, the scales factor are always well determined in order to describe the differential operators. Here we are looking at the problem in a different way. We are interested in the families of line coordinates and the meridian plane is divided into regions. The most important family of curves is just the well known magnetic field lines, projected in the meridian planes. Now new families of local orthogonal lines are defined, which concur to each magnetic axis. Thus, there are as many regions as magnetic axes. Now the differential operators are obtained by using local frames, as the Frenet-frame, in such a way that only the intrinsic parameters of the curves appear. Our system of coordinates is orthogonal and instead of looking for the scale factor we go directly to the differential operators using the curvature of the orthogonal field lines κ_σ together with the well known curvature κ_s of the magnetic field lines projection in the meridian plane. The coordinates in the meridian plane are completed with the usual toroidal angle ϕ. Using this type of coordinates there is a separation of variables in some important equations, i.e. Grad–Safranov. Here we will describe the form of the new equations as the previously mentioned Grad–Safranov, poloidal field equations, and Pfirsch–Schlüter transport in a general case. The application of the results to particular cases as elliptic curves or elliptic curves with triangularity and displacement, will be described elsewhere.

Astrophysics and Space Science **256:** 411–416, 1998.

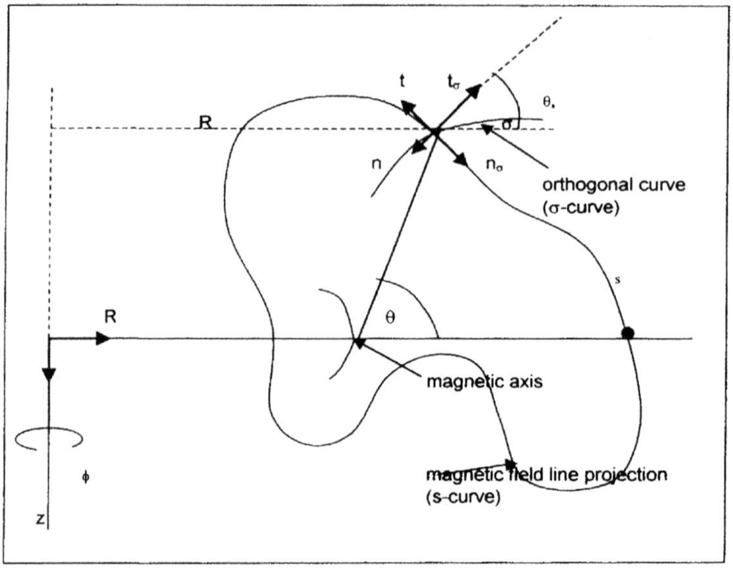

Figure 1. Cross section of a general magnetic surface in a tokamak, showing the system of coordinates described in the text.

2. Theory

Let us consider the family of magnetics surfaces as depending of a parameter λ that identifies the surface, as

$$F(R, \phi, z, \lambda) = 0. \tag{1}$$

Here all the family of magnetics surfaces are defined in this way. The orthogonal lines are now defined by the equation

$$d\mathbf{r} \times \nabla F = 0. \tag{2}$$

In each point we are going to define the following three coordinate lines:

1. the circle with radius R;
2. the intersection of the meridian plane with the magnetic surface at the point; and
3. the line defined by Equation (2).

In the case of toroidal axisymmetry all the preceding lines will be orthogonal in each point. Here we will consider only this case, shown in Figure 1. The following notation will be used: σ = length of arcs along the orthogonal lines (σ-curves); s = length of arcs along meridian magnetic field line projection (s-curves); $\hat{t}$ = unitary tangent vector along the s-curves; $\hat{n}$ = principal normal along the s-curves; $\theta_s = \angle(\hat{t}, \hat{R})$.

The equilibrium equations with toroidal axisymmetry will be:

$$\nabla \times \mathbf{B} = \mu \mathbf{J}, \tag{3}$$

$$\nabla \times (B_p \hat{t} + B_\phi \hat{e}_\phi) = \mu_0 \mathbf{J}_p + \mu_0 \mathbf{J}_\phi, \tag{4}$$

$$\nabla \cdot (B_p \hat{t} + B_\phi \hat{e}_\phi) = 0, \tag{5}$$

$$\nabla = \nabla_\perp + \frac{\partial}{\partial \phi} \hat{e}_\phi. \tag{6}$$

Here $\partial/\partial\phi$ will be zero for functions with toroidal axisymmetry. Let us now introduce the curvatures κ_s and κ_σ respectively of the s-lines and σ-lines (orthogonal curves) In order to proceed further the following important relations can be obtained.

$$\nabla \cdot \hat{t} = \kappa_\sigma - \frac{\sin\theta_s}{R}, \tag{7}$$

$$\nabla \times \hat{t} = \kappa_s \hat{e}_\phi, \tag{8}$$

$$\nabla \cdot \hat{n} = -\kappa_s - \frac{\cos\theta_s}{R}, \tag{9}$$

$$\nabla \times \hat{n} = \kappa_\sigma \hat{e}_\phi, \tag{10}$$

$$\nabla . \hat{e}_\phi = 0, \tag{11}$$

$$\nabla \times \hat{e}_\phi = \frac{\hat{z}}{R}, \tag{12}$$

$$\nabla \hat{t} = \kappa_s \hat{t}\hat{n} + \kappa_\sigma \hat{n}\hat{n} - \frac{\sin\theta_s}{R} \hat{e}_\phi \hat{e}_\phi, \tag{13}$$

$$\nabla \hat{n} = -\kappa_s \hat{t}\hat{t} - \kappa_\sigma \hat{n}\hat{t} - \frac{\cos\theta_s}{R} \hat{e}_\phi \hat{e}_\phi, \tag{14}$$

$$\nabla \hat{e}_\phi = -\frac{1}{R} \hat{e}_\phi \hat{R}. \tag{15}$$

Looking now at the projections of Equation (4) in the Frenet-frame, we arrive at the equations:

– Projection along $\hat{t}$

$$\frac{\partial B_\phi}{\partial \sigma} + B_\phi \frac{\cos\theta_s}{R} = -\mu_0 J p. \tag{16}$$

– Projection along $\hat{n}$ = - $\hat{t}_\sigma$

$$\frac{\partial B_\phi}{\partial s} - B_\phi \frac{\sin\theta_s}{R} = 0. \tag{17}$$

– Projection along $\hat{e}_\phi$

$$\frac{\partial B_p}{\partial s} + B_p \kappa_s = \mu_0 J_\phi. \tag{18}$$

3. Grad–Safranov Equation

Writing down that equation in the form

$$\Delta^{+}(\Psi) = R^2\nabla\cdot\left(\frac{1}{R^2}\nabla\Psi\right) = -[R^2 p'(\Psi) + II'(\Psi)] \tag{19}$$

and considering now the form in the new coordinates, the equation can be written as

$$\frac{\partial^2\Psi}{\partial\sigma^2} + \left(\kappa_s - \frac{\cos\theta_s}{R}\right)\frac{\partial\Psi}{\partial\sigma} = -[R^2 p'(\Psi) + I(\Psi)I'(\Psi)], \tag{20}$$

$$I(\Psi) = RB_\phi. \tag{21}$$

The important point about this form of the Grad–Safranov equation is that now there are only derivatives with respect the orthogonal arc σ. Thus, there is a separation of the variables σ and s.

4. Poloidal Magnetic Field Equation

Developing Equation (5), we obtain

$$\nabla\cdot B = 0 \Rightarrow 0 = \frac{\partial B_\phi}{R\partial\phi} + B_\phi\hat{e}_\phi\cdot\frac{\partial\hat{e}_\phi}{\partial\phi} + \hat{e}_\phi\cdot\nabla_\perp B_\phi + \nabla\cdot\mathbf{B}_p = \nabla\cdot\mathbf{B}_p = 0.$$

From here we arrive at

$$\nabla\cdot\mathbf{B}_p = 0 \Rightarrow \frac{\partial B_p}{\partial s} - \frac{\sin\theta_s}{R}B_p + \kappa_\sigma B_p = 0.$$

The formal integration of this differential equation leads to

$$B_p = B_{p0}\frac{R_0}{R}\exp\left(-\int_0^s \kappa_\sigma ds\right)$$

$$B_p = B_{p0}\frac{R_0}{R}\exp\left(-\int_0^{\theta_s}\frac{\kappa_\sigma}{\kappa_s}\,d\theta_s\right).$$

The analysis of the equation for the cases of elliptic curves with triangularity and displacement will be presented elsewhere (Martín and Haines, 1998).

5. Transport in the Collisional Regime (Pfirsch–Schlüter Transport)

Transport analysis can also be performed with the new coordinates. Looking at the collisional regime, the average normal flow velocity v_n will be given by

$$v_n = \frac{1}{LR_0} \int d\theta R ds \left(\frac{\mathbf{E} \times \mathbf{B}}{B^2} - \eta_{\perp} \frac{\nabla p}{B^2} \right) \cdot (-\hat{n}), \tag{22}$$

where L is the perimeter of the s-curves. Now considering the parallel component of Ohm's Law, the following equation is obtained.

$$B_p \frac{\partial \Phi}{\partial s} = \eta_n j_n B,$$
$$-\nabla \Phi = \mathbf{E}. \tag{23}$$

Writing down the continuity equation for the current

$$\frac{\partial}{\partial s}\left(\frac{j_n}{B}\right) = -\frac{B_\phi}{B_p} \frac{\partial p}{\partial \sigma} \frac{\partial}{\partial s}\left(\frac{1}{B^2}\right), \tag{24}$$

$$\frac{\partial}{\partial s}\left(\frac{j_n}{B}\right) = \frac{B_\phi R}{B_{p0} R_0} \frac{\partial p}{\partial \sigma} \exp\left(\int_0^s \kappa_\sigma ds\right) \frac{\partial}{\partial s}\left(\frac{1}{B^2}\right). \tag{25}$$

These are the general equations to be presented here. To proceed further we must write down explicitly the parametric form of Equation (1). One of the simplest cases to treat is that of elliptic curves (Leontovich, 1975), which in our case can be conveniently defined as: $R = R_0 + \lambda b \cos\theta$, $z = \lambda a \sin\theta$. Thus, the curvature of the orthogonal lines will be:

$$\kappa_\sigma = \frac{(1-\gamma^2)\sin\theta\cos\theta}{\lambda a (\cos^2\theta + \gamma^2 \sin^2\theta)^{3/2}}, \tag{26}$$

$$\gamma = b/a. \tag{27}$$

The details of this treatment will be presented elsewhere.

6. Conclusions

The transport and equilibrium analysis in tokamak can be simplified using a description for magnetic surfaces as a family of nested surfaces depending on a parameter λ. Considering now the vector field derived through the gradient operator, one family of line coordinates is defined as the current lines of this gradient vector field. In the case of toroidal symmetry, an orthogonal system of coordinates is obtained considering the orthogonal curves, defined above, together with the usual magnetic field projections and the azimuthal circles.

Depending on the number of magnetic axis of the toroidal configuration, we will find the same number of orthogonal families as the magnetic axis.

In the description presented here there is no need to find the scale factors of the system of coordinates. Instead, the curvatures of the magnetic and orthogonal field lines appear as the important point functions.

The equilibrium equations are now written using these curvatures. Several relations and differential equations are obtained for the poloidal and toroidal magnetic fields. Grad–Safranov equations with separation of variables are also obtained. Equations for the poloidal magnetic field along a magnetic field line and Pfirsch–Schlüter current are finally considered in a general way, but specific treatment will be presented elsewhere.

References

D'Haeseler, W.D., Hitchen, W.N.G., Callen, J.D. and Shohet, J.L.: 1991, *Flux Coordinates and Magnetic Field Structure*, Springer-Verlag, Berlin/New York.

Leontovich, M.A.: 1975, *Reviews of Plasma Physics*, Vol. 6; Soloviev, L.S., *Hydromagnetic Stability of Closed Plasma Configurations*, Consultants Bureau, New York, pp. 261–264.

Martín, P. and Haines, M.G.: 1998, *Physics of Plasmas* **5**, 410.

Wesson, J.: 1987, *Tokamaks*, Clarendon Press, Oxford.

ROTATION OF WEAKLY COLLISIONAL PLASMAS IN TOKAMAKS, OPERATED WITH ALFVÉN WAVES

V.S. TSYPIN[1], A.G. ELFIMOV[1], A.S. DE ASSIS[2] and C.A. DE AZEVEDO[1]
[1]*Instituto de Física, Universidade do Estado do Rio de Janeiro, 20550-013, Maracanã, Rio de Janeiro, Brazil;* [2]*Instituto de Matemática, Departamento de Matemática Aplicada, Universdade Federal Fluminense, Niterói, Rio de Janeiro, Brazil*

Abstract. In this paper we consider rotation induced by kinetic Alfvén waves in weakly collisional plasma of slightly rippled tokamaks in plateau and banana regimes. Rippled magnetic field of tokamaks retards the plasma rotation in toroidal direction. Here, we are going to find the quasistationary poloidal and toroidal plasma velocities and the radial electric field taking into account the complete form of ponderomotive forces.

Alfvén waves (AW) are usually used for the current-drive and rf heating of plasmas in thermonuclear facilities (especially in tokamak) (Hershkovitz et al., 1994). But AW were never considered to be the cause of plasma rotation in closed thermonuclear facilities both theoretically and experimentally. The effect of plasma rotation is very important because of the possibility to control low-to-high confinement transition (L-H transition), in tokamaks (Tsypin et al., 1996).

Now, rotation of the bulk plasma, as parallel effect at AW plasma heating or AW current-drive, can be supposed as the wasteful using of this energy. That is the reason why the problem of the investigation of weakly collisional plasma rotation induced by rf waves in these facilities is very important.

To find quasistationary quantities of the plasma rotation velocities and radial electric field, we proceed from one fluid motion equations of the plasma:

$$F_\theta^\pi + F_\theta^h = 0, \qquad F_\zeta^\pi + F_\zeta^h = 0. \tag{1}$$

The poloidal F_θ^π (along the poloidal angle θ) and toroidal F_ζ^π (along the toroidal angle ζ) viscous forces are balanced by the rf poloidal F_θ^h and toroidal F_ζ^h forces. These forces have the form (see, e.g., Mikhailovskii and Tsypin, 1984):

$$F_\theta^\pi = \frac{3}{2r}\left\langle \pi_\parallel \frac{\partial \ln B}{\partial \theta}\right\rangle^0, \qquad F_\zeta^\pi = \frac{3}{2R}\left\langle \pi_\parallel \frac{\partial \ln B}{\partial \zeta}\right\rangle^0, \tag{2}$$

where $\pi_\parallel$ is the parallel (along the magnetic field) plasma viscosity, r and R are the minor and major torus radii respectively, $\langle \ldots \rangle \equiv (2\pi)^{-1} \oint (\ldots) d\zeta$ denotes an averaging over ζ and $(\ldots)^0 \equiv (2\pi)^{-1} \oint (\ldots) d\theta$ denotes an averaging over θ.

Astrophysics and Space Science **256:** 417–420, 1998.

The toroidal magnetic field is supposed to have the form $B = B_0(1 + \varepsilon\cos\theta + \delta\cos N\zeta)$, where ε is the inverse aspect ratio, δ and N are the parameters of the rippled torus.

Here, we assume excitation of kinetic AW which are localized near the conversion point because of strong electron dissipation. That is the reason why the plasma dynamics can be well operated in the appropriate place. We suppose that the dispersion relation of these waves has the simplified cylinder geometrical optics approximation

$$N_\perp^2 \approx \frac{\varepsilon_\parallel}{\varepsilon_\perp}(\varepsilon_\perp - N_\parallel^2), \quad \varepsilon_\perp = 1 + \sum_\alpha \frac{\omega_{p\alpha}^2}{\omega_{c\alpha}^2 - \omega^2},$$

$$N_\perp = \frac{ck_\perp}{\omega}, \quad N_\parallel = \frac{ck_\parallel}{\omega}, \tag{3}$$

where $k_\perp$ and $k_\parallel$ are the perpendicular and parallel to the magnetic field components of the wave vector respectively, $\omega_{c\alpha}$ is the cyclotron frequency, $\omega_{p\alpha}$ is the Langmuir frequency, $\varepsilon_\perp$ and $\varepsilon_\parallel$ are the components of the permeability tensor. For kinetic AW, the Landau damping is contained in the longitudinal component of the permeability tensor $\varepsilon_\parallel$ including the damping on trapped particles. We can find the absorbed power of KAW in the weakly collisional plasma

$$P_w = \frac{1}{4}(\mathbf{E}\cdot\mathbf{j}^* + \text{c.c.} \approx \frac{\omega}{8\pi}\,\text{Im}\,\varepsilon_\parallel |E_\parallel|^2. \tag{4}$$

From this equation we can see that we do not need to find the specific expression for the $\varepsilon_\parallel$ if we express the rf forces via the absorbed power P_w. We can compare the theoretical results with the experimental ones taking into account the experimental value of the absorbed power.

As the next step we can find the expressions for components of the rf forces

$$F_\theta^h \approx \frac{P_w}{\omega}(k_\theta - h_\theta k_\parallel), \quad F_\zeta^h \approx \frac{P_w}{\omega}(k_\zeta - h_\zeta k_\parallel), \tag{5}$$

where $k_\theta = m/r$, $k_\zeta = -n/R$, m and n are the poloidal and toroidal wave numbers and $\mathbf{h} = \mathbf{B}/B$. Obtaining Equation (5), we use the relation

$$\text{Re}\,N_\perp\ \text{Im}\,N_\perp \approx \frac{(\varepsilon_\perp - N_\parallel^2)}{\varepsilon_\perp}\frac{\text{Im}\,\varepsilon_\parallel}{2}, \tag{6}$$

which follows from Equation (3). Finally, we can find components of the rf forces

$$F_\theta^h \approx \frac{h_\zeta k_b}{\omega}P_w, \quad F_\zeta^h \approx -\frac{h_\theta k_b}{\omega}P_w, \quad k_b = k_\theta - h_\zeta - k_\zeta(h_\theta), \tag{7}$$

i.e., the longitudinal component of rf force is approximately equal to zero.

Here we consider plateau and banana regimes of weakly collisional plasmas. For the plateau regime the conditions are

$$\frac{\nu_\alpha}{\varepsilon^{3/2}} \geq \frac{v_{T\alpha}}{qR} \geq \nu_\alpha, \quad \frac{\nu_\alpha}{\delta^{3/2}N} \geq \frac{v_{T\alpha}}{R}. \tag{8}$$

where $v_{T\alpha}$ is the thermal speed of species α and q is the safety factor. Using the expressions for viscous forces, repeatedly found, we have these forces in the form of (Mikhailevskii and Tsypin, 1984)

$$F_\theta^\pi = -\frac{\sqrt{\pi}}{2} M_i n_i \frac{q v T_i}{R}\left(U_{\theta i} + \frac{1}{2} U_{Ti}\right),$$

$$F_\zeta^\pi = -\frac{\sqrt{\pi}}{2} M_i n_i \frac{N\delta^2 v_{Ti}}{R} U_{\zeta i}, \tag{9}$$

where $U_{\theta i}$ is the ion poloidal velocity, U_ζ is the toroidal velocity and $U_{Ti} = (c/e_i B_0)\partial T_i/\partial r$ is the ion drift velocity connected with temperature gradient.

Plasma rotation and viscous forces of a tokamak plasma with rippled magnetic field in the conditions of the banana regime,

$$\frac{v_{T\alpha}}{qR} \geq \frac{\nu_\alpha}{\varepsilon^{3/2}}, \quad \frac{v T_\alpha}{R} \geq \frac{\nu_\alpha}{\delta^{3/2} N}, \tag{10}$$

had been studied in (Connor and Hastie, 1973). Using results of these papers we can write

$$F_\theta^\pi = -M_i n_i v_{i*} \frac{q^2}{\varepsilon^2}\left\{(c_u + c_t)U_{\theta i} + (2.37c_t - 1.17c_u)U_{T_i} - \frac{\varepsilon}{q} c_t U_{\zeta i}\right\}. \tag{11}$$

Now we can estimate the rotation velocities and the radial electric field for some tokamaks. We show it for Phaedrus-T tokamak (Hershkovitz et al., 1994), where the AW have been used for the AW plasma heating and current drive, and for the tokamak JET, where AW are intensively studied. The Phaedrus-T had the next core plasma parameters: $T_i \approx 400 \div 600$ eV, $n_0 \approx 2 \cdot 10^{12} \div 10^{13}$ cm^{-3}, $z_{\rm eff} \approx 2$, $B_0 = 0.68$ T, $\tilde{B}_\perp/B_0 = 3 \cdot 10^{-3}$, $r = 26$ cm, $R = 92$ cm, $w = 4 \cdot 10^7$ rad^{-1}, $m = 1 \div 3$, $n = 4 \div 9$, $N = 20$, $\delta \approx 0.05$, $P_w \approx 0.65$ W/cm^3. From these data and Equation (8) we see that the core plasma in this tokamak is in the plateau regime. Using Equations (8) and (9), we find

$$U_{T_i} \approx -6 \cdot 10^5 \text{ cm/s}, \quad \Delta U_{\theta i} \approx 3 \cdot 10^4 \text{ cm/s},$$

$$\Delta U_{\zeta i} \approx -3 \cdot 10^5 \text{ cm/s}, \quad \Delta \bar{E}_r \approx -4 \text{ V/cm}, \tag{12}$$

i.e., these values are of the order or less than those without the rf field. Δ means the rf parts of these quantities.

We use the plasma parameters for JET which is during the plasma heating and current drive by the low hybrid waves: $T_i \approx 2.5 \cdot 10^3$ eV, $n_0 \approx 2.8 \cdot 10^3$ cm^{-3}, $z_{\rm eff} \approx 2$, $B_0 = 3.2$ T, $r = 95$ cm, $R = 285$ cm, $\omega = 4 \cdot 10^7$ rad^{-1}, $m = 1 \div 3$, $n = 4 \div 9$, $N = 20$, $\delta \approx 0.05$, $P_w \approx 0.7$ W/cm^3. From Equation (10) we see that the JET core plasma is in a banana regime. Thus from Equations (10) and (11) we have

$$U_{T_i} \approx -7 \cdot 10^4 \text{ cm/s}, \quad U_{\theta i} \approx 1.5 \cdot 10^5 \text{ cm/s}, \quad U_{\zeta i} \approx -1.3 \cdot 10^5 \text{ cm/s},$$

$$\bar{E}_r \approx -50 \text{ V/cm}, \quad \Delta U_{\theta i} \approx -1.3 \cdot 10^4 \text{ cm/s},$$
$$\Delta U_{\zeta i} \approx -1.3 \cdot 10^5 \text{ cm/s}, \quad \Delta \bar{E}_r \approx 0.0 \text{ V/cm}, \tag{13}$$

i.e., these values are of the order or less than those without considered as positive effect as the rf field does not strongly perturb the ion motion in the core plasma. If is a strong toroidal plasma rotation without external forces found in experiments, as predicted theoretically, toroidal plasma rotation induced by AW will be negligibly small.

The above analysis shows that using the Alfvén waves for tokamak plasma heating and current drive is accompanied by an increased ion poloidal and toroidal rotation. These quantities have been estimated for the Phaedrus-T tokamak, where the main goal was to study AW plasma heating and current drive, and for JET, where AW are under intensive study. The evaluations for experimental data of these tokamaks show that the quasistationary ion rotation velocities and radial electric field are weakly changed by AW. This effect can be considered as positive, because the opinion exists that AW current drive in the core tokamak plasma will practically be absent because of the drag of ions by the electrons and by rf forces. The additional equilibrium problems do not appear in this case.

Acknowledgement

This work was supported by CNPq (National Counsel of Scientific and Technological Development), FAPERJ (Rio de Janeiro Fund of Investigation's Support) and UERJ (Rio de Janeiro State University) (Brazil).

References

Connor, J.W. and Hastie, R.J.: 1973, *Nuclear Fusion* **13**, 221.

Hershkovitz, N. et al.: 1994, in: *Proceedings of the 15th IAEA Conference on Plasma Physics and Controlled Fusion*, Seville, IAEA-CN-60/A315-P-14, (to be published).

Mikhailovskii, A.B. and Tsypin, V.S.: 1984, *Sov. J. Plasma Phys.* **10**(2), 142.

Tsypin, V.S., Elfimov, A.G., de Azevedo, C.A. and de Assis, A.S.: 1996, Alfvén waves and rotation of collisional plasmas in tokamaks, *Phys. Plasmas* (to be published).

CURRENT DRIVE BY ALFVÉN WAVES IN ELONGATED CROSS SECTION TOKAMAK

V. TSYPIN[1], A. ELFIMOV[2], M. TENDLER[3], F. NEKRASOV[1], C. DE AZEVEDO[1] and A. DE ASSIS[4]

[1] *State University of Rio de Janeiro, 20550-013, Brazil;* [2] *University of Sao Paulo, Cx. Postal 20516, Sao Paulo, Brazil;* [3] *The Alfvén Laboratory, Royal Institute of Technology, 10044 Stockholm, Sweden;* [4] *Fluminense Federal University, Niterói, Rio de Janeiro, Brazil*

Abstract. In this paper, an attempt is made to clarify some general aspects of the Alfvén current drive problem for magnetic traps. For large aspect ratio tokamaks, with an elongated cross-section, some general formulas concerning the untrapped particles dynamics and their input to the Landau damping of the Alfvén waves are presented. They are supposed to be used for the further development of the Alfvén current drive theory for tokamaks.

The problem of the noninductive current-drive in cylindrical plasma model and in circular cross-section tokamaks had already been discussed intensively (Elfimov et al., 1992; Nascimento et al., 1994). At the beginning of the study of this problem it has been clear that there are significant difficulties in using of the current-drive in toroidal magnetic traps, especially in a tokamak reactor. For the Alfvén waves, there is an opinion that the efficiency of the current-drive drops as a result of waves absorption by the trapped particles. Now, the theory development of the Alfvén wave (AW) current drive is very important in the course of the future experiments on the TCA/Br tokamak (Nascimento et al., 1994). In this paper, an attempt is made to clarify some general aspects of this problem. For large aspect ratio tokamaks, with an elongated cross-section, some general formulas in the model approximation concerning the untrapped and trapped particles dynamics and their input to the Landau damping of the Alfvén waves, are presented. They are supposed to be used for the further development of the Alfvén current drive theory for tokamaks.

To derive general expressions for the noninductive current induced by AW we proceed from the surface-averaged Ohm's law (Elfimov et al., 1992). The angle dependence of the magnetic field, we get from the expression $B_0 = \sqrt{g_{ik}B^iB^k} = B_s(1 + \varepsilon^* \cos\theta + A\varepsilon \cos^2\theta/2)$, where g_{ik} is the metric tensor, $A = \varepsilon^*(\exp(2\eta) - 1)q^2$, $q = \phi'\chi'$ is the safety factor. For the current calculation, we proceed from the drift kinetic equation supposing $\partial f/\partial\zeta = -inf$, $f = f_0 + f$, $W = v^2/2$, $\mu = v_\perp B_s/2B_0$, $\lambda = \mu/W$

Astrophysics and Space Science **256:** 421–424, 1998.

$$\frac{\partial f}{\partial \theta} - inqf - \frac{2\pi i\omega\sqrt{gB_0}}{v_\parallel \chi'} f - \frac{2\pi\sqrt{gB_0}}{v_\parallel \chi'} C = \frac{2\pi i\omega\sqrt{gB_0}}{v_\parallel \chi'} \frac{e}{T} v_\parallel E_\parallel f_0, \tag{1}$$

where f_0 is the Maxwellian distribution function. We only take the right-hand side of Equation (1) as a function $E_\parallel$ as we further consider the kinetic Alfvén waves.

$$v_\parallel = \pm\sqrt{2W}\sqrt{1 - \lambda\left(1 + \varepsilon^* \cos\theta + \frac{\varepsilon^* A}{2}\cos^2\theta\right)},$$

$$a_{1,2} = \frac{-1 \pm \sqrt{1 + \frac{2a}{\varepsilon^*}\left(\frac{1}{\lambda - 1}\right)}}{A}, \tag{2}$$

Here, we suppose the parameter $A \leq 1$. The untrapped particles are under the conditions $0 \leq \lambda \leq \lambda_1$ and $\theta \leq 2\pi$. The first region of the trapped particle (as in a circular cross-section tokamak) is $\lambda_1 \leq \lambda \leq \lambda_2$ and $\theta_1 \leq \theta \leq \theta\pi - \theta_1$, where $\theta_1 = \arccos(a_1)$ and

$$\lambda_1 = \lambda_{cr}(0) = \frac{1}{1 + \varepsilon^* + \frac{\varepsilon A}{2}}, \qquad \lambda_2 = \lambda_{cr}(\pi) = \frac{1}{1 - \varepsilon^* + \frac{\varepsilon^* A}{2}}. \tag{3}$$

In the new variables, the rf current can be written in the form

$$j_{\parallel\alpha} = 2\pi e_\alpha \frac{B_0}{B_s} \int_0^\infty d\lambda (f_\alpha^+ - f_\alpha^-), \tag{4}$$

where f^+ and f^- are the distribution functions of the particles, moving along and against $\mathbf{h}_0$. The interval $0 \leq \lambda \leq \lambda_2$ is supposed to be divided in two intervals, for untrapped and trapped particles, $0 \leq \lambda \leq \lambda_1$, $\lambda_1 \leq \lambda \leq \lambda_2$. Now, we can solve the ordinary differential Equation (1), allowing, for the dependence on the angle θ Equations (1), (2). For the untrapped particles with $0 \leq \lambda \leq \lambda_1$ and $0 =\leq \theta \leq 2\pi$ the integration constant is determined from the condition $f_{\alpha u}^\pm(0) = f_{\alpha u}^\pm(2\pi)$ and for the trapped particles with $\lambda_1 \leq \lambda \leq \lambda_2$ and $\theta_1 \leq \theta \leq 2\pi - \theta_1$, from $f_{\alpha t}^+(\theta_1 = f_{\alpha t}^-(\theta_1)$, $f_{\alpha t}^+(2\pi - \theta_1) = f_{\alpha u}^-(2\pi - \theta_1)$. Then, we find the Fourier component of the distribution function $f_{\alpha u}^\pm = \Sigma f_{\alpha u m_0}^\pm \exp(i m_0 \theta)$, $Q_\alpha = \Sigma Q_{\alpha m} \exp(i m\theta)$.

For the untrapped particles, the substitution can be used

$$\xi_u = -\arcsin\sqrt{\frac{(1 - a_2)(1 + \cos\theta)}{2(\cos\theta - a_2)}},$$

$$\theta = \arccos\left[-\frac{(1 - a_2) + 2a_2 sn^2(Ku)}{(1 - a_2) - 2cn^2(Ku)}\right], \tag{5}$$

where $u = F(\xi_u, k)/K(k) + 1$, $F(\xi_u, k)$ are the elliptic and complete elliptic integrals, respectively; $sn(Ku)$ is the Jacobian elliptic function, the sine-amplitude.

For the trapped particles "t", we use the substitutions in Equation (2)

$$\xi = \arcsin\sqrt{\frac{2(a_1 - \cos\theta)}{(1+a_1)(1-\cos\theta)}}, \qquad \theta = \arccos\left[\frac{2a_1 - (1+a_1)sn^2(Ku)}{2-(1+a_1)sn^2(Ku)}\right], \quad (6)$$

where $u = F(\xi_t, 1/k)/K(1/k)$.

Let us first make the further simplifications of the untrapped particle distribution function. The analysis shows that the resonant points of this function are described by the equation $\beta_m \pm \alpha_u/\pi = s$, where $s = 0, \pm 1, \pm 2, \ldots, \beta_m = m - nq$. Note that for a wide range of the values λ and A the function $\theta = \theta(u)$, Equation (6), is close to the straight line $\theta = \pi u$, i.e. $(\theta - \pi u)/\theta \ll 1$. That is why at the first approximation we can assume that $\theta = \pi u$, i.e. we take into account only the bounce-resonances with $s = 0$. After some calculations, we find the rf distribution function of the untrapped particles $f^{\pm}_{\alpha u m}$. As it can be seen from Equation (7), the bounce-resonances for the trapped particles t are described by the equation $\alpha_t = \pi s/2$ with $s = 0, \pm 1, \pm 2$ and so on. Taking this into account, after analyzing of the dependence $\theta = \vartheta(u)$ for the further calculation simplification we can suppose that for the trapped particles "t" $\theta = -(\pi - \theta_1 2)\cos(\pi u/2) + \pi$.

Using the model substitutions instead of the exact quantities Equation (6) gives the accuracy not more than several percents for practically all trapped particles except for the narrow region near the separatrix between untrapped and trapped particles where the accuracy go up to 20%. Equation (6) gives us the possibility to find the functions $f^{\pm}_{\alpha t m_0}$.

As it is well known (Elfimov et al., 1992) the noninductive current induced by KAW can be expressed through the absorbed power $P_w \approx \frac{\omega}{8\pi}\,\mathrm{Im}\,\varepsilon_\parallel |E_\parallel|^2$ in this case. Using Equations (2) and (6) we can find the expression for the oscillating currents

$$j_{\parallel\alpha m} = -\frac{4ie_\alpha^2 qRE_{\parallel m}}{T_\alpha}\int_0^\infty f_{0\alpha} W dW$$

$$\times\left(\int_0^{\lambda_1}\frac{\alpha_u d\lambda}{\beta_m^2 - \alpha_u^2/\pi^2} - \frac{\pi^2}{4\beta_m^2}\int_{\lambda_1}^{\lambda^2}\sum_{s=-\infty}^{\infty}\frac{s^2 J_s^2(x1m)}{\alpha_t + s\pi/2}\,d\lambda\right), \qquad (7)$$

where $k^2 = 2(a_1 - a_2)/(1+a_1)(1-a_2)$, $S = 2\omega qR/\sqrt{2\varepsilon\lambda W A(a_1 - a_2)}$, $\alpha_u = kK(k)S$.

As far as we are interested in the imaginary part of the permeability tensor $\varepsilon_{\parallel m}$, after the integration on the variable W we have

$$\mathrm{Im}\,\varepsilon_{\parallel m} = \frac{2\sqrt{\pi}\omega_{pe}^2 qR}{\omega\beta_m v_{Te}}[F_u(A) + F_t(A)], \qquad (8)$$

where

$$F_u(A) = \int_0^{\lambda 2} d\lambda f_u^4[-f_u^2(\lambda)], \quad f_u(\lambda) = \frac{\alpha_u \sqrt{W}}{\pi \beta_m v_{Te}}, \quad f_t(\lambda, s) = \frac{2\sqrt{2W}\alpha_t}{\pi s v_{Te}}, \tag{9}$$

$$F_t(A) = \frac{1}{\beta_m} \sum_{s=1}^{\infty} s \int_{\lambda 1}^{\lambda 2} d\lambda f_t^4(\lambda, s) J_s^2(x_{1m}) \exp[-f_t^2(\lambda, s)]. \tag{10}$$

Now, we can estimate the relative absorption of AW by untrapped and trapped particles using the next tokamak plasma parameters: $c_A/v_{Te} = 0.2$, where $c_A = B_0/\sqrt{4\pi n_0 M_i}$ is Alfvén velocity; $\beta_m = m - nq = 6$. The dependence of absorbed power by different kind of particles $F_u(A)$ and $F_t(A)$ Equation (8) on the tokamak inverse aspect ratio $l_1/2R$ in the center of the semiminor axis can be analyzed.

At $\varepsilon/2 = 0$, we have the cylinder expression for absorbed power. At $\varepsilon/2 \approx 0.065$, AW power, absorbed by untrapped and trapped particles, is approximately the same, $F_u(\varepsilon^*) \approx F_t(\varepsilon^*)$. The dependence of the ratio of absorbed powers $k(\varepsilon, A) = F_t(\varepsilon, A)/[F_u(\varepsilon, A) + F_t(\varepsilon, A)]$ as a function of the parameters ε and A for $c_A/v_{Te} = 0.2$, $\varepsilon^*/2 = 0.125$, $\beta_m = m - nq = 6$ is weak for the tokamak low aspect ratio.

References

Elfimov, A.G., Kirov, A.G. and Sidorov, V.P.: 1992, in: A.G. Litvak (ed.), *High-Frequency Plasma Heating*, American Institute of Physics Translation Series, New York, p. 239.

Nascimento, I.C. and Galvaõ, R.M.O. and TCA/Br Team: 1994, in: *Proc. of Xth Int. Conf. on Plasma Phys.*, Contr. Pap. Brasil, Foz do Icuacu, Vol. 1, p. 69.

FEEDBACK STABILIZATION OF BURN CONDITIONS IN A THERMONUCLEAR REACTOR USING ARTIFICIAL NEURAL NETWORKS: PRELIMINARY RESULTS

J.E. VITELA and J.J. MARTINELL
Instituto de Ciencias Nucleares, Universidad Nacional Autónoma de México, 04510 México, D.F.

Abstract. In this work we present some preliminary results concerning the use of Artificial Neural Networks (ANN) for the feedback stabilization of a thermonuclear reactor at nearly ignited burn conditions. An algorithm to train ANN for this purpose, involving the method of backpropagation through time and the conjugate gradients technique, is discussed. The training results show that ANN can be used successfully for the stabilization of burn conditions in thermonuclear reactors which are unstables to thermal excursions.

1. Introduction

It is expected that thermonuclear reactors for power production will operate under ignited or near-ignited burn conditions. Unfortunately, the corresponding nominal operating points for this type of tokamak reactors will be inherently unstable, i.e. any perturbation will drive the system away from the operating point, if they are designed to operate in the low temperature region where the fusion rate increases with temperature. Here, a nonlinear model is used to represent the time evolution of electron density, the relative density of alpha particles and the temperature of the plasma, where a CDA-ITER scaling for the energy confinement time is adopted. The control variables used to suppress any excursions from the nominal operating point include the concurrent modulation of the D-T refueling rate, the injection of a neutral He-4 beam and an auxiliary heating power. Since any feasible control action is constrained to lie within a set of allowable values, these variables are bounded between maximum and minimum levels. For this purpose a feedforward multilayer artificial neural network with sigmoidal activation function is trained using a Backpropagation Through Time algorithm (BPTT).

2. Physical Model and the ANN Training Algorithm

A zero-dimensional model is adopted in this work to represent a thermonuclear system, in which ions and electrons are assumed to share the same temperature at all times. The plasma is assumed to be solely composed by D-T in equal proportions, alpha particles and electrons. No high-Z impurities are considered. In addition,

Astrophysics and Space Science **256:** 425–429, 1998.

the model assumes instantaneous thermalization of the alpha particles produced by the fusion reactions, and both energy and particle transport are taken into account through an ITER scaling for the energy confinement time τ_E, an alpha particle confinement time τ_α and a D-T fuel confinement time τ_P. Bremsstrahlung is the only radiation loss mechanism included. The evolution of the system is determined by the following set of coupled nonlinear differential equations:

$$\frac{d}{dt}\,n_{\rm DT} = S - 2\left(\frac{n_{\rm DT}}{2}\right)^2 \langle\sigma v\rangle - \frac{n_{\rm DT}}{\tau_P}\,, \tag{1}$$

$$\frac{d}{dt}\,n_\alpha = S_\alpha + \left(\frac{n_{\rm DT}}{2}\right)^2 \langle\sigma v\rangle - \frac{n_\alpha}{\tau_\alpha}\,, \tag{2}$$

$$\frac{d}{dt}\left[\frac{3}{2}\,(n_e + n_{\rm DT} + n_\alpha)T\right] = P_{\rm aux} + Q_\alpha\left(\frac{n_{\rm DT}}{2}\right)^2 \langle\sigma v\rangle$$

$$+\ \eta j^2 - A_b Z_{\rm eff}^2 n_e^2 T^{1/2} - \frac{3}{2}\,(n_e + n_{\rm DT} + n_\alpha)\frac{T}{\tau_E}\,; \tag{3}$$

which correspond to the D-T and alpha particle densities and the energy density balance equations, respectively. In these equations, S_f represent the refueling rate, S_α the neutral He-4 injection rate, and $P_{\rm aux}$ the auxiliary heating power density; $Q_\alpha = 3.5$ Mev is the energy carried by the fusion alpha particles, $\langle\sigma v\rangle$ is the D-T reactivity, A_b and η are the coefficients corresponding to the bremsstrahlung radiation losses and the ohmic heating by the plasma current using the neoclassical parallel resistivity, respectively. Here we will assume $\tau_P = 3\tau_E$ and $\tau_\alpha = 7\tau_E$. With the energy confinement time given by the following scaling law,

$$\tau_E = 0.082\, I_p^{1.02}\, R^{1.6}\, B_0^{0.15}\, A_i^{0.5}\, \kappa_x^{-0.19}\, P_{\rm net}^{-0.47}\,; \tag{4}$$

where $P_{\rm net} = V_{\rm core}(P_{\rm aux} + P_\alpha + P_{oh} - P_{br})$. Using the quasineutrality condition $n_e = n_{\rm DT} + 2n_\alpha$, the above set of equations can be transformed into a set of equations for the electron density n_e, the relative fraction of alpha particles $f = n_\alpha/n_e$ and the plasma temperature T. The ignited steady state condition, i.e. where $P_{\rm aux} = 0$ and $S_\alpha = 0$, is obtained for $S_o = 4.16 \times 10^{18}$ m^{-3} sec^{-1} with $n_0 = 9.8 \times 10^{19}$ m^{-3}, $T_0 = 8.28$ Kev and $f_0 = 0.0624$, values that will be assumed to constitute the desired operating point for the ignited tokamak reactor of this work.

The purpose of the ANN is to provide the sequence of control actions that drives the system from an arbitrarily selected initial state $\mathbf{z}_1$ within a given region in phase space, toward a target state $\mathbf{z}_t$, without violating the constraints imposed on the state and/or the control variables. It is desired to stabilize the dynamical system, through the use of a feedforward ANN with sigmoidal activation functions to provide a control law of the form $\mathbf{u} = \mathbf{u(z)}$, such that any initial perturbation in the state variables is suppressed by the neural network, returning to a nominal

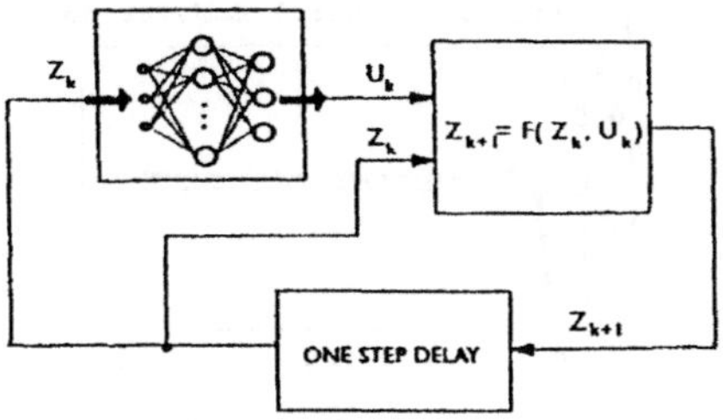

Figure 1.

operating point in the vecinity of the target state. The joint ANN-dynamical system configuration is illustrated in Figure 1, where the output of the neural network **u** is associated with the control variables S_f, S_α and P_{aux} through the following transformation equations: $S_f = S_0[1 + k_1(2u_1 - 1)]$, $S_\alpha = f_0 n_0 k_2(2u_2 - 1)^2$ and $S_{\mathrm{aux}} = 1.5 n_0 T_0 k_3(2u_3 - 1)^2$; with the following arbitrary values $k_1 = 1$, $k_2 = 0.1$ and $k_3 = 0.1$. The following steps summarize the iterative algorithm used to train a feedforward neural network as a dynamic system controller:

- Step 1: An admissible volume of the phase-space (which in our case is a region around the steady state ignited conditions) is divided in a number P of cells.
- Step 2: Randomly select the initial values of the set of weights $\{\mathbf{w}\}$ that specify the ANN (see Figure 1).
- Step 3: Generate a set of P trajectories by randomly selecting an initial state from each one of the cells in phase space, and then allowing the ANN-dynamical system to evolve in time until a pre-specified maximum number of time steps is reached, the state of the system has moved out of a certain region, or satisfied some other criteria.
- Step 4: Convergence is achieved when the entire set of P trajectories satisfies the convergence criterion $|z_i - z_{it}| \leq \varepsilon$, for all i, where ε is a pre-specified error range and the subindex i corresponds to the components of the state-space vector **z**. Otherwise proceed to the next step.
- Step 5: Using BPTT, calculate the gradient of the error, $\nabla\mathcal{E}$, produced by all the different trajectories and update the weights using the method of conjugate gradients.
- Step 6: Repeat steps 3–5 until the training of the ANN is successfully completed, i.e. when the entire set of P trajectories, each of which starts from a different cell, reaches the target $\mathbf{z}_t$ within the error range ε.

3. Training Results

After a preliminary training session, an ANN was obtained and here we show the resulting transient behavior given the following initial conditions: the electron

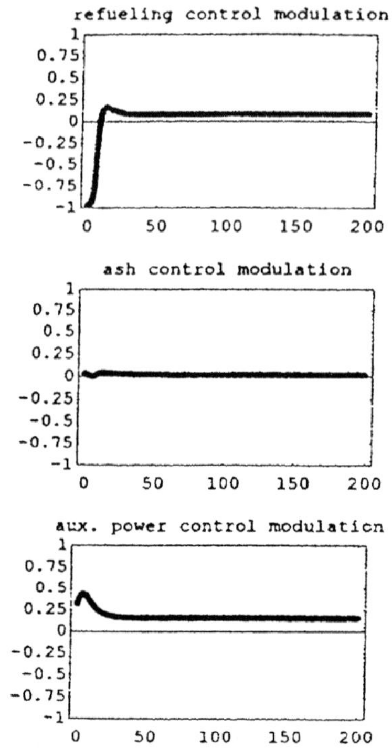

Figure 2.

density is $1.15 \times n_0$, the relative fraction of alpha particles is $0.9 \times f_0$, and the plasma temperature is $0.8 \times T_0$. Figure 2 shows the time behavior of the control variables $2u_1 - 1$, $(2u_2 - 1)^2$, and $(2u_3 - 1)^2$, provided by the neural network while in Figure 3 it is shown the time behavior of the normalized state variables, $z_1 = n_e/n_0$, $z_2 = f_\alpha/f_0$ and $z_3 = T/T_0$, resulting from the control actions. It is observed that the simultaneous modulation in the injection rate of the "cold" D-T fuel and the auxiliary heating power within the first 20 sec into the transient, is able to control the transient returning the state variables within a vecinity of the target state. In this particular example the injection of alpha particles is negligible. Eventually the entire system relaxes to a state with a D-T refueling rate $S = 1.09\, S_0$, slightly higher than the desired ideal value S_0; and a value for the auxiliary heating power density of $\sim$ 15% its maximum allowable.

4. Conclusions

In this work we have shown some preliminary results concerning the use of an ANN for the stabilization of a thermonuclear reactor at nearly ignited burn conditions. Although satisfactory learning has yet to be obtained, the promisory results presented here shows that ANNs can be used successfully to stabilize a nonlinear

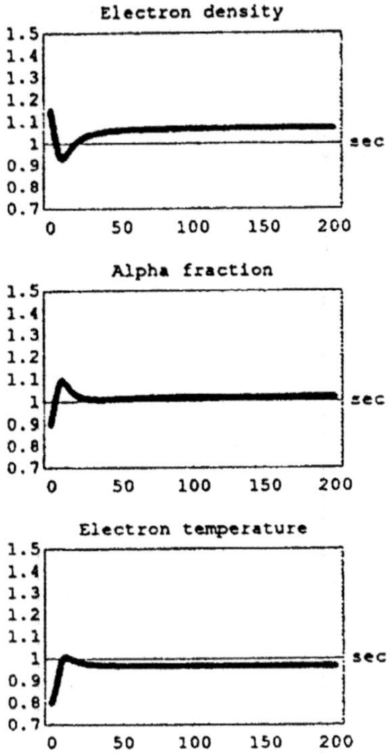

Figure 3.

thermonuclear system for temperature and density departures significatively far away from their nominal operating values.

Acknowledgments

The authors acknowledge partial financial support from DGAPA-IN101696 (UNAM) and CONACYT-3155P-A9607 projects.

References

Hui, W., Bamieh, B.A. and Miley, G.H.: 1994a, Robust burn control of a fusion reactor by modulation of the refueling rate, *Fusion Technol.* **25**, 318.

Hui, W., Bamieh, B.A. and Miley, G.H.: 1994b, Robust burn control of a fusion reactor with modulation of refueling rate, *Fusion Technol.* **26**, 1151.

Rumelhart, D.E., Hinton, G.E. and Williams, R.J.: 1986, Learning internal representations by error propagation, in: D.E. Rumelhart and J.C. McClelland (eds.), *Parallel Distributed Processing: Exploration in the Microstructure of Cognition*, Vol. I, MIT Press, pp. 319–362.

Werbos, P.: 1990, Backpropagation through time: What it does and how to do it, *Proc. IEEE* **78**(10), 1550.

EFFECT OF INDUCED TOROIDAL ROTATION ON POLOIDAL ROTATION AND ION HEAT CONDUCTIVITY OF TOKAMAK EDGE PLASMAS

V.S. TSYPIN[1], D.KH. MOROZOV[2], J.J.E. HERRERA[2], J.J. MARTINELL[2], M. TENDLER[3], I.F. POTAPENKO[1], A.S. DE ASSIS[4] and C.A. DE AZEVEDO[1]

[1]*Institute of Physics, State University of Rio de Janeiro, 20550-013, Maracanã, Rio de Janeiro, Brazil;* [2]*Instituto de Ciencias Nucleares, Universidad Nacional Autonoma de Mexico, a Postal 70-543, 04510 Mexico;* [3]*The Alfvén Laboratory, Fusion Plasma Physics, Royal Institute of Technology, 10044 Stockholm, Sweden;* [4]*Institute of Mathematics, Group of Applied Mathematics, Fluminense Federal University, Niterói, Rio de Janeiro, Brazil*

Abstract. The transport processes in edge (collisional) plasmas of tokamaks with smooth profiles of macroscopic plasma parameters and induced poloidal and toroidal plasma flows, are considered. The toroidal and poloidal velocities of particles, the radial electric field and the ion heat flux are derived. It is shown that forces, induced by radio frequency waves, plasma turbulence or neutral beam injection, can be used to control the poloidal and toroidal plasma velocities, as well as ion heat conductivity, in a wide range of these values.

The possibility of controlling plasma poloidal flows is very attractive, as the low-to-high confinement transition (L–H transition) in tokamaks depends on sheared plasma poloidal flows (Callen, 1992). External radio frequency waves, plasma turbulence or biased electrodes are usually considered to control the plasma poloidal flows. Neutral beam injection can also strongly affect poloidal plasma velocities and transport processes in tokamaks, as has been shown both for collisional plasmas, as well as for the weakly collisional tokamak plasma.

It has recently been shown that the kinetic (Graddock and Diamond, 1991; Tsypin et al., 1996a) or slow (Tsypin et al., 1996b) Alfvén waves (AW) can be used to create the localized sheared poloidal flows in the edge plasmas of tokamaks. Plasma rotation, induced by AW, was found out experimentally in the Phaedrus-T tokamak (Wukitch et al., 1996). Calculations, performed in (Tsypin et al., 1996a, 1996b) are in satisfactory coincidence with the experimental results (Wukitch et al., 1996).

We generalize these results for slightly rippled tokamaks with external forces, affecting the plasma, where the collisional parameter $b = \nu^{*2}$ is large enough for the ion-electron heat exchange and toroidal perturbations of the ion and electron temperatures to play an essential role in the ambipolarity condition. We analyze the possibility of controlling the plasma poloidal and toroidal velocities as well as

Astrophysics and Space Science **256:** 431–436, 1998.

ion heat fluxes by external forces. We suppose that results obtained in this paper can be used in the future experiments on the TCA/Br tokamak (Nacimento et al., 1994).

Let us consider a slightly rippled tokamak, in which the angle dependence of the magnetic field is given by $B = B_0(1 + \varepsilon\cos\theta + \delta\cos N\zeta)$, where θ and ζ are the poloidal and toroidal angles, ε is the inverse aspect ratio, δ is the ζ-modulation depth of the magnetic field, and N is the number of ripples. To find the poloidal and toroidal plasma velocities, we use the ambipolarity condition, in the well-known form that follows from the current continuity equation $\nabla\mathbf{j} = 0$,

$$\langle j^r \rangle = 0, \tag{1}$$

where

$$\langle\cdots\rangle = \int_0^{2\pi}\int_0^{2\pi}(\cdots)\sqrt{g}\,d\theta d\zeta \Big/ \int_0^{2\pi}\int_0^{2\pi}\sqrt{g}\,d\theta d\zeta.$$

From MHD equations we find

$$\left(M_i n\frac{d\mathbf{V}_i}{dt} + \nabla p + \nabla\cdot\hat{\pi} - \mathbf{F}^h\right)_{\theta,\zeta} = -\frac{\sqrt{g}}{c}\,j^r B^{\zeta,\theta}, \tag{2}$$

where $\mathbf{F}_h$ is an external force being exerted on the plasma. In the general case, this force can originate from any kind of rf wave, plasma turbulence or neutral beam injection. In particular, this force has been calculated for the case of slow Alfvén waves (Craddock and Diamond, 1991; Tsypin et al., 1996a). We take the plasma viscosity in the form (Tsypin et al., 1996b)

$$(\nabla\cdot\hat{\boldsymbol{\pi}})_\theta = \frac{3}{2}\pi_\parallel\frac{\partial}{\partial\theta}\ln B - \frac{1}{2}\frac{\partial\pi_\parallel}{\partial\theta}, \quad (\nabla\cdot\hat{\boldsymbol{\pi}})_\zeta = -\frac{3}{2}\pi_\parallel\frac{\partial}{\partial\zeta}\ln B + \frac{\partial\pi_\parallel}{\partial\zeta}. \tag{3}$$

Other quantities, depending on the angles θ and ζ can be found from the well-known MHD equations.

We take the parallel viscosity tensor $\pi_\parallel$ in the form derived in (Mikhailovskii and Tsypin, 1971):

$$\pi_\parallel = -\frac{2}{3}\frac{p_i}{\nu_i}(0.96\beta - 0.59\gamma), \tag{4}$$

where

$$\beta = \frac{3}{r}\left\{\varepsilon U_{\zeta i}\frac{\partial}{\partial\zeta}\ln B - U_{\theta i}\frac{\partial}{\partial\theta}\ln(\sqrt{g}\,n^{2/3}B) + U_{T_i}\frac{\partial}{\partial\theta}\ln\frac{B}{n}\right\}, \tag{5}$$

$$\gamma = -\frac{3}{r}\left\{0.34U_{\theta i}\frac{\partial}{\partial\theta}\ln n + U_{T_i}\left(1.36\frac{\partial}{\partial\theta}\ln B - 0.84\frac{\partial}{\partial\theta}\ln n\right)\right\}. \tag{6}$$

Then, the ion heat flux Γ_{T_i} can be derived by integrating the temperature evolution equation over the plasma volume

$$\Gamma_{T_i} = \langle q_i^r \rangle, \tag{7}$$

where after the integration over the angle θ, we obtain

$$\Gamma_{T_i} = -\frac{2nT_i\nu_i}{M_i\omega_{ci}^2}\frac{\partial T_i}{\partial r} - \frac{5cB_s}{4\pi\varepsilon e_i R}\int_0^{2\pi} d\theta \frac{p_i}{B^2}\frac{\partial T_i}{\partial\theta} . \tag{8}$$

As can be seen from Equations (5) and (6), it is necessary to calculate the ion poloidal velocities $U_{i\theta}$ and $U_{e\theta}$.

Now we can find the plasma rotation velocities using the ambipolarity conditions Equation (1), the expressions for perturbed temperatures, and the ion viscosity Equations (4–6):

$$U_{i\zeta} = 1.39\frac{\nu_i R^2}{M_i n_0 v_{T_i}^2 (N\delta)^2} F_\zeta^h, \tag{9}$$

$$\begin{aligned} U_{i\theta} &= G_{u1}(\alpha, b)U_{T_i} + G_{u2}(\alpha, b)\left(\frac{\varepsilon}{q}\frac{j_\parallel}{e_i n_0} + U_p\right) \\ &\quad + 1.39\frac{\nu_i R^2}{M_i n_0 v_{T_i}^2} G_{u3}(\alpha, b) F_\theta^h, \end{aligned} \tag{10}$$

where

$$G_{u1}(\alpha, b) = -\frac{f_2(\alpha, b)}{f_1(\alpha, b)}, \qquad G_{u2}(\alpha, b) = 1.35\frac{M_e}{M_i}\frac{\alpha b^2}{f_1(\alpha, b)},$$

$$G_{u3}(\alpha, b) = \frac{d(b)}{f_1(\alpha, b)},$$

$$f_1(\alpha, b) = d(b)\left(1 + \frac{2}{3}\alpha\right)(1 + 0.19\alpha) + 0.18\alpha^2 b\left(1 + 15.2b\frac{M_e}{M_i}\right),$$

$$f_2(\alpha, b) = d(b)\left(1 + \frac{2}{3}\alpha\right)(1.83 + 1.52\alpha) - 0.9\alpha b,$$

$$d(b) = 1 + 2.2b\sqrt{M_e/M_i} .$$

Functions $G_{u1}(\alpha, b)$, $G_{u2}(\alpha, b)$, $G_{u3}(\alpha, b)$ are plotted for the deuterium plasma case in Figure 1. We can conclude, analyzing this figure, that toroidal rotation does not allow the strong variations of poloidal rotation, when forces induced by radio frequency waves, plasma turbulence or neutral beam injection are present. The

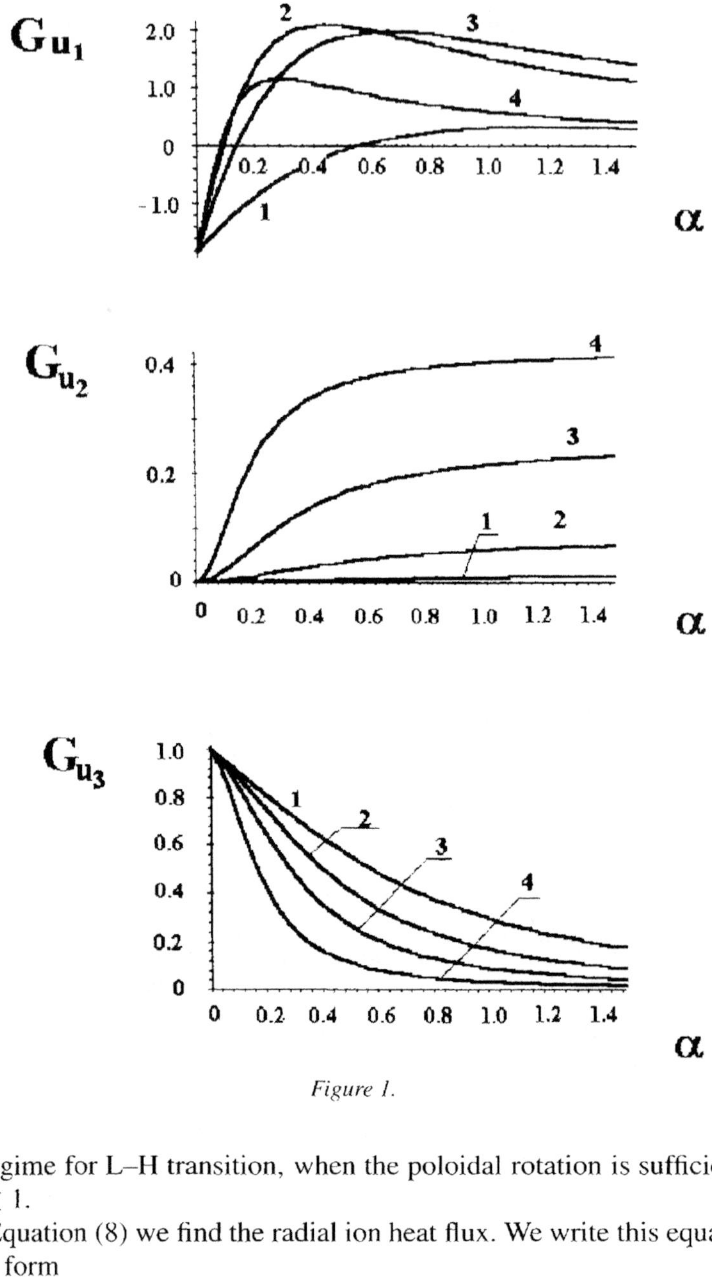

Figure 1.

relevant regime for L–H transition, when the poloidal rotation is sufficient, occurs when $\alpha \ll 1$.

From Equation (8) we find the radial ion heat flux. We write this equation in the Shafranov form

$$\Gamma_{T_i} = -\frac{2nT_i\nu_i}{M_i\omega_{ci}^2}\frac{\partial T_i}{\partial r}\left\{1 + 1.6q^2\left[G_{T_1}(\alpha, b) + G_{T_2}\frac{1}{U_{T_i}}(\alpha, b)\right.\right.$$
$$\left.\left.\times\left(\frac{\varepsilon}{q}\frac{j_\parallel}{e_i n_0} + U_p\right) - 1.39\frac{\nu_i R^2}{U_{T_i}M_i n_0 v_{T_i}^2}G_{T_3}(\alpha, b)F_\theta^h\right]\right\}, \qquad (11)$$

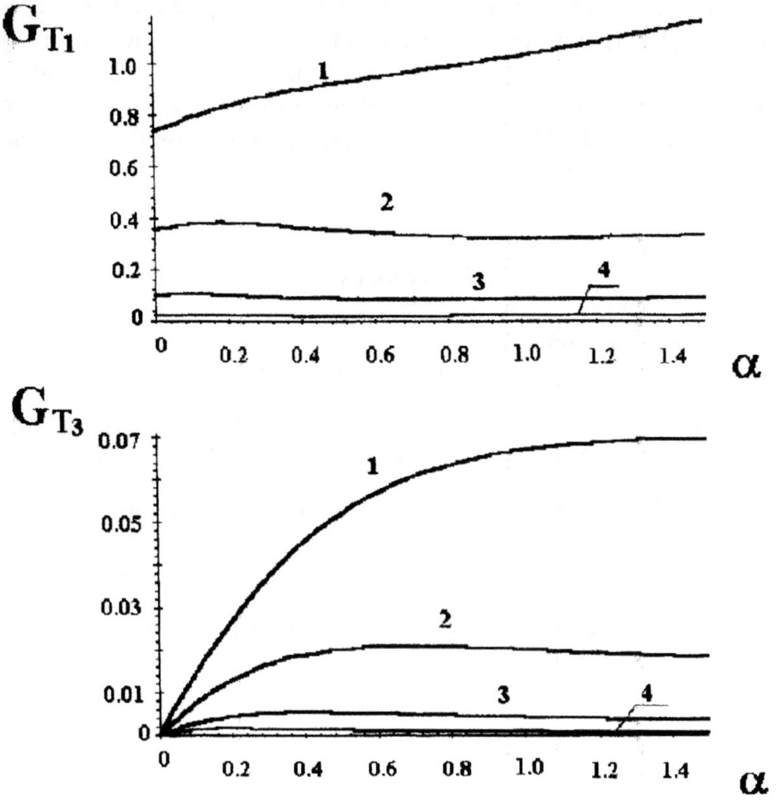

Figure 2.

where

$$G_{T_1}(\alpha, b) = \frac{(1+\alpha/2) f_3(\alpha, b)}{f_1(\alpha, b)},$$

$$G_{T_2}(\alpha, b) = 0.76 \frac{M_e}{M_i} \frac{\alpha b(1+\alpha/2)}{f_1(\alpha, b)} \left[\left(1 + \frac{2}{3}\alpha\right)(1+0.19\alpha) - \frac{0.18\alpha^2 b}{d(b)} \right],$$

$$G_{T_3}(\alpha, b) = \frac{\alpha(1+7.6bM_e/M_i)(1+\alpha/2)}{5 f_1(\alpha, b)},$$

$$f_3(\alpha, b) = \left(1 + \frac{2}{3}\alpha\right)\left[1 + 0.19\alpha + \frac{\alpha}{5}(1.83 + 1.52\alpha)\left(1 + 7.6b\frac{M_e}{M_i}\right)\right] + 1.4 \frac{M_e}{M_i} \frac{\alpha^2 b^2}{d(b)}.$$

If we suppose that b satisfies the inequality $1 < b < \sqrt{M_i/M_e}$, we obtain the previously known expression for the radial ion heat flux.

As proper analysis shows, the quantity $G_{T_2}(\alpha, b)$ is negligible under the condition $b \leq M_i/M_e$. The quantities $G_{T_1}(\alpha, b)$ and $G_{T_3}(\alpha, b)$ are plotted in Figure 2. As can be seen from Figure 2, the neoclassical contribution in the radial ion heat flux, as a function of α is approximately of the same order for $b \geq \sqrt{M_i/M_e}$, dropping with increasing b. The influence of induced forces on the radial heat flux is usually small for $b \geq \sqrt{M_i/M_e}$.

References

Callen, J.D.: 1992, *Phys. Fluids* **B4**, 2142.

Graddock, G.G. and Diamond, P.H.: 1991, *Phys. Rev. Lett.* **67**, 1537.

Mikhailovskii, A.B. and Tsypin, V.S.: 1971, *Plasma Physics* **13**, 785.

Nacimento, I.C., Galvaõ, R.M.O., Tuszel, A.G., Degasperi, F.T., Ruchko, L.F. and TCA/Br Team: 1994, in: *Proc. of Xth Int. Conf. on Plasma Physics*, Contr. Pap. Brasil, Foz do Iguacu, Vol. 1, p. 69.

Tsypin, V.S., Elfimov, A.G., de Azevedo, C.A. and de Assis, A.S.: 1994a, *Physics of Plasmas* **3**, 4606.

Tsypin, V.S., Elfimov, A.G., dc Azevedo, C.A. and de Assis, A.S.: 1994b, *Physics of Plasmas* **3**, 2818.

Wukitch, S., Litwin, C., Harper, M., Parker, R. and Hershkovitz, N.: 1996, *Phys. Rev. Lett.* **77**, 294.

LOW-ANGLE THOMSON SCATTERING EXPERIMENT FOR DETERMINATION OF PLASMA ELECTRON DENSITY AND TEMPERATURE

D.O. CAMPOS, L.A. BERNI, M. MACHIDA and S.A. MOSHKALYOV
Instituto de Física "Gleb Wataghin" (DEQ-Plasma), Universidade Estadual de Campinas, UNICAMP, 13083-970, Campinas, SP, Brasil

Abstract. In this work we show results of measurements of the electron density n_e and temperature T_e by using small-angle Thomson scattering in a theta-pinch plasma. The method of measurement and analysis developed here is based on integration of only two ranges of the scattering profile. This method simplifies the experimental set up, because it needs only two detectors for simultaneous determination of n_e and T_e, instead of the commonly 5–7 detectors used in 90° experiments.

The results obtained for low-angle scattering are compared with the measurements at 90° using single pass and multipass under the same plasma conditions.

1. Introduction

Thomson scattering has been largely used in pulsed plasma devices, especially in fusion devices, for determination of the electron densities (n_e) and temperatures (T_e), providing measurements with local and temporal resolution. The most common scattering geometry is based on 90° scattering, i.e., the scattered light is observed at an angle θ perpendicular to the incident laser beam direction. T_e and n_e are obtained, generally, after fitting of the theoretical spectrum with the experimental curve (Sheffield, 1975), measured by one (single channel) or more detectors (multichannel) positioned on the exit of a spectrometer.

As discussed in a preliminary work (Campos et al., 1995), the form of the electron component of the scattered spectrum depends on the parameters n_e, T_e and on the observation angle θ.

The influence of T_i, the ion temperature, can play a significant role only in a narrow region near the maximum of the spectrum (ion component), which is avoided in the measurements because the stray light coming from the laser also appears in this region. At low angles, the whole spectrum becomes narrow and more intense in its central region. Figure 1 shows the scattered spectra for a ruby laser light ($\lambda = 694.3$ nm) by a plasma with fixed density and temperature, at different observation angles.

In this work we show results of measurements of n_e and T_e by using small-angle Thomson scattering in a theta-pinch plasma. The method of measurement and analysis developed here is based on integration of only two ranges of the scattering profile. This method simplifies the experimental set up, because it needs

Astrophysics and Space Science **256:** 437–441, 1998.

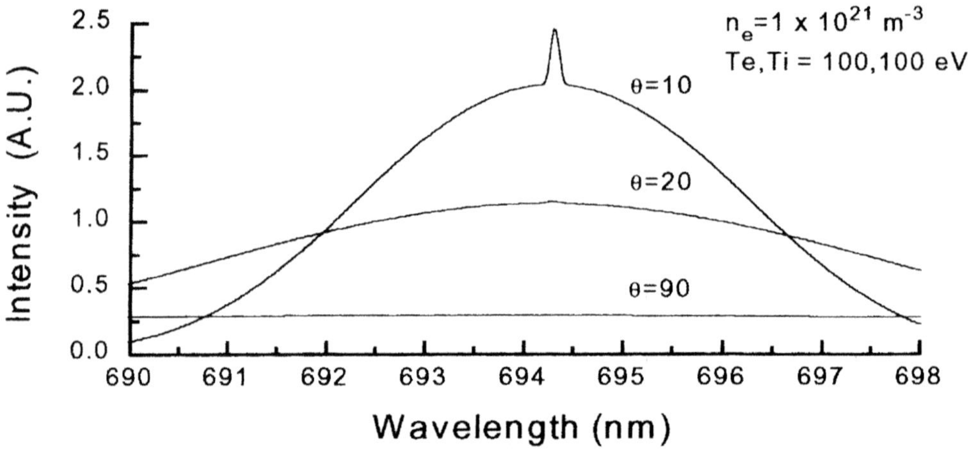

Figure 1. Thomson scattering spectra for different observation angles (θ) under the same plasma conditions.

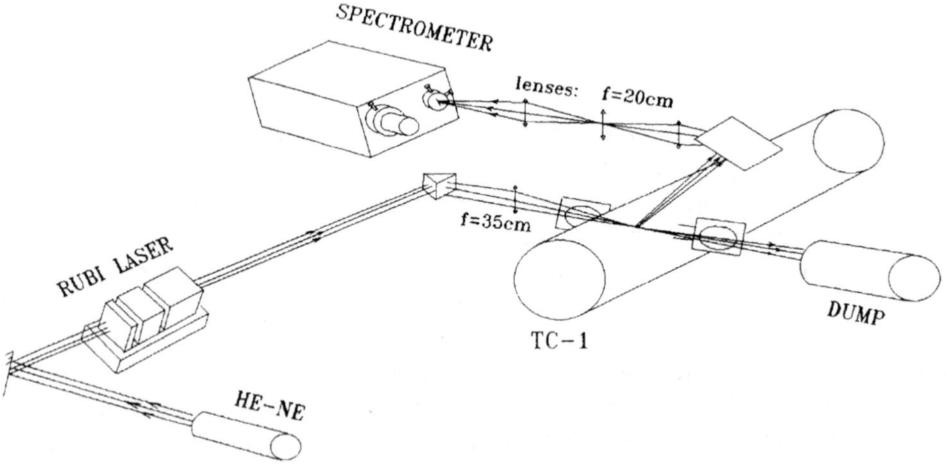

Figure 2. Low-angle Thomson scattering set up.

only two detectors for simultaneous determination of n_e and T_e, instead of the 5–7 detectors commonly used in 90° experiments. The results obtained for low-angle scattering are compared with the measurements using 90° single pass and multipass (Berni et al., 1996) under the same plasma conditions.

2. Experiment

Figure 2 shows the experimental set up for the low-angle experiment. The plasma is generated in the compact torus TC-1 UNICAMP. This system incorporates a ruby laser in the Q-switched mode, with an energy pulse of 3 J and a duration of

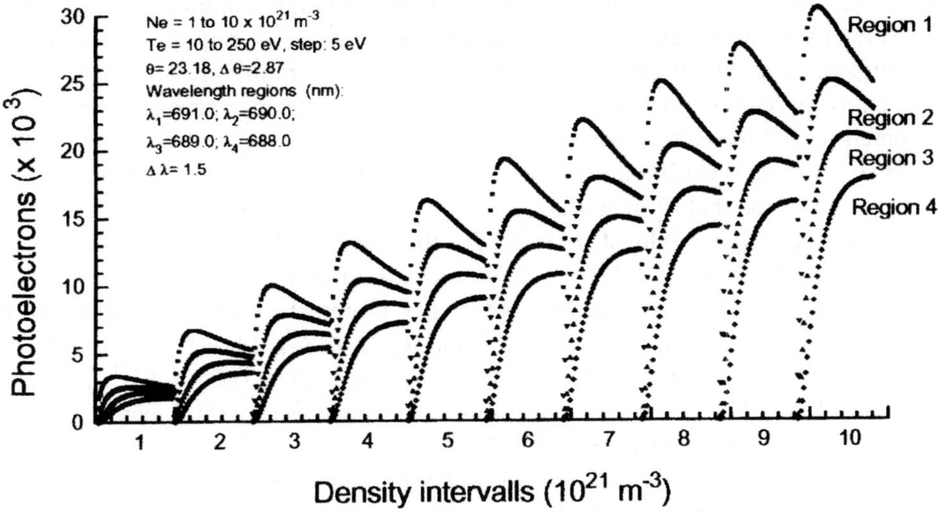

Figure 3. Number of photoelectrons for different wavelength regions. The electron density changes in intervals of 1×10^{21} m^{-3} along the horizontal axis. For each density interval the temperatures changes in steps of 5 eV.

40 ns, and beam divergence of approximately 2 mrad. The 90° scattered light is collected using a special optical system consisting of a spectrometer (Spex 0.75 m, 1.1 nm/mm) and a photomultiplier (RCA 7265). The laser power was monitored using a photodiode. The He-Ne laser shown in Figure 2 is used for alignment purposes and the dump collects the laser light coming from the plasma region to prevent elevation on the stray light level.

The method used here for determination of the electron density and temperature involves measurements of the spectrum in only two different wavelength regions ($\lambda_i \pm \Delta\lambda_i$, where λ_i is the central wavelength in the selected region), which are selected by means of a numerical code. This selection takes into account the expected density and temperature ranges, and the observation angle. Figure 3 shows as an example the number of scattering photoelectrons for different wavelength regions. Observe that the electron density is represented in intervals of 1×10^{21} m^{-3}, it does not change continuously along the horizontal axis. For each density interval the temperature is changed in steps of 5 eV, originating the curves. The calculation of the number of photoelectrons, takes into account the integral of the intensity along $\lambda_i \pm \Delta\lambda_i$ and also along the observation solid angle Ω, which implies in changes on θ and on the azimuthal angle φ inside Ω. The integration inside Ω is especially important at low scattering angles, when the θ variation can lead to significant differences on the scattering spectrum. The intensity calibration was carried out using Rayleigh scattering in neutral nitrogen gas (Berni et al., 1996).

The determination of n_e and T_e is done by means of a numerical code which compares the measured intensities of two selected wavelength regions with the respective prediction (like that of Figure 1). The standard deviations of a sequence of 5–6 intensity measurements lead to a range of possible n_e and T_e values in this comparison, from which the averaged density and temperature are obtained.

The results obtained for the low-angle Thomson scattering measurements were compared to 90° measurements in single pass and multipass configuration (Campos et al., 1996a, 1996b) under the same plasma conditions, i.e., at maximum compression of the theta-pinch discharge.

For the scattering angle of $\theta = 22.2°$, the solid angle $\Omega = 0.01$ sr (which implies in a variation of $\pm\Delta\theta = 2.86°$ around θ) and the wavelength regions chosen to be (690.0 ± 1.1) nm and (689.0 ± 1.1) nm, the following values were obtained: $n_e = (1.2 \pm 0.4) \times 10^{21}$ m^{-3} and $T_e = (48 \pm 11)$ eV. The results of the 90° measurements are: (a) for single pass $n_e = (1.7 \pm 0.2) \times 10^{21}$ m^{-3} and $T_e = (50 \pm 8)$ eV; and (b) for the multipass experiment (with 5 passes) $n_e = (1.5 \pm 0.2) \times 10^{21}$ m^3 and $T_e = (47 \pm 11)$ eV.

As expected, the results obtained from the low-angle experiment agree, within the respective errors, with the results of the 90° experiment in single and multipass. Five points along the spectrum were measured and the conventional fitting procedure described in Section 1 was used for the determination of n_e and T_e for the 90° Thomson scattering. The observed errors have as main source the low reproducibility of the plasma discharge of the theta-pinch device, added to the bremstrahlung emission and electrical noises.

3. Conclusion

A low-angle Thomson scattering experiment has been installed and tested successfully for the determination of electron density and temperature on a theta-pinch device.

The results obtained from this set up were compared to the more conventional 90° Thomson scattering in single pass and multipass configurations, under the same plasma conditions, showing a good agreement.

In this way, the reliability of the method proposed for low-angle scattering was demonstrated. One advantage in using this method lies in the simplification of the data acquisition and analysis system, which uses only two regions of observation, making only two detectors necessary for simultaneous measurements in both wavelength regions, instead of five or more detectors necessary for the 90° case. Furthermore, the detector calibration procedure also becomes simplified and the implantation of the system experiences a cost reduction.

Acknowledgements

This work was supported by the Fundação de Amparo à Pesquisa do Estado de São Paulo (FAPESP), Financiadora de Estudos e Projetos (FINEP), and Conselho Nacional de Desenvolvimento Científico e Tecnológico (RHAE/CNPq).

References

Berni, L.A., Campos, D.O., Machida, M., Moshkalyov, S.A. and Lebedev, S.V.: 1996, Molecular Rayleigh scattering as calibration method for Thomson scattering experiments, *Brazilian Journal of Physics* **26**(4), 755.

Campos, D.O., Machida, M., Berni, L.A., Moshkalyov, S.A. and Lebedev, S.V.: 1995, The feasibility study of a low-angle Thomson scattering for determination of plasma electron density and temperature, in: *3o. Encontro Brasileiro de Física dos Plasmas – Fifth Brazilian Plasma Astrophysics Workshop*, Águas de Lindóida, SP, 4–6 December 1995.

Campos, D.O., Machida, M., Berni, L.A., Kantor, M., Moshkalyov, S.A. and Lebedev, S.V.: 1996a, Experimental study of conventional and multipass Thomson scattering diagnostics, *Japanese Journal of Applied Physics* **35**(12A), 6273.

Campos, D.O., Machida, M., Berni, L.A., Kantor, M., Moshkalyov, S.A. and Lebedev, S.V.: 1996b, The application of a multipass system for Thomson scattering diagnostics in magnetically confined plasmas, *Brazilian Journal of Physics* **26**(4), 742.

Sheffield, J.: 1975, *Plasma Scattering of Electromagnetic Radiation*, Academic Press, New York.

GAS EMBEDDED COMPRESSIONAL Z-PINCH IN H_2 AND D_2

L. SOTO
Comisión Chilena de Energía Nuclear, Casilla 188-D, Santiago, Chile

H. CHUAQUI, M. FAVRE, R. SAAVEDRA and E. WYNDHAM
Facultad de Física, Pontificia Universidad Católica de Chile, Casilla 306, Santiago 22, Chile

M. SKOWRONEK and P. ROMEAS
Université Pierre et Marie Curie, Paris, France

R. ALIAGA-ROSSEL and I. MITCHELL
The Blackett Laboratory, Imperial College, London, U.K.

Abstract. Recent experiments in a gas embedded compressional Z-pinch are presented. The experiments have been carried out in H_2 and D_2, using a pulse power generator capable of delivering a $dI/dt > 10^{12}$ A/s. The pinch is initiated by a focused laser pulse, which is coaxial with a cylindrical DC microdischarge. This configuration results in double column pinch at early times, which at current rise evolves into a gas embedded compressional Z-pinch. Diagnostics used are Rogowskii coil, single frame holographic interferometry and holographic shadowgraphy, visible streak camera images from which, current, density, line density, pinch radius and plasma motion are obtained. The pinch is characterised by a maximum on axis density which is much higher than the expected value from filling pressure, with a Bennett temperature of 75 eV at 180 kA.

1. Introduction

The gas embedded Z-pinch configuration was considered for some time as the ideal solution to the inherent problems associated with a conventional compressional Z-pinch, namely the limited maximum density achievable and discharge stability. Early experimental work on these discharges showed that the electron density was lower than that to be expected from filling gas pressure. Furthermore, after a few tens of nanoseconds the discharge becomes unstable under $m = 1$ mode, leading to a helical plasma structure. Theoretical studies have shown that for a gas embedded Z-pinch and $m = 1$ mode is to be expected, whereas an $m = 0$ mode ought to be present in a conventional compressional Z-pinch. This is indeed fully confirmed by the experimental evidence obtained so far. A number of experiments have been carried out in gas embedded Z-pinch using our small pulse power generator (GEPOPU). Experimental results obtained confirm that, after an initial expansion, $m = 1$ instabilities leading to a helix develop, eventually disrupting the plasma column. A gas embedded Z-pinch configuration which addresses the density and

Astrophysics and Space Science **256:** 443–446, 1998.

stability issues has been reported (Soto et al., 1994a). In this configuration a central conduction channel, together with a coaxial current path is initially established. Compression up to twice the expected value from filling pressure is observed, as well as, enhanced stability. It has been conjectured that the stability observed can be explained by resistive effects and finite Larmor radius effects (Chuaqui et al., 1993; Soto et al., 1994b). In the present work studies in the gas embedded compressional Z-pinch covering a wider pressure range than previously reported in H_2 are presented, as well as, preliminary results in D_2. The comparison between discharges in H_2 and D_2 provides some insight into the relevance of the dynamics in the plasma evolution.

2. Experimental Apparatus

The experiments were carried out on GEPOPU, a generator capable of delivering currents up to 200 kA to a 1.5 Ω impedance load for 120 ns. The value of dI/dt of the current ramp was approximately 2×10^{12} A/s. A DC microdischarge is established between two stainless steel conical hollow electrodes with 2 mm diameter, separated by 10 mm. A few nanoseconds before the application of the main voltage from the driver, a pulsed Nd-YAG laser (20 ns, 200 mJ at 1.06 μm) is focused through the anode onto the cathode. As the main voltage is applied, the laser generated plasma acts as an electron source to provide preionization on a central plasma column. Initially, this conduction channel is highly resistive. The combined preionization scheme produces two parallel concentric conductive paths.

An Nd-YAG frequency doubled 6 ns laser pulse was used for optical diagnostics to obtain simultaneous single shot image-plane holographic interferometry and holographic shadowgraphy. Visible streak camera provides radial and axial plasma motion. Single frame image converter camera are used to corroborate results obtained with the streak camera. From these diagnostics the electron density profile, $n_e(r)$, the electron line density, N, and the external Z-pinch radius, a, are obtained with good temporal and spatial resolution. The total current, $I(t)$, and the external voltage, $V(t)$, were also measured with a Rogowskii coil and a capacitive divider. Filling pressures of 1/3 and 1/6 atm were used in H_2 and D_2, although not all diagnostics were employed for all conditions.

3. Results and Discussion

Results obtained from single frame interferometry indicate that the axial electron density becomes larger than 4×10^{25} m^{-3} at 75 ns after the current start. From the corresponding shadowgram it can be seen that the refractivity is high at the pinch centre, which is in agreement with the fact that no fringes are visible at the centre of the interferogram. At later times, 180 ns, near the peak current of 130 kA, the

Figure 1. Radial streak photograph in which the different expansion rates are observed, having separate coaxial initial current channels coalescing at later times.

electron density at centre is 3.8×10^{25} m^{-3}. The line density under these conditions is of the order of 5×10^{19} m^{-1}, which significantly higher than reported previously. This difference is due to a higher electron density at the pinch centre and a much better defined plasma edge density profile. The Bennett temperature can thus be estimated at only 40 eV, which is consistent with the negative results obtained with soft X-ray diagnostics used to measure temperature. The Bennett temperature for the higher current case studied, 180 kA, is estimated at 75 eV.

From axial streak photographs it is observed that the initial laser spark at the centre has independent dynamics, which is consistent with a metallic laser spark in expansion. The spark remains near the electrode. These observations agree with previous interferometric results (Soto et al., 1994a) which show that the maximum spark size is 2 mm at 50 ns. From radial streak photographs, as shown in Figure 1, an initial 10 ns fast expansion phase is observed, followed by about 50 ns expansion of the central channel of 2×10^4 m/s. The expansion of the internal wall of the annular plasma is of the same order, whereas the external wall has a slower expanding velocity of 6×10^3 m/s. The reduced expanding rate of the external wall compared to the central channel is presumably due to the magnetic field generated by the central current channel. The fact that the laser initiation does make a difference is an indirect evidence that there is an important fraction of the current flowing near the axis. At the end of this phase the two initial plasmas coalesce into one with an apparent internal structure, a brighter central region which is maintained during the following 110 ns. The corresponding expansion rates for D_2 at 1/3 atm are $\sim 1.2 \times 10^4$ m/s and $\sim 4 \times 10^3$ m/s, lower than the values obtained for H_2. The expansion rate is reduced slightly less than expected from the mass difference. This can be understood by comparing the currents for both cases, which are smaller for the D_2 case, as it would be expected in view of the fact that the resistivity would be higher for the latter case.

Further work is required for discharges in D_2, a complete set of electron density and line density are required to do a full analysis of the influence of a higher mass gas. Future work being considered includes experiments with smaller initial radius

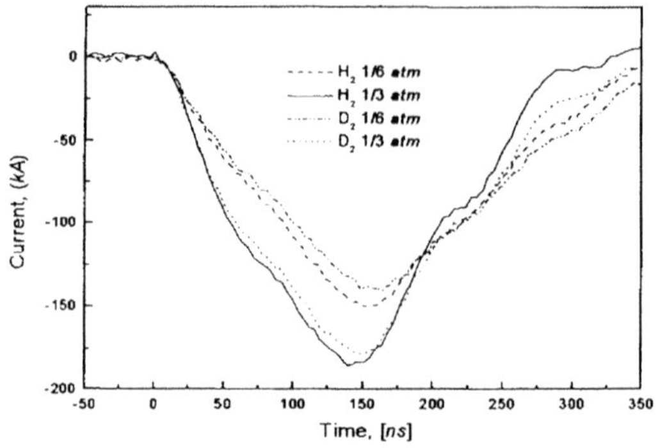

Figure 2. Current traces for H_2 and D_2 at 1/3 and 1/6 atm. The driving voltage is the same for all conditions.

for the annulus, lower background pressure and up to 1.2 MA peak current in the LLAMPÜDKEÑ generator (Chuaqui et al., 1998).

Acknowledgements

This work has been partially funded by FONDECYT, grants 1950048, 1960555 and 2960008, ECOS/CONICYT programme and Imperial College (MAGPIE project). R. Saavedra holds a CONICYT scholarship.

References

Chuaqui, H., Soto, L., Favre, M., Wyndham, E.S. and Skowronek, M.: 1993, in: *Proc. III International Conference of Dense Z-Pinches*, London, p. 27.

Chuaqui, H., Wyndham, E., Friedli, C. and Favre, M.: 1998, *Laser Part. Beams* **15**, 241.

Soto, L., Chuaqui, H., Favre, M. and Wyndham, E.: 1994a, *Phys. Rev. Lett.* **732**, 2891.

Soto, L., Chuaqui, H., Favre, M. and Wyndham, E.: 1994b, in: *Proc. 1994 Int. Conf. on Plasma Physics*, Vol. 1, Contributed Papers, 216, Iguazu Falls, Brazil.

INVESTIGATION OF VARIATION OF FOCUS ACTION WITH THE LENGTH OF THE INNER ELECTRODE FOR A PLASMA FOCUS

A.V. GHOLAP
Department of Applied Physics, National University of Science and Technology, Bulawayo, Zimbabwe

M. MATHUTHU and T.G. ZENGENI
Department of Physics, University of Zimbabwe, Harare, Zimbabwe

Abstract. A 2.3 kJ plasma focus was operated at different pressures with argon, nitrogen and air fillings. It was observed that good focusing action can be obtained by reducing the length of the inner electrode from 150 to 40 mm, when inner electrode has negative polarity. Under the present experimental conditions the optimum length was found to be 40 mm.

1. Introduction

The focusing action in the plasma focus device is affected by the material of internal electrode (IE) (Rout et al., 1993), the insulator sleeve length and material (Zakaullah et al., 1989; Beg et al., 1992) and the polarity of internal electrode. The strong focusing action is important from the point of view of increasing neutron yield (Decker et al., 1982). Hence it is required to optimize the system parameters taking into consideration that the focus dynamics has two separate phases viz., the rundown phase and the radial collapse phase.

Thus, optimization can be achieved by varying the duration of the rundown phase by varying electrode length of internal electrode which has negative polarity and pressure of the filling gas.

2. Experimental Setup

The schematic of the plasma focus device is shown in Figure 1a. The inner and outer electrode system was mounted on a brass plate. The inner electrode was made of brass rod of diameter 20 mm while the outer electrode system consists of eight brass rod of diameter 8 mm fixed around the inner electrode in the form of a squirrel cage. The distance between the inner electrode and outer electrode system was 60 mm. The length of the electrodes was varied from 150 to 40 mm to investigate the focusing action. The chamber was evacuated by a two stage rotary pump reaching an ultimate base pressure of 10^{-3} Torr. The pressure of the filling gas was measured by mechanical diaphragm pressure gauge.

Astrophysics and Space Science **256:** 447–450, 1998.

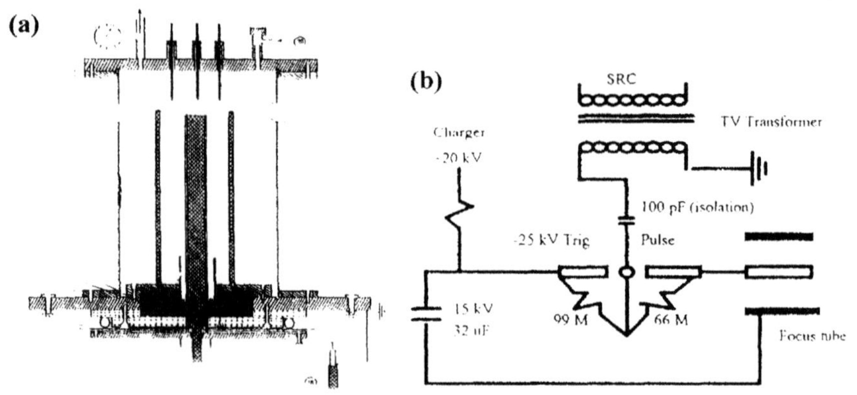

Figure 1. (a) Schematic of the plasma focus device. (b) Experimental layout for the plasma focus.

Energy is stored in a 32 μF and 27 nH Maxwell Capacitor by charging it to 12 kV. The capacitor transfers the energy to the plasma chamber when switched through a swinging cascade spark gap which is in turn switched by a 25 kV negative pulse with a rise time of 1 μs and jitter time of about 50 ns. The spark gap is mounted directly above the capacitor using parallel plate configuration (Lee et al., 1988). Experimental layout is shown in Figure 1b.

3. Current and Voltage Signals

To display the current signals a self integrated Rogowski coil of 400 turns of enamelled copper wire wound around the anode of the focus was connected to 100 MHz digital scope. To avoid reflections in the system the coil was terminated with 50 Ω resistance with 1:10 attenuator. The calibration constant of the coil was 11.6 kA/V. A resistive voltage probe with a rise of time of 15 ns was strapped across the anode and the cathode plate. Voltage signal was observed on 100 MHz digital scope. Three different sets of electrodes having lengths of 150, 80 and 40 mm were investigated using argon, air and nitrogen gases.

4. Results and Discussions

The typical Rogowski coil and high voltage signals are shown in Figure 2a. The signals were obtained for a particular gas say argon by varying pressure. The full width half maximum in ns and the amplitude of high voltage spike was scanned from the high pressure to low pressure for different electrode lengths. A good focus was characterized by the sharper HV spike, the narrower FWHM and the deeper current signal dip. Figures 2b, 2c and 2d shows the plots of HV spikes against gas

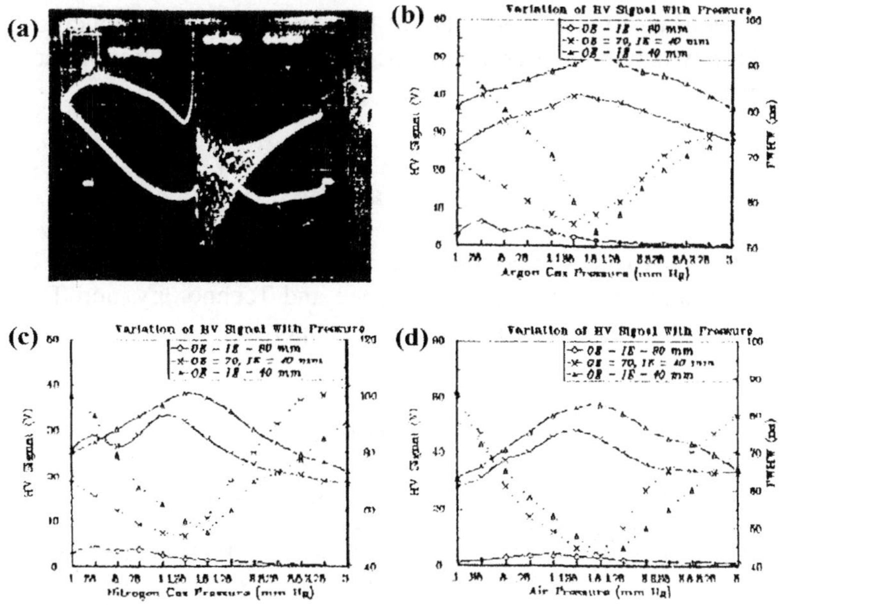

Figure 2. (a) The current and voltage signals Upper Trace voltage: 4 kV/div. Lower trace current: 75 kA/div. Time base 1 μs/div. (b) Pressure vs. HV signal (ARGON). (c) Pressure vs. HV signal (N_2). (d) Pressure vs. HV signal (AIR).

pressures together with FWHM of the corresponding signal for argon, nitrogen and air.

Argon showed good focusing action in the range of pressure from 0.1 to 2.5 Torr, nitrogen in the range 0.1 to 2.0 Torr and air in the range 0.25 to 2.0 Torr. At these pressures no focusing action was registered at 150 mm length of the electrode. However, the focusing was stronger when electrode length was 40 mm than when it was 80 mm.

The FWHM show that good focus action is shifted towards higher pressure range as the electrode length is reduced. For argon and air the lower pressure range for focusing action is decreased from 0.3 mm of Hg (Smith and Gholap, 1988) with positive polarity to 0.1 mm of Hg with negative polarity while for air it is decreased from 0.5 mm of Hg to 0.25 mm of Hg while the upper pressure range is increased from 2 to 2.1 mm of Hg for argon and from 1.1 to 2.0 mm of Hg for air, respectively.

The reduction in the length of electrode for good focusing action in present set up may be explained on the basis of the reduction in the time required to travel the axial distance in the rundown phase due to shorter length of electrode. Moreover, at 150 mm length of electrode no focusing was observed. This may be due to (i) the much longer time required for the formation of the sheath and lift off and (ii) the longer axial rundown phase. Moreover, in this configuration more impurities may be ablated from the electrode and insulating material. Thus, the longer time required to form focus could be compensated by reducing the axial rundown phase which in the present case achieved by the reduction in the length of the centre electrode.

Acknowledgements

The authors wish to express their gratitude to the Research Boards of the University of Zimbabwe, National University of Science and Technology, and Third World Academy of Sciences for providing financial support.

References

Beg, F.N., Zakaullah, M. and Murtaza, G.: 1992, *Physics Scripta* **46**, 152–154.

Decker, G., Kies, W. and Pross, G.: 1982, *Phys. Lett.* **89**, 393–396.

Lee, S., Tou, T.Y., Moo, S.P., Eissa, M.A., Gholap, A.V., Kwek, K.H., Mulyodrono, S., Smith, A.J., Surayadi, M., Usada, W. and Zakaullah, M.: 1988, *Am. J. Phys.* **56**(1), 62–68.

Rout, R.K., Shyam, A., Chitra, V. and Bansal, A.: 1993, *Indian J. Pure & Appl. Phys.* **29**, 678–681.

Smith, A.J. and Gholap, A.V.: 1988, in: Lee and Sakanaka (eds.), *Proc. of Symp. on Small Scale Lab. Plasma Expts*, , World Scientific, pp. 105–116.

Zakaullah, M., Baig, T.J., Beg, S. and Murtaza, G.: 1989, *Phys. Lett. A* **137**(12), 39–43.

NUMERICAL STUDY OF THE FIRST COMPRESSION OF A TWO-DIMENSIONAL PINCH

L. BILBAO and D. GRONDONA
Instituto de Física del Plasma, CONICET – Universidad de Buenos Aires, Ciudad Universitaria – Pab. I, 1428 Buenos Aires, Argentina

Abstract. For studying the first compression phase of a Plasma Focus machine, we developed a 2-fluid, time-dependent, 2-D code. The code includes two species (ions and electrons), two temperatures, radiation, thermal conductivity, electrical conductivity, magnetic diffusion, and viscosity and is coupled to an external circuit. In order to get reasonable initial conditions for the pinch phase, a simple 1-D model is used for the run-down phase. Under this somewhat ideal Plasma Focus discharge condition, relatively simple scaling laws are obtained. In particular, for a given machine, the pressure (p_o) of maximum thermonuclear production scales with the charging voltage (V_o) as $p_o \propto V_o^2$; the thermonuclear production (Y_n) at these pressures increases with the current (I_o) as $Y_n \propto I_o^4$ and the mean energy per particle is almost independent of all parameters (note that this means that the converging speed should be almost the same regardless of the machine at the optimum conditions). These laws show a reasonable agreement with experimental data and can be explained with a dimensional analysis.

1. Introduction

In a Plasma Focus discharge after the current sheath reaches the top of the electrodes, its motion becomes radially inward, starting the pinch phase. A shock wave, that travels ahead the magnetic piston, moves towards the axis and later on reflects off the axis, starting the first compression phase. The magnetic piston converges until it is reached by the reflected shock. This is the time of maximum compression. The following expansion of the magnetic piston continues until the development of an $m = 0$ instability produces several compression (the second compression). For studying the first compression phase, we developed a 2-fluid, time-dependent, 2-D code. The code includes two species (ions and electrons), two temperatures, radiation, thermal conductivity, electrical conductivity, magnetic diffusion, viscosity and is coupled to an external circuit (Bernal et al., 1998). The code is used in order to get the behavior of the pinch as a function of the parameters. In the present version is not intended for simulated an actual Plasma Focus device, because neither the breakdown nor the run-down phases are included. In order to get reasonable initial conditions for the pinch phase, a simple 1-D model is used for the run-down phase. Under this somewhat ideal Plasma Focus discharge condition, relatively simple scaling laws are obtained.

Astrophysics and Space Science **256:** 451–457, 1998.

TABLE I
List of parameters.

Case	a (cm)	b (cm)	l (cm)	C (uF)	L_o (nH)	V_o (kV)
BA	1.8	3.60	12.0	10.50	31.0	$20 \div 40$
NE	3.3	6.25	21.0	225.00	20.0	$15 \div 25$
SP	2.5	7.00	6.5	1.56	40.0	$120 \div 220$
FR	8.0	12.00	56.0	1250.00	16.0	$20 \div 40$

2. Parameters

Different configurations were studied. They are characterized by the following geometrical and electrical magnitudes: anode length l and radius a, cathode radius b, bank capacity C, external inductance L_o and charging voltage V_o. Also the initial filling pressure (p_o) and the gas type is given. In what follows we will assume Deuterium gas. Four different configurations were studied (see Table I).

Other related variables used are: L_c external plus gun inductance

$$L_c = L_o + 2 \times 10^{-9}\, l \, \ln(b/a)\,, \tag{1}$$

I_m maximum current (using L_c)

$$I_m = V_o \sqrt{C/L_c}, \tag{2}$$

I_d current at the end of the coaxial stage (taken as the initial current for the collapse), I_p current at the pinch time, (I^2/a) proportional to a magnetic force per unit length (I refers to I_m, I_d or I_p) and $(p_o\, a)$ proportional to a mass per unit length (p_o is the filling pressure in mbar). Note that the ratio $\left(I^2/a\right)/(p_o\, a)$ gives a mean energy per particle (temperature).

3. Simplified Coaxial Stage

In order to get reasonable initial conditions for the compression stage, a simplified snow-plow model was used for reproducing the coaxial stage. Two coupled equations were numerically solved: one for the axial position (z) and the other for the circuit as follows:

$$\frac{dz}{dt} = \frac{100\, I}{a\sqrt{p_o}}\,, \tag{3}$$

and

$$\frac{d\,(LI)}{dt} = V_o - RI, \tag{4}$$

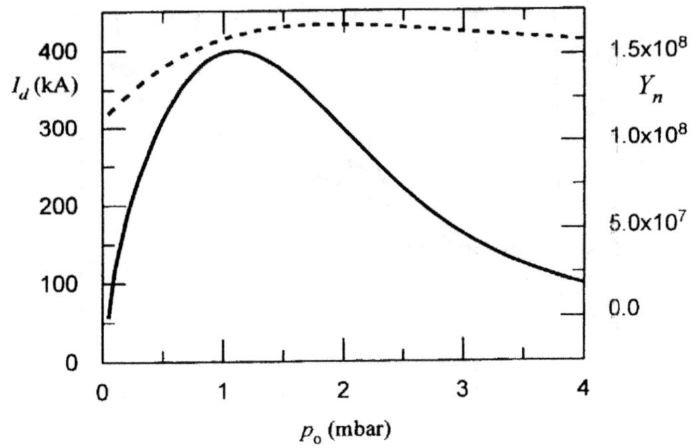

Figure 1. Neutron production (solid) and current (dashed) vs. filling pressure, case BA, 30 kV.

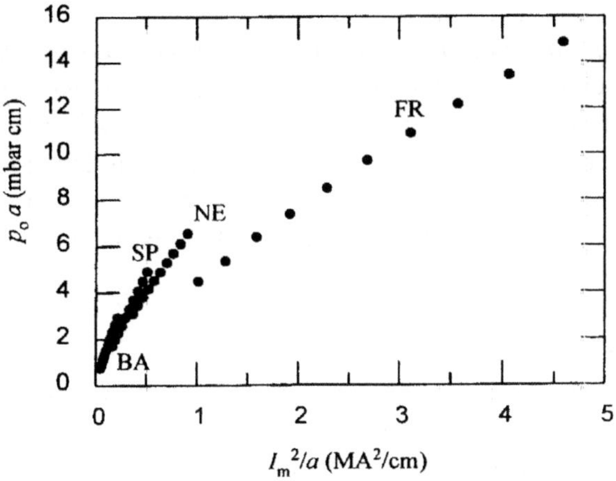

Figure 2. Pressure of maximum neutron production vs. maximum current (or charging voltage).

with

$$L = L_o + 2 \times 10^{-9}\, z\, \ln(b/a)\,. \tag{5}$$

When $z = l$ the current is taken as the initial current for the compression (I_d). In Figure 1 the current I_d is plotted as a function of p_o for the case BA, $V_o = 30$ kV.

4. Results

The calculations proceed as follows. For each case and for each possible value of V_o the initial filling pressure is browsed until the maximum neutron yield is obtained. They represent the optimized regime for a given case and voltage.

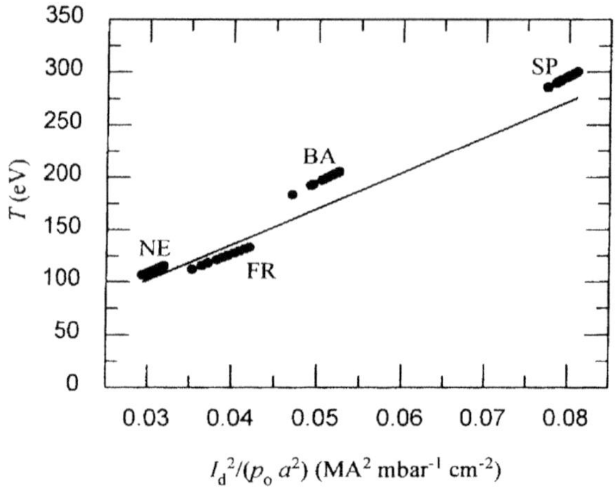

Figure 3. A comparison between the ion temperature (scatter) and a dimensional analysis (solid).

In Figure 2 the $(p_m\, a)$ is plotted against (I_m^2/a), that is the pressure of maximum production vs. the charging voltage or, alternatively, the mass per unit length vs. force per unit length, showing two facts: firstly, that $p_m \propto V_o^2$ and secondly, that the mean energy per particle for a given machine is almost independent of the initial energy bank. This can be seen in Figure 3 where the mean temperature obtained with the 2D code is given together with the analytical expression

$$T = 3.4 \times 10^3 \frac{I_d^2}{p_o\, a^2}\,, \tag{6}$$

obtained as the ratio between the force and mass per unit length.

The neutron production during the first compression follows a $Y_n \propto I^4$ law. In Figure 4 the calculated neutron output is plotted. For comparison a mean experimental law is also given.

Another interesting feature is the ratio between the electron and the ion temperature. Taking into account the fact that the ions are heated up by the shock wave the cooling effect of the electrons will more import if the relative duration of the pinch is longer. The relative equipartition time can be estimated as the ratio between the duration of the pinch and the electron-ion equipartition time, that is

$$\frac{\Delta t}{\tau_{eq}} \propto \frac{(a/v)}{(T^{3/2}/n)} \propto \frac{(p_o\, a)^3}{(I^2/a)^2}\,. \tag{7}$$

In Figure 5 the ratio T_e/T_i obtained with the code at the end of the first compression phase are plotted against the above relative equipartition time. As expected, the ions are cooler when the relative equipartition time is longer. In Figure 6 the ratio T_e/T_i is plotted as a function of the bank energy. This suggests that the ratio (T_e/T_i) increases with the bank energy. This means that high energy machines are less efficient in neutron production than smaller machines. However, this fact

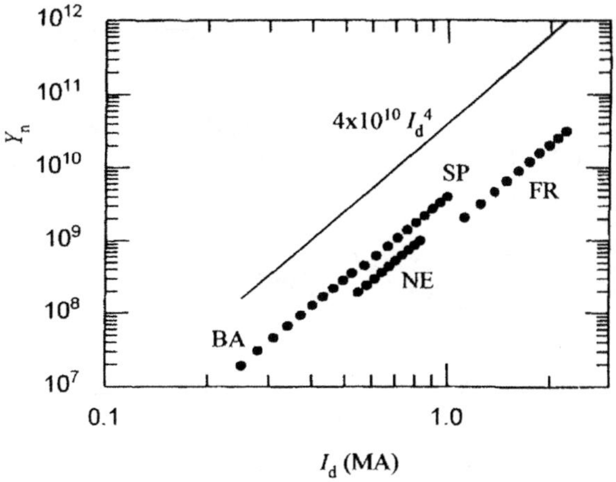

Figure 4. First compression (scatter) and mean total measured neutron yield (solid) vs. current.

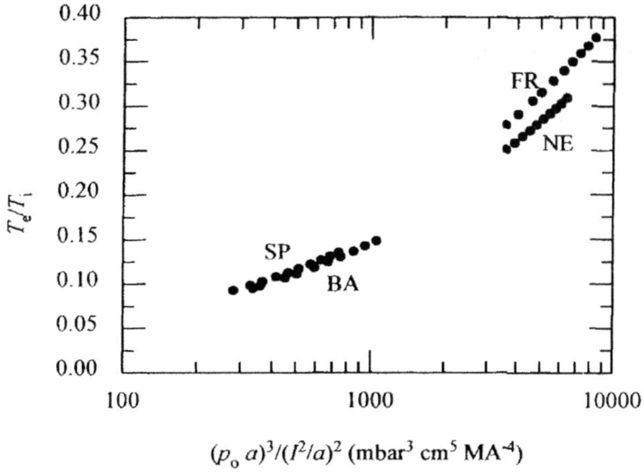

Figure 5. Electron-ion temperature ratio and a dimensional analysis.

cannot explain the experimentally observed, drastic drop in neutron production over 500 to 800 kJ bank energy.

5. Conclusions

Many of the measured properties in Plasma Focus devices are present in this "ideal" non-cylindrical pinch.

Although the neutron production of the first compression is smaller than the measured one, they have some common scaling laws: the neutron production with the current as $Y_n \propto I^4$ and the pressure of maximum neutron production with the charging voltage as $p_m \propto V_o^2$.

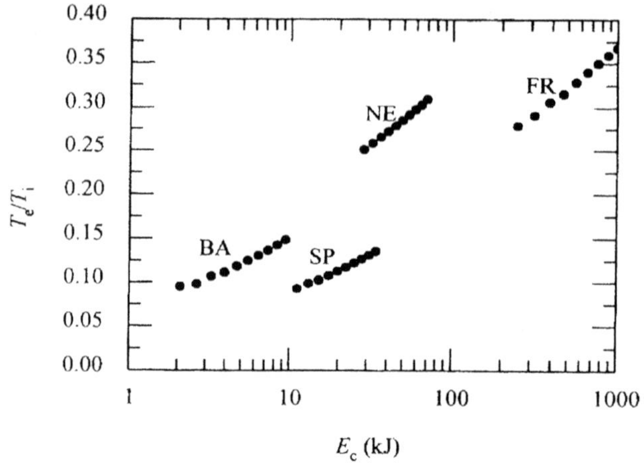

Figure 6. Electron-ion temperature ratio vs. initial bank energy

The equipartition time seems to play an important role on the neutron production. Larger machines are less efficient than smaller ones. This can be seen from Equation (7) and using that the velocity is

$$v^2 \propto \frac{(I^2/a)}{(p_o\, a)}\,; \tag{8}$$

therefore

$$\frac{\Delta t}{\tau_{eq}} \propto \frac{(p_o\, a)}{v^4}\,. \tag{9}$$

Usually, the larger the machine the larger are the values of $(p_o\, a)$ and the only way to keep the electrons cold is by increasing v (see Figure 6, case SP).

The cooling effect of the equipartition time cannot explain the well known drop in neutron production that happens at high energies. Other explanations are needed. For example, it was suggested that the "good" and "bad" breakdown regions are separated by the line $p \propto V_o$ (Gourlan et al., 1978) that always will be "crossed" at high energies by the pressure of maximum production $p_m \propto V_o^2$.

Due to the fact that no other model for the neutron production can explain all these properties in a broad range of energies, further studies should be conducted in this direction. For instance, studying the successive stages, including more realistic simulations for the breakdown and/or the coaxial stage.

Acknowledgment

Work partly supported by grants EX034 (University of Buenos Aires) and PIA6641 (CONICET), Argentina.

References

Bernal, L., Bilbao, L., Grondona, D. and Temporal, M.: 'Two-dimensional finite volume Lagrangian numerical code'.

Gourlan, C., Kroegler, H., Maisonnier, C., Rager, J.P., Robouch, B., Bertalot, L., Gentilini, A., Arcipiani, B., Pedretti, E. and Steinmetz, K.: 1978, Present status of the Frascati 1 MJ Plasma Focus Programme, in: V. Nardi, H. Samlin and W.H. Bostick (eds.), *Proc. 2nd Int. Conf. on Energy Storage, Compression and Switching*, Plenum Press, New York, p. 221.

2-D CODE FOR SIMULATING A HOLLOW GAS PUFF Z-PINCH

L. BERNAL
Laboratorio de Fluidodinámica y Plasma, Universidad de Mar del Plata, Funes y Peña, 7600 Mar del Plata, Argentina

L. BILBAO
Instituto de Física del Plasma, CONICET – Universidad de Buenos Aires, Ciudad Universitaria – Pab. I, 1428 Buenos Aires, Argentina

Abstract. We compare the dynamics of plasma in a hollow gas puff Z-pinch device, obtained with Schlieren pictures, with the predictions of a 2-D MHD Lagrangian code. We show that agreement between the model and experiment require an axially variable line density and width of the injected gas.

1. Introduction

The concept of gas puff Z-pinch has been widely studied (Marrs et al., 1985; Harcs et al., 1985; Hammel, 1984). A hollow gas puff Z-pinch (HGP) can be produced by injecting an annular gas jet, usually from the cathode to the anode (Shiloh, 1978; Burkhalter et al., 1979; Shiloh et al., 1978; Stallings et al., 1979; Degnan et al., 1981; Clark et al., 1982; Li and Yang, 1991). Although both cylindrical foils and annular gas jets are similar experiments, in an actual gas puff devices the density varies both radially and axially. Starting from a uniform thin layer distribution, different shapes and density distribution may be produced leading to many different situations. A priori, it is not clear which parameter determines the dynamics. For instance, the dependence with axial coordinate of the linear density and thickness and/or the radial distribution. Further, the optimization of a given device is performed in order to get a maximum for a particular property (for example, X-ray production) and not always a cylindrical pinch gives this maximum. We have investigated the different dynamics arising from many combination of these parameters in order to explain the dynamics observed in a low-energy HGP, called NOVA (Acuña et al., 1992, 1994), at the University of Mar del Plata (Argentina). Using a 2D, time dependent, two fluid (ions and electrons), two temperature model (Bernal et al., 1998) we try to clarify the influence of the initial distribution on the dynamics of the collapse.

Astrophysics and Space Science **256:** 459–465, 1998.

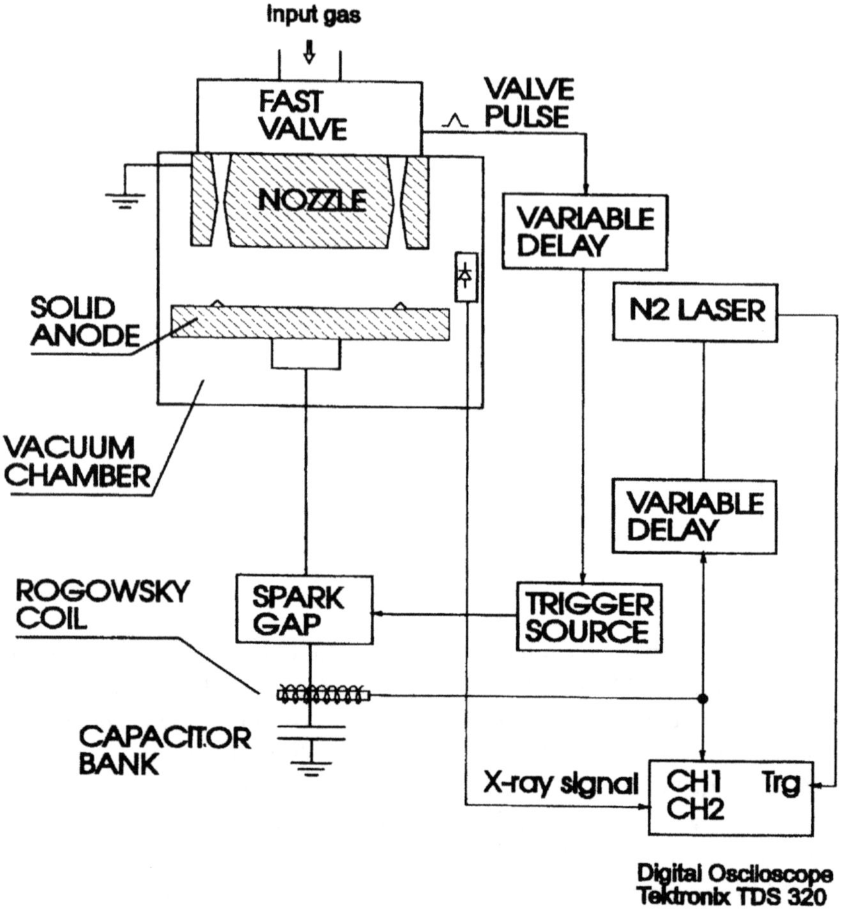

Figure 1. Hollow gas puff Z-pinch experiment setup.

2. Experimental setup

The NOVA Z-pinch has 2 kJ of stored bank energy (working at 8.5 kV) and 80 nH of inductance. Working in the hollow gas puff mode, a jet of neutral, high pressure (1 bar) gas is injected through a annular nozzle with 32 mm of inner radius and 35 mm of outer radius, see Figure 1. The electrode separation was 18 or 23 mm. The main diagnostics were a Rogowsky coil (for recording dI/dt), a PIN-diode (BPX 65) with a 200 μm of Be for soft X-ray detection and Schlieren pictures using a N_2 laser. Using density probes a global idea of the axial and radial dependence of the neutral gas density was obtained. Depending on the used gas and on the pressure at the fast valve we get different axial velocities ranging from 100 to 200 m/s. Depending on the delay of the breakdown the gas front can reach and bounce off the anode producing some sort of mass accumulation of neutral gas

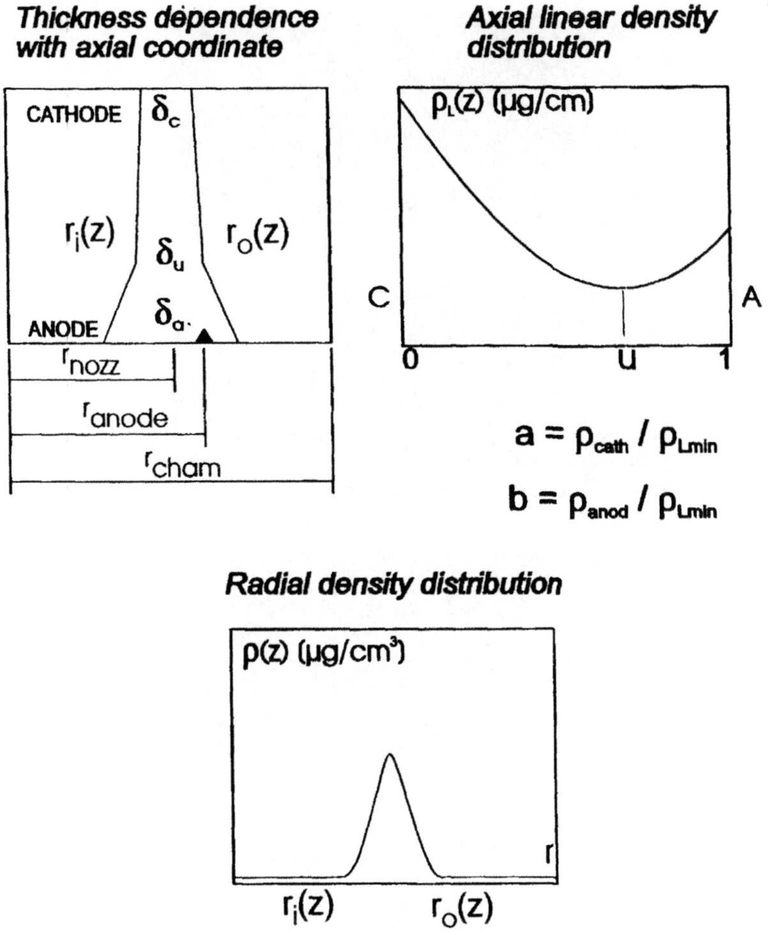

Figure 2. Definition of the parameters for the gas distribution.

on this electrode. Under this particular initial conditions and working with Argon a high producing soft X-ray radiation regime was found. In this case the combination of a relatively long injection time together with the presence of a solid anode produces a particular mass distribution along the axis, having a minimum linear density at a given distance from the anode because the accumulation in the anode is produced simultaneously with the increasing density in the outlet (cathode).

Taking into account the radial expansion the density can be represented by six parameters as described in Figure 2. The combination of these factors produce a rather complicated dynamics. Schlieren pictures with an N_2 laser show this particular dynamics. As can be seen from Figure 3 a highly non-cylindrical compression is obtained. For a given energy this is the best regime for soft X-ray radiation production in this device. It is clear from these results that a simple snowplow model can not give a good description. Although the dynamics is mainly determined by

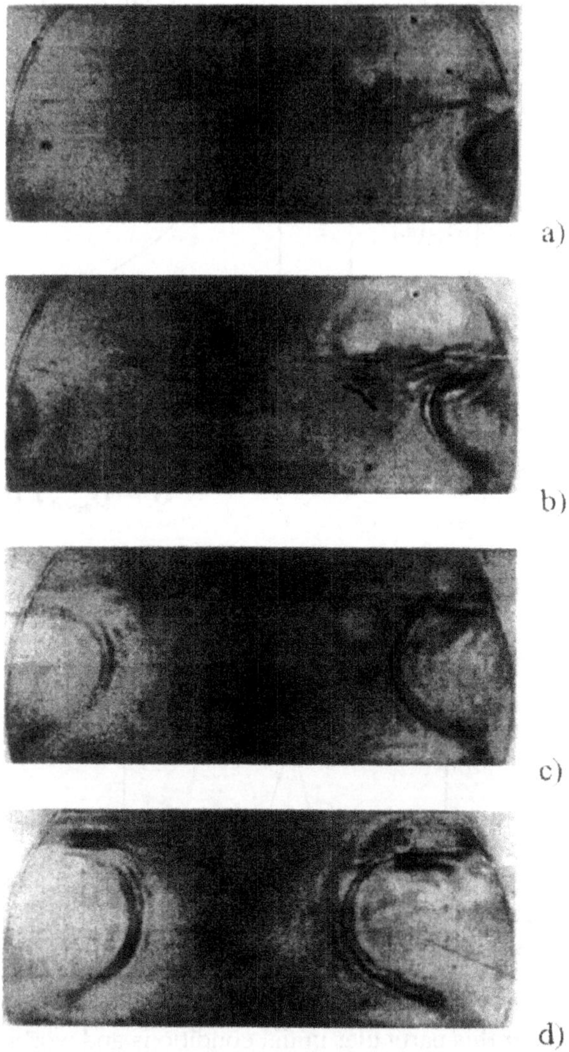

Figure 3. Schlieren sequence (from different shots) for: (a) −415 ns, (b) −325 ns, (c) −205 ns and (d) −150 ns.

the momentum equation, we have included other phenomena like thermal diffusion in order to evaluate the relative weight of these factors.

3. Results

For reproducing the results of the NOVA experiment different values of a, b, δ_a, δ_u, δ_c and $\rho_{l\,\min}$ were tested. The closer simulation to the experimental data was

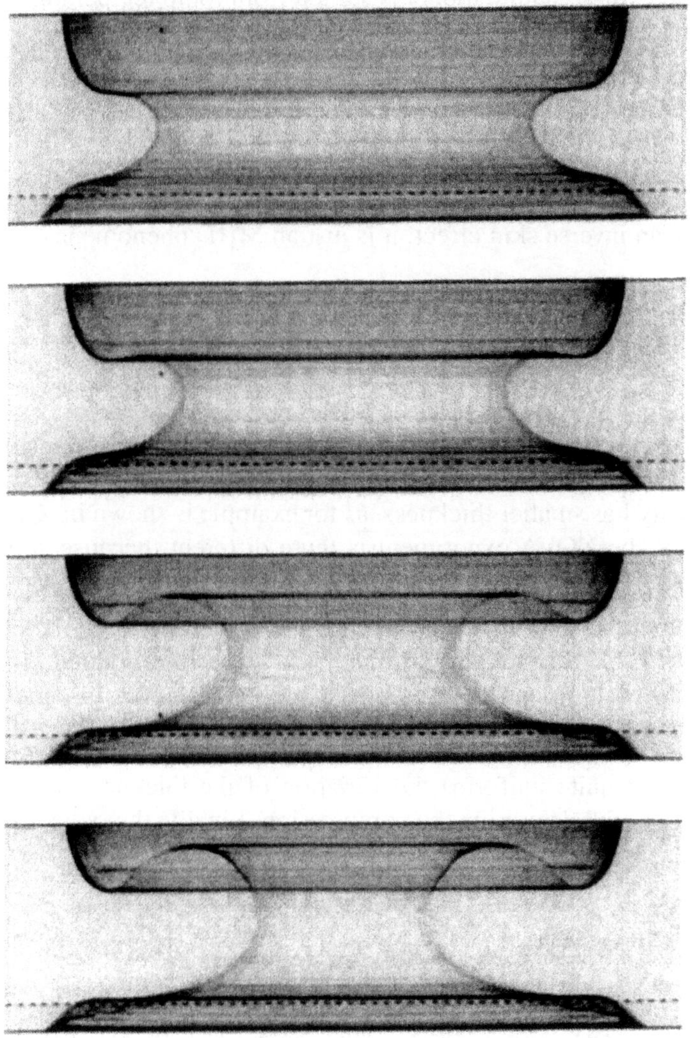

Figure 4. Calculated Schlieren sequence for the same times as in Figure 3.

obtained using $a = 10$ and $u = 0.4$. In Figure 4 we show the calculated evolution of the discharge. Very similar shapes were obtained using just the ideal MHD part of the code, indicating that diffusive processes play a secondary role during convergence. It is important to say that this statement applies to the convergence and is not intended for describing the beginning of the discharge. It seems reasonable that magnetic diffusion plays a role during the first stages of the discharge, until the Joule heating heats the plasma and prevents the diffusion for going on. Further, the plasma model is not adequate for describing the breakdown stage due to the presence of free electrons being accelerated by the electric field. Anyway, in some cases if the hydrodynamic model is started from room temperature, multiple shells or incomplete implosion of the gas is observed, in a similar way of the results reported in (Degnan et al., 1981). Starting from a given current distribution with an adequate temperature, having a sort of equilibrium, we can see a frozen structure propagating towards the axis. Here frozen is used in the way of indicated magnetic flux conservation. Due to hydrodynamic compressions the current distribution can suffer strong variations, further, the current can be reversed in a part of the sheath. This is not an inverse skin effect, it is just an MHD phenomena.

4. Conclusions

The numerical calculation shows that the main behavior of a gas puff Z-pinch is controlled by two parameters: axial distribution of the linear density and thickness variations along the axis. Usually they act in opposite way, because the higher linear density has smaller thickness, as for example is shown in (Clark et al., 1982). The case of the NOVA experiment is quite different, because some sort of mass concentration is produced in the electrode in front of the outlet. This means that the minimum linear density is reached in a point located between the electrodes. This produces a PF-like discharge with increased radiation production. The evolution of such a distribution can be reproduced with a 2D code. In conclusion, as long as the thickness of the 'liner' is small (as compared to the radius of the discharge) the dynamics is dictated by the axial distribution of the linear density. In case this is uniform (or quite uniform) the variation of the thickness as a function of the axial position will determine the compression. Finally, the radial distribution plays a secondary role.

Acknowledgments

This work is partly supported by grants EX034 (University of Buenos Aires) and PIA6641 (CONICET), Argentina.

References

Acuña, H., Bernal, L. and Pouzo, J.: 1992, in: J. Herrera (ed.), *Proc. 5th. LAWPP*, Ciudad de México, México, p .36.

Acuña, H., Bernal, L. and Pouzo, J.: 1994, in: P.H. Sakanaka, E. del Bosco and M.V. Alves (eds.), *Proc. 1994 International Conference on Plasma Physics*, Foz do Iguaçu, Brazil, p. 125.

Bernal, L., Bilbao, L., Grondona, D. and Temporal, M.: 'Two-dimensional finite volume Lagrangian numerical code'.

Burkhalter, P.J., Shiloh, J.H., Fisher, A. and Cowan, R.D.: 1979, *J. Appl. Phys.* **50**, 4532.

Clark, W. et al.: 1982, *J. Appl. Phys.* **53**, 5552.

Degnan, J.H., Reinovsky, R., Honea, D.L. and Bengston, R.D.: 1981, *J. Appl. Phys.* **52**, 6550.
Hammel, B.A.: 1984, Ph.D. Thesis, University of Colorado.
Harcs, J.D., Marrs, R.E. and Fortner, R.J.: 1985, *J. Appl. Phys.* **18**, 1985.
Li, C.R. and Yang, T.C.: 1991, *J. Phys. D: Appl. Phys.* **24**, 48.
Marrs, R.E., Dietrich, D.D., Fortner, R.J., Levine, M.A. et al: 1985, *Appl. Phys. Lett.* **42**, 946.
Shiloh, J.H.: 1978, Ph.D. Thesis, University of California, Irvine.
Shiloh, J.H., Fisher, A. and Rostoker, N.: 1978, *Phys Rev. Lett.* **40**, 515.
Stallings, C., Childers, K., Both, I. and Schneider, R.: 1979, *Appl. Phys. Lett.* **35**, 524.

OPTICAL MEASUREMENTS IN AN EXPLODING WIRE EXPERIMENT

L. BIBBO, N. GIOVAMBATISTA, P. GÓMEZ, M. OLIVIERI, C. TIBALDI,
L. BERNAL and J. POUZO
Laboratorio de Fluidodinámica y Plasma, Universidad de Mar del Plata, Funes y Peña, 7600 Mar del Plata, Argentina

L. BILBAO
Instituto de Física del Plasma, CONICET – Universidad de Buenos Aires, Ciudad Universitaria – Pab. I, 1428 Buenos Aires, Argentina

Abstract. We present optical measurements (Shadow and Schlieren pictures) of a 12.5 Joule exploding wire experiment. After performing a large number of shots some sort of time evolution can be reconstructed from the pictures. The pictures were taken during the dark phase (zero current) and the plasma phase. Although the electrical signals are quite reproducible from shot to shot (allowing a good time correlation and indicating that the global dynamics is reproducible) we observed that the individual details are quite different, probably due to the non-uniformity of the wire. We observed a uniform expansion (the volume increases almost linearly with time) with the presence of surface instabilities and there is no evident difference in the density distribution between the dark and the plasma phases. Finally, we derive a mean radius and velocity evolution with time and an approximate energy balance.

1. Introduction

We present optical measurements (Shadow and Schlieren pictures) of an exploding wire experiment performed with a 12.5 Joule capacitor working at 10 KV, delivering about 10 KA of peak current, using a copper wire of 45 μm diameter and 30 mm length. The current signal is obtained from a Rogowski coil and integrated using a passive integrator. We also registered the visible light emitted from the wire with a fast pin diode (BPX 65) placed at 100 mm from the axis wire (see Figure 1). For the optical measurements we use an N_2 laser (1 ns UV pulse) synchronized to the main discharge by means of a secondary Rogowski coil (see Figure 2).

The electrical signals are quite reproducible from shot to shot (allowing for a good time correlation) indicating that the global dynamics is reproducible. After performing a large number of shots some kind of time evolution can be reconstructed from the pictures. In Figures 3 and 4 both Schlieren and Shadow sequences are given, together with the corresponding time correlation. As it can be seen, the pictures show either the dark phase (zero current) and the plasma phase.

Astrophysics and Space Science **256:** 467–472, 1998.

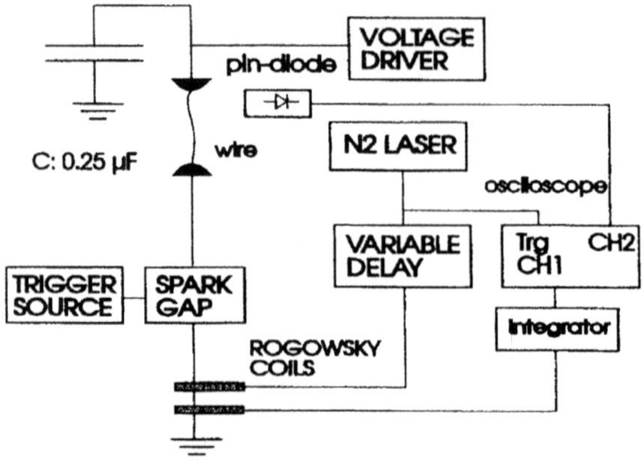

Figure 1. Experimental setup.

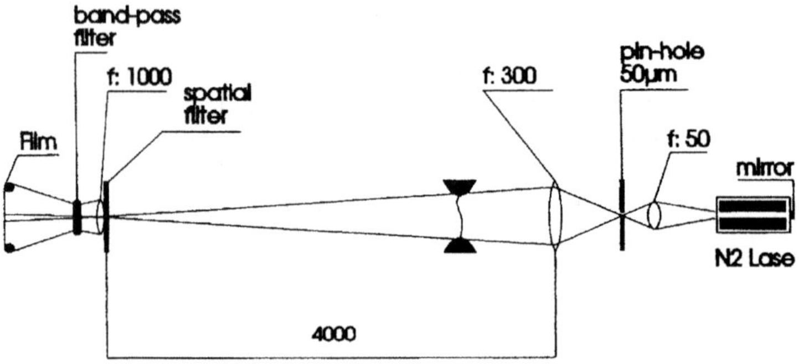

Figure 2. Optical arrangement.

2. Discussions and Conclusions

From the experimental results it is possible to infer a simple model of the process, which can be divided into three different stages. In the initial stage, due to Joule heating, the wire is heated up. From the measured current and voltage, the energy deposition due to Joule heating is $E_J \approx 5$ J. Note that the energy required to vaporized the wire is $E_V \approx 2$ J. For this reason, after a period of time (t_o) that depends on the radius of the filament, the wire is converted into a dense gas with an internal energy $E_I \approx 3$ J. Due to the low conductivity of the gas, the current is interrupted.

The second stage (dark pause) is characterized by an almost adiabatic expansion because there is no heating (due to the lack of current) nor losses (no radiation is detected). A simple model of this expansion is obtained assuming that the expanding column of gas works against the atmosphere at supersonic velocity. Therefore, the pressure (p) and the velocity (v) at the boundary are related as

$$p = \rho_o v^2,$$

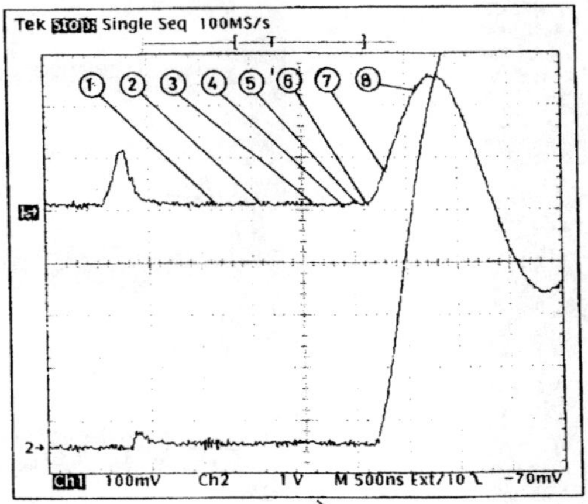

a)

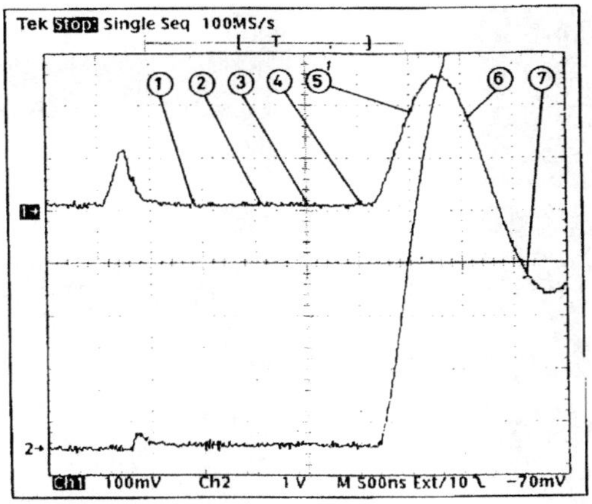

b)

Figure 3. Current (UB) and photodiode (LB) signals and time correlation for (a) Schlieren and (b) Shadow.

where ρ_o is the mass density of the atmosphere.

Using the adiabatic relationship

$$pV^\gamma = C = \text{const.}$$

and $V = \pi h r^2$, where V is the volume and γ the adiabatic exponent, the velocity becomes

$$v = \frac{dr}{dt} = \frac{C^{1/2}}{(\pi h)^{\gamma/2} r^\gamma \rho_o^{1/2}} .$$

Integrating from the beginning of the expansion ($t = t_o$ and $r \approx 0$), the radius evolution is

$$r = \left[C'(t - t_o)\right]^{1/(\gamma+1)}$$

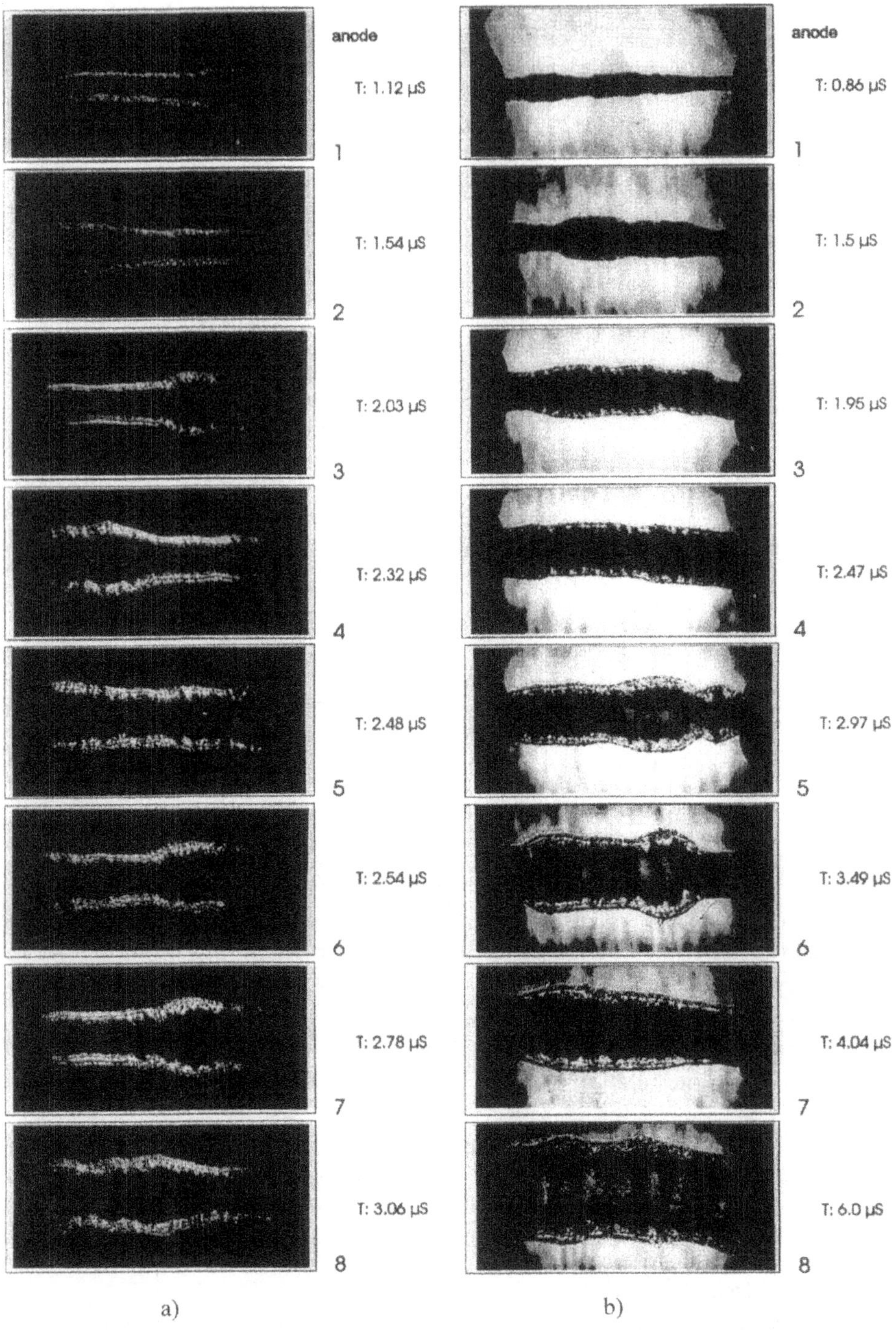

Figure 4. Time evolution of the wire: (a) Schlieren and (b) Shadow. (See Figure 3 for the time correlation with the current.)

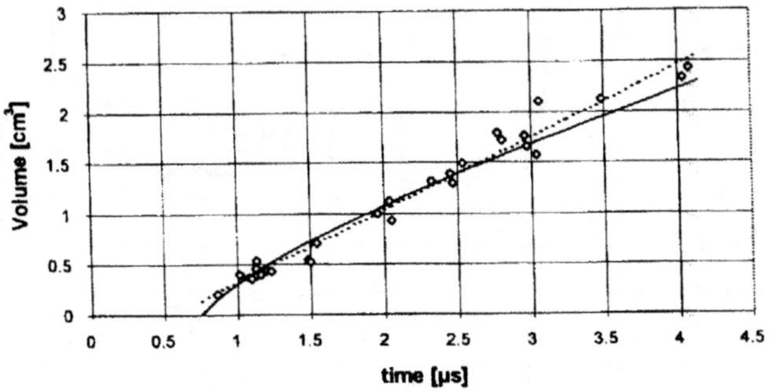

Figure 5. Volume evolution: calculated (solid) and experimental (scatter). Also an uniform expansion is given (dotted).

with

$$C' = \frac{(\gamma + 1)\, C^{1/2}}{(\pi h)^{\gamma/2}\, \rho_o^{1/2}}.$$

Therefore,

$$V = C''\,(t - t_o)^{2/(\gamma+1)}\ ,$$

where $C'' = \pi h C'^{(2/(\gamma+1))}$. From the experimental data the best fit using $\gamma = 5/3$ is

$$C'' = 0.92\ \mathrm{cm}^2 \mu\mathrm{s}^{-3/4} \quad \text{and} \quad t_o = 0.77\ \mu\mathrm{s}$$

and the related constants are

$$C' = 0.045\ \mathrm{cm}^{8/3} \mu\mathrm{s}^{-1} \quad \text{and} \quad C = 1.2 \times 10^{-5}\ \mathrm{g\ cm}^4 \mu\mathrm{s}^{-2}.$$

In Figures 5 and 6, this volume fit as well as those for the radius and velocity are plotted (solid line) together with the experimental points, showing a good agreement.

Using the above constants, the pressure of the gas is

$$p = \frac{540\ \mathrm{atm}}{[r\ (\mathrm{mm})]^{10/3}}\ ;$$

this means that during the dark pause (from $t \approx 0.8\ \mu$m, $r \approx 1.5$ mm to $t \approx 3\ \mu$m, $r \approx 4$ mm) the pressures vary from 140 to 5.3 atm, well above the atmospheric pressure. The energy spent during the expansion, obtained from the work performed, is ≈ 2.5 J, in good agreement with the available energy at the beginning of the dark pause, that is ≈ 3 J.

The third stage (plasma phase) begins when the falling density reaches the value of spontaneous breakdown and the current is reestablished. Note that from the Schlieren and Shadow pictures it is not possible to discriminate the dark pause

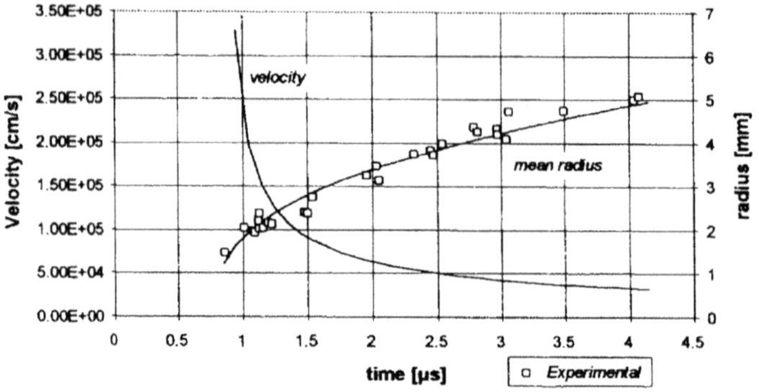

Figure 6. Velocity and mean radius evolution: calculated (solid) and experimental (scatter).

from the plasma phase. During this stage and depending on the strength of the current a pinching phase can be produced. This is not the present case. From the pictures a continuous expansion is observed. This behavior can be understood by comparing the pressure of the gas with the magnetic pressure. The maximum value of the magnetic pressure (obtained using the peak current of 10 KA) is

$$p = \frac{15\ \text{atm}}{[r\ (\text{mm})]^2}\ ,$$

that is, smaller than the internal pressure as long as $r \leq 15$ mm. Beyond this point due to the fact that the Joule heating will increase the internal energy and simultaneously will decrease the energy of the circuit, the magnetic pressure will never overcome the internal pressure preventing a pinch phase.

Acknowledgments

Work partly supported by grants EX034 (University of Buenos Aires) and PIA6641 (CONICET), Argentina.

References

Bernal, L. and Bilbao, L.: 1998, *Astrophys. Space Sci.* **256**, 459.
Bilbao, L.: 1990, *J. Comp. Physics* **91**, 361.
Bilbao, L. and Grondona, D.: 1998, *Astrophys. Space Sci.* **256**, 451.
Bilbao, L. and Linhart, J.G.: 1996, *Plasma Physics Reports (Fizika Plazmy* **22**, 457.

STUDIES OF PLASMA DYNAMICS IN A SMALL PLASMA FOCUS OPERATING IN HYDROGEN-ARGON MIXTURES

M. FAVRE, P. SILVA, H. CHUAQUI and E. WYNDHAM
Facultad de Física, Pontificia Universidad Católica de Chile, Casilla 306, Santiago 22, Chile

P. CHOI and C. DUMITRESCU-ZOITA
LPMI, Ecole Polytechnique, Palaiseau 91128, France

Abstract. We report on experimental observations in PFP-I, a small 3.8 kJ plasma focus, which is operated in Hydrogen-Argon mixtures to investigate the effect of parameter modifications on the overall performance of the device. An extensive array of diagnostics is been used, which includes voltage and current probes in the external circuit, a novel small magnetic probes array located along the cathode rods, filtered PIN diodes located side and end on, and filtered multi-pinhole and slit-wire X-ray camera. Extended operating range from below 0.2 Torr upwards has been achieved. Hot spot formation has been investigated as a function of H_2-Ar mixing ratio. Hot spot sizes around 150 μm in the soft X-ray region, have been inferred from the slit-wire measurements. Pin-hole time integrated X-ray pictures and time resolved PIN diode measurements have been used to determine characteristic hot-spot temperatures in the 350 to 450 eV range.

1. Introduction

Many of the interesting transient phenomena which are characteristic of dense magnetized plasmas can be investigated in small Plasma Focus devices, at very low operational cost. In particular, hot-spot formation is a common feature in a variety of transient discharges of the Z-pinch type, such as vacuum spark, gas puff, exploding wire, composite pinch and plasma focus (Negus and Peacock, 1975; Shiloh et al., 1978; Zakharov et al., 1984). Hot-spot phenomena have attracted renewed interest in view of potential applications as soft X-ray radiation sources. Although hot-spots have been investigated in different experiments, the physical mechanisms involved in their generation and time evolution, are still not clear. We present experimental results on the investigation of hot-spot formation in PFP-I, a small 3.8 kJ Plasma Focus device operating at constant mass density in Hydrogen-Argon mixtures, at pressures from below 0.2 Torr upward. Constant mass load operation allows effects due to Argon admission on radial collapse and subsequent hot-spot formation to be decoupled from pure dynamic effects. The main diagnostics for hot-spot investigations is a combination of multi pinhole and slit-wire (Chuaqui et al., 1994) X-ray photography, with filtered PIN diodes. Pin-hole and slit-wire images are recorded on HP5 film. The simultaneous recording of pin-hole and slit-wire images allows both the temperature and characteristic size of the hot-

Astrophysics and Space Science **256:** 473–478, 1998.

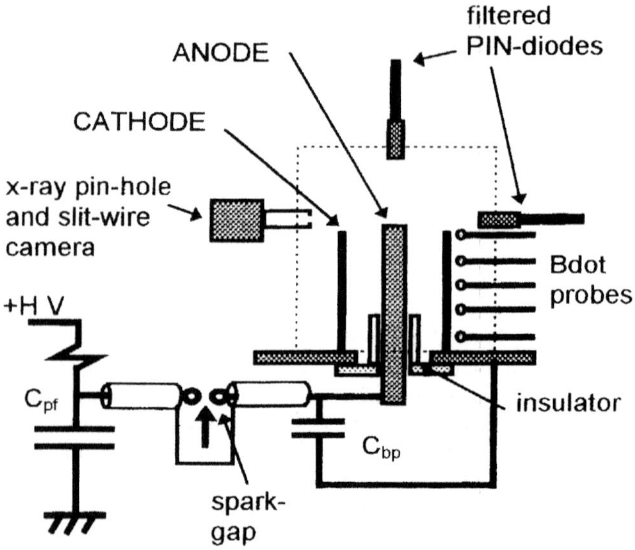

Figure 1. Schematics of the plasma focus device and associated diagnostics.

spots to be determined. Typical size for the hottest emitting region, at temperatures between 350 and 450 eV, is found to be around 50 μm, with a typical duration of the high temperature phase of less than 20 ns.

2. Experimental Apparatus

A schematic of the experimental set-up is shown in Figure 1. A low inductance, 9 channel triggered spark-gap is used to transfer the charge from the 9 μf, 30 kV capacitor bank to the focus. 10 ns transmission lines, with an effective impedance of 4.2Ω, connect the spark-gap to the bank and focus. The electric circuit design has been optimized to enhance high frequency oscillation of the voltage at initial breakdown, in order to improve breakdown efficiency and uniform current sheath initiation. Typical peak current is 140 kA at 20 kV charging.

The discharge is formed between a hollow anode, 1.2 cm radius and 14 cm length, and six cathode rods, symmetrically arranged in a 3.2 cm radius circle. The insulator is a Pyrex sleeve, 5 cm long. The focus is operated in H_2-Ar mixtures, at pressures from 0.2 Torr upwards. An array of small magnetic probes located along the cathode rods is used to study the current sheath structure during the run down phase of the discharge. The probes are electrostatically shielded and placed behind the cathode rods, allowing non-perturbing measurements to be made. Experimental studies of current sheath evolution based on measurements with the magnetic probe array are presented in a companion paper (Favre et al., 1998). An array of filtered PIN diodes (BPX65) located side and end-on, is used in conjunction with a multi-

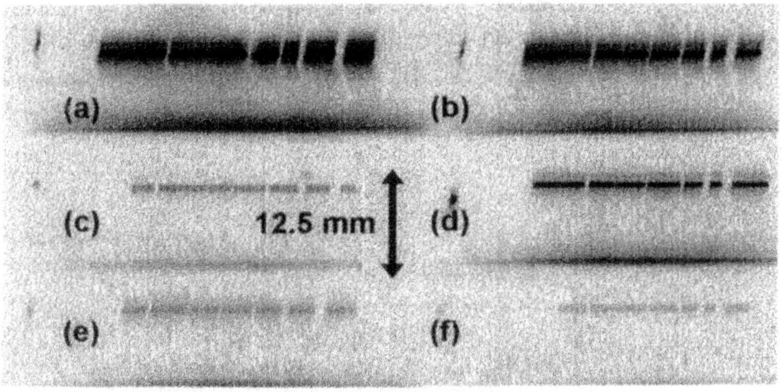

Figure 2. Pin-hole and slit-wire X-ray images of a single shot in 80% H_2 + 20% Ar. (a) to (f) refer to the filter set (see main text).

pinhole and slit-wire X-ray camera. Six pin-holes of 100 and 200 μm diameter are used with six 100 μm slits. X-ray images are recorded in HP5 film, using an adapted Nikonos camera. The same filter combination is used in PIN diodes and pin-hole slit-wire camera. Matched pairs of filters were used in order to discriminate the Argon line emission at around 4 and 20 Å. Thin wires of 25 to 250 μm diameter were used to estimate the characteristic hot-spots size.

3. Experimental Results

Characteristic experimental data corresponding to a single shot at 15 kV, with a mixture of 80% Hydrogen plus 20% Argon, is presented in the next set of figures. Figure 2 shows the combined pin-hole slit-wire images. The different filters, as labeled in the figure, are: (a) 12 μm Beryllium, (b) 16 μm Beryllium, (c) 12.5 μm Titanium, (d) 17 μm Kim-fol, (e) 15 μm Aluminium, and (f) 3 μm Silver plus 24 μm Kim-fol. 100 μm pin-holes were used with the Beryllium filters and 200 μm pin-holes elsewhere. The pin-hole images are seen on the left hand side of the corresponding slit-wire image. In all cases, the diffuse band below the slit-wire image corresponds to anode radiation. Both images, pin-hole and slit-wire, in filters (a), (b), (d) and (e) show a hot-spot embedded in a plasma column. Filters (c) and (f) do not resolve the plasma column. In all cases the image is pin-hole limited, which indicates that the characteristic size of the hot-spot is below 100 μm over the filter transmission bands.

Figure 3 shows densitometric scans across the slit-wire images. Six different wires were used, as shown in the figure: (1) 100 μm Copper, (2) 25 μm Tungsten, (3) 50 μm Tin-Copper alloy, (4) 80 μm Tungsten, (5) 160 μm Tin-Copper alloy, and (6) 250 μm Silver. Labels “a” to “f” identify the corresponding filter, as in Figure 2. The fact that the 25 μm diameter wire produces a well defined absorption

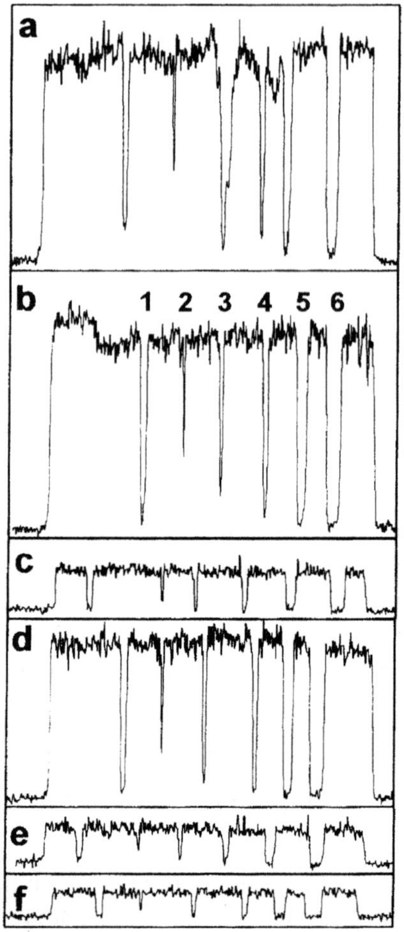

Figure 3. Densitometric scans across slit-wire images of Figure 2.

peak is a good indication that the characteristic size of the hot-spot is of the same order of magnitude.

Figure 4 shows voltage, time derivative of the discharge current, and PIN diode signals. Several pulses of X-rays are observed, following the drop in dI/dt. The X-ray signals are a combination of hot-spot, plasma column and beam target anode radiation. The end-on detector, with a Nickel filter, is used to identify strong electron beam induced emission. It can be seen that a strong beam induced X-ray pulse is associated with the second spike in the voltage trace, with minor excursions coinciding with the first and third voltage spike.

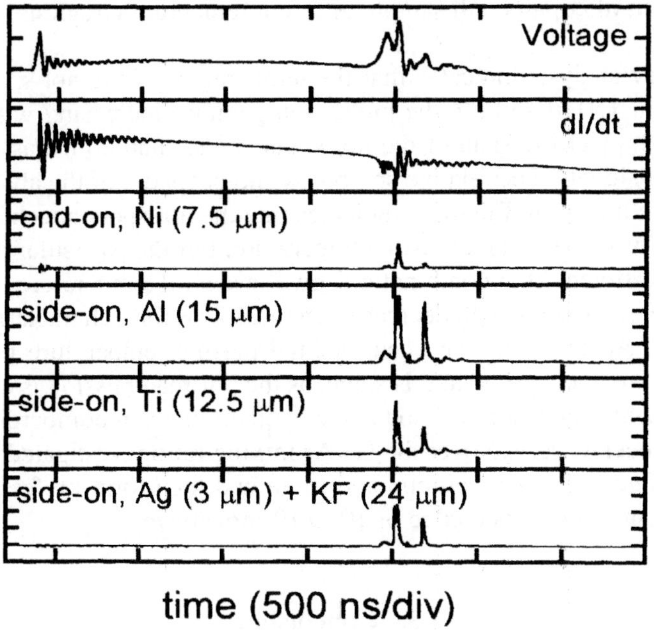

Figure 4. Voltage, time derivative of sicharge current, and PIN diode signals. Same shot as in Figure 2.

4. Discussion

X-ray emission is a transient phenomenon associated with the pinching phase of the focus discharge. It is a combination of thermal radiation, which includes bremsstrahlung, recombination, line, and electron beam target emission. To analyze the X-ray images and PIN diode signals we have used XRAYFIL (Dumitrescu-Zoita, 1996), a code that calculates the emission spectrum for a given plasma using a collisional radiative model of the level population and convolutes the resulting spectrum with the transmission characteristics of the filters used, as well as the response function of the corresponding detector. Level population for the dominant species, in this case Argon, is derived from a separate code, RATION (Lee et al., 1989). The calculation is repeated over a set of plasma temperatures and densities. Comparison of the measured ratio of signal, optical density or voltage, trough the different filters with the calculated values, allows plasma temperature to be inferred from the experimental data. In a time integrated X-ray image, the information over the film is a superposition of the emission at different stages of the pinch evolution, with temperature and density changing continuously in time. To overcome this difficulty, pairs of matched filters are chosen, which maximize detection in a particular band of the expected emission spectrum. A change in temperature will then result in the emission shifting from one pair of filters to another, thus allowing the plasma temperature to be effectively determined at different stages of the pinch evolution. Similar considerations apply to the space integrated observations, in which different filter pairs will sample different temperature regions of the pinched plasma.

Preliminary analysis indicates that the temperature of the hot-spots evolves in the range from 150 to 450 eV, the higher temperature increasing with Argon percentage. The typical life span of the hot-spots is less than 20 ns, as inferred from the PIN diode signals. The relative sharpness and deepness of the absorption peaks in the slit-wire images in Figure 3, indicates that the hot-spots are characterized by an inner core of higher characteristic temperature. For the particular case shown in Figure 2, the different filter combinations gives a central temperature of 440 eV, in a $\sim$50 μm diameter core, which decreases up to 220 eV for a characteristic diameter of $\sim$150 μm. Accurate measurements of the plasma temperature using the PIN diode signals are not possible at this stage, as the plasma emission is not collimated and includes emission from the anode, due to electron beam bombardment. Similar behavior has been observed over the H_2-Ar mixing ratio investigated. The number of hot-spots fluctuates between one and three and they form preferentially either next to the front of the anode edge or 10 to 12 mm above.

5. Conclusion

We have presented preliminary observations of hot-spots in plasma focus operating in H_2-Ar mixtures with a combination of space and time resolved X-ray diagnostics. Typical sizes for the hot-spots have been established and a characteristic temperature range has been measured. Further analysis of the existing data is in progress and a detailed characterization of the hot-spot size as a function of temperature and mixing ratio will be published elsewhere.

References

Chuaqui, H. et al.: 1994, in: *Proc. ICPP*, Foz de Iguazu, Brasil, Vol. III, p. 29.
Dumitrescu-Zoita, C.: 1996, Ph.D. Thesis, Université de Paris Sud.
Favre, M. et al.: 1998, *Astrophys. Space Sci.* **256**, 337.
Lee, R. et al.: 1989, *J. Quzut. Spectrosc. Radiat. Transfer* **32**, 91.
Negus, C.R. and Peacock, N.J.: 1975, *J. Appl. Phys. D* **12**, 91.
Shiloh, J. et al.: 1978, *Phys. Rev. Lett.* **40**, 515.
Zakharov, S.M. et al.: 1984, *Sov. J. Plasma Phys.* **10**, 303.

TIME RESOLVED STUDIES OF A PULSED HOLLOW CATHODE CAPILLARY DISCHARGE

P. CHOI and C. DUMITRESCU-ZOITA
LPMI, Ecole Polytechnique Palaiseau 91128, France

M. FAVRE, J. MORENO, H. CHUAQUI and E. WYNDHAM
Facultad de Física, Pontificia Universidad Católica de Chile, Casilla 306, Santiago 22, Chile

M. ZAMBRA
Comisión Chilena de Energía Nuclear, Casilla 188-D, Santiago, Chile

Abstract. Experimental studies on a fast pulsed capillary discharge are presented. The discharge operates in a 0.8 mm inner diameter capillary, at 12 kV applied voltage. On axis discharge initiation is achieved by means of the hollow cathode effect. A short, less than 10 ns, XUV pulse is produced. Preliminary time resolved spectroscopic studies indicate that a hot, fast evolving short duration capillary plasma is produced.

1. Introduction

The capillary discharge is known as a high brightness source in the VUV and soft X-ray region (McCorkle, 1981). It has also recently been demonstrated to be a suitable medium for discharge based X-ray laser studies (Rocca et al., 1994). The successful operation of a capillary discharge as a high temperature plasma radiation source depends partly on the initial formation of a conducting plasma column away from the wall and partly in the ability to provide a high rate of current rise into a high impedance load. The rapid formation of the ionization path is a necessary condition for the preparation of the current channel while the rate of current rise determines if and when the plasma column will expand or contract when it is heated by the input current. In this paper we report on time resolved studies on a fast capillary discharge which operates in the nanosecond regime with current in the kA region into a sub-millimetre diameter capillary. To achieve the rapid formation of the discharge column, the initial ionization path is prepared through the transient hollow cathode effect (Choi et al., 1987). This discharge is characterized by a rapid voltage collapse accompanied by the growth of an intense electron beam leading to a very high current density channel on axis even at near vacuum operating conditions. The electron beam facilitates the formation of an on-axis ionization channel, which develops throughout a moving virtual anode (Choi et al., 1993). As a result of this, the capillary discharge could be accurately synchronized to an external triggering event by increasing the ionization growth processes within the

Astrophysics and Space Science **256:** 479–484, 1998.

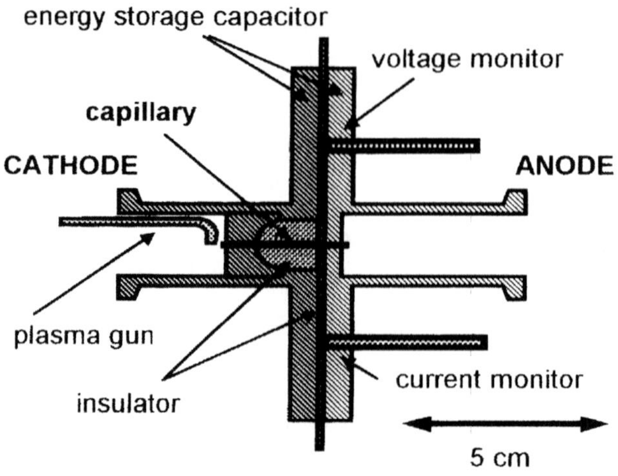

Figure 1. Experimental set-up.

hollow cathode region (Chuaqui et al., 1989). To produce the high rate of current rise required, it is necessary to operate at high voltage, with low circuit inductance. A novel high voltage discharge geometry with very low inductance is used, which integrates the energy storage medium directly onto the discharge electrodes of the capillary system. The discharge chamber is arranged on the axis of a pair of parallel plate electrodes, separated by a thin dielectric to form a local energy storage capacitor. No additional switches are included, in order to maintain the inductance at a minimum. The local energy storage capacitor is pulse charged throughout a primary capacitor using a triggered switch. The long formative time in low pressure discharge enables a high over-voltage to be established across the electrodes and for the discharge to be triggered externally. By minimizing the circuit inductance, a very fast rising current of $> 10^{12}$ A.s^{-1} can be obtained at voltage below 30 kV.

2. Experimental Apparatus

A schematic of the capillary discharge set-up is shown in Figure 1. A pair of 90 mm diameter brass electrodes form the anode and cathode of the discharge, as well as the parallel plate capacitor. An alumina capillary, 0.8 mm diameter, 30 mm long, is located on axis between the two electrodes. On the cathode side, an additional plate is used to provide the optimum geometry of the aperture for the operation of the transient hollow cathode assisted ionization. The insulation that separates and isolate the two electrodes is also the dielectric medium for the capacitor, and so there is a trade off between the requirement for a material with high breakdown strength and high dielectric constant. A primary DC charged capacitor of 5 nF is used to pulse charge the storage capacitor within 100 ns. A novel set of current and voltage monitor has been integrated into the ground side of the electrode which

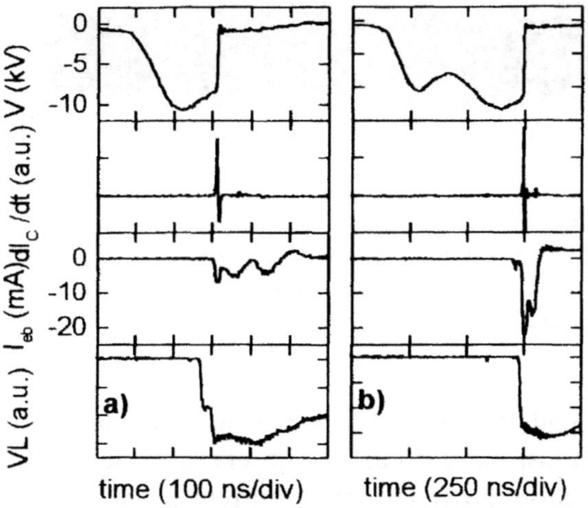

Figure 2. Characteristic signals: (a) triggered discharge, (b) self-breakdown discharge.

forms part of the energy storage capacitor, in order to measure the voltage and current local to the discharge. At a charging voltage of 25 kV, current above 10 kA is produced through the capillary, with a 10–90% current rise time of below 5 ns. The discharge is triggered by means of an auxiliary discharge from a small cable plasma gun, located a few mm behind the cathode aperture, with a delay of 80 to 120 ns. Best results have been obtained by pumping the cathode region through the capillary, thus allowing a higher pressure in the hollow cathode region, while maintaining a lower pressure in the capillary to produce a lower line density plasma. The experiments have been performed in Argon at pressures between 80 and 200 mTorr in the hollow cathode region and a few mTorr at the Anode region. Apart from the voltage and current measurements, the plasma emission is studied in the visible with PIN diodes and photomultipliers, in the VUV region with time resolved spectroscopy and in the soft X-ray region with filtered PIN diodes and XRD. For time resolved spectroscopic studies, a VUV grazing incidence spectrometer with a three frame gated intensified microchannel plate is used. In these observations the spectrometer was fitted with a 600 groves/mm gating, with a useful spectral range from 70 to 240 Å. The framing time and frame separation can be adjusted by selecting the cable length to the high voltage pulser. A 5 ns gating time was used in these measurements. Electron beams are measured using a Faraday cup and a scintillator-photomultiplier assembly.

3. Experimental Results

Figure 2 shows characteristic single shot signals for (a) a triggered discharge at 12 kV, and (b) self-breakdown discharge, both at 160 mTorr. A cable delay of 110 ns is used for the auxiliary triggering discharge. The voltage rises to its max-

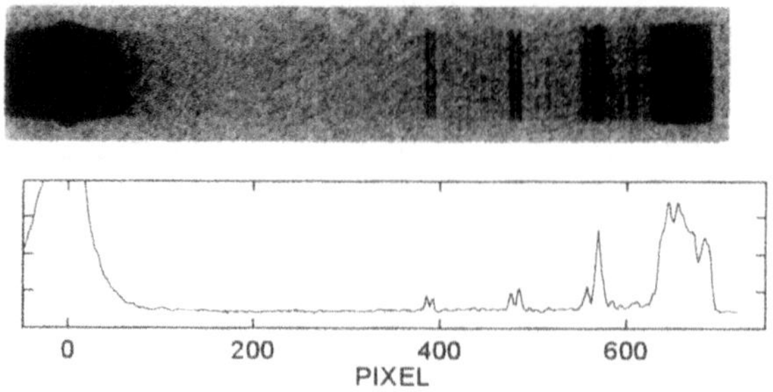

Figure 3. Time integrated spectrum at 210 mTorr.

imum value in around 100 ns, and collapses in less than 10 ns at breakdown. At voltage collapse, a fast current pulse is observed, which is time correlated with an electron beam pulse. After the initial current pulse, the voltage recovers and grows up to $\sim$ 1 kV. Additional electron beam pulses after breakdown are driven by the residual high voltage. The visible light emission is monitored axially, from behind the cathode. The first step observed in the visible corresponds to the auxiliary discharge. The second step coincides with the onset of the main discharge across the capillary. At later times the visible emission follows the oscillation of the main discharge current, as inferred from a software integration of dI/dt. The main feature of the triggered discharge are highly reproducible over the pressure range investigated.

When the self-breakdown, untriggered discharge is compared with the triggered one, the characteristic delay to breakdown in this particular shot increases by around 500 ns. An initial low current electron beam pulse is observed prior to electric breakdown. A fast current pulse is also observed at voltage collapse, which again coincides with the fast rise in visible light emission. At both triggering conditions, XRD and PIN diode observations show an initial fast, less than 10 ns width, VUV pulse which coincides with the main discharge current pulse. A high energy component, above 5 keV, in the electron beams has not been detected, presumably due to the relatively low applied voltage. The positive excursion in the electron beam signals is probably due to discharge plasma reaching the Faraday cup collector plate. In general, the time delay to breakdown is not reproducible in the untriggered discharge, and an appreciable jitter is observed. Figure 3 shows a time integrated spectra obtained over 50 shots, at 210 mTorr in the HCR. Figure 4 shows two frames of a single shot spectrum obtained at 320 mTorr. The exposure time and frame separation is 5 ns. Both the time integrated and the single shot spectra exhibit a well defined line structure, with no continuum emission over the 20–240 Å spectral range. An accurate calibration of the observed spectral range, with line species identification is currently under way.

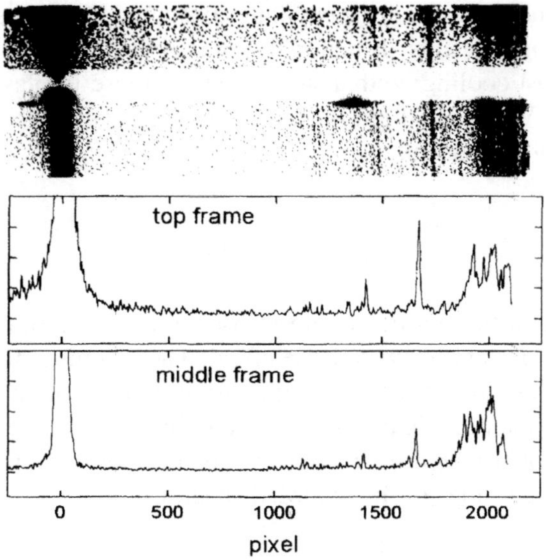

Figure 4. Single shot spectra at 320 mTorr.

4. Discussion

Transient hollow cathode discharges (THCD) are characterized by a rapid voltage collapse accompanied by the growth of an intense electron beam leading to the formation of an on-axis ionization channel, which develops throughout a moving virtual anode (Rocca et al., 1994). The electron beams originate in a plasma region inside the hollow cathode region, which develops spontaneously following the application of the high voltage across the electrodes. Initially the electron beams are of high energy, of the order of the external potential, and low characteristic current. At later times, just before electric breakdown, the electron beams are of lower energy but much higher current. Apart from the high energy component, these features are clearly identified in the sequence of events leading to electric breakdown in the pulsed capillary discharge. Auxiliary triggering inside the HCR provides an additional source of electrons to enhance the electron beam emission. A single electron beam pulse is observed and electric breakdown is achieved in a time much shorter than in self-breakdown operation. In the late case, the formation of the electron beam relays purely on self-sustained ionization processes inside the HCR, and the two periods observed, before and at the time of electric breakdown, are characteristic of a self-breakdown THCD operating on a much lower aspect ratio (Favre et al., 1992). The spectroscopic data clearly shows the fast evolution of the capillary plasma, as well as the lack of significant continuum radiation over the wave length investigated. The plasma channel is heated up very rapidly, in the sub-10 ns time scale, within a few ns after current rise. The partial voltage recovery after the fast current pulse is a good indication that the initial discharge channel has been initiated along the capillary axis, away from the wall. Expansion of the plasma channel results in material being ablated from the capillary wall, thus leading to a fast cooling, with a subsequent increase in plasma resistivity. As a result of this, VUV to XUV light is emitted only in a very short pulse associated with the duration of the current.

5. Conclusion

Our results show that the pulsed capillary discharge can be initiated successfully with the assistance of the hollow cathode effect. The resulting discharge is characterized by a short, less than 5 ns half width current pulse, with an associated XUV light pulse. The enhanced HCE due to the auxiliary discharge allows the XUV light pulse to be synchronized to external events. Further work is being carried out to identify the relevant lines of the recorded spectra.

Acknowledgments

This investigation has been funded by FONDECYT project 1950798 and a CNRS-CONICYT collaboration program. J. Moreno holds a postgraduate scholarship from CONICYT.

References

Choi, P. et al.: 1987, *IEEE Trans. Plasma Sci.* **15**, 428.
Choi, P. et al.: 1993, *Phys. Lett.* **63**, 2750.
Chuaqui, H. et al.: 1989, *Appl. Phys. Lett.* **55**, 1065.
Favre, M. et al.: 1992, *IEEE Trans. Plasma Sci.* **20**, 53.
McCorkle, R.A.: 1981, *Appl. Phys. A* **A26**, 261.
Rocca, J.J. et al.: 1994, *Phys. Rev. Lett.* **73**, 2192.

CURRENT SHEATH STUDIES IN A SMALL PLASMA FOCUS OPERATING IN HYDROGEN–ARGON MIXTURES

M. FAVRE, P. SILVA, H. CHUAQUI and E. WYNDHAM
Facultad de Física, Pontificia Universidad Católica de Chile, Casilla 306, Santiago 22, Chile

P. CHOI
LPMI, Ecole Polytechnique, Palaiseau 91128, France

Abstract. We present preliminary time and space resolved studies of current sheath formation in Plasma Focus discharges, using a novel array of non-invasive magnetic field probes. The experiments are performed in a Mather type plasma focus, operating at 2 kV. The discharge is formed between a hollowed anode and six symmetrically arranged cathode rods. The array of small magnetic probes is located along the cathode rods. The probes are of millimeter size. They are shielded behind the rods, as to minimize capacitive coupling to the anode electrode, and allow non-perturbing measurements to be made. A simple analytical model of current sheath evolution is used to analyze the probe signals. The experiments have been performed in pure Hydrogen and Hydrogen with Argon mixture, at pressures from below 0.2 Torr upwards. The effect of the Argon mixture on the current sheath structure is investigated with the probe array. It is found that at constant mass density operation, the increase in the percentage of Argon results in a thinner sheath, with steeper current profile.

1. Introduction

It has been long recognized that overall performance in plasma focus operation depends critically on initial breakdown and current sheet formation. Although several studies on initial breakdown or current sheath formation and evolution have been reported (Donges et al., 1980; Ananin et al., 1978), most of the measurements rely on invasive probes or time integrated optical observations (Kwek et al., 1990; Bruzzone et al., 1991), which, as a whole, do not provide accurate enough information for quantitative studies. In particular, signals from mm size probes inserted into the plasma are difficult to interpret due to plasma shielding and skin depth effects. The structure of the current sheath depends strongly on the operating pressure and the presence of impurities. In this work we present preliminary time and space resolved studies of current sheath formation in Plasma Focus discharges in Hydrogen-Argon mixtures, using an array of non-invasive magnetic field probes.

2. Experimental Apparatus

The experiments are performed in a Mather type plasma focus, which is powered by a 9 μf capacitor bank, operating at 20 kV. The discharge is formed between a

Astrophysics and Space Science **256:** 485–490, 1998.

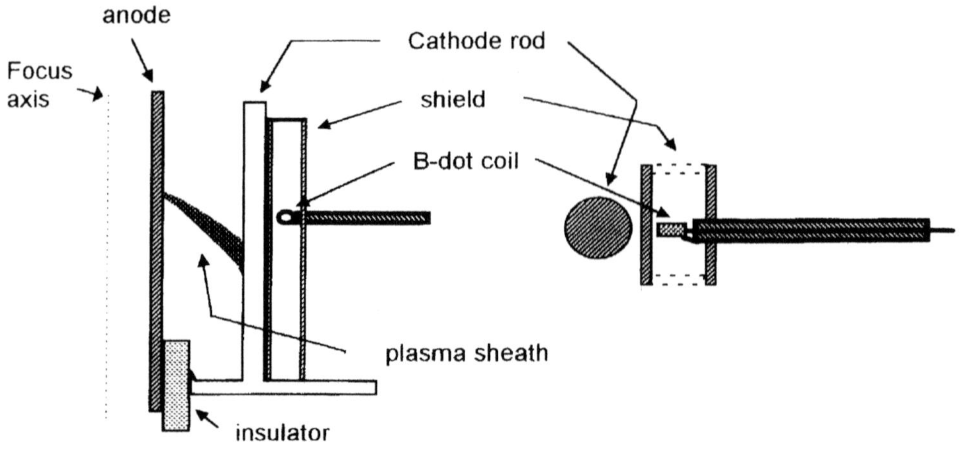

Figure 1. Schematic of the B-dot probe.

hollowed anode and six symmetrically arranged cathode rods. Further details on the experimental set-up are presented in a companion paper in these proceedings (Favre et al., 1998). The array of small magnetic probes is located along the cathode rods. The probes are made of five turns of thin copper wire, forming a 1.5 mm diameter coil. They are electrostatically shielded and located behind the rods, away from the centre anode, as to minimize capacitive coupling to the anode electrode and plasma effects.

A schematic of the probe design is shown in Figure 1. Such arrangement allows non-perturbing measurements to be made. The probes have been used in two different configurations: as a symmetric array at a given height along the focus electrodes, to study sheath uniformity and symmetry, and arranged in a row of up to six probes along one of the cathode rods, to study space and time evolution of the current sheath during the run down phase of the focus discharge. The array probe signals are recorded simultaneously, with 2 ns time resolution. The experiments have been performed in pure Hydrogen and Hydrogen with Argon admission, at pressures from below 0.2 Torr upwards. Other available diagnostics include voltage and current probes in the external circuit, filtered PIN diodes located side and end on, and a multi-pinhole X-ray camera.

A simple analytical model has been used to interpret the probe signals. The model assumes a zero thickness sheath, which moves at a given angle with respect to the anode. The magnetic fields due to the anode, cathode and sheath currents are calculated analytically, for different positions along the axial direction. A typical current sheath velocity, which is inferred from the actual probe array data, is used to obtain a time dependent magnetic field, at a given probe location. A sinusoidally rising current is assumed. A moving finite thickness sheath is obtained by superposition of several axially displaced zero thickness sheaths. Actual probes signals are compared with the time derivative of the magnetic field signal obtained with the

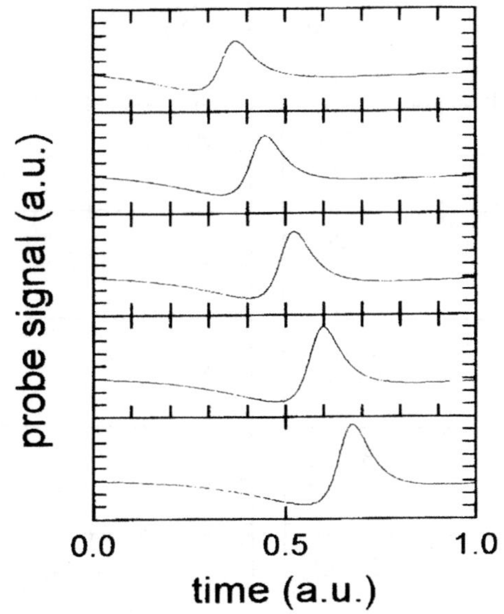

Figure 2. Calculated probe signal for a zero thickness sheath moving at 60° with respect to the anode.

model. Figure 2 shows the time evolution of a five probe array, for a zero thickness sheath, which forms a 60° angle with respect to the anode. The calculated signals allows to correlate the peak of the probe signal with the passage of the central part of the sheath and the width at the base with the actual thickness of the sheath.

3. Experimental Results

Figure 3 shows characteristic signals obtained with shots at 1.3 Torr, pure Hydrogen (Figure 3a), and 0.27 Torr, with a mixture of 80% Hydrogen/20% Argon (Figure 3b), which corresponds to the same mass density as in the pure Hydrogen case. V is the voltage, dI/dt is the time derivative of the discharge current, PIN-A is a PIN diode located axially, at 40 cm from the focus region, with a 2 μm Cu filter, and B1 to B5 are signals from the B-dot probe array, with individual probes located along a single cathode rod, at different axial positions, starting at 4.5 cm from the base plate, with each one separated by 1.0 cm. Superimposed on all traces, on an enlarged vertical scale, is the discharge current. Maximum current is $\sim$ 140 kA. The current trace indicates that at this mass density, focus occurs at peak current. Strong focus action at both pressures is inferred from the current dip, voltage spike and X-ray signal in the PIN diode.

Table I shows characteristic sheath thickness and velocity for different Hydrogen-Argon mixing ratios, at constant gas mass density an decreasing pressure. $V_{B1\rightarrow B3}$ and $V_{B3\rightarrow B5}$ are the mean sheath velocity between positions of probes B1

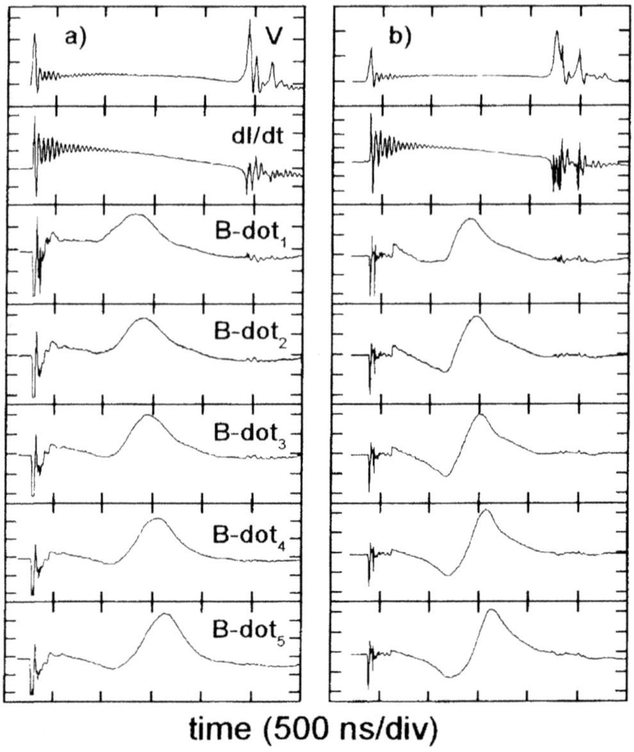

Figure 3. Electric, X-ray and B-dot signals at: (a) 1.3 Torr, pure H_2, and (b) 0.27 Torr, 80%H_2 + 20%Ar.

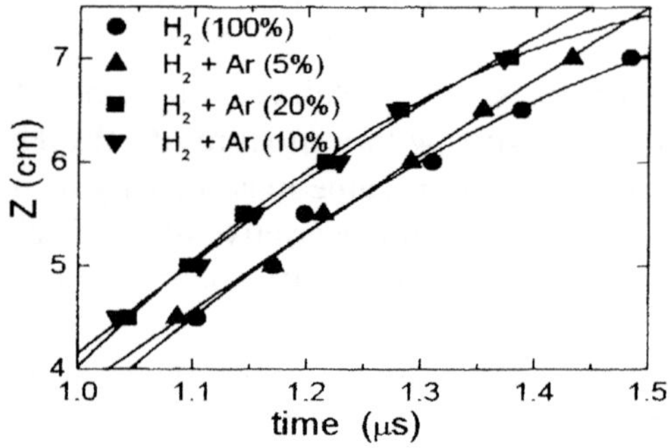

Figure 4. Axial position of the current sheath, as a function of Argon percentage.

and B3, and probes B3 and B5, respectively. ΔB1 and ΔB5 are the sheath thickness at probes B1 and B5.

TABLE I
Characteristic sheath velocity and thickness at different H_2/Ar ratios.

GAS	P (Torr)	$V_{B1 \rightarrow B3}$ (cm/μs)	$V_{B3 \rightarrow B5}$(cm/μs)	ΔB1(mm)	ΔB5(mm)
H_2, 0%Ar	1.30	7.9	8.3	7.3	4.8
H_2, 5%Ar	0.66	7.9	8.7	7.3	3.8
H_2, 10%Ar	0.45	6.4	10.2	5.9	4.2
H_2, 20%Ar	0.27	6.4	12.4	5.8	4.3

4. Discussion

Three conditions on the current sheath should be satisfied in order to achieve an efficient high energy density pinch at the compression stage: (i) uniform initial breakdown across the insulator surface, (ii) formation of a thin, axially symmetric and uniform sheath which propagates axially during the rundown phase, with high snowplough efficiency, and (iii) current sheath reaching the axis at the time when the discharge current is maximum. Condition (i) is determined mainly by the geometric features of the electrode-insulator arrangement and the electrical characteristics of the electric circuit associated with the focus device. In particular, efficient breakdown is found to be enhanced if an initial high frequency, high voltage oscillation waveform is applied. Further discussions on the effect of initial high frequency, high voltage oscillation are presented elsewhere (Choi et al., 1996). If the condition (i) is accomplished, condition (ii) depends on the ionization and radiation properties of the particular gas or gas mixture being used and the applied voltage. Condition (iii) depends on the electrode geometry, mass density of the working gas and rate of rise of the discharge current. The voltage and dI/dt signals shown in Figure 3 exhibit a high frequency oscillation which is a result of the design features of the electric circuit of the device. Under these conditions, uniform initial breakdown and current sheath initiation are achieved, as it is inferred from the regular, single humped time evolving B-dot signals. Further confirmation of efficient initial breakdown has been obtained with observations using three symmetrically arranged B-dot probes, which show identical signals. Figure 3 also shows that keeping the mass density constant at an optimized value for a given applied voltage, the formation of the dense pinch at focus compression can take place at peak current, regardless of the mass mixture ratio and gas pressure. This has been observed over a wide H_2/Ar mixing ratio, up to pressures below 0.2 Torr. The simple model of B-dot probe signals presented above allows some particular characteristics of the axial evolution of the current sheaths at different H_2/Ar to be extracted from the experimental data. Despite the limitations of the model, which assumes a flat current sheath, perpendicular to the electrodes, instead of a curved structure due to the 1/r dependence in the magnetic field, the main features of the

B-dot probe signals are reproduced by the model, as it can be seen when comparing Figure 1c with B3 in Figure 3a. The peak in the probe signal corresponds to the center of the current sheath, whereas the two minima observed at each side of the central peak define the width of the sheath. Mean sheath velocities shown in Table I indicate that at low Ar content, the initial velocity tends to be lower than at higher Ar content, whereas at the final stages, before radial collapse, the characteristic axial velocity increases with increasing Ar content. On the other hand, the initial sheath thickness is found to be smaller at higher Ar content, becoming more or less independent of the gas composition near the end of the axial phase. Although the time from initial breakdown to axial pinch formation is the same for both conditions in Figure 3, the B-dot probe array signals show clear differences in the shape and time evolution of the corresponding current sheaths. Besides an increase in sheath velocity and a decrease in sheath thickness as the sheath propagates axially, the B-dot probe signals in Figure 3a exhibit a clear time symmetry around B3. This is not observed in Figure 3b, where the signal evolves continuously, with a changing time dependent profile. To first approximation, this signal behavior can be interpreted as an indication of a transition from a near constant sheath current density profile when operating in pure hydrogen, to a time dependent, increasingly steeper sheath current density profile, as a result of the enhanced ionization due to the argon admixture. Further interpretation of the B-dot probe data will require an improved analytic model of the probe signals, which includes more realistic finite thickness effects and curvature.

Acknowledgments

This work has been partially funded by a doctoral research grant from FONDECYT. P. Silva acknowledges a scholarship from CONICYT.

References

Ananin, S.I. et al.: 1978, *Sov. J. Plasma Phys.* **4**, 175.
Bruzzone, H. et al.: 1991, *Meas. Sci. & Tech.* **2**, 1195.
Choi, P. et al.: 1996, in: *Proc. BEAMS'96*, Prague, Czech Republic, p. 570.
Donges, A. et al.: 1980, *Phys. Letters* **76A**, 391.
Favre, M. et al.: 1998, *Astrophys. Space Sci.* **256**, 337.
Kwek, K.H. et al.: 1990, *IEEE Trans. Plasma Sci.* **18**, 826.

HARD X-RAY DIAGNOSTIC OF Z-PINCH DISCHARGES

A. ROBLEDO-MARTINEZ
Dept. Energia, Universidad Autonoma Metropolitana, Azcapotzalco DF, 02200 Mexico City

R. ALIAGA-ROSSEL, I.H. MITCHELL, J.P. CHITTENDEN, A.E. DANGOR and M.G. HAINES
The Blackett Laboratory, Imperial College, London SW7, U.K.

Abstract. An experimental technique for the detection and time-resolved characterisation of hard X-radiation from a Z-pinch is described. The detectors employed consist of combinations of plastic scintillators and photomultipliers tubes coupled through optical fiber. The scintillators were fitted with thick aluminium and lead filters and cross filtering is then used to obtain the energy of the X-rays. It was found that the X-rays can reach energies of up to 4 MeV, much higher than the applied voltage. The observed radiation intensity varied with the disruption time lag.

1. Introduction

Hard X-rays are a common feature in plasma focus, terawatt fiber z-pinches, vacuum diodes and gas-puff z-pinch discharges. It has been noted in these investigations that the onset of the radiation coincides with a dip in the current signal and is associated with a pinch of the plasma. The pinch produces beams of energetic electrons, with energies that often are in excess of the energy that would otherwise be imparted on them by the applied voltage alone. When the e-beams hit the metallic electrodes or wall chamber hard X-rays are emitted. This regime of the discharge is often referred to in the literature as the disruption.

The electrons are stopped by the metal wall and emit radiation through the bremsstrahlung mechanism. This radiation has a continuum spectrum plus some low-energy characteristic lines (which are not relevant to the range of energies involved here). The intensity of the continuous X-radiation emitted is a function of the energy of the electron beam and the chemical nature of the target electrode. Thus the intensity $I(E)$ in the energy interval $E + dE$, per incident electron in a monoenergetic beam, is given by Kramers' formula as:

$$I(E)dE = kZ(E_0 - E), \quad E \leq E_0, \tag{1}$$

where E_0 is the energy of the electron, k a constant and Z the atomic number of the target electrode. This spectrum has been shown to be adequate for electrons with energies in the range from tens to hundreds of keV (Dyson, 1990). Measurements made at higher energies (5.3 MeV) showed that Equation (1) also provides a good fit to the measured spectrum (O'Dell et al., 1968; Ferdinande et al., 1971). Figure 1

Astrophysics and Space Science **256:** 491–497, 1998.

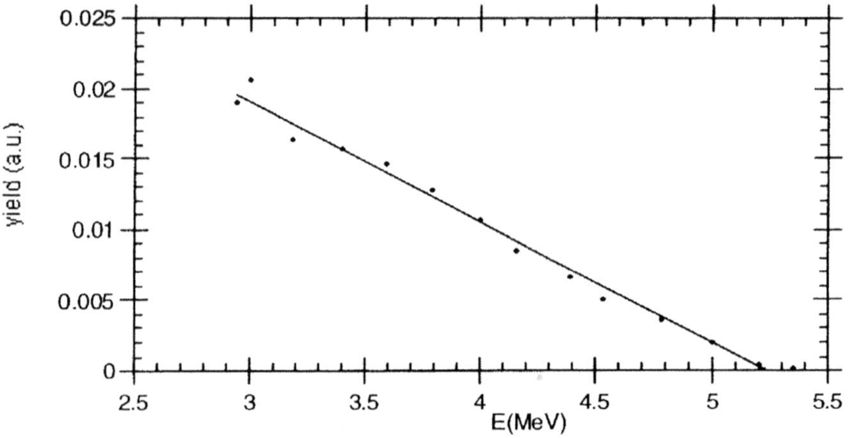

Figure 1. X-ray spectrum of a 5.3-MeV e-beam. Experimental data from O'Dell et al. (1968), corrected by Ferdinande et al. (1971).

shows the hard X-ray spectrum emitted by a monoenergetic 5.3 MeV beam; note that the experimental data fits Equation (1) reasonably well.

2. Experimental Set-Up

2.1. Diagnostics

For the experiments, MAGPIE, a terawatt, 2.4 MV, 336 kJ generator (Mitchell et al., 1996) was used. Its design allows the current to flow into the load in about 150 ns. Two different loads were used in the experiments, 33 μm diameter carbon fibers and aluminum 25μm diameter fibers. In both cases the length of fiber was 25 mm. A complete suite of diagnostics including soft and hard X-rays detectors, laser probes, streak and frame imaging, optical spectrometry as well as pinhole and streak photography were employed. Figure 2 shows a top view of the discharge chamber and the layout of the main diagnostics (the pinch axis is perpendicular to the plane of the drawing).

2.2. Hard X-Ray Detectors

The hard X-ray detectors used were developed specifically for use in electrically noisy environments. They are based on the conventional photomultiplier (PMT) and scintillator combination where problems can arise from placing photomultipliers close to the source of noise. However, placing the photomultipliers at some distance from the scintillator introduces a different problem because the wavelength of the scintillator emission is greatly attenuated by fibre optics required to link the two together. This problem is overcome by wrapping a fluorescent optical

TABLE I
Filters employed.

Detector No.	Filter
1	3 mm Al
2	3 mm Al plus 2.7 mm Pb
3	3 mm Al plus 5.4 mm Pb

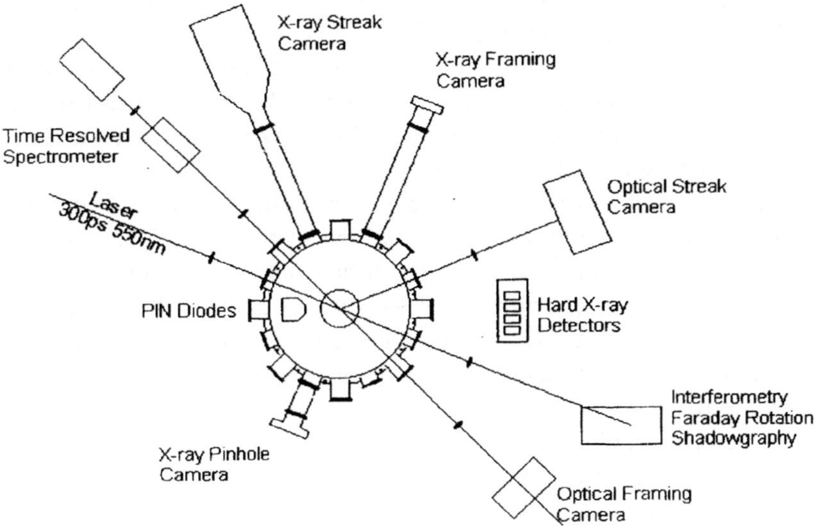

Figure 2. Layout of the diagnostics and the discharge chamber in MAGPIE.

fiber around a block of the scintillator, NE102A in this case. The output from the NE102A, due to the incident X-radiation, is stepped down in wavelength to the green by the fluorescent fibre. This is then coupled to a standard multi-mode optical fiber and relayed to a photomultiplier tube in the screened room (Choi and Aliaga-Rossell, 1994). The PMT output is then recorded on the digital oscilloscope. An array of three of these detectors is mounted outside the vacuum chamber at about 1m away from the pinch. Apart from the viewing aperture, they are screened by 5 cm of lead. Each detector was fitted with a different filter, as shown in Table I. The combined detector response is in the range 200 keV to 6 MeV.

When an X-ray pulse is incident upon a filtered detector the resulting energy-integrated signal reflects the energy response of both the filters and the detectors.

If the pinch is assumed to emit radiation with a given spectral shape $I(E)$ then the detected signal amplitude $v(t)$ is given by (Knoepfel, 1979):

$$v(t) = K \int_{0}^{E_0} I(E, t) R(E) f(E) dE, \tag{2}$$

where $R(E)$ is the filter transmission, E_0 the energy of the e-beam, K a constant, $f(E)$ the scintillator's spectral response and $I(E, t)$ the emission spectrum in Equation (1). The filter transmission was calculated using the XRD program. The scintillator response as a function of energy was obtained from available experimental data (Roulston and Naqvi, 1957).

The voltage output expected from each detector for a given energy of e-beam is calculated by numerically integrating Equation (2) using the relevant filter transmission. This is carried out for various energies between 300 keV and 4 MeV. Figure 3 shows the calibration curves employed in the measurement of the X-ray energies. Curves E_{12} and E_{13} give the calculated X-ray energy as a function of signal ratio for detectors 1–2 and 1–3, respectively (see Table I).

The main source of error in the present measurements comes from the experimental determination of the signal ratio which in turn is a function of: (a) the different gain of the detectors and (b) the (6-bit) A-D conversion rate. The value of K in Equation (2) is determined by the overall gain of the system after the scintillator which depends upon the losses due to the coupling of fiber and the PMT and the gain of the PMT. In practice the value of K is different for each detector. In order to obtain the relative gains a series of tests were carried out fitting identical filters to the detectors and comparing the measured signals. A linear regression was applied to the resulting data giving the relative gains of detectors 1 and 2 and of 1 and 3. For signal ratios closer to unity a small error in the ratio is reflected in a significant error in the energy value due to the steepness of the curve. For smaller amplitudes the sampling error becomes a significant source of error.

3. Results and Discussion

Hard X-rays were observed at about 120–200 ns after current start and always coincided with a dip in the di/dt signal. Figure 4 shows how the onset of X-radiation (which is shown as negative intensity for the sake of convenience) coincides with the dip. Optical streak images show that the plasma at this time is in a series of density islands with a tenuous plasma in between. Measurements made with ancillary PIN diodes show that the hard X-ray emission originates in the anode.

The intensity of the radiation and its energy varied from shot to shot but some general trends could be established: (a) aluminium fibers gave lower energy maxima than carbon, (b) the intensity of the radiation is proportional to the magnitude of the dip in the dI/dt signal, (c) earlier disruptions are more intense than late

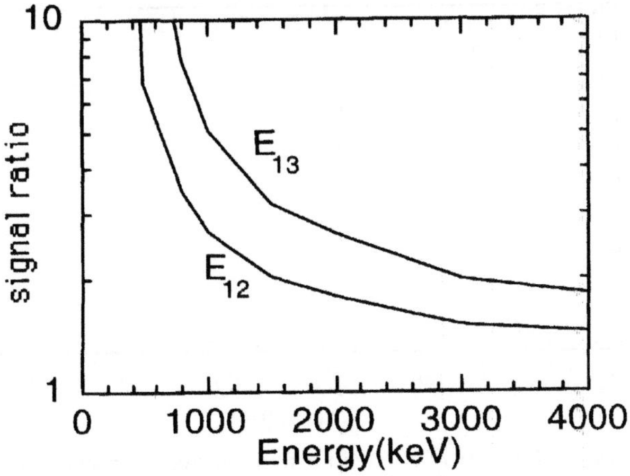

Figure 3. Detector calibration curves. E_{12} and E_{13} correspond to the signal ratios of detectors 1–2 and 1–3, respectively.

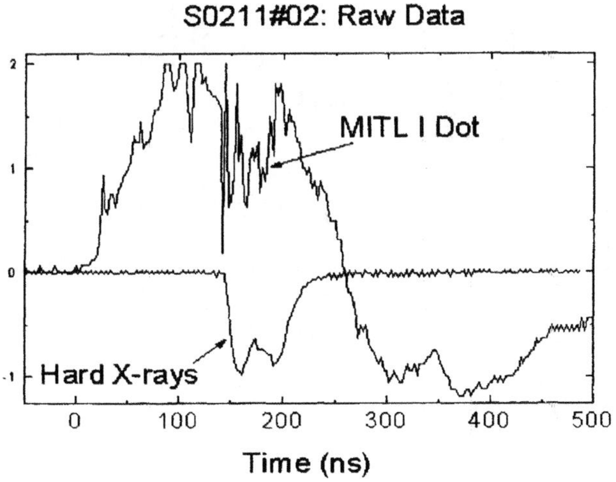

Figure 4. Onset of X-rays and dip in di/dt signal. Current starts at $t = 0$, thus the onset delay shown is 120 ns.

ones. Figure 5 shows the intensity of hard X-rays plotted as a function of delay for 9 carbon shots. This figure shows that, except for one case, the intensity decreases with increasing onset delay (as measured from current start).

Figure 6 shows the time-resolved X-ray energy measurement for an aluminum shot. The measured energies have typical values of 1–2 MeV. The energy of the X-rays in carbon shots was higher (almost 4 MeV) especially at the beginning of the pulse but then decreased to a more typical value of 1.5–2 MeV. Sometimes the energy also increased again in the tail of the pulse. Aluminium shots, on the

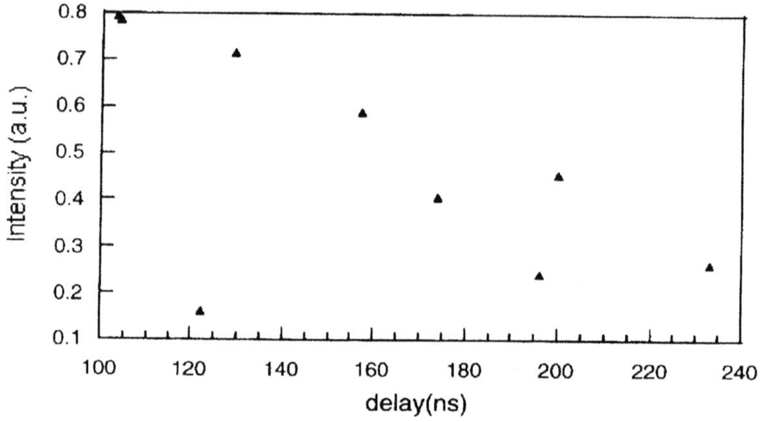

Figure 5. Intensity of X-radiation as a function of onset delay.

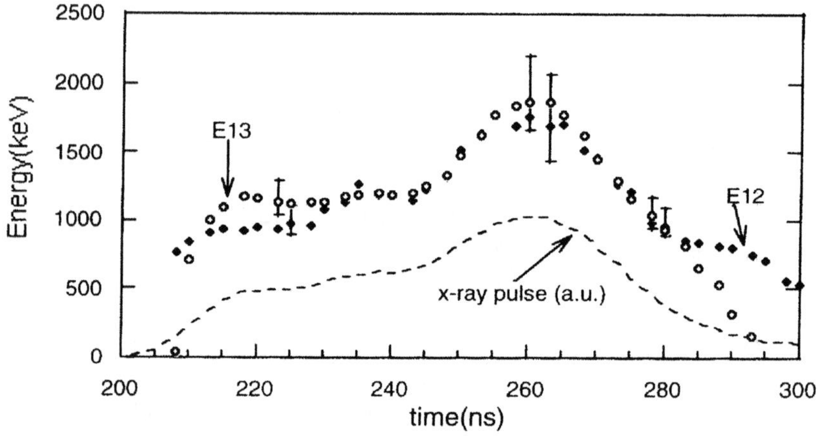

Figure 6. Energy measurement of hard X-rays from a 25 μm-aluminum fiber pinch. $t = 0$ corresponds to current start.

contrary, always reached the maximum energy at or near the intensity peak. Figure 6 shows that, within the experimental error, the energy measured with detectors 1–2 is in good agreement with that of 1–3. This means that the assumption of a monoenergetic beam implicit in Equation (1) is reasonable (Robledo et al., 1997).

During the disruption two mechanisms are thought to be at work: the transient necking off of the pinched plasma to form density islands could produce an energetic singular ion beam with off-axis oppositely moving guiding centre ions carrying the equal and opposite momentum; electrons will similarly be energised but their guiding centre motion off-axis will give a slow motion to the anode. The low density plasma between the density islands will probably have an anomalous resistivity, and the high voltage across this region will also permit a runaway diode action.

4. Conclusions

The intensity and energy of the hard X-rays produced by a z-pinch discharge were experimentally measured. The emission takes place when the plasma undergoes disruption. This is associated with the breakup of the plasma column into density islands. It is thought that this necking off of the plasma provides the accelerating mechanism for the e-beams that ultimately produce the radiation. The measured energy of the beams is in the range 1–4 MeV, much higher than the applied voltage.

Acknowledgments

This work was funded by the EPSRC-UK. A. Robledo-Martinez thanks the Mexican Science Foundation (Conacyt) and SN1 for financial support.

References

Choi, P. and Aliaga-Rossell, R.: 1994, *Rev. Sci. Instrum.* **65**, 3034.

Dyson, N.A.: 1990: *X-Rays in Atomic and Nuclear Physics*, 2nd ed., Cambridge University Press, London.

Ferdinande, H., Knuyt, G., Van de Vijver, R. and Jacobs, R.: 1971, *Nucl. Instr. Meth.* **91**, 135.

Knoepfel, H.: 1979, in *Diagnostics for Fusion Experiments*, Varenna, Pergamon Press, Oxford, p. 111.

Mitchell, I.H., Bayley, J.M., Chittenden, J.P., Worley, J.F., Dangor, A.E. and Haines, M.G.: 1996, *Rev. Sci. Instr.* **67**, 1533.

O'Dell, A.A., Sandifer, C.W., Knowlen, R.B. and George, W.D.: 1968, *Nucl. Instr. Meth.* **61**, 340.

Robledo, A., Mitchell, I.H., Aliaga-Rossel, R., Chittenden, J.P. Dangor, A.E. and Haines, M.G.: 1997, *Phys. Plasmas* **4**, 490.

Roulston, K.I. and Naqvi, S.I.H.: 1957, *Nucleonics* (October issue), 86.

THERMAL DECOMPOSITION OF ILMENITE IN A NON-TRANSFERRED ARC THERMAL PLASMA FLOW REACTOR

M. MANRIQUE, T. FIGUEIRA and J. GÓMEZ
Department of Materials Science, Simón Bolívar University, Caracas, Venezuela

P.R. TAYLOR
Department of Metallurgical and Mining Engineering, College of Mines, University of Idaho, Moscow, Idaho, 83843, U.S.A.

Abstract. The thermal decomposition of ilmenite concentrates in a non-transferred arc thermal plasma flow reactor was investigated. The plasma torch was operated using argon-helium and argon-nitrogen mixtures as plasma gases. The effect of power, feed rate, and particle size was studied. Products were characterized by XRD, SEM, TEM and ICP techniques. Free energy minimization plots were used to predict the thermodynamic behavior of the system under different operating conditions. The results of this investigation showed a significant segregation process where titanium-rich products are collected in the reactor, while iron-rich products are collected in the quenching box and filter. The average size of the product is in the submicron range. Titanium nitride was formed and condensed in the reactor zone when using argon-nitrogen mixture as plasma gas.

1. Introduction

Titanium is usually produced from two raw materials, rutile (natural TiO_2) and ilmenite, being the ilmenite the preferred raw material in many processes due to its lower cost and availability. In Venezuela, there are several ilmenite deposits and some of them has been quantified. TiO_2 can be produced by decomposition of ilmenite at high temperatures using plasma furnaces in reducing environments like carbon, hydrogen and methane (Skrivan et al., 1974, Down et al., 1982; Ryoda et al., 1989; Allen, 1987) which result in a consolidated material which must be grounded and chemically treated in order to obtain pure TiO_2. The present study describes the fundamental thermodynamics and the experimental results obtained on the thermal decomposition of ilmenite in a non-transferred arc thermal plasma flow reactor.

2. Thermodynamics

Free energy minimization plots for the Fe-Ti-O system have been developed using the CSIRO Thermopackage (Turnbull and Wadsley, 1988). Plasma gases are also taken into account in these thermodynamic calculations, but are not shown

Astrophysics and Space Science **256:** 499–503, 1998.

in the plots due to their higher content. Figure 1 shows the thermodynamic equilibrium for the system Fe-Ti-O in an argon-helium atmosphere. This plot shows that at temperatures above 5000 K, which are found inside the plasma jet, only the monoatomic gases Fe, Ti, O are stable. In the temperature range 3600–4800 K, the monoatomic gases become unstable and recombine to form new species. The possible reaction that may take place in this zone during the reaction process is:

$$Ti(g) + O(g) = TiO(g). \quad (1)$$

In the temperature range 2800–3600 K, titanium monoxide gas and iron gas become unstable and recombine to form new species. Some of the possible reactions that may take place in this zone are:

$$TiO(g) + O(g) = TiO_2(g), \quad (2)$$

$$Fe(g) + O(g) = FeO(g), \quad (3)$$

$$O(g) = 0.5O_2(g), \quad (4)$$

$$TiO_2(g) = 0.25Ti4O7(l) + 0.125O_2(g). \quad (5)$$

In the temperature range 1600–2800 K, the oxides become unstable and recombine to form ilmenite. Some of the possible reactions that may take place in this zone are:

$$0.25Ti_4O_7(l) + Fe(g) + 1.25O(g) = FeTiO_3(s), \quad (6)$$

$$0.25Ti_4O_7(l) + Fe(g) + 0.625O_2(g) = FeTiO_3(s), \quad (7)$$

$$TiO_2(g) + Fe(g) + O(g) = FeTiO_3(s), \quad (8)$$

$$TiO_2(g) + Fe(g) + 0.5O_2(g) = FeTiO_3(s). \quad (9)$$

At temperatures below 1750 K ilmenite decomposes to form FeO(l) and $FeTi_2O_5$(s), which remains stable down to room temperature. Figure 1 shows that ilmenite may be thermally decomposed at temperatures above 2600 K. If quenched rapidly the product would be Fe(s) and a mixture of titanium oxides. The iron could then be preferentially leached to obtain an oxide powder that could then be oxidized to TiO_2. The product would be very fine sized.

3. Experimental System

The experimental system consists of a tubular reactor, non-transferred arc thermal plasma torch, rapid quench section, filter, powder feeder and data acquisition system. A detailed description can be found elsewhere (Taylor et al., 1995). The solid

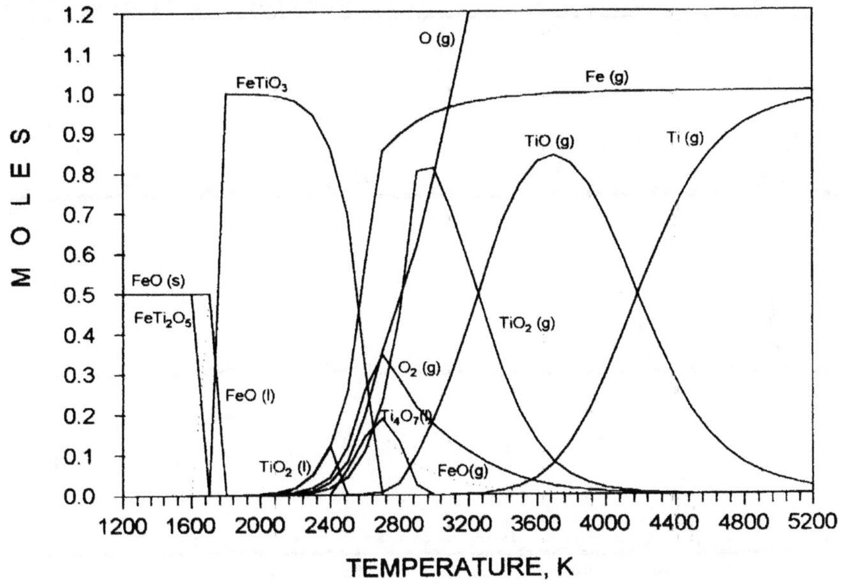

Figure 1. Free-energy minimization plot for the Fe-Ti-O system in argon-helium atmosphere.

product is collected from the quench section, the filter and also from the inside wall of the reactor. Mixtures of argon-nitrogen and argon-helium were used as plasma gas. The ilmenite powder and argon as carrier gas are fed into the reactor using a powder feeder, through a water cooled injection probe inserted at an angle near the plasma jet. The gaseous stream leaving the quench section is passed through a filter before venting to the stack.

4. Experimental Procedure

The ilmenite concentrate used was from the Agua Mena creek heavy minerals deposit in Venezuela. The sample was ground previously in a ceramic jar mill to obtain the required grain size. The plasma torch was started with argon, and after stabilizing, nitrogen was added in the required amounts to reach the operating power. The reactor was preheated until it reached thermal equilibrium before starting the feeding. The feeding time was 15 minutes for most of the experiments. After the feeding was stopped the whole system was shut down and allowed to cool under an argon atmosphere. The product was removed when the system reached room temperature.

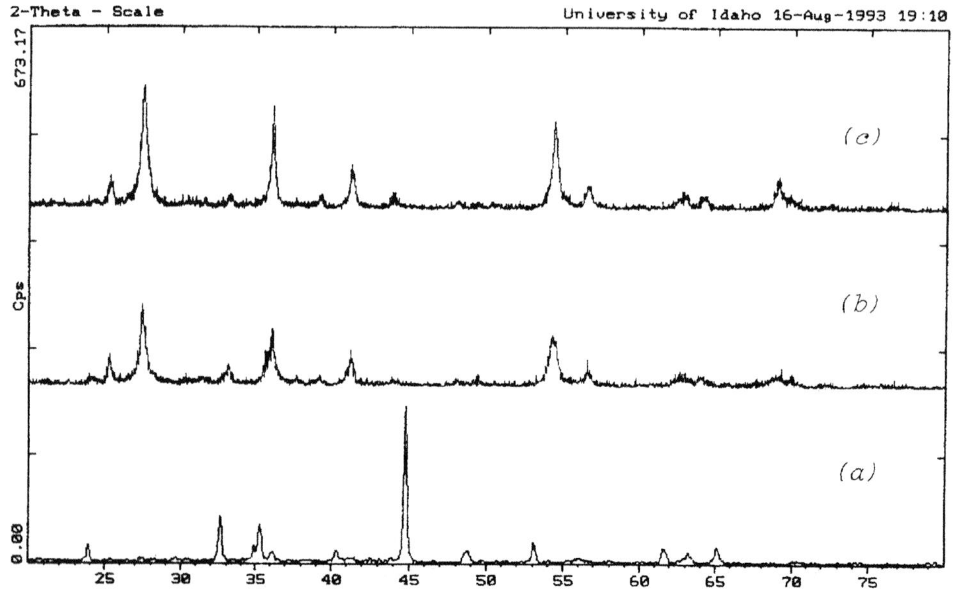

Figure 2. X-ray diffraction pattern: (a) collected powder in quench chamber, (b) sample 5a leached in HCl, and (c) sample 5a leached in H_2SO_4.

5. Results and Discussion

The solid products obtained in each run were characterized physically and chemically. Titanium oxides-metallic iron segregation is one of the main characteristics of these experiments. As a result of this phenomena titanium oxides tend to remain in the reactor while metallic iron is preferentially found in the quenching box and filter products due to its higher volatility. Titanium nitride-metallic iron segregation was observed in those experiments carried out under argon-nitrogen atmosphere. The X-ray diffraction pattern of the solid product collected in the quenching box for one of the test is shown in Figure 2a. This plot shows the diffraction patterns for metallic iron and unreacted ilmenite. Figures 2b and 2c show the diffraction pattern for titanium dioxide (anatase and rutile) found after leaching of the product collected in the reactor with hydrochloric acid and sulphuric acid respectively. The size of these products were in the micron and submicron range. Samples from the different products were analyzed using SEM and TEM techniques.

Figure 3 shows a TEM micrograph of the powder collected in the quenching box after leached with hydrochloric acid. This micrograph shows two different sizes, spherical particles of size 1–2 microns corresponding to unreacted ilmenite and very small particles in the submicron range formed by titanium suboxides. Samples collected in the reactor show diffraction patterns of titanium dioxide and metallic iron being this in less amount.

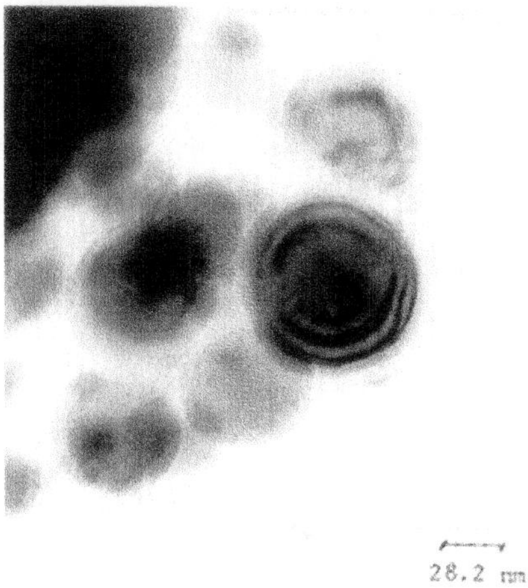

Figure 3. TEM micrograph of the collected powder at the quench chamber.

6. Conclusions

Submicron size mixtures of titanium dioxide and metallic iron powder were obtained by thermal decomposition of ilmenite concentrates in a non-transferred arc thermal flow reactor using argon and helium as plasma gases. The titanium dioxide and metallic iron were segregated during the plasma processing due to the greater volatility of iron during processing, this may be a promising method to produce relatively pure titanium dioxide directly from ilmenite concentrates. Titanium nitride and metallic iron were obtained in the reactor zone when using argon-nitrogen atmosphere. This mixture may be a precursor to the development of a ceramic matrix composite with the desirable hard properties of TiN in an iron matrix. Free energy minimization plots were used to analyze the stability of the different species that may be present in each stage of the thermal decomposition process and were used to confirm the stability regions of the products formed.

References

Allen, B.: 1987, The smelting of ilmenite in a D.C. transferred-arc plasma furnace with a molten anode configuration, *Mintek*, M304.

Down, M. et al.: 1982, Titanium product collection in a plasma reactor, *U.S. Patent* **4**, 356, 029.

Ryoda, S. et al.: 1989, Direct smelting of ores or metal compounds by plasma arc, *PCT Int. Appl.*, 664–680.

Skrivan, J.F. et al.: 1974, Process for the benefication of titaniferous ores utilizing hot wall continuous plasma reactor, *U.S. Patent* **3**, 856, 918.

Taylor, P.R. et al.: 1995, Formation of titanium carbide from ilmenite concentrates in a thermal plasma reactor, *Plasma Chemistry and Plasma Processing* **15**, 545–557.

Turnbull, A.G. and Wadsley, M.W.: 1988, *The CSIRO Thermochemistry System*, Version 5, IMEC, Australia.

SPATIAL AND TEMPORAL BEHAVIOR OF A LASER GENERATED TITANIUM PLASMA

FRANCISCO LIRA and RICARDO CASTELL
Laser Spectroscopy Laboratory, Physics Department,
Universidad Simón Bolívar, Caracas 1080, Venezuela

Abstract. Optical emission studies have been performed on the laser ablation of Titanium metallic targets. The high intensity radiation of a pulsed N_2 laser (337.1 nm; 8 J/cm^2) was focused onto the target to generate a short living (150 ns) plasma cloud that expands away from the surface into vacuum at high velocities. Time resolved measurements were taken as a function of the distance from the studied surface with a spatial rsolution of 20 μm, by collecting the emitted light in a direction perpendicular to the plasma expansion axis. A simple diffusion model with a "source" term is proposed to demonstrate the interaction of the laser beam with the plasma plume, as derived from analysis of the space-time behavior of several UV and visible atomic transitions.

1. Introduction

Plasmas induced by the incidence of laser radiation on metallic surfaces show interesting features when the energy of the laser photons coincides with the energy of a resonant transition in the atomic or molecular system. Despite very extensive experimental work has been reported (Künnemeyer and Kock, 1983; Walkup et al., 1986), many effects of laser beam interaction with the metallic surface and the generated plasma are still not satisfactory. Furthermore, none of them has considered so far the spatial and temporal evolution of the transitions emitted during the interaction. The high degree of ionization reached when duration of the laser pulse exceeds the lifetime of the upper level has been explained (Measures and Cardinal, 1981) by relating the strong increase in the rate of energy accumulation with a sharp decrease of the breakdown threshold when resonance between the radiation and the atomic system was established.

In this work we have used an N_2-laser capable of producing nanosecond pulses of intense UV radiation to study its interaction with the generated plasma. The experiments reported here were done under vacuum. Time measurements with a resolution of $\pm$ 2 ns for both neutral (Ti I) and ionized (Ti II) species were taken as a function of the distance from the surface with a spatial resolution of 20 μm.

Astrophysics and Space Science **256:** 505–509, 1998.

2. Experimental Setup

The N_2-laser used in our experiments was constructed in the Institut für Experimentalphysik V, Ruhr-Universität, Bochum, Germany and described previously (Cubbedo and Curry, 1973; Rebhan et al., 1980). It is capable of producing 15–20 ns FWHM pulses of 337.1 nm radiation with energies between 15 and 20 mJ for a gas mixture of 95% N_2 and 5% SF_6. The laser radiation was focused onto the target by a quartz lens with a focal length of 250 mm producing a sharp 0.4×0.3 mm spot. Spectroscopically pure titanium was used as target material with its surface perpendicular to the laser beam axis. The plasma produced was imaged 1:1 onto the entrance slit (30 μm) of a 1-meter monochromator (McPherson-2051) by two plano-convex UV lenses mounted on a precise translator stage with a three-mirror system that allowed us to rotate 90°, and scan the image of the plasma across the slit in a direction perpendicular to the target surface. The experimental setup is shown schematically in Figure 1. A spatial resolution, in the direction normal to the surface, of about ± 15 μm at the target was obtained. The selected transition was detected by a photomultiplier (RCA 1P28A) coupled to the monochromator. The current pulses from the photomultiplier were monitored with a digitizing oscilloscope (HP 54502 A) with a maximal sampling rate of 400 MHz. The oscilloscope was triggered by a current pulse from a Si photodiode (EG&G FND – 100A) which had a rise time less than 1 ns and was illuminated by one edge of the laser beam. Each laser pulse produced a time-resolved spectrum that was stored in a computer using a interface board (AT-GPIB/NI-488). Intensity variations in the laser line were averaged by taking ten consecutive laser shots. The contour curves finally were obtained by scanning the slit of the monochromator in steps of 0.02 mm starting from the metallic surface.

3. Experimental Results

From the spectra of the titanium plasmas produced in vacuum ($p = 2 \times 10^{-5}$ mbar) and in air (normal pressure) taken between 360 and 380 nm was possible to identify and select the strongest and best isolated Ti II lines near the laser transition. For the region between 630 and 640 nm we selected one Ti I line at 632.4 nm. In the case of the plasmas produced in vacuum, the Ti I lines were very weak in comparison to close-by Ti II lines which were therefore well separated. In the case of plasmas produced in air, the line profiles were broader following the Stark broadening mechanism (Andreic et al., 1993) and a strong continuum showed up as previously reported (Henc-Bartolic et al., 1994). Two Ti II lines, the transition (a^2P-z^2P^0) at 336.6 nm and the transition (a^4F-z^4G^0) at 336.48 nm were selected for the measurements in vacuum.

The typical behavior of the Ti I transitions can be seen in Figure 2 for the line at 632.4 nm. It can be observed that near the metallic surface the line intensity

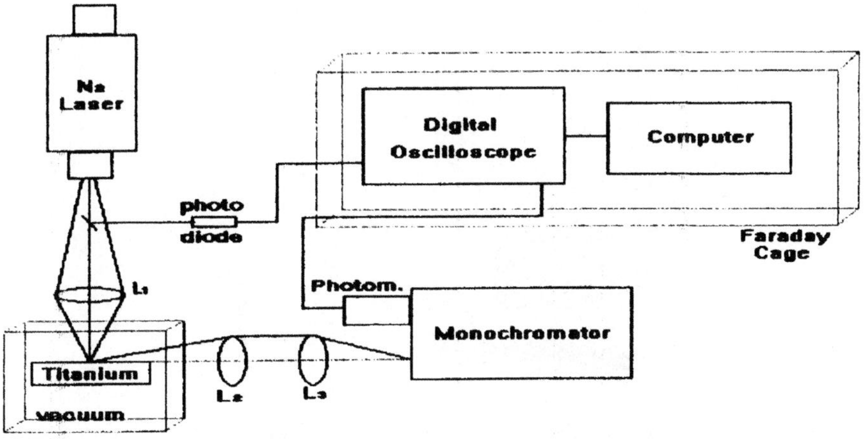

Figure 1. Schematic of the experimental setup for the time and spatial evolution measurements. A three-mirror system (not shown) rotates the plasma image 90° in front of the entrance slit of the monochromator.

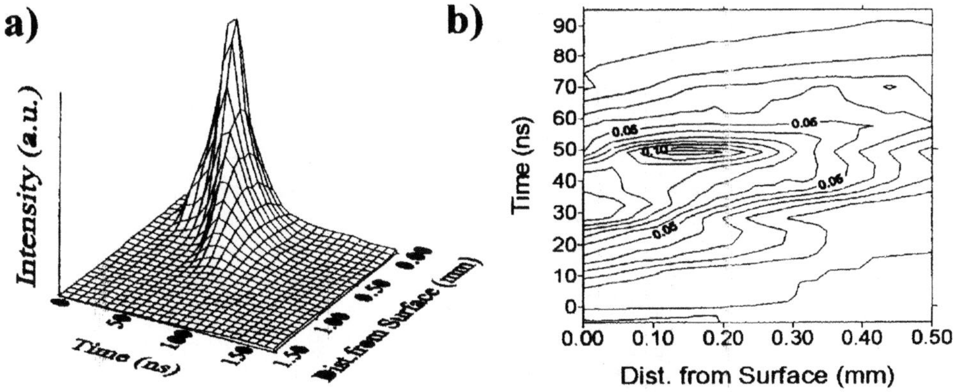

Figure 2. (a) Intensity distribution and (b) contour curves for the Ti I line at 632.4 nm emitted from the laser generated plasma in vacuum (6 × 10^{-6} mbar) as a function of time and distance from the metallic surface.

follows the laser intensity with a delay of ~ 10 ns but away from the surface (~ 0.15 mm) such a delay is larger (~ 20 ns) while the intensity of the transition has increased. For the Ti II transition (a^2P-z^2P^0) at 336.6 nm, close to the laser line and with a lower level that does not decay further by photon emission, the contour curves (Figure 3) show a different feature: its intensity follows rather close the laser reaching its maximum value at the same time or even before the laser pulse does, for distances up to ~ 0.40 mm from the surface.

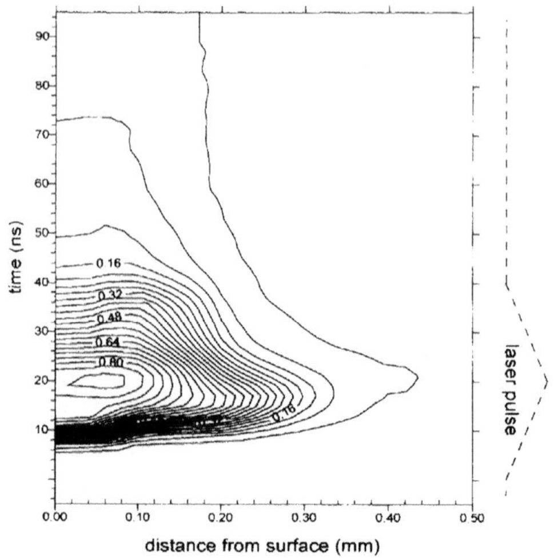

Figure 3. Contour curves for the intensity of the Ti II line at 336.6 nm emitted by the plasma in vacuum (2×10^{-6} mbar) as a function of time and distance from the surface. Laser pulse time history is shown.

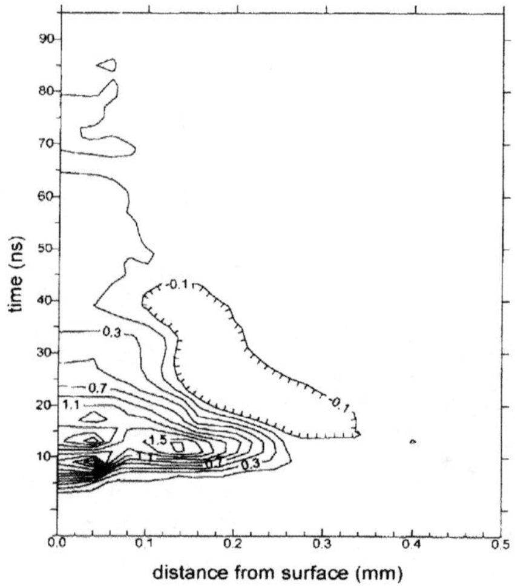

Figure 4. Calculated "source" term for the Ti II line at 336.6 nm in vacuum. Laser-plasma interaction is observed during the first half of the laser pulse.

4. Discussion and Conclusions

A model based only upon the interaction between the laser beam and the metallic surface, as the source of the plasma emitting species, could not be fitted to the experimental results. An additional process that populates the upper states of the observed transitions has to be considered. Assuming that the intensity of emission

(E [s^{-1} m^{-1}]) depends from the upper state density (ρ [m^{-1}]) as: $E = \rho/t$, and applying the continuity equation, one can write:

$$\frac{Q}{\eta + 1} = \tau\frac{dE}{dt} + \tau\frac{d(\bar{V}.E)}{dx} + E,$$

where τ [s] is the lifetime of the transition, $\bar{V}$ [m/s] is the mean velocity of the expanding atoms at a given distance and time from the surface, η is the ratio between the number of atoms in the upper level that decay to other levels to the ones that decay through the specific transition, and Q [s^{-1} m^{-1}] is a source term responsible for the additional population increase of the upper state. In order to estimate $\bar{V}$ we have considered that, in vacuum, the atoms are emitted from the surface with a Maxwellian distribution, move away thereafter with a constant velocity, and that the amount of ablated atoms is proportional to the laser (gaussian) intensity at each time.

The results obtained with this model for the "source" term, $Q/\eta + 1$, from the data for the Ti II transition at 336.6 nm are shown in Figure 4. It can be seen that the "source" term becomes relevant only while the laser pulse lasts, reaches its maximum value at a distance of $\sim$ 0.135 mm from the surface and peaks in time even before the laser pulse does. This evidences the interaction laser-plasma by the absorption of the radiation that pumps the upper level and prevents the laser from further reaching the titanium surface. During the second half of the laser pulse, the plasma starts to cool down and the recombination process becomes dominant as can be inferred from the spatial and temporal behavior of the Ti I transition at 632.4 nm.

From our results we conclude that the incidence of the laser radiation creates in vacuum a plasma that expands with an average velocity $\sim$ 8 Km/s, and thereafter the duration of the laser pulse radiation allows its interaction with the atoms in the upper energy levels by introducing an additional source of plasma radiation for the Ti II transitions. On the other hand, the source of the Ti I radiation, is mainly due to the recombination process and not to direct excitation from the laser.

References

Andreic, Z., Henc-Bartolic, V. and Kunze H.-J.: 1993, *Physica Scripta* **47**, 405.
Cubbedo, R. and Curry, S.M.: 1973, *IEEE J. of Quant. Elect.* **QE-9**, 499.
Dreyfus, R.M., Kelly, R. and Walkup, R.E.: 1986, *Appl. Phys. Lett.* **49**, 1478.
Gingelbach, R.M. and Wentzek, P.L.G.: 1991, *Appl. Phys. Lett.* **58**, 1597.
Henc-Bartolic, V., Andreic, Z. and Kunze H.-J.: 1994, *Physica Scripta* **50**, 368.
Herman, J., Boulmer-Leborgne, C. and Dubreuil, B.: 1990, *Appl. Surf. Sci.* **46**, 2291.
Künnemeyer, R. and Kock, M.: 1983, *J. Phys. B: At. Mol. Phys.* **16**, L607.
Measures, R.M. and Cardinal, P.G.: 1981, *Phys. Rev.* **A23**, 804.
Rebhan, U., Hildebrandt, J. and Skopp, G.: 1980, *Appl. Phys.* **23**, 341.
Rothenberg, E. and Koren, G.: 1984, *Appl. Phys. Lett.* **44**, 664.
Walkup, R.E., Jasinski, J.M. and Dreyfus, R.M.: 1986, *Appl. Phys. Lett.* **48**, 1690.

ION-GAS INTERACTION IN A VACUUM ARC AT INTERMEDIATE PRESSURES

D. GRONDONA, H. KELLY and A. MÁRQUEZ
Instituto de Física del Plasma (CONICET), Departamento de Física, Facultad de Ciencias Exactas y Naturales (UBA), Ciudad Universitaria Pab. I, 1428 Buenos Aires, Argentina

Abstract. Measurements of the ion flux as a function of the pressure in a plasma arc are presented. We report data on Ti and Pt cathodes using N_2 and Ar as the filling gases. Values of an effective collision cross section derived from the attenuation measurements are inferred.

1. Introduction

It is a well known fact that in plasma arcs operated at moderate currents (below a few kA, without erosion of the anode) the cathode is an intense source of metallic ions through the cathodic spot mechanism (Kutzner and Craig Miller, 1992). In a vacuum arc (metallic plasma), the ionic current fraction reaches $\approx 8\%$ of the total current (Kimbling, 1974; Davis and Craig Miller, 1969). However, when the arc is operated in a gaseous environment, the ionic flow is attenuated due to its interaction with the neutral particles. The study of this interaction is important both from the technological applications point of view (because reactive gases are employed to produce chemical reactions), but also to develop theoretical models taking into account the presence of the neutral gas. The data reported in the literature (Kimbling, 1974; Meunier and Drouet, 1987; Meunier and Douyon de Azevedo, 1992) include the study of Cu, Ag and C cathode materials and N_2, He, H_2, Ar and SF_6 gases. Here we report data on Ti and Pt cathodes using N_2 and Ar as the filling gases.

2. Experimental Arrangement

The arc was produced by discharging an electrolytic capacitor bank ($C = 0.075$ F) connected to a series inductor-resistor ($L = 2$ mH, $R = 0.33$ Ω) which critically damped the discharge. The vacuum chamber was a 25 cm long and 10 cm diameter stainless steel cylinder. The electrodes were introduced through the ends of the cylinder, and these ends faced the positioned electrodes. The gap between anode and cathode was set at 5 mm. The chamber was pumped down to $\approx 10^{-5}$ mbar with an oil diffusion pump; then its was closed and the filling gas was introduced through a needle valve to the desired pressure p. The arcs were performed in N_2 and Ar with p values in the range 0.1–3.5 mbar, using Ti and Pt cathodes. The

Astrophysics and Space Science **286:** 511–515, 1998.

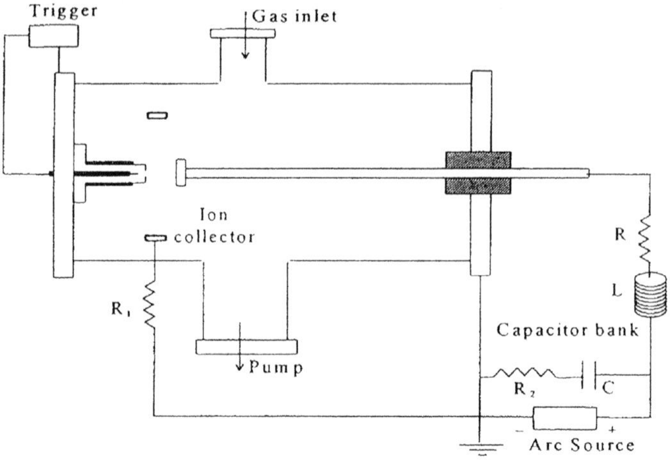

Figure 1. Experimental arrangement.

arc was ignited by applying a short ($\sim \mu$s) high-voltage pulse to a trigger electrode contained in the cathode. The operating voltage was in the range 220–280 V, giving a current pulse (measured through a low-value resistor, $R_2 = 0.008 \pm 0.001 \Omega$) with an amplitude of 400–500 A and HAFW duration of 35 ms.

The ion flux at an angle of 90° with respect to the cathode normal was collected by a probe consisting of a conducting ring (with a radius of 2.3 cm and a thickness of 3 mm) surrounding the cathode, and electrically connected to it via a low value resistor ($R_1 = 0.9 \pm 0.05\ \Omega$). As the potential difference between the probe and the plasma is ≥ 10 V (the plasma potential is close to the anode potential, which is ≈ 25 V in this case) whereas the electron temperature at the probe position is $\leq$ 3–4 eV, nearly no electrons will reach the surface, and the probe current (i_p) will correspond to ion saturation currents. In Figure 1 a sketch of the experimental arrangement is shown.

For a discharge performed in pure vacuum, it is possible to relate i_p with the total ion current (I_i) ejected from the cathode. This can be done by taking into account the angular distribution of the ion flux experimentally determined in vacuum (Kutzner and Craig Miller, 1989) and the geometry of the experiment. With the additional assumption of uniformity of the ion emission from the cathode surface, it is obtained:

$$I_i(A) = 9\, i_p(A). \qquad (1)$$

In this way one can determine I_i for different p values and by analyzing the function $I_i/I = f(p)$ a cross section for the ion absorption in the neutral gas can be inferred. It must be noted that two underlying assumptions are involved in the above described analysis: (i) the presence of the background gas does not change the geometrical collecting efficiency of the probe derived from vacuum conditions; and (ii) the ion flux ejected from the cathode surface remains the same as for vacuum conditions (effects of redeposition or cathode "poisoning" are not considered).

3. Results

In Figure 2 I_i/I as a function of p for a Ti cathode (N_2 and Ar gases) and a Pt cathode (N_2 gas) are presented. Each point in the graph represents an average over 3–10 shots performed under the same experimental conditions. For each case the data can be fitted by an exponential law of the type:

$$I_i/I\ (\%) = A + B\, e^{(-\alpha\ p(\text{mbar}))}, \tag{2}$$

where $A = 0.15$, $B = 4.1$, $\alpha = 1.4$ for the Ti-N_2 case; $A = 0.42$, $B = 3.3$, $\alpha = 1.0$ for the Ti-Ar case; and $A = 0$, $B = 2.1$, $\alpha = 0.55$ for the Pt-N_2 case. From Equation (2) and assuming that the ions have traveled through the neutral gas a distance equal to the difference between the radii of the probe and the cathode (≈ 2 cm), it can be inferred an effective cross section (σ) for ion absorption in the gas. The obtained values for σ are: $(3.0 \pm 0.1) \times 10^{-17}$ cm^2 for the Ti-N_2 case; $(2.1 \pm 0.2) \times 10^{-17}$ cm^2 for the Ti-Ar case; and $(1.2 \pm 0.5) \times 10^{-17}$ cm^2 for the Pt-N_2 case, respectively.

4. Discussion and Final Remarks

The values of σ obtained in this work are very similar among them, in spite of the differences between the masses of the ions and the reactive or inert character of the background gas. Moreover, in (Meunier and Douyon de Azevedo, 1992) a value of 2×10^{-17} cm^2 was obtained for σ for the C^+ ions/H_2 molecules interaction, which is also similar to those derived in this work. This fact suggests that the main process responsible for the ion attenuation in the background gas is a charge-exchange of the ion with a gas particle followed by a fast recombination of the resulting gaseous ion in its own gas. This process takes place for a low kinetic energy of the cathodic ions, because the cross section for energy transfer from elastic collisions of the ions with the neutral particles is very high (for instance, for Ti^+ ions propagating in Ar gas, it reaches $\approx 5 \times 10^{-15}$ cm^2 (Meyer et al., 1981)). It is interesting to note that, for all the cases investigated, the limit of I_i/I (%) from expression (2) when $p \rightarrow 0$ does not reach the expected vacuum value ($= 8$). This effect can be understood by noting that the presence of the neutral gas not only produces an ion absorption but also changes the spot mechanism: it is a well known fact that the cathodic erosion decreases as p increases (Kimbling, 1974), and hence one should expect that the ion flux emerging from the cathode is also a decreasing function of the gas pressure. To take into account this effect on the value of I_i/I, detailed measurements of the erosion rate as a function of p are required. Although we have

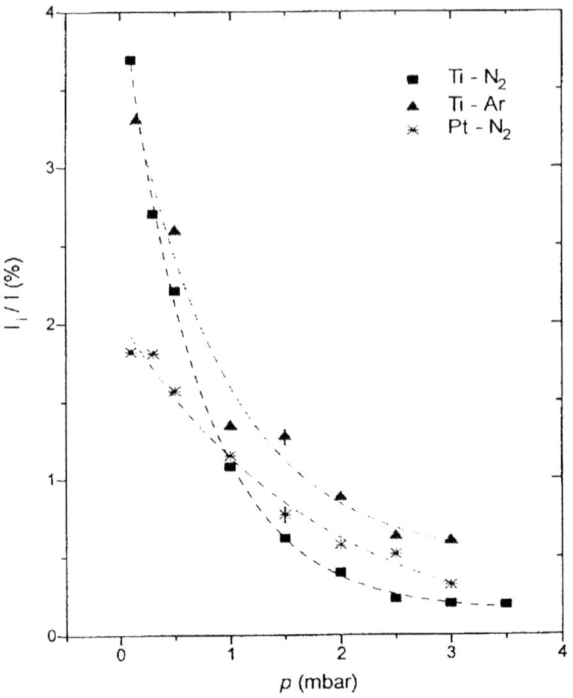

Figure 2. Normalized ion flux as a function of the filling pressure.

not performed these measurements for all the cases investigated yet, preliminary results (Bruzzone et al., 1996) obtained for a limited set of p values for the Ti-N_2 case have allowed us to estimate that the value of σ is smaller for only a factor of ~ 2 than that obtained from Figure 2. From a practical point of view, the presence of a background gas (although it is necessary to produce coatings of compounds) reduces the ion flux and hence the coating rate, thus setting a practical limit on the value of the cathode-substrate distance for each p.

Acknowledgment

Work supported by grants of the Buenos Aires University and CONICET.

References

Bruzzone, H., Kelly, H., Márquez, A., Lamas, D., Ansaldi, A. and Oviedo, C.: 1996, *Plasma Sources Sci. & Technol.* **5**, 582.
Davis, W.D. and Craig Miller, H.: 1969, *J. Appl. Phys.* **40**, 2212.
Kimbling, C.W.: 1974, *J. Appl. Phys.* **45**, 5235.
Kutzner, J. and Craig Miller, H.: 1989, *IEEE Trans. Plasma Sci.* **PS-17**, 688.

Kutzner, J. and Craig Miller, H.: 1992, *J. Phys. D* **25**, 686.
Meunier, J.L. and Douyon de Azevedo, M.: 1992, *IEEE Trans. Plasma Sci.* **PS-20**, 1053.
Meunier, J.L. and Drouet, M.C.: 1987, *IEEE Trans. Plasma Sci.* **PS-15**, 515.
Meyer, K., Schuller, I.K. and Falco, C.M.: 1981, *J. Appl. Phys.* **52**, 5803.

TiN PRODUCTION IN A GLOW DISCHARGE

P.J. ARANGO, A. DEVIA, A.F. ROJAS, C.A. PEÑA, J.A. ORTIZ and E. RESTREPO
Departamento de Física, Laboratorio de Física de Plasma, Universidad Nacional de Colombia, Sede Manizales, A.A. 127 Manizales, Colombia

Abstract. We have been studying direct current glow discharges aiming at the production of different paint of coatings on steel substrates. In a set of preliminary experiments, long lasting (hours) discharges performed in a Nitrogen atmosphere between a stell anode (holding the substrates) and a titanium cathode separated by a small gap (few mm) produced very thin coatings of the substrates whose colour depend on the parameter of (interelectrodes gap, gas pressure, discharge current and time duration). In some conditions, the coatings showed a golden colour, typical of the Nti compound. A discussion of the results is given.

1. Introduction

In the industrial devices that produced coatings, a high-current electrical discharge is generate between conducting electrodes in a chamber to low pressure. The current is conducted by a ionized vapour of the electrode material. The cathode throws plasma jets toward the anode. Emission of current depends on the cathode material (Harris, 1980), on the pressure of environment gas and the gap interelectrodes (Kimblin, 1974; Boxman and Goldsmith, 1989). The plasma jet contains ions flowing away from the cathode with a mean charge state depending on the cathode material (Boxman and Goldsmith, 1989; Bruzzone et al., 1996). Putting a substrate in a location where it intercepts the plasma jet in nitrogen ambient is possible the production of nitride films (Naoe and Yamanaka, 1970; Osipov et al., 1978; Aksenov et al., 1977; Bruzzone et al., 1994).

2. Experimental Arrangement

The discharge is produced in a stainless steel cylindrical chamber 6.35 mm (1/4 in) thick, 200 mm diameter and 300 mm long. The cathode is composed by a titanium bar 30 mm diameter and 200 mm long. The anode is formed by a circular 345 steel plate 30 mm diameter and 1 mm thick, supported by a sample-holder also of steel. The distance of separation between electrodes is 2 mm and the pressure varies between 1 and 1×10^2 mbar. in a nitrogen environment. Figure 1 is a sketch drawing of the system used.

The steel substrates are submitted to mirror polishing process and cleaning with a chemical abrasive before coating.

Astrophysics and Space Science **256:** 517–521, 1998.

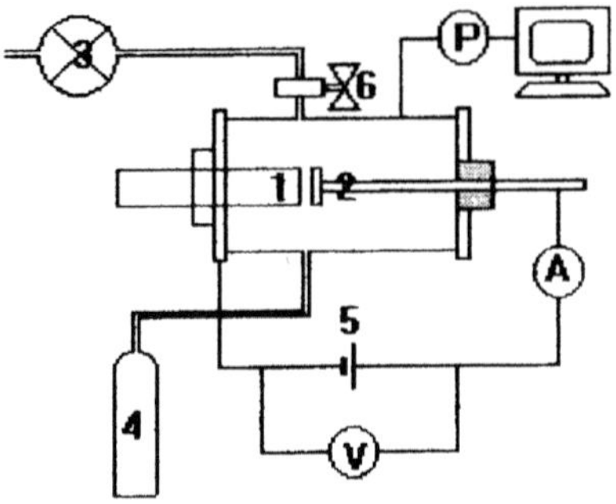

Figure 1. Sketch arrangement of the system. 1: cathode, 2: anode, 3: turbomolecular pump, 4: nitrogen source, 5: D.C. source, 6: electrovalve.

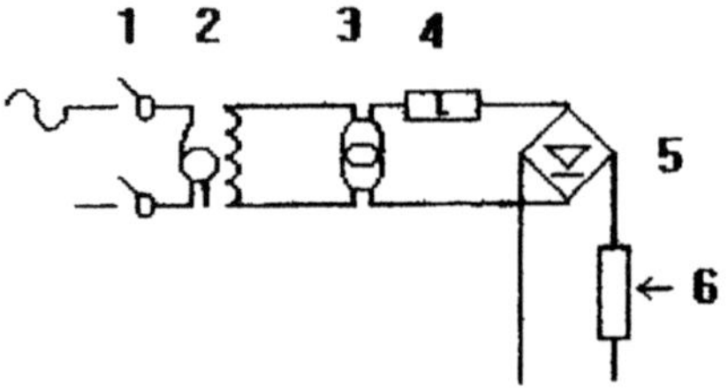

Figure 2. D.C. source circuit. 1: switch, 2: self transformer, 3: transformer, 4: air core coil, 5: complete wave rectifier, 6: limiting resistance.

Samples are subject to a glow discharge for a period of time varying between 4 to 12 hours. The evacuation system of the chamber is a TCP 015 Balzers Pfeiffer turbomolecular pump. The pressure monitoring is through a TPR pirani and IKR could cathode gauges and TPG 300 modular gauge. Vacuum is provided again and is filled up to work pressure in order to start discharging.

The generation circuit of discharge is composed by a 0–100 V self transformer, a 150 V.A. transformer, an air coil and a limiting resistance of 15 kΩ, 225 W.

Figure 3 is a photograph of the system used in the experiment. Foreground shows the chamber, above it are pirani and cold cathode gauges in connection with the modular gauge that is showed at the lower right. The modular gauge is connected by an interface to the computer (background) for pressure monitoring in function of time. Under the chamber is the turbomolecular pump.

3. Results

Although quantitative measures were not made in the treated samples, a visual inspection shows that there is a titanium nitrate coating observed in an area equivalent to the transversal section of the cathode with characteristics depending on the working conditions. Figure 4 shows photographs of different coatings.

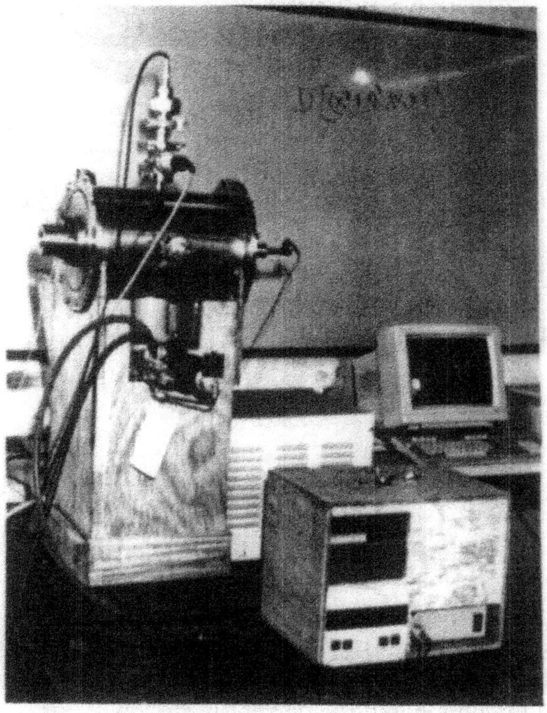

Figure 3. Photograph of the experimental system.

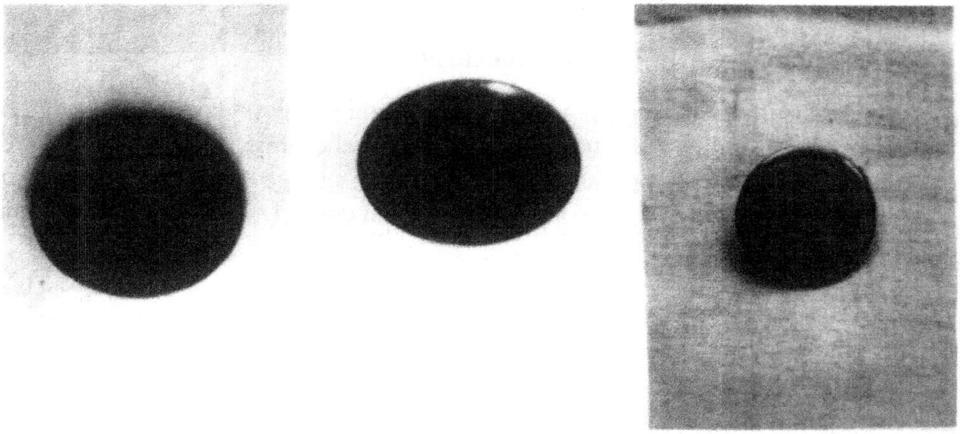

Sample	Pressure (mbar)	Current (mA)	Voltage (V)	Time (hours)
1	10^2	50	450	12
2	10^1	48	500	7
3	1–1.5	30	450	6

Figure 4. Aspects of different coatings.

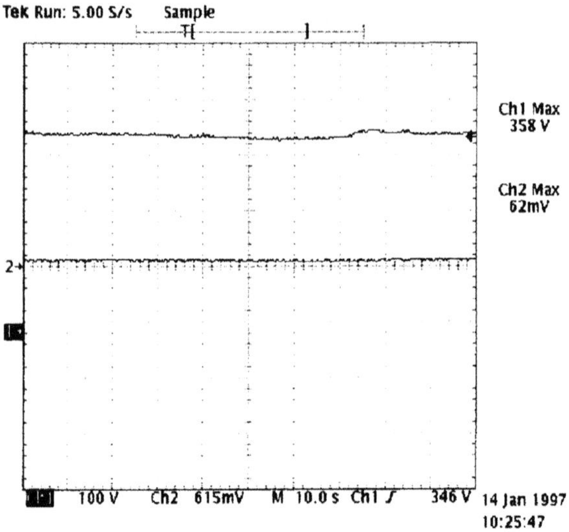

Figure 5. V and I Vs t in a typical discharge.

Figure 5 is a voltage and current monitoring in function of time of the experiment typical discharge made with a Tektronix TDS 254A oscilloscope. The upper curve corresponds to voltage and the current is the lower curve multiplied by 0.25.

4. Conclusions

In these experiments TiN coatings are produced on stainless steel surfaces of the kind of 7 cm^2 in a more or less uniform way. The final coating finish (color, texture, etc.) depends on the work pressure, on the time of discharge to which the sample in submitted and must depend also on the voltage and current of discharge (Bruzzone et al., 1994).

Acknowledgements

We are grateful to the Consejo de Ciencia y Tecnología (COLCIENCIAS) for financial support. We are grateful, also, to Instituto de Física del Plasma de la Universidad de Buenos Aires (INFIP) and, in particular, to Professors H. Bruzzone and A. Márquez for their technical assistance.

References

Aksenov, I.I., Belous, V.A., Padalka, V.G. and Khoroshikh, V.M.: 1977, *Fiz. Khim Obrab. Mat.* **6**, 89.

Boxman, R.L. and Goldsmith, S.: 1989, *IEEE Trans Plasma Sci.* **17**, 705.
Bruzzone, H., Kelly, H., Márquez, A., Lamas, D., Ayllón, E. and Topolevsky, R.: 1994, in: *International Conference on Plasma Physics*, p. 273.
Bruzzone, H., Kelly, H. Márquez, A., Lamas D., Ansaldi, A. and Oviedo, C.: 1996, *Plasma Source Sci. Technol.* **5**, 582.
Harris, L.P.: 1980, *Vaccum Arcs de J.M. Lafferty*, Wiley, New York.
Kimblin, C.W.: 1974, *J. Appl. Phys.* **45**, 5235.
Naoe, M and Yamanaka, S.: 1970, *Japan J. Appl. Phys.* **9**, 293.
Osipov, V.A., Padalka, V.G., Sablev, L.P. and Stupak, R.I.: 1978, *Prib. Tekh. Eksp.* **6**, 173.

CORRELATION BETWEEN THE POPULATION OF ACTIVE SPECIES IN LOW PRESSURE PLASMAS AND THEIR REACTIONS WITH ORGANIC COMPOUNDS

PEDRO PATIÑO
Escuela de Química, Facultad de Ciencias, Universidad Central, P.O. Box 47102, Caracas 1020A, Venezuela

Abstract. The oxidation of four alkanes, two olefins and twelve aromatic compounds in the liquid phase with either high voltage or radio frequency glow discharges has been studied. The results of the liquid-plasma interactions of this work considerably differ from those in homogeneous gas phase. Neither fragmentation products in the traps nor polymers on the reactor walls have been detected. The selectivity of product formation is far superior to plasma oxidations in gas phase, making these techniques attractive for preparative chemistry. Total conversion, i.e., mass transformed against initial mass of substrate, has been studied as a function of the temperature of the liquid and the oxygen flow rate in the reactor. Optical emission spectroscopy has been utilized to detect and monitor the active species in the plasma. At the conditions for maximum oxidation no evidence of $O_2(^1\Delta_g)$ has been observed. A correlation between the behavior of the $O(^3P)$ population in the discharge and the reactivity of the different compounds, as a function of the oxygen flow rate in the reactor, is presented.

1. Introduction

The use of glow discharges in organic synthesis has been widely reported. The interactions of plasmas with gases and solids have been studied extensively in the last two decades, leading to numerous syntheses (Suhr, 1983) and methods for surface modifications (d'Agostino, 1990). In these processes often the selectivity is limited, mainly for two reasons: in gas-phase reactions the primary products are normally hot and may undergo consecutive reactions before they are thermalized. In plasma-solid interactions, the products remain in the reactor and are exposed to the plasma for a considerable time. This again might cause various consecutive reactions. The study of plasma-liquid interactions, with a few exceptions (Hickling and Newnes, 1959, 1961; Plasman et al., 1965), had been neglected for a long time. Because such systems provide rapid quenching of the hot primary products, they should be far more selective than reactions in the gas phase. Compared with plasma-solid interactions they are superior because, in stirred systems, products are continuously eliminated from the surface layer, thus reducing consecutive reactions. Following works on oxidation of liquid hydrocarbons by means of low pressure oxygen plasmas (Iocacca et al., 1989, 1992; Ropero et al., 1992), we report here results from the interactions of such plasmas with n-pentane, n-hexane,

Astrophysics and Space Science **256:** 523–527, 1998.

n-heptane and n-octane (Patiño et al., 1995), 1-dodecene, 1-tetradecene, toluene, ethylbenzene, n-propyl-benzene, i-propylbenzene, t-butylbenzene, and seven additional di- and tri-alkylbenzenes (Patiño et al., 1996). A correlation between the behavior of the population of $O(^3P)$, the active species in the plasma, and the reactivity of the different compounds, as a function of the oxygen flow rate in the reactor, is presented.

2. Experimental

The reactions have been carried out making the cold plasma reach the surface of the liquid hydrocarbon. The reactor is a double walled glass vessel of ca. 300 ml. In the case of the high voltage system, the plasma is produced by means of a home-made low frequency AC power source capable of supplying up to 7 kV and 200 mA. In the radio frequency system the plasma is produced by means of a Branson IPC 100 generator with a fixed frequency of 13.4 MHz and variable power up to 500 watts. Substrate (1.5–3.0 ml) is placed at the bottom of the reactor and this is cooled down by means of a methanol-liquid nitrogen bath, the temperature being controlled within $\pm 1°$. In order to study the population of the active species, the light from the discharge has been focused onto the entrance slit of a 2.0 m Jarrel–Ash Czerny–Turner scanning spectrometer, detected by an RCA 4832 photomultiplier that fed a lock-in amplifier, and finally recorded on a Cole–Parmer pen recorder. Products of the reactions are identified and measured by means of GM-MS.

3. Results and Discussion

Applied power has been in the range of 8–30 W in the high voltage system and 35–100 W in the radio frequency device. Time of reaction has been 30 to 60 minutes. This has been rather short in order to avoid successive reactions. The most important products have been: secondary alcohols and the corresponding ketones for alkanes; epoxides and aldehydes for α-olefins; and phenols for aromatic compounds. $O(^3P)$ being the main reactive species in the plasma, the formation of these products has been explained by Suhr et al. (1984). Reactivity has been studied as a function of temperature of the liquid and oxygen flow rate, this ranging from 15 to 54 mMol/h in high voltage and 15 to 220 mMol/h in radio frequency. The optimum conversions are 7 to 80%, alkanes being the less reactive species and olefins the most. Maximum reactivity has been achieved at a different oxygen flow rate for each compound, the conversion decreasing towards both lower and higher rates. The results are shown in Figure 1 for n-pentane.

At low rates the production of active species in the plasma decreases for there are fewer oxygen molecules that can suffer electronic impact. At the side of higher pressures, due to the higher concentration of starting gas molecules in the plasma,

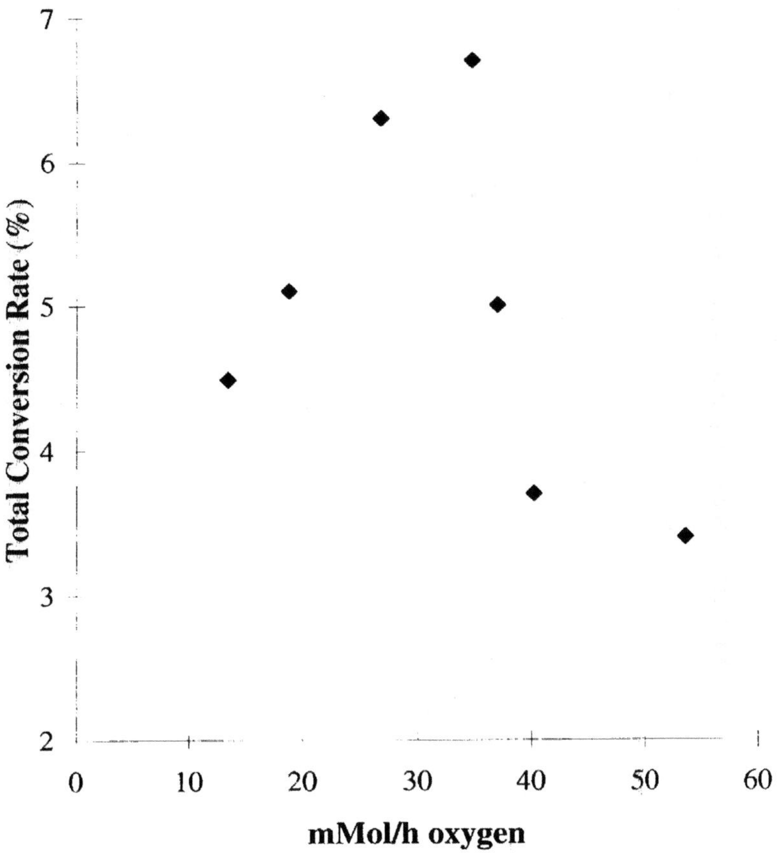

Figure 1. Conversion rate of n-pentane vs. oxygen flow rate. The temperature of the liquid was −129°C and the power of the source was 20 W.

the frequency of collisions between these species and the electrons increases. This implies that the electrons do not have time to be sufficiently accelerated by the electric field and hence most of them do not acquire high enough energy to produce the dissociation of such molecules, thus producing less fragmentation. The region of optimum flow rate is obviously a compromise between the two tendencies. This behavior has been demonstrated by following the relative population of the active species in the plasma, $OI(2p^3P)$. It is produced by means of the process

$$O_2 + e^- \rightarrow O + O + e^- (\text{O may be } ^3P, ^1D, ^1S, \text{etc.}) \quad (1)$$

$O(^3P)$ concentration can be correlated to the $3p\ ^3P \rightarrow 3s\ ^3S$ ($\lambda = 844.6$ and 844.7 nm) transition. The optical emission signal of OI has been recorded and plotted *versus* the oxygen flow rate in the reactor, as shown in Figure 2.

The behavior has shown to be very similar to the variation of the reactivity with the oxygen flow rate. Walkup et al. (1984) have measured the OI concentration by means of the two photon laser induced fluorescence technique. The atomic oxygen

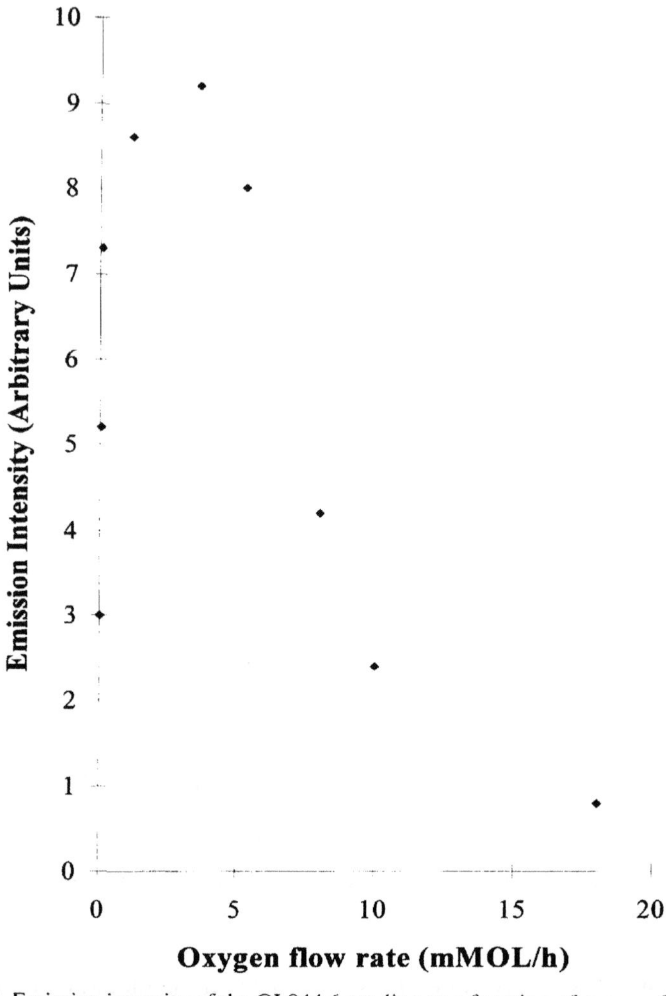

Figure 2. Emission intensity of the OI 844.6 nm line as a fucntion of oxygen flow rate.

population has been monitored against the CF_4 percentage in a plasma produced through O_2-CF_4 mixtures. From 0 to 10% of CF_4 the OI density increases due to the increasing amount of electrons that the easy to ionize CF_4 molecules supply to the plasma. At higher carbon tetrafluoride concentrations the OI density decreases and is negligible at about 70% of CF_4. Clearly, as the percentage of carbon tetrafluoride increases, the probability of electron collisions with O_2 molecules decreases and hence so the production of OI does. Under the same conditions used for $O(^3P)$, an attempt to monitor the population of $O_2(^1\Delta_g)$ has been made. This can be correlated to the intensity of the 0,0 band of the O_2 atmospheric system $b^1\sum_g^+ \rightarrow X^3\sum_g^-$ at 759.4 nm. Nevertheless, we have recorded the resolved structure of this band at oxygen pressures higher than 2 torr ($\sim$ 270 mMol/h), it has not been observed at the range of pressures of this study. This means that $O_2(^1\Delta_g)$ is not a relevant species in the discharge. This has been confirmed by the fact that none peroxide has been detected among the products of the oxidation of the olefins treated in this study, as expected from the reaction of $O_2(^1\Delta_g)$ with such compounds (Suhr et al., 1984). This confirms that $O(^3P)$ is, if not the unique, the most relevant oxidising agent in the plasma.

Acknowledgements

Thanks to CDCH-UCV for supporting this research and Dr. Jimmy Castillo for his help.

References

d'Agostino, R.: 1990, *Plasma Deposition, Treating, and Etching of Polymers*, Academic Press, Boston.

Hickling, A. and Newnes, G.R.: 1959, *J. Chem. Soc.*, 368.

Hickling, A. and Newnes, G.R.: 1961, *J. Chem. Soc.*, 5177, 5186.

Iacocca, D., Ropero, M., Patiño, P. and Suhr, H.: 1989, in: *Proceed. of the 9th International Symposium on Plasma Chemistry*, IUPAC, Pugnochiuso, Italy, p. 762.

Iacocca, D., Patiño, P., Ropero, M. and Hernández, F.E.: 1992, *Rev. Brasileira de Aplic. de Vácuo* **11**, 126.

Pasman, H.J., Vlugter, J.C. and Brenking, C.J.: 1965, *Brennstoffchemie* **46**(6), 241; **46**(9), 41; **46**(11), 14.

Patiño, P., Hernández, F.E. and Rondón, S.: 1995, *Plasma Chem. Plasma Process.* **15**, 159.

Patiño, P., Ropero, M. and Iacocca, D.: 1996, *Plasma Chem. Plasma Process.* **16**, 563.

Ropero, M., Patiño, P., Armas, F., Káram, A. and Iacocca, D.: 1992, *Acta Cient. Ven.* **43**, 280.

Suhr, H.: 1983, *Plasma Chem. Plasma Process.* **3**, 1.

Suhr, H., Schmid, H., Pfeundschuh, H., and Iacocca, D.: 1984, *Plasma Chem. Plasma Process.* **4**, 285.

Walkup, R., Saenger, K. and Selwyn, C.S.: 1984, *Materials Research Society Symp. Proceed.* 30, Boston, MA.

SPECTRAL BROADENING OF THE 3463 Å LINE IN A ZIRCONIUM PLASMA GENERATED BY LASER IRRADIATION

H. RIASCOS L.
Departamento de Física, Universidad Tecnológica de Pereira, A.A. 97, Pereira, Colombia

Abstract. In a vacuum chamber a plasma is produced by shining a pulsed Nitrogen laser, 20 mJ and 20 ns (Nordstrom, 1995), focused in an area of 0.1 mm^2 over a high purity polycrystalline Zirconium surface. This is accomplished by an optical system coupled to a monochromator, with a 0.1 Å resolution power, by detecting the visible and ultraviolet radiation for Zr I and Zr II. For the transition $a^2H_{4.5} \rightarrow y^2G^0_{3.5}$, measurements are performed at different distances with respect to the impact area, to determine the Doppler and Stark effects dependence on the laser fluency (J/cm^2). The Zr II ions velocity is determined in order to estimate the Doppler broadening effect in the fitting of the Voigt profile for the spectral line studied. These measurements are performed in a shoot to shoot way.

1. Introduction

This work presents the spectroscopy study of the 3463 Å spectral line produced when a non-commercial pulsed Nitrogen laser is focused onto a high purity flat solid Zirconium surface. The laser energy could be as high as 20 mJ depending on the Nitrogen purity. The pulse duration depends on the gas used and the laser alignment. The pulse found lasted between 8 and 20 ns. This laser is capable, by changing the capacitor's charge, of producing power densities around 108 W/cm^2 (Krohin, 1972; Mulser et al., 1973).

The target, placed on a X–Y table, was made up of a 95% purity 10 cm^2 Zirconium plate, the size of the target's focal spot was 0.1 mm by 1.0 mm. The line of vision to observe the emission spectra, in the 3400 to 3800 Å was taken parallel to the target's surface and perpendicular to the laser's beam incidence axis. In this spectral range, the emission lines found correspond to the ionized Zr II atom (Riascos, 1994).

To detect the emitted radiation, a 1.0 m monochromator, with a 0.1 Å resolution, was used. The calibration of the monochromator was achieved with a Mercury lamp. The emission lines relative intensities were integrated using a fast integrator, a Boxcar, and then stored in a microcomputer running a program on Q-Basic, see Figure 1. It should be noticed that in the visible range, the signal to noise ratio of the relative intensity of the emission lines, is small, making these measurements less reliable (Riascos, 1994). The intensity was found to decrease with the number of firings in the same target's region. The diagnostic technique developed is based

Astrophysics and Space Science **256:** 529–532, 1998.

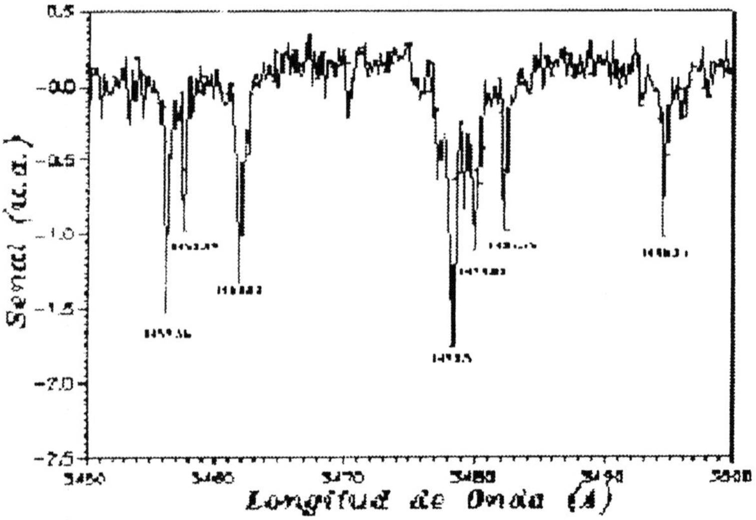

Figure 1. This figure shows the 3450 to 3500 Å spectral range. A large number of lines can be observed in this spectrum corresponding to Zr I (the neutral atom) as well as Zr II (the atom singly ionized).

on the spectral broadening due to the Stark effect produced by the plasma electrons (Griem, 1974; Kepple and Griem, 1982).

2. Results

The 3463.02 spectral line was chosen for a more detailed study because it is isolated and intense. The line's profile was measured with a 10 μm width in the opening slits before and after the monochromator and at 2000 V in the photomultiplier, a 10^{-6} mbar pressure and a pulse energy of 8.8 mJ, see Figure 2.

The observed intensity distribution was calculated measuring the area under the curve. The table below shows the main features of this line.

λ (Å) Zr II	Transition	gf	Intensity (u.r.)	$\Delta\lambda_{1/2}$ (Å)
3463.02	$y^2G^0_{3.5} \rightarrow a^2H_{4.5}$	9.3	750	0.19

2.1. Role of the Stark and Doppler Effects in the Line's Broadening

The physical situation is the following: along the monochromator's line of vision each plasma element emits radiation. The signal detected is the integral along the

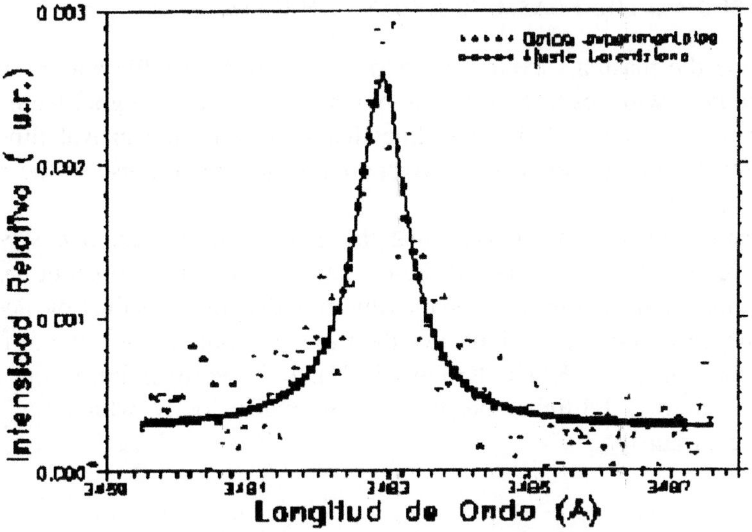

Figure 2. This figure shows the profile of the 3463.02 Å line at a pressure of 10^{-6} mbar and a laser energy of 13.8 mJ. The crosses correspond to the data and the squares to a Lorentzian fit.

line of vision. The plasma is considered optically thick. We assume that the plasma behaves as a plasma in Local Thermal Equilibrium (LTE)) (McWhirter, 1962).

To determine the average line's width when the Doppler and Stark effects contribute, the line's profile will be calculated using Voigt's profile. In our case only the Stark effect affected the broadening and using the analysis program Table curve, a fit to a Lorentzian was performed. The contribution due to the Doppler effect was negligible as can be shown from the following relation:

$$\Delta\lambda_{D1/2} = 7.16 \times 10^{-7} \lambda_0 \sqrt{\frac{T}{\mu}} ,$$

where λ_0, the central wavelength, is given in Å, T, the plasma's temperature, in K and μ is the atomic weight. The table below shows the contribution due to the Doppler and Stark effects. As can be seen the Doppler is smaller than the Stark effect.

Comparison between broadening contributions from Stark and Doppler effects.

λ (Å)	$\Delta\lambda_{S1/2}$ (Å)	$\Delta\lambda_{D1/2}$ (Å)	%
3463.02	0.19	0.028	14.7

2.2. Measuring the Plasma's Expansion Rate

To determine the plasma's expansion velocity, the line's profile was measured at different heights with respect to the target's surface. These heights were 0.4, 0.8 and 1.2 mm. The slope of the best line that represents the arrival time of the ions at different heights versus the corresponding heights, represents the plasma's expansion velocity.

To measure the time we assume that the gate of the boxcar opens when the plasma is created. Then the time difference between the gate's opening and when the signal appears inside the gate (the maximum of the signal) is the time needed for the ions to emit at a certain height. For the times measured, we found a plasma's expansion rate of 14.50 Km/s (Krohin, 1972). It is worth pointing out that the velocity measured is 9.8 times the plasma's thermal velocity, which may suggest an explosive expansion.

References

Griem, H.R.: 1974, *Spectral Line Broadening by Plasmas*, Academic Press, New York.
Kepple, P.C. and Griem, H.R.: 1982, *Physical Review* **26**(1), 484.
Krohin, O.N.: 1972, in: F.T. Arecchi and E.O. Schulz-Du Bois (eds.), *Laser Handbook*, p. 1371.
Mulser, P., Sigel, R. and Witkowski, S.: 1973, *Physics Reports* **6**(3), 187.
Nordstrom, R.J.: 1995, *Applied Spectroscopy* **49**, 1490.
Riascos, H.: 1994, M.Sc. Thesis, Universidad Simon Bolivar, Caracas.
McWhirter, R.W.P.: 1962, in: R.H. Huddlestone and S.L. Leonard (eds.), *Plasma Diagnostic Techniques*, Academic Press, New York, p. 201.

A COMPACT HIGH-VOLTAGE PULSE GENERATOR FOR PLASMA APPLICATIONS

V.A. SPASSOV*, J. BARROSO and M. UEDA
National Institute for Space Research (INPE), Associated Plasma Laboratory (LAP), P.O. Box 515, 12201-970 Sao Jose dos Campos, S.P., Brazil

L. GUERGUIEV
Wayne State University, 3832 Eldridge, Detroit, MI 48212, U.S.A.

Abstract. The design and construction of a compact high-voltage pulse generator for providing input electron beam power for the LAP/INPE 32 GHz gyrotron and for treatment of metal and polymer materials by plasma immersion ion implantation (PIII) are described. The generator was built on a circuit category of Pulse Forming Network (PFN), consisting of nine LC sections with $L = 270\ \mu$H and $C = 2.5$ nF.

The instrument was designed to produce a flat 30 kV, several Amps pulse in 15 μs pulse length with pulse repetition frequency (PRF) of 8 to 100 Hz. By means of a resonant charging inductance it is possible to gain an output voltage with a factor of 1.8 higher than the voltage supplied by the pulse generator. The generator is fed with sine-wave, constant current source, and a 60 kV, 15 mA switching power supply.

1. Introduction

The INPE Plasma Laboratory Gyrotron Project was devised with the main purpose of designing, constructing and operating a high power microwave radiation source. In order to achieve this goal, it was necessary to employ and handle special technologies as, for instance, high voltage switching and pulsed power techniques. In particular, for the gyrotron experiment a high-voltage pulse modulator was built at INPE in a circuit category known as a hard-tube pulser to provide the input electron beam power. In this system, a large capacitor bank (2.0 μF, 100 kV) is discharged through a high-voltage tetrode (TH 5188) connected in series with the gyrotron. Although this experiment is still in operation, its circuit complexity has led to great difficulty in maintenance and servicing. To overcome such a drawback, the construction of another kind of modulator, called line-type pulser, was proposed. In addition, the cost involved in building a line-type pulser is significantly lower than for a hard-tube one.

The Plasma Immersion Ion Implantation (PIII) technique was developed during the late 80s for surface treatment of nonplanar components in contrast with the

* Visiting Scientist, Faculty of Physics, Sofia University, 5 J. Bourchier Blvd., Sofia 1126, Bulgaria.

Astrophysics and Space Science **256:** 533–538, 1998.

traditional method of implantation using ion accelerators. There are two basic limitations of the ion implantation technique based on accelerators: line-of-sight and high cost of equipment and processing. PIII technique overcomes these limitations by extracting directly the ions of interest from the plasma where the samples to be irradiated are immersed. The application of negative high-voltage pulses to a target immersed in plasma and the subsequent sheath expansion in the presence of high voltage allows three dimensional implantation of the samples. Preliminary results of INPE PIII research have shown successful N implantation in Al sample, but indicated the necessity of better plasma optimisation and confirmed the need of higher PRF and negative high-voltage pulse generation.

Both needs of high-voltage pulse generator (tuneable from 0 to 30 kV, pulse duration more than 10 μs), and higher PRF were the motivation of the design and its construction. Our efforts have been concentrated on several important objectives:

(1) Design of a line-type pulse generator (pulse duration of 15 μs, tuneable pulse voltage in the range 0–30 kV for nonlinear loads such as gyrotron and PIII reactor.

(ii) Design of resonant charging inductance for line-type pulse generator.

(iii) Design of high-voltage tuneable sine-wave constant current switching power supply (0–60 kV) for line-type high-voltage pulse generator.

2. Pulse Generator

The line-type pulser is commonly known as a pulse forming network (PFN), as it has a lumped constant transmission line which serves not only as a source of electrical energy during the pulse occurrence but also as a pulse-shaping element. The PFN in this generator basically consists of inductors and capacitors set in a specific configuration in order to provide desired pulse shape. Figure 1 shows the electrical diagram of the pulser circuit. In this scheme, the PFN capacitors are charged from DC power supply (60 kV, 15 mA). The charging operation is completed when the equilibrium voltage on the network is nearly equal to the power supply voltage V_0. The energy stored in the capacitance elements is discharged into the load upon closing a fast switch tube (thyratron CX 1171).

The PFN output impedance Z_0 can be determined as $Z_0 = (L/C)^{1/2}$, where L and C are the inductance and capacitance values of each cell, respectively. If the load is matched to the PFN output impedance, a rectangular voltage pulse with duration time $\delta = N(LC)^{1/2}$ and with amplitude $V = 0.5V_0$ occurs onto the load after closing the thyratron, where N is the number of LC cells in PFN and V_0 is the PFN charging voltage (Glasoe and Lebacqz, 1948).

The triggering of the thyratron CX 1171 is driven by positive voltage pulse of 300 V and a negative cutoff voltage pulse of 450 V in consecutive order to the thyratron grid g_2 while on the grid g_1 a constant positive voltage 100 V is applied.

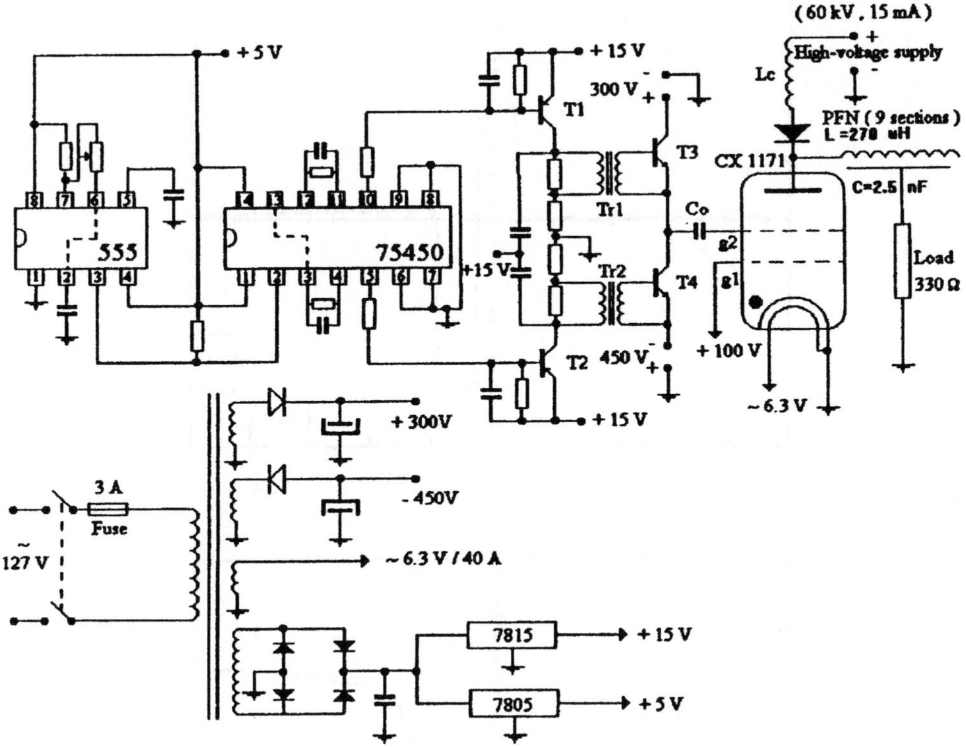

Figure 1. Circuit of the high-voltage generator based on thyratron CX 1171.

By means of integrated timer 555, a tuneable oscillator generates rectangular pulses, which are applied to the input of the circuit 75450. Two independent pulse sequences, alternatively changing in time, are obtained and drive the fast transistor switches T_3, T_4 by driver transistors T_1, T_2 and pulse transformers T_{r_1}, T_{r_2}, respectively. The capacitor C_0 is included to prevent switch circuit from thyratron high voltage.

The instrument tests have shown a high stability in the high-voltage pulse amplitude and in continuous work (more than 1 hour). The shape of a grid pulse and output high-voltage pulse across the resistance load of 330 Ω are shown in Figures 2 and 3, respectively.

3. Resonant Charging Inductance

Utilizing computer simulation, it was found that by using a charging inductance L_c of 1.39 H and PRF of 20.8 Hz it is possible to gain an output voltage with factor 1.8 higher than the charging voltage of the pulse generator. For better insulation, the inductor must be immersed in transformer oil during the experiment.

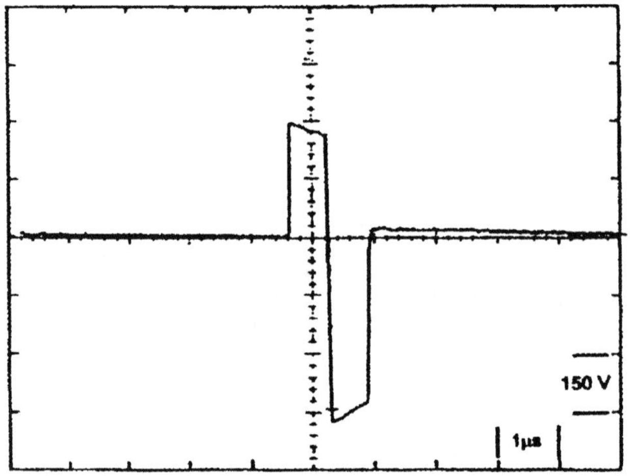

Figure 2. Grid voltage waveform for start-stop pulse of the thyratron CX1171.

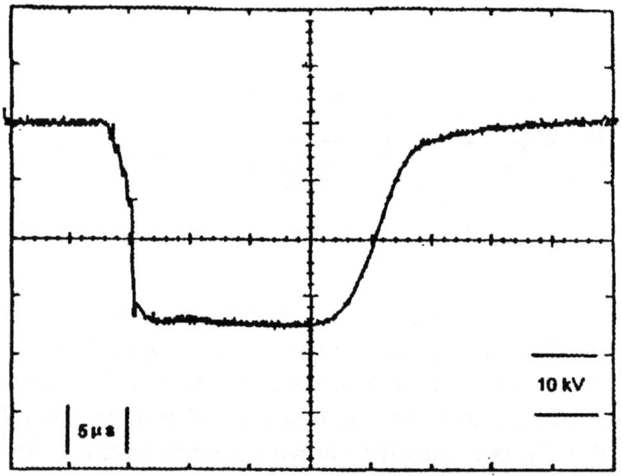

Figure 3. Pulse output waveform of pulse generator on a 330 Ω load.

4. High-Voltage Power Supply

A practical circuit diagram of a high-voltage, tuneable (0–60 kV, 15 mA), sine-wave, constant current source switching power supply, designed using results from (Chaltakov, 1988), is shown in Figure 4. It consists of:

(i) A generator of equidistant rectangular pulses, based on the timer 555. It is possible to tune the frequency of the generator by the potentiometer R_{B1} (fine) and R_{B2} (coarse) (Figure 4a).

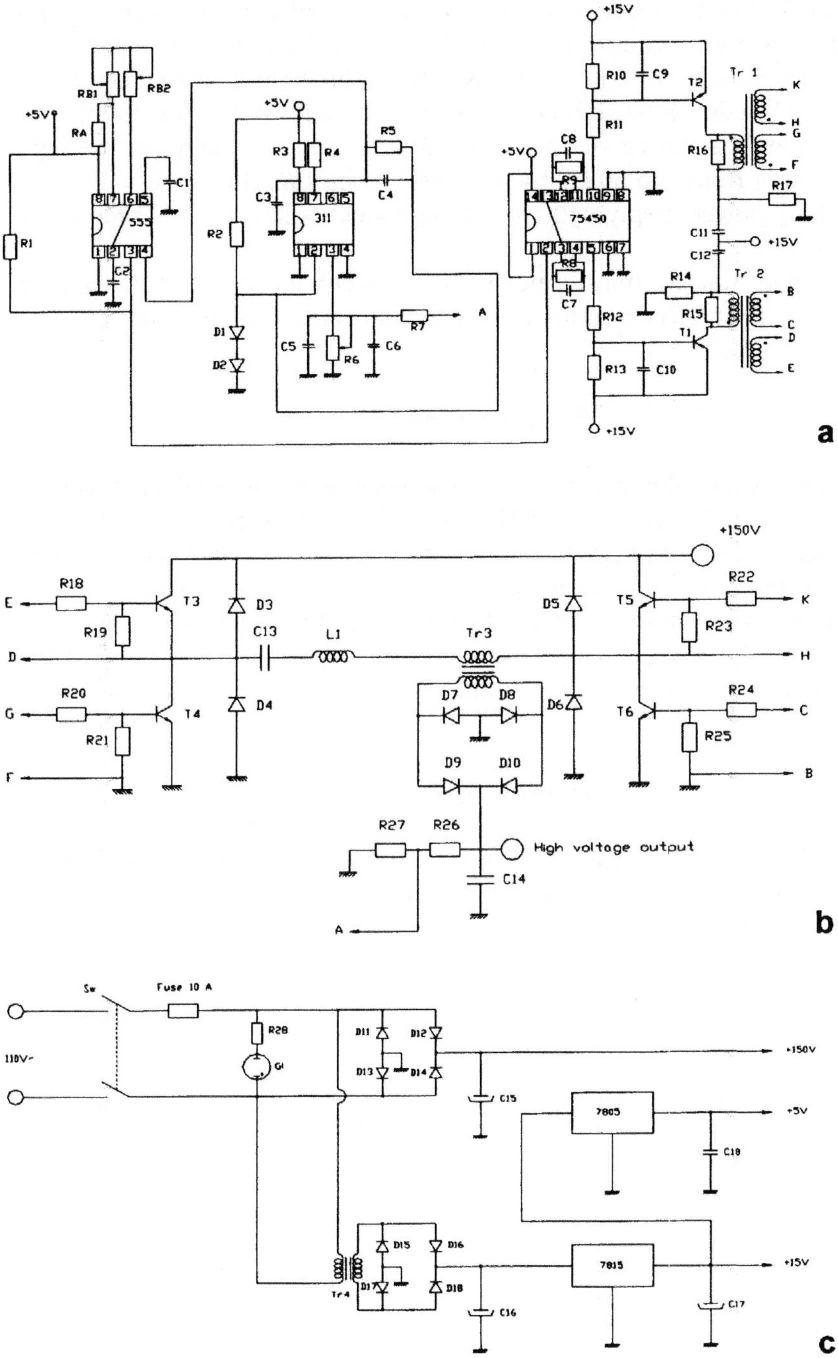

Figure 4. Circuit of the high-voltage tuneable switching power supply.

(ii) Two independent, but alternatively changing in time, pulse sources, based on the circuit 75450, the driver transistors T_1, T_2 and pulse transformers T_{r_1}, T_{r2} direct the transistor switches T_3, T_6 and T_4, T_5, respectively (Figure 4a).

(iii) A resonance circuit $L_1C_{13}T_{r_3}$, which is between the transistor switches T_3, T_6 and T_4, T_5. The diodes D_3–D_6 are preventive measures used (Figure 4b).

(iv) An error amplifier based on the circuit 311 is used for tuning of the high voltage output from 0 to 60 kV by the potentiometer R_6 (Figure 4a).

(v) An auxiliary power supply with three independent voltages: 150, 15 and 5 V (Figure 4c).

The high voltage transfer transformer T_{r3}, diode rectifiers D_7–D_{10} and charging capacitor C_{14} are assembled in additional block, immersed in transformer oil.

5. Conclusion

The performance test of the high-voltage pulse generator was conducted using a resistance load of 330 Ω. PRF up to 25 Hz was achieved at the maximum charging voltage. Further increase of the PRF results in decrease of the maximum attainable PFN voltage, as the full voltage is not available to the capacitors. An efficiency of 90% was obtained for charging voltages higher than 10 kV, due to the negligible switching losses. The major losses appear to come from the resistivity and the leakage inductance of the transfer transformer.

Acknowledgments

One of the authors (V.A. Spassov) would like to thank the Brazilian Council for Science and Technological Development (CNPq) and FAPESP (S. Paulo) for a visiting-scientist scholarship.

References

Chaltakov, I.V.: 1988, *J. Physics E; Sci. Instruments* **21**, 663.
Glasoe, G.N. and Lebaqz, J.V.: 1948, *Pulse Generators*, McGraw-Hill Book Company, New York.

LASER ABLATION IN THE SYNTHESIS OF ZIRCONIUM OXIDE

RICARDO CASTELL[1] and THIERRY POIRIER[2]
[1] *Physics Department and* [2] *Surface and Interface Research Group, Universidad Simón Bolívar, Apdo. 89000, Caracas 1080-A, Venezuela*

Abstract. Results of an experimental study of the oxidation of a zirconium plasma generated by laser ablation of a metallic target under low pressures of oxygen are presented. The crystallographic identification of the collected product by electronic diffraction shows metastable ZrO cubic phase. The oxidation phenomena are interpreted by thermochemical simulation suggesting a model in accordance with other cases of synthesis of metastable oxides at high temperatures, such as plasma reactors.

1. Introduction

Laser ablation has proven to be an important technique for thin film deposition and micromachining of metals, ceramics (Geohegan, 1993) and polymers. Its high ability to removal of coatings also makes it a very interesting application for cleaning surfaces, such as decontamination of radioactive surfaces or laser ablation of aircraft skin (Reitz and Rawers, 1993). An interesting property of lasers is the possibility of high velocity heating and quenching the materials. Such thermal treatments can allow the formation of metastable phases in the target (for example, with surface martensitic transformations). Nevertheless, the quickest thermal treatments are realized from the laser plume (see Figure 1a) which results from the target-laser interaction (Kools et al., 1992). Temperatures between 6500 and 12000 K may be obtained (Kools et al., 1992; Butt et al., 1994), and the pressure of the plume reaches relatively high pressures (on the order of tens of atmospheres) compared with the backing gas pressure, particularly in the very early stage of evolution. As the plume expands and the plume pressure subsequently drops, the partial pressures and relative pressures must change significantly. Reitz and Rawers (1993) report an additional mechanism of ablation: a microscopic expansion occurs on the surface, producing spallation of a thin surface layer (see Figure 1b). Butt et al. (1994) calculated gas phase compositions above the surface of a $ZrC_{0.95}$ target at 9000 K, as a function of plume pressure (from 0.01 to 50 atmospheres): the Zr neutral gas phase is the dominant species. When the plume is made by laser-ablating a metal within a reactive atmosphere, such as oxygen, a plume-oxygen reaction may occur and create an oxide phase of the metal. Gilgenbach et al. (1994) and Ching et al. (1995) have detected the formation of AlO molecules produced by KrF laser-ablated Al atoms in oxygen atmospheres at low pressures. We present here a

Astrophysics and Space Science **256:** 539–545, 1998.

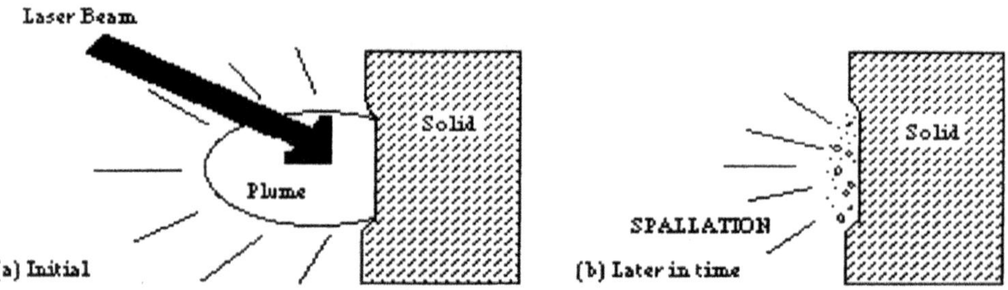

Figure 1. Laser-target interaction: formation of the plume, then spallation (Reitz and Rawers, 1993).

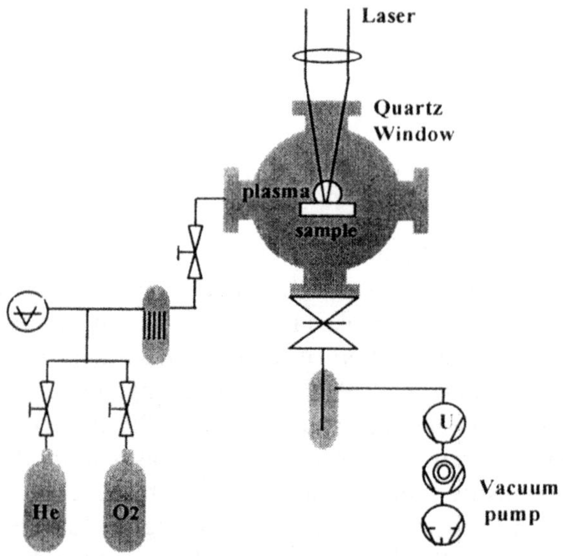

Figure 2. Laser ablation system (normal incidence configuration).

study of the formation of zirconium oxide by reaction of ultraviolet laser-ablated zirconium atoms in oxygen atmospheres at room temperature.

2. Experimental Procedure

The experimental setup is depicted in Figures 2 and 3. The ablation source is a pulsed N_2-SF_6 laser (Rebban et al., 1980) with a pulse duration of 20 ns which focused at the target surface delivers an energy density of about 6 J/cm^2. Helium and oxygen atmospheres are made by injecting small volumes of gas through a needlevalve and controlling the pressure by a Leybold Heraeus manometer. The behaviour of laser-target interaction is visually observed, with a normal incidence laser exposure, under various atmospheres and pressures (study No. 1). The configuration for characterizing the material yielded from the plume-atmosphere reaction

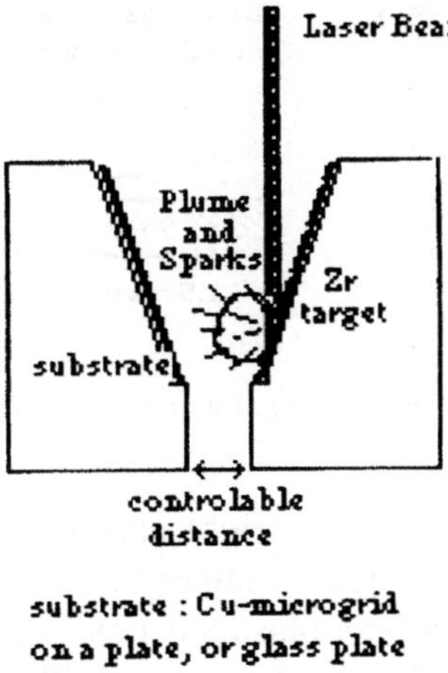

Figure 3. Deposition system for collecting laser-ablated materials.

is realized to collect a significant quantity of material, therefore a 30° incidence laser radiation is used on the target material (Figure 3), in front to the substrate. Visible layers are easily detected on transparent glass substrates. Due to the small quantities collected, classical X-Ray Diffraction (XRD) techniques do not allow a good characterization of the synthesized phases (study No. 2). Thus, the cristallographic study is done by Transmission Electronic Microscopy (TEM, Philips EM420), using the Electronic Diffraction mode. The material is collected onto a TEM sample holder, which consists of a copper microgrid, covered by a thin carbon layer. This implicates special precaution to deposit the material, which is propulsed with high velocity (Kools et al., 1992) and high temperature from the plume to the microgrid, without destroying the carbon layer. A distance of about 1 cm, with a reduced time of exposure (about 180 shots) is necessary.

3. Results and Discussion

3.1. Study No. 1: Plume-Atmosphere Reaction

In all atmospheres, the 2–3 first laser impacts create a plasma with an intense bright yellow light, which disappears afterwards, presenting then a less bright plasma. It is due to a surface contamination of the Zr target. Ghaderi (1988) used this property of laser ablation to remove and analyse the surface contamination of materials. Morant et al. (1989) show that it corresponds to previous exposure of pure zirconium metal in oxygen at room temperature creates a 2 nanometer ZrO_2 surface layer. Such a layer has a lower reflectivity and thermal conductivity than the Zr metal, therefore its coefficient of emission has to be higher. Ching et al. (1995) report similar effects on Al_2O_3 compared to Al metal. In a Helium atmosphere, a

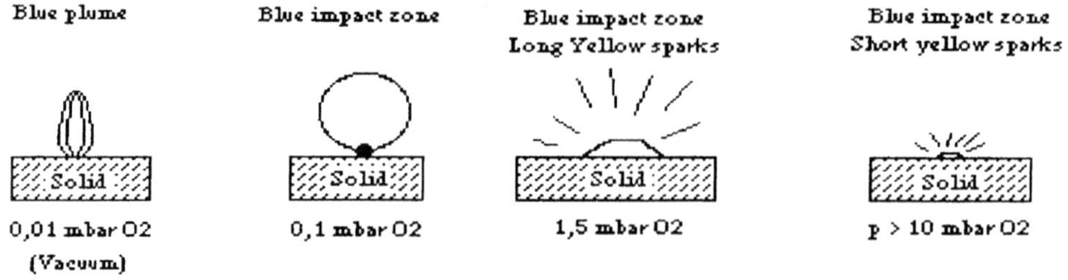

Figure 4. Laser-target interaction in different atmospheres.

dark blue plasma is observed, whose height appears to decrease and stabilize at pressures above 0.1 mbar. In vacuum, a rather large blue plasma is observed, and in oxygen atmospheres a bright blue plasma is observed, and after 50 shots, bright yellow sparks are ejected from the target material, with a maximum of intensity and path length at 1.5 mbar of oxygen (Figure 4). At lower pressures, the sparks tend to disappear, whereas at pressures higher than 4 mbar, their intensity and path length tend to lower and stabilize at 100 mbar. In the Helium atmosphere, the extension of the plume is limited as the pressure of the backing gas increases. Butt et al. (1994) report that the plume expands and its pressure and temperature decrease quickly, and they confirm this observation. In the case of the oxygen atmosphere, the fact that the first 50 laser shots do not yield sparks may be explained by the original high smoothness and resultant reflectivity of the target. The efficiency of the laser increases with the roughness of the surface. The existence of sparks in oxygen atmospheres, which are not observed in Helium, is typical of an oxidation phenomenon of zirconium metal. Oxygen acts as a backing gas, i.e. its pressure (above 10 mbar) limits the intensity and path length of the projected sparks, and acts as a reactive medium, as sparks are observed above 0.1 mbar.

3.2. Study No. 2: Synthesis of Cubic Metastable ZrO

A 1.5 mbar oxygen pressure is chosen to collect more oxide particles from laser ablation. The distance (about 1 cm) between the target and the copper microgrid allows as well to protect the carbon layer of the latter, as to select the material to be received: the microgrid stays out of the range of expansion of the plume (chemical nature not well determined), but stays within the range of the sparks (zirconium oxide). A study made by Electronic Diffraction (TEM) shows the presence of Cubic ZrO zirconium suboxide (Figure 5). Such a component is metastable according to the CSIRO thermodynamic tables (Turnbill and Wadsley, 1988). Referring to Figures 1a and 1b, reported by Reitz and Rawers (1993) about laser-target interaction, an interesting way of determining whether ZrO is synthesized directly from the Zr plume-oxygen interaction or from the spallation and oxidation of solid Zr metal in the reactor is comparing the exposure in the present case and in the case of a classical oxidation of Zr metal at room temperature (Morant et al., 1989): Morant

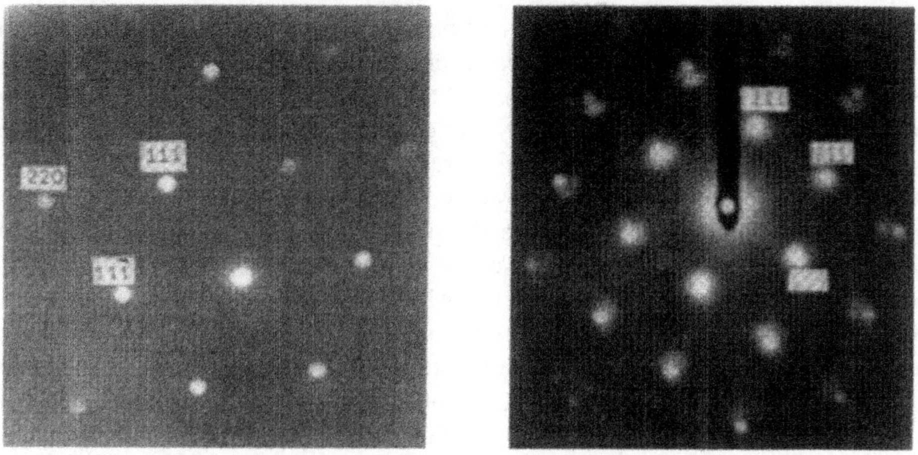

Figure 5. TEM electronic diffraction patterns of cubic ZrO yielded under 1.5 mbar oxygen. Zone axis [−1 1 0] and [0 −1 1].

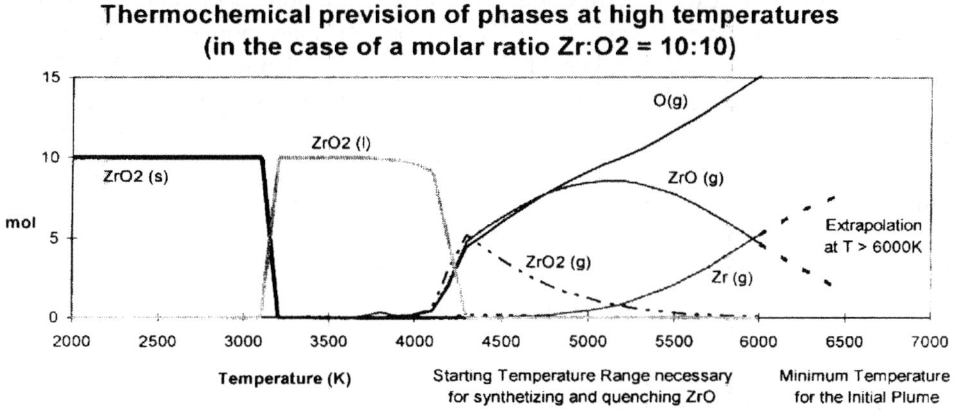

Figure 6. Thermodynamically stable phases in a plasma at several temperatures: in the laser plume, only Zr(g) can exist.

et al. report that oxygen chemisorption on metallic zirconium at room temperature occurs at exposures lower than about 1–2 L (1 L $= 10^{-6}$ torr.s) with formation of ZrO at 25 L. In the present case, the maximum exposure time is about $200 \cdot 10^{-9}$ s (Castell, 1995), thus the exposure may be calculated as follows: *maximum exposure during laser ablation* $= 200 \cdot 10^{-9} * 1.5 \cdot 10^{-3}/(760 * 10^{6}) = 4 \cdot 10^{-7}$ L. The ratio of exposures of Zr in oxygen that are able to synthesize the ZrO phase is the following:

$$\textit{exposure}\ (25°\text{C})/\textit{necessary exposure during laser ablation}$$
$$= 25/4 \cdot 10^{-7} = 62.5 \cdot 10^{6}.$$

Such a high ratio cannot justify an oxidation of spalled solid zirconium in oxygen atmosphere, although solid Zr may reach temperatures of 2000°C. Thus it is possible to exclude the former hypothesis, and to explain the oxidation mechanism of zirconium by examining the thermodynamic behaviour of a (gaseous) Zr plasma in an oxygen atmosphere. There are several thermodynamic computer packages (such as CSIRO, Thermochemistry System, SOLGASMIX, STEPSOL, and Outokumpu) available for comprehensive thermodynamic calculations. Taylor et al. (1994) used the CSIRO software, based on the free energy minimization (FEM) calculation, to study the behaviour of the Al-O system at very high temperatures. They show interesting results, such as the significant presence of Al suboxides, until 5000 K. The Zr-O system has to be studied in the conditions of the plume, i.e. at temperatures of 6500–12000 K (Kools et al., 1992; Butt et al., 1994) and several oxygen quantities, in order to predict how the expanding plume may react with oxygen. The Outokumpu software indicates that the existence of ZrO_2 is thermodynamically impossible above 6500 K. The CSIRO software indicates that the major gas phases above 6000 K are Zr(g) and ZrO (g) (Figure 6). The ZrO_2 phase is not present at these temperatures. An extrapolation of the curves, at any relative pressure of oxygen, leads to 100% gaseous Zr phase above 7000 K. Such a temperature is in the lower range of the laser plasmas (Kools et al., 1992; Butt et al., 1994). It is interesting to note that the quenching velocity in laser plasmas may be extremely high: a minimum quenching velocity V in our case would be: $V = (T_p - T_e)/t$, where T_p is the lowest plasma temperature (6500 K), T_e the minimum temperature of light emission (about 1000 K), and t the maximum time of existence of light emission (Castell, 1995) by the laser ablation ($200 \cdot 10^{-9}$ s). Thus, it can be estimated that $V = 25 \cdot 10^9$ Kelvin/second. Moreover, the plasma temperature is superior to the temperature of existence of the ZrO gas phase, thus the latter cannot be formed yet, as the plume is quenched. As gaseous ZrO is not formed at the laser plasma temperature, it can neither appear during quenching, nor yield the solid cubic ZrO metastable phase. The rapid ejection of sparks from the plume can be attributed to the highly pyrophoric property of zirconium (Webster, 1980), which generates an explosive oxidation of Zr particles in oxygen (which is at room temperature). Therefore, the ultrarapid condensation and oxidation of Zr in an oxygen-poor atmosphere, is the mechanism proposed for the formation of cubic ZrO metastable phase from the laser plume:

$$\text{Zr (gas)} \rightarrow \text{Zr (condensed) then Zr (condensed)} + 0.5\ O_2 \rightarrow \text{ZrO (cubic)}.$$

The formation of a metastable solid suboxide from a plasma at lower temperatures has been obtained by Pirzada (1991), during the plasma decomposition of silica components: quenching gaseous SiO yielded the solid SiO metastable phase, i.e.: SiO (gas) $\rightarrow$ SiO (solid) As Pirzada worked at the temperatures of existence of the SiO gaseous phase, his results stay in accordance with the present model.

4. Conclusions

Metastable Cubic ZrO has been synthesized by laser-ablation of pure metallic zirconium in a 1.5 mbar atmosphere. An optimal backing gas pressure has been determined for yielding a higher rate of Zirconium oxide sparks. The synthesis of ZrO results from the high temperature combination between gaseous Zr and oxygen, and from their high velocity quenching. A deposition at a low distance from the plume allows the observation of both metallic Zr splats and encrusted grains of Zr Oxide.

Acknowledgments

The support of the Decanato de Investigaciones of the Universidad Simon Bolivar (USB) is acknowledged. Milton Manrique (USB) is gratefully acknowledged for his contribution to this work.

References

Geohegan, D.B.: 1993, *MRS Proc.* **285**, 27.

Reitz, W. and Rawers, J.: 1993, A review: Laser ablation and its effects on surface removal', *Surface Modification Technologies VI*, The Minerals, Metals & Materials Society.

Kools, J.C.S., Baller, T.S. and De Zwart, S.T.: 1992, Gas flow dynamics in laser ablation deposition, *J. Appl. Phys.* **71**(9), 1.

Butt, D., Wantuck, P. and Sappey, A.: 1994, Laser diagnostics of zirconium carbide vaporization, *J. Am. Cer. Soc.* **77**(6), 1411–1417.

Gilgenbach, R.M., Ching, C.H. and Lash, J.S.: 1994, Laser diagnostic experiments on KrF laser ablation plasma-plume dynamics relevant to manufacturing applications, *Phys. Plasmas* **1**(5), 1619–1625.

Ching, C.H., Gilgenbach, R.M. and Lash, J.S.: 1995, 'Detection of AlO molecules produced by KrF laser-ablated Al atoms in oxygen gas and plasma environments, *J. Appl. Phys.* **78**(5), 3408–3410.

Rebban, U., Hildebrandt, J. and Skopp, G.: 1980, *Appl. Phys.* **23**, 341.

Ghaderi, S.: 1988, Analysis of ceramics by laser ablation Fourier transformation mass spectrometry, in: Evans, C. (ed.), *Ceramic Transactions*, Vol. 5, Advanced Characterization Techniques for Ceramics, Am. Cer. Soc., pp. 73–84.

Morant, C., Sanz, J.M. and Galan, L.: 1989, An XPS study of the interaction of oxygen with zirconium, *Surf. Sci.* **218**, 331–345.

Turnbill, A.G. and Wadsley, M.W.: 1988, *The CSIRO Thermochemistry System, version 5*, IMEC, Australia.

Castell, R.: 1995, personal data.

Taylor, P.R., Pirzada, S.A. and Manrique, M.: 1994, Thermodynamic predictions for material processing in a plasma reactor using solid oxide feed materials, *Metall. and Mat. Trans. B.* **25B**, 713–720.

Webster, R.T.: 1980, Zirconium, in: Cubberly, W. (ed.) *Metals Handbook*, Vol. 2: *Properties and Selection: Non Ferrous Alloys and Pure Metals*, 9th ed., ASM, pp. 826–831.

Pirzada, S.A.: 1991, Silicon carbide synthesis and modelling in a non-transferred arc thermal plasma reactor, M.Sc. Thesis, University of Idaho, p. 102.

GPSR Compliance

The European Union's (EU) General Product Safety Regulation (GPSR) is a set of rules that requires consumer products to be safe and our obligations to ensure this.

If you have any concerns about our products, you can contact us on ProductSafety@springernature.com

In case Publisher is established outside the EU, the EU authorized representative is:

Springer Nature Customer Service Center GmbH
Europaplatz 3
69115 Heidelberg, Germany

Batch number: 10755248

Printed by Printforce, the Netherlands